Green Energy and Technology

For further volumes:
http://www.springer.com/series/8059

Zhen Fang
Editor

Pretreatment Techniques for Biofuels and Biorefineries

Springer

Editor
Prof. Dr. Zhen Fang
Biomass Group
Xishuangbanna Tropical Botanical Garden
Chinese Academy of Sciences
Kunming, Yunnan
People's Republic of China

ISSN 1865-3529 ISSN 1865-3537 (electronic)
ISBN 978-3-642-32734-6 ISBN 978-3-642-32735-3 (eBook)
DOI 10.1007/978-3-642-32735-3
Springer Heidelberg Dordrecht London New York

Library of Congress Control Number: 2012953996

Printed on acid-free paper

Springer is part of Springer Science+Business Media (www.springer.com)

Preface

Pretreatment is the first and the most crucial step for effectively using biomass and for developing new routes to produce biofuels and value-added products. Pretreatment is a process intensive step and, for example, it is the single most expensive processing step in cellulosic ethanol production, making up approximately 20–40 % of the product cost. Although there are many research articles that focus on pretreatment techniques, it was felt by the authors that there was a lack of a comprehensive source where one could turn to understand the many possible methods and their range of application.

This text includes 19 chapters contributed by world-leading experts on pretreatment methods for biomass. It gives an extensive coverage for different types of biomass (e.g. molasses, sugar beet pulp, cheese whey, sugarcane residues, palm waste, vegetable oil, straws, stalks and wood), for different types of pretreatment approaches (e.g. physical, thermal, chemical, physical–chemical and biological) and for methods that show subsequent production of biofuels and chemicals such as sugars, ethanol, extracellular polysaccharides, biodiesel, gas and oil. In addition to traditional methods such as steam, hot-water, hydrothermal, diluted acid, organosolv, ozonolysis, sulfite, milling, fungal and bacterial, microwave, ultrasonic, plasma, torrefaction, pelletization, gasification (including biogas) and liquefaction pretreatments, novel techniques (e.g. nano- and solid-catalysts, organic electrolyte solutions and ionic liquids) are introduced and discussed.

Each chapter was strictly reviewed externally by experts in biofuels listed in the Acknowledgement. The chapters are categorized into seven parts:

- Part I: Biopretreatment
- Part II: Thermal Pretreatment
- Part III: Chemical Pretreatment
- Part IV: Physicochemical Pretreatment
- Part V: Gasification, Liquefaction and Biogas
- Part VI: Novel Pretreatment Techniques
- Part VII: Treatment of Different Types of Biomass

This book offers a review of state-of-the-art research and provides guidance for future paths for developing pretreatment techniques of biomass for biofuels in the fields of biotechnology, microbiology, chemistry, materials science and engineering. It is our intention to provide a systematic introduction to pretreatment techniques. It is an accessible reference book for students, researchers, academicians and industrialists in biorefineries.

Acknowledgement

First and foremost, I would like to thank all the contributing authors for their many efforts to insure the reliability of the information given in the chapters. Contributing authors have really made this project realizable.

Apart from the efforts of authors, I would also like to acknowledge the referees listed below for carefully reading the book chapters and giving constructive comments that significantly improved the quality of the book:

Prof. Nicolas Abatzoglou (Université de Sherbrooke, Canada), Dr. Muhammad T. Afzal (Univ. of New Brunswick, Canada), Dr. Taku Michael Aida (Tohoku Univ., Japan), Dr. Zainal Alimuddin Zainal Alauddin (Universiti Sains Malaysia), Prof. Jacques E. Amouroux (LGPPTS/UPMC/ENSCP, France), Dr. Anju Arora (Indian Agricultural Research Institute), Dr. Hassane Assaaoudi (McGill Univ., Canada), Dr. Hassan Azaizeh (Tel Hai College, Israel), Prof. Nicolas Brosse (Nancy Univ., France), Prof. Rafael B. Mato Chaín (Universidad de Valladolid, Spain), Dr. Pascale Champagne (Queen's Univ., Canada), Prof. Hongzhang Chen (Chinese Academy of Sciences), Prof. Andrzej G. Chmielewski (Warsaw Univ. of Technology), Prof. Jae-woo Chung (Gyeongnam National Univ. of Science and Technology, South Korea), Dr. Daniel Ciolkosz (Pennsylvania State Univ.), Dr. Louis J. Circeo (Applied Plasma Arc Technologies, LLC, Atlanta), Dr. Rudolf Deutschmann (Lakehead Univ., Canada), Dr. Tim Dumonceaux (Agriculture and Agri-Food Canada), Dr. Animesh Dutta (Univ. of Guelph, Canada), Prof. Elena Efremenko (The M.V. Lomonosov Moscow State Univ.), Prof. Xu Fang (Shandong Univ., China), Prof. Toshi Funazukuri (Chuo Univ., Japan), Prof. Pag-asa Gaspillo (De La Salle Univ., Philippines), Dr. Anli Geng (Ngee Ann Polytechnic, Singapore), Prof. Ashwani K. Gupta (Univ. of Maryland), Prof. Rick Gustafson (Univ. of Washington), Prof. Michikazu Hara (Tokyo Institute of Technology), Prof. Adriaan van Heiningen (Univ. of Maine), Dr. Kazuhiko Ishikawa (Advanced Industrial Sciencr and Technology, Japan), Dr. Keikhosro Karimi (Isfahan Univ. of Technology, Iran), Prof. Hansie Knoetze (Stellenbosch Univ., South Africa), Prof. Gunnur Kocar (Ege Univ., Turkey), Prof. Ramesh Chander Kuhad (Univ. of Delhi South Campus, India), Dr. Christopher Lan (Univ. of Ottawa), Mr. Jean-Rémi Lanteigne (École Polytechnique de Montréal, Canada), Dr. Jean-Michel Lavoie (Université de Sherbrooke, Canada), Dr. Jianjun Li (Chinese Academy of Sciences), Dr. Yebo Li (Ohio State Univ.), Prof.

Yin Li (Chinese Academy of Sciences), Dr. Lin Lin (Jiangsu Univ., China), Prof. Yun Liu (Beijing Univ. of Chemical Technology), Dr. Poupak Mehrani (Univ. of Ottawa), Dr. Felipe Alatriste Mondragon (Instituto Potosino de Investigación Cientifica y Tecnologica, Mexico), Dr. Antonis Mountouris (National Technical Univ. of Athens), Dr. Naim Najami (The Academic Arab College of Education, Israel), Prof. Yonghao Ni (Univ. of New Brunswick), Prof. Lucia García Nieto (Universidad de Zaragoza, Spain), Dr. Abdul-Sattar Nizami (Univ. of Toronto), Dr. Melek Özkan (Gebze Institute of Technology, Turkey), Prof. Igor Polikarpov (Universidade de São Paulo, Brazil), Prof. Xinhua Qi (Nankai Univ., China), Dr. Wensheng Qin (Lakehead Univ.), Dr. Armando T. Quitain (Kumamoto Univ., Japan), Dr. R. Michael Raab (Agrivida, Massachusetts), Dr. Mala Rao (National Chemical Lab., India), Prof. Joseph P. Roise (North Carolina State Univ.), Dr. Guus van Rossum (Univ. of Twente, the Netherlands), Prof. Roger Ruan (Univ. of Minnesota), Prof. Elio Santacesaria (Complesso Universitario di Monte S.Angelo, Italy), Dr. Anton Sonnenberg (Wageningen UR, the Netherlands), Dr. Andy Soria (Univ. of Alaska), Dr. Wolfgang Stelte (Technical Univ. of Denmark), Dr. Chia-Hung Su (Ming-Chi Univ. of Technology, Taiwan, ROC), Dr. Lee Keat Teong (Universiti Sains Malaysia), Mr. Xiao-fei Tian (Chinese Academy of Sciences), Prof. Montserrat Zamorano Toro (Universidad de Granada, Spain), Mr. Satriyo Krido Wahono (Indonesian Institute of Sciences), Dr. Haisong Wang (Chinese Academy of Sciences), Dr. Chunbao (Charles) Xu (Western Univ., Canada), Dr. Jing Yang (Southwest Forestry Univ., China), Dr. Wennan Zhang (Mid Sweden Univ.), Dr. Xiao Zhang (Washington State Univ.), Prof. Xiao-yu Zhang (Huazhong Univ. of Science and Technology, China), Dr. Y.-H. Percival Zhang (Virginia Tech), Dr. Xuebing Zhao (Tsinghua Univ., China), Dr. Junyong Zhu (USDA Forest Service).

I am also grateful to Ms. June Tang (Associate Editor, Springer) for her encouragement and guidelines during my preparation of the book.

Finally, I would like to express my deepest gratitude towards my family for their kind cooperation and encouragement, which help me in completion of this project. Especially, I would like to dedicate this book to my Mother (Ms. Shubi Yu) for her love and support during her final illness. She passed away at 9:30 am on April 9, 2012 in Taining, Fujian.

May 30, 2012 Zhen Fang
Kunming

Contents

Editor's Biography

Prof. Dr. Zhen Fang is leader and founder of biomass group, Xishuangbanna Tropical Botanical Garden, Chinese Academy of Sciences. He is also an adjunct full Professor of Life Sciences, University of Science and Technology of China. He is the inventor of "fast hydrolysis" process. He is specializing in thermal/biochemical conversion of biomass, nanocatalyst synthesis and its applications, pretreatment of biomass for biorefineries. He obtained his PhDs from China Agricultural University (Biological & Agricultural Engineering, 1991, Beijing) and McGill University (Materials Engineering, 2003, Montreal).

Contributors

Mania Abdollahi-Neisiani Department of Chemical Engineering, Ecole Polytechnique de Montréal, C.P. 6079, succ. Centre Ville, Montréal, H3C 3A7, Canada
e-mail: mania.abdollahineisiani@polymtl.ca

Murni Melati Ahmad Chemical Engineering Department, Universiti Teknologi PETRONAS, Bandar Seri Iskandar, 31750 Tronoh, Perak, Malaysia

Razol Mahari Ali Management and Humanities Department, Universiti Teknologi PETRONAS, Bandar Seri Iskandar, 31750 Tronoh, Perak, Malaysia

Felipe A. F. Antunes Department of Biotechnology, School of Engineering of Lorena, University of São Paulo, Lorena-12.602.810, Brazil

Azlin Suhaida Azmi Department of Biotechnology Engineering, Faculty of Engineering, International Islamic University Malaysia, Jalan Gombak, 53100 Kuala Lumpur, Malaysia

Sobhi Basheer TransBiodiesel Ltd, Nazareth Street 79, P.O. Box 437, Shefar-Am 20200, Israel

Alexander N. Bratsev Institute for Electrophysics and Electric Power RAS (IEE RAS), Dvortsovaya nab., 18, St.-Petersburg, 191186, Russia

Anuj K. Chandel Department of Biotechnology, School of Engineering of Lorena, University of São Paulo, Lorena-12.602.810, Brazil
e-mail: anuj.kumar.chandel@gmail.com

Jamal Chaouki Department of Chemical Engineering, Ecole Polytechnique de Montréal, C.P. 6079, succ. Centre Ville, Montréal, H3C 3A7, Canada
e-mail: jamal.chaouki@polymtl.ca

Hong-zhang Chen National Key Laboratory of Biochemical Engineering, Institute of Process Engineering, Chinese Academy of Sciences, Beijing 100190, China
e-mail: hzchen@home.ipe.ac.cn

Jing Dai Renewable Materials Program, Department of Forest, Rangeland and Fire Sciences, University of Idaho, Moscow, Idaho 83844-1132, USA

Pratibha Dheeran Biotechnology Area, Indian institute of Petroleum, Dehradun, India

Ehiaze Augustine Ehimen Department of Energy Technology, Aalborg University, Pontoppidanstræde 101, DK-9220 Aalborg Ø, Denmark

Zhen Fang Biomass Group, Chinese Academy of Sciences, Xishuangbanna Tropical Botanical Garden, 88 Xuefulu, Kunming, Yunnan province, 650223, China
e-mail: zhenfang@xtbg.ac.cn

Ellen C. Giese Department of Biotechnology, School of Engineering of Lorena, University of São Paulo, Lorena-12.602.810, Brazil

Motonobu Goto Bioelectrics Research Center, Faculty of Engineering, Kumamoto University, 2-39-1 Kurokami, Kumamoto 860-8555, Japan
e-mail: mgoto@kumamoto-u.ac.jp

Feng Guo Biomass Group, Chinese Academy of Sciences, Xishuangbanna Tropical Botanical Garden, 88 Xuefulu, Kunming, Yunnan province 650223, China

Takahisa Hinata Hokkaido Central Agricultural Experiment Station, Hokkaido Naganuma, Japan

Jessica Hoffmann Department of Energy Technology, Aalborg University, Pontoppidanstræde 101, DK-9220 Aalborg Ø, Denmark

Jens Bo Holm-Nielsen Department of Energy Technology, Aalborg University, Pontoppidanstræde 101, DK-9220 Aalborg Ø, Denmark

Fang Huang School of Chemistry and Biochemistry, Institute of Paper Science and Technology, Georgia Institute of Technology, Atlanta, GA 30332-0440, USA

Toshiyuki Imai Green Plan. Co. Ltd, Osaka Sakai, Japan

Rouzbeh Jafari Department of Chemical Engineering, Ecole Polytechnique de Montréal, C.P. 6079, succ. Centre Ville, Montréal, Canada H3C 3A7
e-mail: rouzbeh.jafari@polymtl.ca

Yoshiaki Kimura Hokkaido Central Agricultural Experiment Station, Hokkaido Naganuma, Japan
e-mail: kimura-yoshiaki@hro.or.jp

Sachin Kumar Sardar Swaran Singh National Institute of Renewable Energy, Jalandhar-Kapurthala Road, Wadala Kalan, Kapurthala-144601, Punjab, India
e-mail: sachin.biotech@gmail.com

Vadim A. Kuznetsov Institute for Electrophysics and Electric Power RAS (IEE RAS), Dvortsovaya nab., 18, St.-Petersburg, 191186, Russia

Pak Sui Lam Biomass and Bioenergy Research Group, Clean Energy Research Center, Department of Chemical and Biological Engineering, University of British Columbia, 2360 East Mall, Vancouver, BC, Canada V6T 1Z3
e-mail: wilsonlam82@yahoo.com

Jean-Remi Lanteigne Chemical Engineering Department, École Polytechnique de Montréal, C.P. 6079, succ. Centre Ville, Montréal, QC, Canada H3C 3A7

Jean-Philippe Laviolette Department of Chemical Engineering, Ecole Polytechnique de Montréal, C.P. 6079, succ. Centre Ville, Montréal, Canada H3C 3A7
e-mail: jean-philippe.laviolette@polymtl.ca

Sean Lim Lay PETRONAS Research Sdn. Bhd., Lot 3288 & 3289, Off Jalan Ayer Itam, Kawasan Institusi Bangi, 43000 Kajang, Selangor

Baoqiang Liao Dept of Chem. Eng., Lakehead University, Thunder Bay, ON, Canada P7B 5E1

Armando G. McDonald Renewable Materials Program, Department of Forest, Rangeland and Fire Sciences, University of Idaho, Moscow, Idaho 83844-1132, USA
e-mail: armandm@uidaho.edu

Mas Fatiha Mohamad Biomass Processing Laboratory, Green Technology MOR, Universiti Teknologi PETRONAS, Bandar Seri Iskandar, 31750 Tronoh, Perak, Malaysia

Ladan Jafari Naimi Biomass and Bioenergy Research Group, Clean Energy Research Center, Department of Chemical and Biological Engineering, University of British Columbia, 2360 East Mall, Vancouver, BC, Canada V6T 1Z3

Nandhagopal Narayanaswamy Sardar Swaran Singh National Institute of Renewable Energy, Kapurthala, India

Ivy dos Santos Oliveira Department of Biotechnology, School of Engineering of Lorena, University of São Paulo, Lorena-12.602.810, Brazil

Ebru Toksoy Öner IBSB—Industrial Biotechnology and Systems Biology Research Group, Marmara University, Bioengineering Department, Goztepe Campus, 34722 Istanbul, Turkey
e-mail: ebru.toksoy@marmara.edu.tr

Sergey D. Popov Institute for Electrophysics and Electric Power RAS (IEE RAS), Dvortsovaya nab., 18, St.-Petersburg, 191186, Russia

Victor E. Popov Institute for Electrophysics and Electric Power RAS (IEE RAS), Dvortsovaya nab., 18, St.-Petersburg, 191186, Russia

Armando T. Quitain Graduate School of Science and Technology, Faculty of Engineering, Kumamoto University, 2-39-1 Kurokami, Kumamoto, 860-8555, Japan

Arthur J. Ragauskas School of Chemistry and Biochemistry, Institute of Paper Science and Technology, Georgia Institute of Technology, Atlanta, GA 30332-0440, USA
e-mail: arthur.ragauskas@ipst.gatech.edu

Lasse Rosendahl Department of Energy Technology, Aalborg University, Pontoppidanstræde 101, DK-9220 Aalborg Ø, Denmark
e-mail: lar@et.aau.dk

Philip G. Rutberg Institute for Electrophysics and Electric Power RAS (IEE RAS), Dvortsovaya nab., 18, St.-Petersburg, 191186, Russia
e-mail: rc@iperas.nw.ru

Mitsuru Sasaki Graduate School of Science and Technology, Faculty of Engineering, Kumamoto University, 2-39-1 Kurokami, Kumamoto, 860-8555, Japan

Wei Shi Dept of Chem. Eng., Lakehead University, Thunder Bay, ON, Canada P7B 5E1

Silvio Silvério da Silva Department of Biotechnology, School of Engineering of Lorena, University of São Paulo, Lorena-12.602.810, Brazil
e-mail: silvio@debiq.eel.usp.br

Shahab Sokhansanj Biomass and Bioenergy Research Group, Clean Energy Research Center, Department of Chemical and Biological Engineering, University of British Columbia, 2360 East Mall, Vancouver, BC, Canada V6T 1Z3

Alexander V. Surov Institute for Electrophysics and Electric Power RAS (IEE RAS), Dvortsovaya nab., 18, St.-Petersburg, 191186, Russia

Ahmed Tafesh TransBiodiesel Ltd, Nazareth Street 79, PO Box 437 Shefar-Am 20200, Israel
e-mail: atafesh@transbiodiesel.com

Hideyuki Takenaka Hokkaido Central Agricultural Experiment Station, Hokkaido Naganuma, Japan

Xiaofei Tian Biomass Group, Chinese Academy of Sciences, Xishuangbanna Tropical Botanical Garden, 88 Xuefulu, Kunming, Yunnan province, 650223, China

Saqib Sohail Toor Department of Energy Technology, Aalborg University, Pontoppidanstræde 101, DK-9220 Aalborg Ø, Denmark

Zahra Tooyserkani Biomass and Bioenergy Research Group, Clean Energy Research Center, Department of Chemical and Biological Engineering, University of British Columbia, 2360 East Mall, Vancouver, BC, Canada V6T 1Z3

Yoshimitsu Uemura Chemical Engineering Department, Universiti Teknologi PETRONAS, Bandar Seri Iskandar, 31750, Tronoh, Perak, Malaysia

Shilpi Verma Department of Chemical Engineering, Indian Institute of Technology, Roorkee, India

Charles (Chunbao) Xu The Institute for Chemicals and Fuels from Alternative Resources, Faculty of Engineering, Western University, London, ON, Canada N6A 5B9
e-mail: cxu6@uwo.ca

Seiichi Yasui Zukosha. Co. Ltd, Hokkaido Obihiro, Japan

Suzana Yusup Chemical Engineering Department, Universiti Teknologi PETRONAS, Bandar Seri Iskandar, 31750 Tronoh, Perak, Malaysia
e-mail: drsuzana_yusuf@petronas.com.my

Jun-ying Zhao National Key Laboratory of Biochemical Engineering, Institute of Process Engineering, Chinese Academy of Sciences, Beijing 100190, China

Michael Zviely Virdia (Formerly HCL CleanTech), Research & Development Department, Herzlyia 46733, Israel
e-mail: michael.zviely@virdia.com

Part I
Biopretreatment

Chapter 1
Biological Pretreatment of Lignocellulosic Biomass for Enzymatic Saccharification

Nandhagopal Narayanaswamy, Pratibha Dheeran, Shilpi Verma and Sachin Kumar

Abstract Biological delignification is an attractive approach for pretreatment of lignocellulosic biomass. This approach is very cost effective, low-energy requirement, environment friendly, low formation of toxic materials such as furfural, hydroxymethylfurfural, etc. Biological approach has been demonstrated using direct microorganism as well as using enzymes extracted from microbes. The microbial treatment includes fungi such as white-rot fungi, brown-rot fungi and soft-rot fungi, and bacteria. Both of brown-rot and soft-rot fungi principally degrade the plant polysaccharides with minimal lignin degradation, while white-rot fungi are capable of complete mineralization of both the lignin and the polysaccharide components. This chapter presents a brief review of the relevant and updated literature on biological pretreatment of lignocellulosic biomass. Various approaches used by different researchers for biological delignification of lignocellulosic biomass, including microbial and enzymatic approaches, mode of action, effect of biological pretreatment on lignocellulosic biomass, effect of biological pretreatment on enzymatic hydrolysis, have been included in this chapter. The chapter also provides a glimpse of the gaps, which need to be studied.

Keywords Lignocelluloses · Pretreatment · Lignin · Biological delignification · Fungi · Bacteria

1.1 Introduction

In view of environmental and fossil fuel security concern, the future energy economy will probably be based on a broad range of alternative energy resources such as wind, water, sun, nuclear fission as well as biomass. Extensive use of fossil fuels in the

S. Kumar (✉) · N. Narayanaswamy
Sardar Swaran Singh National Institute of Renewable Energy, Jalandhar-Kapurthala Road, Wadala Kalan, Kapurthala 144601, Punjab, India
e-mail: sachin.biotech@gmail.com

P. Dheeran
Biotechnology Area, Indian institute of Petroleum, Dehradun, India

S. Verma
Department of Chemical Engineering, Indian Institute of Technology, Roorkee, India

Z. Fang (ed.), *Pretreatment Techniques for Biofuels and Biorefineries*,
Green Energy and Technology,
DOI 10.1007/978-3-642-32735-3_1,

last century has greatly depleted the energy reserves. Presently, the petroleum-based fuels—gasoline, aviation turbine fuels, and diesel—all liquid fuels, and compressed natural gas (CNG) are almost excessively used in the transportation sector. The increasing rate of consumption of fossil fuels has raised severe problems including the issues of depletion of energy resources, increase in fuel prices, and global climate change. The major attraction of the use of renewable energy fuels is the reduction of environmental impacts that are associated with the use of the fossil fuels [1]. Therefore, an imperative technology is required to ward off the apprehensive problems of meager fossil fuels and its negative impact on environments. Finally, researchers are looking for the economical way to produce alternative fuels and energy preferably from abundantly available biodegradable or eco-friendly and renewable raw materials such as biomass or renewable resources such as sun, wind, water, etc. These resources have a vital role and equal contribution in the energy sector [2].

Among the potential bioenergy resources, lignocellulosic biomass has been identified as a cheap and effective feed-stock for the production of biofuels such as bioethanol, biobutanol, and biogas. Lignocellulosic biomass is available about 180 million tons per year from agriculture and other sources [3, 4]. Lignocellulose is the most abundant renewable and natural resource, which have a promising role in renewable energy sector and have fetched many researchers toward a new road map to the biofuels production. Biofuels such as ethanol, butanol, hydrogen, biogas, etc. from lignocellulosic biomass and non-food sources have caught worldwide attention. The lignocellulosic biomass has increased its attention because these raw materials do not compete with food crop and is less expensive than conventional feed-stock like sugarcane, corn, etc. In general, lignocellulosic feed-stocks are observed as promising alternative sources because it consist massive amount of carbohydrates [5]. All lignocellulosic biomass predominantly comprise cellulose, hemicellulose, and lignin, but in a different ratio with respect to distinct biomass [6]. However, lignocellulosic materials are naturally recalcitrant and have more complex structure [3, 5]. Lignocellulosic biomass for the production of biofuels includes forest residues such as wood; agricultural residues such as sugarcane bagasse, corn cob, corn stover, wheat, and rice straws; industrial residue such as pulp and paper processing waste; municipal solid wastes; and energy crops such as switch grass [7–11].

Cellulose is the most abundant organic compounds on the earth and this polysaccharide consists of linear chain of several hundred to 10 thousands recurring D-glucose units with molecular formula $(C_6H_{10}O_5)\,n$, linked by $\beta(1 \rightarrow 4)$ glycosidic bonds. Cellulose is a structural component of a primary cell wall in green plants and algae. Naturally, cellulose can be found in two different forms in the plant materials, consists of parts with crystalline structure and amorphous structure. The crystalline celluloses are well organized, which are tightly bundled and bound together by strong inter chain hydrogen bonds while this is less pronounced in amorphous cellulose.

Hemicelluloses, the second most abundant natural polymer on the earth [3, 12], are the heterogeneous polymers consisting of pentoses (D-xylose, D-arabinose), hexoses (D-glucose, D-mannose, and D-galactose), and sugar acids. Hemicellulose is a connector between cellulose and lignin, and it leads to more rigidity. In hardwood, hemicelluloses are dominantly found as xylan, whereas in softwood glucomannan are most common [3, 12, 13]. Xylans are commonly found as heteropolysaccharide in many plants with backbone chain of 1,4-linked β-D-xylopyranose units. Along with xylose unit, xylan may comprise arabinose, glucornic acid or its 4-O-dimethyl ether, acetic acid, ferulic, and *p*-coumaric acids. Xylan can be simply extracted in an acid or alkaline environment but in the case of glucomannan requires stronger acid or alkaline environment [3, 12]. Hemicellulose is also an economically important natural polymer as it contains ample amount of pentose sugar, which can be used as a substrate in food, pharmaceutical, and biofuels industries.

Lignin, the third largest available biopolymer in nature [3, 12], is a heterogenous and irregular arrangement of phenylpropanoid polymer that reduces the chemical and enzymatic degradation to maintain the recalcitrant and insoluble properties of lignocellulose. Three phenylpropionic alcohols primarily exist as monomers of lignin (i) coniferyl alcohol, (ii) coumaryl alcohol, and (iii) sinapyl alcohol. In general, herbaceous plants such as grasses, rice, and wheat straws have the lowest contents of lignin, while in softwoods lignin content is found to be higher. Lignin is the major rate-limiting component in the carbon recycling reaction, as its oxidation rate is naturally very slow [14, 15]. Furthermore, lignin has an important role in conducting water in plant stems and giving physical strength to the plants.

The main routes to produce fuels from biomass (biofuels) include fermentation of sugars to alcohol, gasification and chemical synthesis, and direct liquefaction. The biological process for converting lignocellulose to biofuels requires: (1) delignification to liberate cellulose and hemicelluloses from the matrix; (2) depolymerization of the carbohydrate polymers to produce free sugars; and (3) fermentation of mixed hexose and pentose sugars [16–19]. All these processes comprise the same main components: hydrolysis of the hemicellulose and the cellulose to monomer sugars, fermentation, and product recovery. The main difference between the process alternatives is the hydrolysis steps, which can either be accomplished by an acid or by enzymes [20].

Lignocellulosic materials need to be saccharified to produce fermentable sugars. This is an intensive process involving a combination of pretreatment and either chemical (acid hydrolysis) or enzymatic hydrolysis [20–22]. In the chemical process, the hydrolysis of sugar polymers in lignocellulosic material is catalyzed by an acid, whereas in the enzymatic process, enzymes are used for hydrolyzing cellulose and hemicellulose to sugar monomers [23–25]. Several factors influence the yields of the monomeric sugars from the lignocellulosic matter and the by-products during hydrolysis. These factors include biomass particle size, liquid-to-solid ratio, type and concentration of acid used, temperature, reaction time, length of the macromolecules, porosity of the biomass, degree of polymerization of cellulose, configuration of the cellulose chain, association of cellulose with other protective polymeric structures within the plant cell wall such as lignin, pectin, hemicellulose, proteins, and mineral elements, etc. [26–28].

Enzymatic hydrolysis offers major advantages over other chemical routes (e.g., acid hydrolysis) such as higher yields, minimal by-product formation, low-energy requirements, mild operating conditions, and low-chemical disposal costs [29]. Hydrolysis of cellulose to glucose in aqueous media catalyzed by the cellulase enzyme suffers from slow reaction rates due to high crystalline structure of cellulose, degree of polymerization, pore volume, acetyl group bound to hemicellulose, surface area, hydrophobicity, and biomass particle size, which make the penetration of enzymes to the active sites very difficult [30–32]. The enzymatic hydrolysis without pretreatment yields sugars which is $<20\,\%$ of the theoretical quantity, whereas $>90\,\%$ of the theoretical quantity of sugars are obtained with enzymatic saccharification after pretreatment [33, 34]. Therefore, pretreatment is a necessary and prudent step to break the crystalline structure of the lignocelluloses, the removal of lignin to expose the cellulose and hemicellulose molecules for efficient enzymatic conversion, and saccharification of feed-stock [5, 31, 35–39].

Physical, physico-chemical, chemical, and biological processes have been studied for the pretreatment of lignocellulosic materials [40–42]. Enzymatic hydrolysis of lignocellulosics can be significantly enhanced by physical, chemical, and biological pretreatments of the lignocellulosic materials to remove and modify the lignin and hemicellulose and to reduce the fiber crystallinity. The physical and chemical pretreatment including grinding, organosolv process involving extraction with hot aqueous ethanol, ozonolysis, acid/alkaline treatment, oxidative delignification, carbon dioxide explosion, hydrogen peroxide, ultrasonic irradiation, ammonia fiber expansion, wet explosion, and acid or SO_2-catalyzed steam explosion, ammonia fiber explosion (AFEX) and biological pretreatment have been followed and optimized up to certain levels [5, 31]. The objective of physical pretreatment or mechanical pretreatment is generally used to reduce the particles size, crystallinity, and degree of polymerization; and consequently it leads to increase the surface area for enzyme and/or chemical accessibility. In thermal pretreatment method, various methods have been investigated such as steam explosion/steam pretreatment, liquid hot water, etc.

Chemical pretreatment is another important technique that has been commonly followed by many industries like paper and pulp industries for few decades. This treatment is mostly used by the researchers, which includes catalyzed steam explosion, acid/alkali treatment, ammonia fiber/freeze explosion (AFEX), ionic liquid pretreatment, organosolv, and pH-controlled hot water treatment. All the above treatments require different chemicals and different operating conditions [3, 43].

Biological pretreatment have been studied elaborately by various researchers because this technique is very cheap, less energy consuming process, and the refulgent area of research. In this method, microorganisms or enzymes are used as catalyst in order to modify lignin and to degrade the hemicellulosic content in the biomass. Several white-rot fungi and brown-rot fungi, such as *Phanerochaete chrysosporium*, *Pleurotus ostreatus*, *Ceriporiopsis subvermispora*, *Postia placenta*, *Phanerochaete carnosa*, *Gloeophyllum trabeum* and *Trametes versicolor* have been studied for pretreatment of biomass such as wheat and rice straws, corn stover and switch grass [31, 44]. An overview of biological pretreatment and its applications are shown in Fig. 1.1.

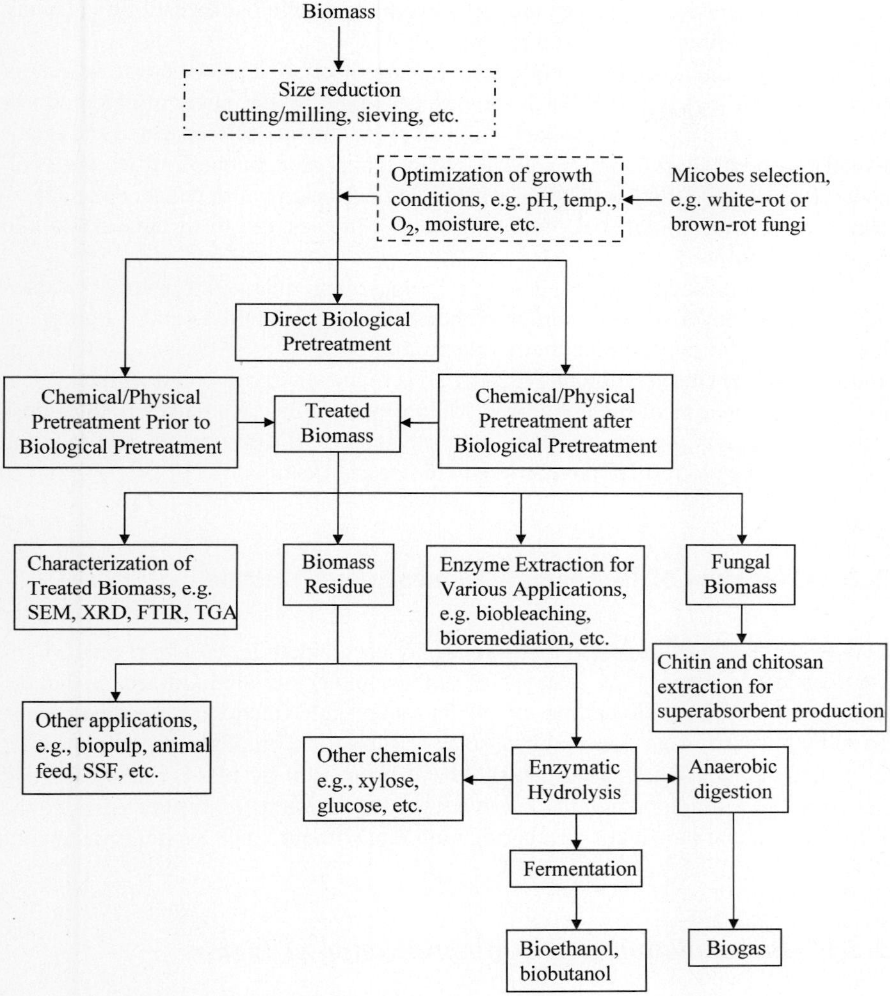

Fig. 1.1 Overview of biological treatment and its applications

All the pretreatment methods, except biological method, require expensive equipment that have demand of high energy depending on which the process to be carried out. Furthermore, these techniques often result in effluent and residue that tremendously have negative impacts on environments, inhibit the enzymatic reaction and the growth of microorganisms, which suppose to ferment the product of enzymatic saccharification [5, 31]. Indeed, biological pretreatment method using white-rot fungi has increased its attention because of the following inherent advantages, (i) safe and environmental-friendly method; (ii) low-energy consumption and cost effective; (iii) selective degradation; (iv) in some cases treated biomass directly can be used for enzymatic conversion or fermentation; (v) increase the cellulose digestibility of many types of forage fiber and agricultural wastes [45].

This pretreatment retains many special features itself, that is why now researchers looked into biological route to achieve desired target. So far, many research papers have already been reported that the biological pretreatment has been tested and established beyond its level. Organo-solvent (ethanol, methanol, butanol, ethylene glycol, *n*-butylamine, etc.) also used along with biological treatment to enhance the degradation of internal lignin seal, removing hemicelluloses and disturbing crystalline nature of cellulose [46].

Despite all these advantages, however, biological pretreatment is a very slow process; and moreover some important components (hemicelluloses and cellulose) of biomass are also consumed either by same microorganism or by some foreign invaders. Low-saccharification rate (35–40 %) is found when compared with chemical and physical treatment methods [47]. Main objective of this chapter is to discuss various biological pretreatment methods, advantages, disadvantages, and to come out with the key to resolve the barriers in biological treatment.

1.2 Overview of Biological Pretreatment Methods

Falkowski et al. [48] reported that lignin may accumulate in terrestrial ecosystems for decades, on longer time scales most of these molecules are oxidized, so that the accumulation of organic carbon in soils is a miniscule fraction of the total carbon fixed by the ecosystem. Lakes may also store substantial amounts of organic matter in sediments. Various microorganisms have been used by many researchers and this zero pollution approach has received good attention as it helps to enhance the fermentation and enzymatic saccharification rate without much capital investment.

1.2.1 Bio-oxidation or Bio-mineralization of Lignin

In the 1920s, a small amount of research was conducted on biodegradation of lignin and few concepts about lignin were recapitulated in the 1930s (i) lignin is among the plant cell wall polymers that is more resistant to biodegradation, but can be degraded, (ii) white-rot fungi degrade lignin in wood, and (iii) could not be delignified (without loss of wood carbohydrates) [49]. Waksman et al. [50] investigated the lignin degradation in compost and soil environment. In 1951, Gottileb and Pelczar reported that white-rot fungus, *Polyporus versicolor*, used Braun's native lignin as the growth substrate. This finding revealed that lignin could be used as a sole carbon and energy source for white-rot fungi. In addition to white-rot fungi, other groups of fungi were found to degrade lignin, partially, namely basidiomycetous litter-decomposing and brown-rot fungi as well as soft-rot fungi in the 1950s [51, 52]. The first lignin compound was studied and reported in the 1960s; [53–56] and biodegradation assays based on ^{14}C-lignin were developed in the 1970s and it was revealed how lignin was optimally degraded under laboratory conditions [57, 58]. The white-rot fungus, *P. chrysosporium*, was used as experimental organism in USA and in other laboratory, *Sporotrichum pelverulentum* was chosen for lignin bio-degradation studies [59, 60]. Previously, *T. versicolor* was a well studied experimental fungus [53, 61].

In the late 1970s and the early 1980s, very important concepts were investigated in the physiology of lignin degradation by *P. chrysosporium*. The most important discoveries were (i) low nitrogen requirement and lignin mineralization takes place during secondary metabolisms; (ii) highest mineralization was found at 100 % oxygen, thus, lignin degradation is oxidative and agitation having detrimental effects in lignin degradation; (iii) veratryl alcohol formation takes place during lignin oxidation [58]. Some other fungi, for example, *Phlebia radiata* [49, 62] were also found to readily degrade lignin and lignin model compounds in a similar way. The first extracellular enzymes involved in lignin degradations were discovered in 1983–1984 [63–65] and named as lignin peroxidases (LiPs). The catalytic mechanism of LiPs, based on initial one-electron oxidation of the lignin model compounds followed by subsequent breakdown reactions via radical cation intermediates, was experimentally verified [66, 67].

In the 1990s, in addition to pivotal studies on catalytic and enzymatic properties of the lignin-modifying peroxidases as well as their molecular biology, the major areas of research were dealt with the potential applications of white-rot fungi and their enzymes in biopulping (biomechanical pulping), pulp bleaching, and other applications. The most promising fungi for biopulping are so-called selective lignin degraders, that is, fungi that degrade larger amounts of lignin relative to carbohydrates such as *C. subvermispora* [49].

Generally, wood basidiomycetous fungi that cause white-rot in wood, called white-rot fungi, are the major lignin degraders in nature, which specifically degrades the lignin in different woody and straw or lignocellulosic biomass such as corn stover, wheat straw, paddy straw, sugarcane trashes, various wood materials, etc. [68]. The concept behind this biological degradation of lignin is secretion of several lignolytic enzymes by these white-rot fungi. Lignolytic enzymes such as LiP, manganese peroxidase (MnP), laccase (Lac), and versatile-peroxidase (VP) have a vital role in many applications like biopulping, biobleaching, biofuels, xylose, and enzyme production [32, 68, 69]. After this breakthrough discovery, many researchers conducted various experiments on molecular biology studies on lignolytic enzyme coding gene, for example, cloning and sequencing of LiP gene [70], cloning and sequencing of MnP gene [71, 72], heterologous expression of Lac [73], homologous expression of peroxidases [49, 74], 3D structure of Lac [75], and cell-free mineralization of ^{14}C-labeled synthetic and natural lignins by MnP [76]. The following microbial diversity has been described by various researchers for lignin degradation:

1.2.1.1 White-Rot Fungi

Wood-rotting basidiomycete fungi are usually divided into white-rot and brown-rot fungi. As mentioned earlier, several white-rot fungi are involved in lignin biodegradation such as *P. chrysosporium*, *C. subvermispora*, *Phlebia subserialis*, *Echinodontium taxodii*, etc. [31, 32, 44, 77]. Majorly, white-rot fungi grow well on hard woods such as birch and aspen. On the other hand, certain species *Heterobasidion annosum*, *Phellinus pini*, and *P. radiata* grow well on soft woods such as pine and spruce [32]. However, the feasibility of biological pretreatment is still in its infancy because of the extremely long treatment time as well as the difficulty in selectively degrading lignin [5, 78, 79].

The growth of fungi on lignocellulosic biomass results in a loss of dry matter. During the fungal growth, all the main components (cellulose, hemicelluloses, and lignin) are consumed in part by the fungus for its growth and metabolic activities. The loss and the selective degradation of lignin is greatly depends upon the strain which is taking the course of degradation. For example, *Flammulina velutipes*, *Fomes marginatus*, and *Laetiporus sulfurous* decompose wheat straw very slowly or poorly, hence, these white-rot fungi are unsuitable for biological delignification. Some other fungi *Ganoderma applanatum*, *Poria* sp., and *Trametes gibbosa* grow well on wheat straw, but they degrade the hemicellulose and cellulose; therefore, these strains are also not found suitable for biodelignification. Although it is very difficult to remove lignin alone from the lignocellulose, some unique fungal species such as *Stropharia rugosoannulata*, *Hapalopilus rutilans*, *P. ostreatus*, *C. subvermispora*, *Lentinula edodes*, and *Pleurotus eryngii* have high affinity with lignin; and they are able to consume lignin faster than non-lignin content of biomass. Therefore, these strains are good delignifier and can be used efficiently in biological pretreatment of lignocellulose [80, 81].

White-rot fungi are more commonly found on angiosperm than on gymnosperm wood species in nature [82]. Generally, syringyl (S) units of lignin are more selectively degraded whereas guaiacyl (G) units are more resistant to degradation. The transmission electron microscopy revealed that *C. subvermispora* and *Pleurotus eryngii* partially removed the middle lamella while *P. radiata* apparently removed the lignin from secondary cell walls, when these fungi were grown on straw [83]. In fibers, the middle lamella contains a high concentration of G lignin, while, secondary walls contain a high proportion of S lignin. Various environmental conditions like cultivation time, pH, nutrient ingredients (nitrogen source), and oxygen level have been optimized by many researchers in order to achieve the maximum degradation of lignin [84]. Lignin degradation by white-rot fungi occurs through the action of lignin degrading enzymes such as peroxidases (LiP and MnP) and phenol oxidase (Lac) [5, 78, 85]. These enzymes are regulated by carbon and nitrogen sources [16]. Almost all white-rot fungi produce Lac and MnP, but only some of them produce LiP [32].

White-rot fungi degrade lignin in biomass with two different mode of degradation, named as selective and non-selective degradation. In non-selective degradation, all three components (lignin, cellulose, and hemicellulose) were almost degraded equally, whereas in selective decay mostly hemicellulose and lignin were degraded [32]. Some white-rot fungi species remove lignin without loss of cellulose from LCCs and cause white-mottled or white-pocket type of rot and those species referred as selective delignifier, for example, *Phellinus nigrolimitatus* [32, 86]. More than

1,500 species of white-rot fungi are able to decompose lignin with little consumption of cellulose [87]. There are also some fungi that are able to degrade the same wood with both types of attack selective and non-selective [49]. Good examples of such fungi are *G. applanatum* and *H. annosum*. The selective delignifiers have a pivotal role in biopulping, biobleaching and bio-fuel industries. However, the ratio of lignin–hemicellulose–cellulose decayed by a selected fungus can differ enormously; and even different strains of the same species, for example, *P. chrysosporium* and *C. subvermispora*, may act in another way on the same kind of wood. *C. subvermispora* was found to be one of the most lignin removers from woody materials, but grew poorly on rice straw [88]. Furthermore, the comparative studies of *C. subvermispora* and *P. chrysosporium* revealed that *C. subvermispora* genetic inventory and expression patterns exhibit increased oxidoreductase potential and less cellulolytic capability relative to *P. chrysosporium* [89]. Some examples of white-rot fungi, which posses selective degradations, are *Pycnoporus cinnabarinus*, *P. ostreatus*, *P. eryngii*, *P. radiata*, *Phlebia tremellosus*, *P. subserialis*, *P. pini*, and *Dichomitus squalens* [32, 86, 90]. The selective delignifiers have a potential role in pretreatment of various lignocelluloses in order to attain the considerable amount of feed-stock for the biofuel production.

Some species remove lignin more readily than carbohydrates [86]. Many white-rot fungi colonize cell lumina and cause cell wall erosion. Eroded zones form as decay progresses and large voids filled with mycelium. This type of rot is referred to as non-selective or simultaneous rot [86]. *T.* (syn. *Coriolus, Polyporus) versicolor* and *Fomes fomentarius* are typical simultaneous-rot fungus [32, 61]. Therefore, the use of non-selective fungi is greatly limited by its non-selective degradation of plant cell walls and it may be used in biological pretreatment to some extent.

1.2.1.2 Brown-Rot Fungi

Most of the brown-rot fungi degrade cellulose and hemicellulose more rapidly than lignin in woods. But the lignin is modified up to certain level and left as modified brown lignin residue, hence collectively called as brown-rot fungi. Many brown-rot fungi such as *Serpula lacrymans*, *Coniophora puteana*, *Meruliporia incrassata*, *Laetoporeus sulphureus*, and *G. trabeum* are used in various investigations [91, 92]. Most of the brown-rot fungi prefer soft-wood to hard-wood as substrate, for example *S. lacrymans* (dry-rot fungus) and *C. puteana* are the most harmful fungi occurring in wood in temperate region.

Brown-rot fungi have a unique mechanism to break down the wood. In contrast to white-rot fungi that de-polymerize the cell wall carbohydrates only to the extent that they utilize degraded product in fungus metabolism, brown-rot fungi accumulates the de-polymerized cell wall cellulose and hemicellulose since the fungus does not utilize all the products in the metabolism [61]. Early in the decay process, these brown-rot fungal hyphae penetrate from one cell to another through existing pores in wood cell walls. The hyphae start penetration from the cell lumen, where they are in close connection with the S3 layer. The brown-rot affects the S2 layer of the wood cell wall first [49].

Although brown-rot fungi consume economically important materials in biomass, the potential biotechnology application of brown-rot fungus is used to produce cattle feed from pine dust through solid-state fermentation. The brown-rotted lignin is used as an adhesive as it reacts more rapidly than native lignin due to increased phenolic-hydroxyl groups, for example, to replace phenol-formaldehyde flake board resin [49]. *G. trabeum* is the most extensively used fungus for treatment of wood chips. For example, Monrroy et al. [46] pretreated bioorganosolv process of *Pinus radiata* wood chips by using bioorganosolv process. They used *G. trabeum* for 3 weeks followed by organosolv treatment with various ratios of ethanol–water mixture at pH 2 and optimized H factor (factor that combine time and temperature in one variable). They found significant improvement in solvent accessibility and H factor was found to be decreased from 6,000 to 1,156 for obtaining 161 g ethanol/kg of *P. radiata* wood. Another example, Ray et al. [93] pretreated Scots pine (*Pinus sylvestris*) sapwood by *C. puteana* for 35 days and they found that glucose release from the wood increased by four to five folds after 10 days exposure with minimum loss of weight (5 %) and maximum sugar release occurred 15 days after exposure to *C. puteana* with 9 % weight loss.

To some extent, brown-rot fungi have similar pathways to degrade the lignocellulose as white-rot fungi. The wood decay mechanisms of both types of fungi rely on radical formation, low pH, and the production of organic acids such as oxalic acid. The radical formation would maximize the solubility of lignin in alkali and the decay process is an oxidation reaction, hence decay can be enhanced by high oxygen supply. However, many proposed mechanisms are not fully proven experimentally [49].

1.2.1.3 Soft-Rot Fungi and Other Microfungi

Blanchette [86] has described two kinds of soft-rot: type I consisting of biconical or cylindrical cavities that are formed within secondary walls and type II refers to an erosion form of degradation. For example, *Daldinia concentrica* is the most efficient fungus of type II group, which primarily affect hardwood. Nilsson et al. [94] found 53 % weight loss in birch wood within 2 months. During early stage of classification of different wood-rotting fungi, *Xylariaceous ascomycetes* from genera such as *Daldinia, Hypoxylon*, and *Xylaria* have often been regarded as white-rot fungi, but today these fungi are categorized to soft-rot fungi as they cause typical type II soft-rot in wood. In coniferous wood (e.g., pine wood), the weight loss was very low and it has been thought that these type of woods have more guaiacyl units in middle lamella, which inhibit the growth of soft-rot fungi [49].

Some microfungi (*Penicillium chrysogenum*, *Fusarium oxysporum*, and *Fusarium solani*) identified in a forest soil sample are able to mineralize grass lignins upto 27 % [49]. However, most of the soft-rot and microfungi consume readily economically important carbohydrates during invading and have very less applications in biological pretreatment.

1.2.1.4 Bacteria and Actinomycetes

In biological pretreatment process, bacteria and actinomycetes are not as efficient as white- and brown-rot fungi. Very few bacteria, such as filamentous bacteria belonging to the genus *Streptomycetes* are well known degraders of lignin, have been studied for pretreatment. These bacteria have been found to have some role in final mineralization of lignin. Non-filamentous bacteria *Pseudomonas* degrade very little amount of lignin. Since these bacteria do not have extra cellular oxidoreductase, which is one of the very essential enzymes for delignification and cannot be utilized in biological pretreatment. Actinomycetes are bacteria which form multicellular filaments; thus, they resemble fungi, also produce extracellular peroxidase as white-rot and brown-rot fungi, for example LiP-type enzyme. *Streptomyces* sp. EC1 produces peroxidase and cell-bound demethylase requiring H_2O_2 and Mn^{2+}, both have been produced at relatively high levels in the presence of Kraft lignin or wheat straw [49]. Bacteria actinomycetes *Streptomyces viridosporus* have also been studied up to some extent [95]. Godden et al. [96] studied activity of peroxidase and catalase in six actinomycetes strains.

Thermophilic actinomycetes have been isolated from a wide range of natural substrates, for example from desert sand and compost. The genera of the thermophilic actinomycetes isolated from compost include *Nocardia*, *Streptomyces*, *Thermoactinomyces*, and *Micromonospora*. Actinomycetes degrade lignin as their primary metabolic activity and at high nitrogen levels compared to white-rot fungi, most of which degrade lignin via their secondary metabolism [97].

The lignin-degrading actinomycete species examined till date have been shown to oxidatively de-polymerize lignin. The primary degradative activity of actinomycetes is solubilization of lignin, with low levels of mineralization compared with the white-rot fungi. The depolymerization reactions produce a modified water-soluble, acid precipitable polymeric lignin as the principal lignin degradation product. The range of actinomycete species capable of metabolizing lignin is still unknown. Moreover, the strains examined thus far solubilize lignin to an acid-precipitable polymeric lignin-like product.

1.3 Enzymes Involved in Lignin Degradation or Mineralization

Enzymes face several challenges in the degradation of macromolecular lignin [49]. As mentioned earlier, this substrate is a large heterogeneous polymer and very difficult to degrade by microbes. Indeed, lignin does not contain enzymatically hydrolysable linkages and is stereo-irregular. For lignin degradation, the enzymes or agents must be oxidative. Many extracellular enzymes involved in lignin degradation are, as mentioned earlier, LiPs (LiPs, ligninases, EC 1.11.1.14), manganese peroxidases (MnPs, Mn-dependent peroxidases, EC 1.11.1.13) and Lacs (benzenediol:oxygen oxidoreductase, EC 1.10.3.2). Further, some accessory enzymes are also involved in hydrogen peroxide production. Glyoxal oxidase (GLOX) and aryl alcohol oxidase

(AAO) (EC 1.1.3.7) belong to this group. LiPs and MnPs are heme-containing glycoproteins, which require hydrogen peroxide as an oxidant [49, 98].

1.3.1 Lignin Peroxidase (LiP)

Tien et al. [63] discovered LiP in the extracellular medium of *P. chrysosporium* grown under nitrogen limitation. The enzyme uses H_2O_2 as co-factor or mediator for activity and is capable of oxidizing and/or cleaving lignin and lignin model compounds. This was supposed to be the key reaction of lignin degradation. Very few fungi are found to produce extracellular LiP [98]. *P. chrysosporium*, *T. versicolor*, *Bjerkhandera* sp., and *T. cervina* are some fungi, which can produce LiPs [32]. Indeed, LiP was found to play only a minor role in lignin degradation by *T. versicolor*, at least as measured by bio-bleaching of kraft pulp [99].

LiPs are monomeric homo-protein and glycol protein belonging to oxidoreductase family, which specifically act on peroxide as an acceptor (peroxidases). These enzymes have molecular weight of 40 kDa and isoelectric points (pI) ranging from 2.8 to 5.3. The absorption spectrum of the native enzyme in *P. chrysosporium* has a very distinct maximum at 406–409 nm due to the presence of a single heme group, where Fe^{3+} pentacoordinates with four heme tetrapyrrole nitrogen and a histidine of LiPs (protoporphyrin IX) [32, 98]. The interaction of LiPs with its substrate follows ping-pong mechanism [100]. As shown in Fig. 1.2, LiPs are oxidized by H_2O_2 to two-electron oxidized intermediates (LiP I) along with iron ions as Fe^{4+} and free radical residues on tetrapyrolle. LiP I then oxidises the donor substrate by one electron, where the donor substrate, VA (3,4-dimethoxybenzyl alcohol, VA) yields second intermediate LiPs complex (LiPs II) in which iron ion is found in same oxidation state, that is, Fe^{+4}, but there is no free radical residue on tetrapyrolle of heme and a radical cation. LiP II then oxidises a second molecules of donor substrate (VA), confers another radical cation and native form of LiP. Here the reformation of native LiP mainly depends upon the LiP II reduction step, which is a rate limiting step in catalytic cycle. Because the reduction of LiP II is a relatively slow process and LiP II is less potent than LiP I complex. Consequently, LiP II complex is long available for reaction again with H_2O_2 leads to inactivation of enzyme and forms LiP III complex (Fig. 1.2), which is characterized as a complex between LiP and superoxide. The catalytic cycle of LiP is described in Fig. 1.2. VA radical cations act as redox mediators and are capable to reduce LiP III complex back to its native form, LiP. In this LiP catalytic cycle reaction, VA radical cations ($VA^{\bullet+}$) are usually restored back after its oxidation reaction with non-phenolic compounds of lignin.

As in this catalytic cycle reaction, VA plays an important role. Three major functions of VA have been investigated so far. Firstly, VA acts as a mediator in electron-transfer reaction. Secondly, VA is a good substrate for compound II, therefore VA is essential for completing the catalytic cycle of LiP during the oxidation of terminal substrates. Furthermore, if the inactive LiP III complex forms, the intermediate $VA^{\bullet+}$ will be capable of reducing LiP III complex back to its native form LiP

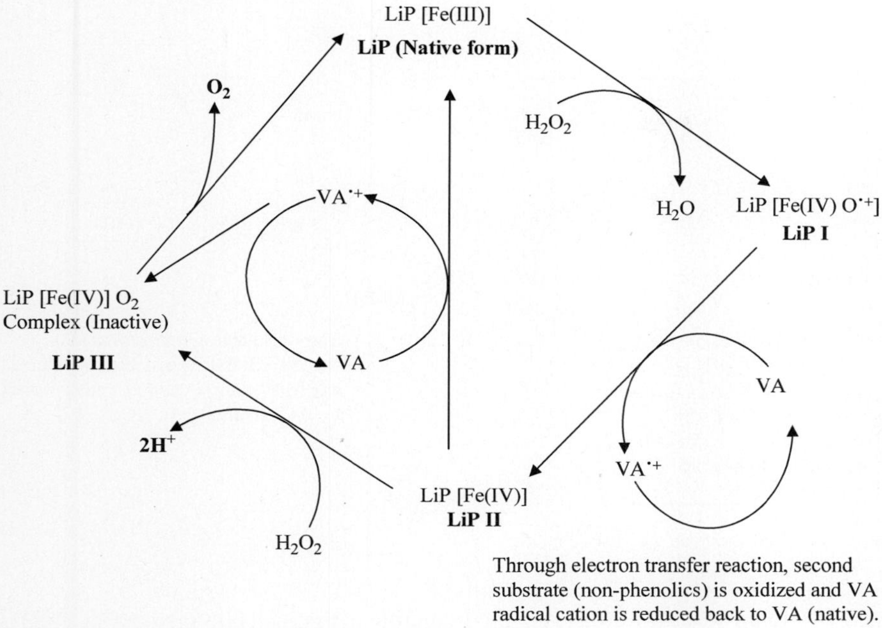

Fig. 1.2 Catalytic cycle of LiP [32, 98]

(Fig. 1.2). Thirdly, VA prevents the H_2O_2-dependent inactivation of LiPs by reducing LiP II complex back to its native form LiP. Almost all the white-rot fungi synthesize VA via *de novo* glucose pathway during early stage of secondary metabolism in parallel with LiP production [98].

LiPs oxidize non-phenolic and phenolic units of lignin by removing one electron and creating free radicals, which lead to chemically decompose the polymer. LiP has been shown to oxidize fully methylated lignin, lignin model compounds as well as various polyaromatic hydrocarbons. LiPs cleave selectively Cα–Cβ bond, aryl Cα bond, aromatic ring opening and demethylation in the lignin molecule [32, 98].

1.3.2 Manganese Peroxidase (MnP)

Manganese peroxidase (EC 1.11.1.13, Mn(II):hydrogen-peroxide oxidoreductase, MnP) also require H_2O_2 as an oxidant in the Mn-dependent catalyzing reaction in which Mn^{2+} is converted to Mn^{3+} by MnP. Mn^{3+} then oxidizes phenolic rings to phenoxyl radicals, which leads to decomposition of compounds. Both LiPs and MnPs are heme-containing glycoproteins [49, 101, 102]. But LiPs are not as widespread as MnPs, and major difference between MnPs and LiPs in lignin degradations are as LiPs generally oxidize nonphenolic lignin substructures and MnPs oxidize phenolic

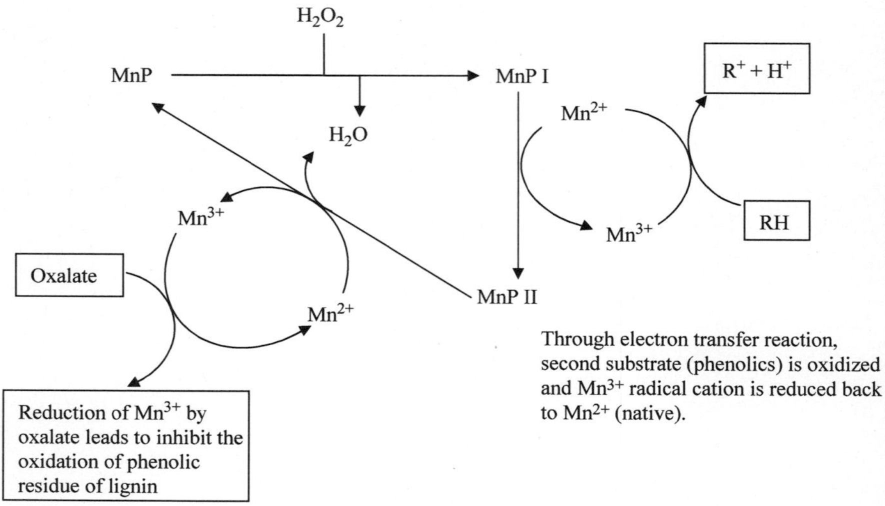

Fig. 1.3 Catalytic cycle of MnP [32, 98]

rings of lignin [49]. MnPs have an important role in lignin depolymerization, chlorolignin, and demethylation of lignin. Therefore, MnPs have a very essential role in biological pretreatment of lignocellulosic biomass. So far, many researchers have reported that *P. chrysosporium*, *Pleorotus ostreatu*, *Trametes* sp., and several other species, which belong to *Meruliaeiae*, *Coriolaceae*, and *Polyporaceae* produce MnP [32].

MnPs contain one molecule of heme as iron protoporhyrin IX and comprise with 357 amino acid residues, three sugar residues (Glc Nac, Glc Nac at Asn 131, and a single mannose at Ser 336), two structural calcium ions, a substrate Mn^{2+} and 478 solvent molecules. For MnP, the acidic amino acids, aspartic acid, and two glutamic acids have been proposed as manganese-binding residues [32, 98]. MnPs act on its substrate almost similar to LiPs action. Thus, the native form of MnP is oxidized by addition of H_2O_2 to form MnP I complex (Fig. 1.3). Then this catalytic cycle involves in the oxidation of Mn^{2+} to Mn^{3+} by MnP I and MnP II complexes. Finally, Mn^{3+} oxidizes the lignin compounds by diffusing into the lignified cell wall and attacks it from inside. Indeed, MnP I can directly involve in the oxidation of phenolic compounds such as 2,6-dimethoxyphenol, guaiacol, and phenolic tetrameric lignin model compounds. This oxidation reaction clearly elucidates that MnP oxidizes the phenolic part of the lignin indirectly via Mn ions. But MnP naturally does not oxidize aromatic compounds of lignin directly as LiP. Because they do not have tryptophan residue, required for electron transfer to non-phenolic substrates [98, 103]. Recently, MnPs have been isolated from *Bjerkandera* sp. BOS55 and *P. eryngii* that are found to be oxidized Mn^{2+} as well as aromatic compounds [98]. Hence, it is very clear that addition of Mn^{2+} may play further enhancement in the bio-oxidation of phenolic compounds of lignin and may induce MnP production in fungi.

1.3.3 Laccase (Lac)

Laccases (Lac, EC 1.10.3.2, benzendiol: oxygen oxydoreductase) belong to blue copper protein or oxidase family. Lac has been found in fungi, bacteria, and plants. The major producers of Lac are of fungi kingdom, whose diversity can be found in soil, phytopathogenic, and freshwater inhabiting ascomycetes and basidiomycetes [104]. Lac is generally larger than peroxidases as it has a molecular weight of approximately 60 kDa and pI 3–6 [49]. Optimum pH for better Lac activity is found to be 3–5 [105]. Lac catalyzes four single-electron oxidations of aromatic amines and phenolic compounds such as phenolic substructure of lignin, which coincide with the reduction of O_2 to H_2O [32, 98]. Indeed, it can also oxidize nonphenolic compounds under certain conditions, for example, 2,2′-azinobis(3-ethylbenzthiazoline-6-sulfonate) (ABTS) [106], 1-hydrobenzotriazole (1-HBT) [107], and violuric acid [108]; natural mediators such as 4-hydroxybenzoic acid, 4-hydroxybenzyl alcohol [109], and 3-hydroxyanthranilate [110]. Therefore, the natural mediator should be produced by organisms for the complete oxidation of lignin.

Lac is produced by almost all the white-rot fungi. Generally, it has several Lac encoding genes and secrete as multiple isoforms [49, 106]. Lac contains four copper atoms of three distinct types per enzyme, and each type has a different role in the oxidation of substrate [98]. Type I copper directly involves in the reaction with the substrate. The type I copper gives a maximum absorbance at a wavelength of 610 nm, which gives to the enzyme a typical blue color. The type II copper and the two type III copper cluster are found in triangular forms. Copper II and III complexes involve in the binding, the reduction of O_2 and the storage of electrons originating from the reducing substrates. The type II copper does not have absorbance in a visible range, while the type III copper has a maximum absorption at 330 nm, hence copper II and III complexes do not have any color [98]. The entire crystalline structure of Lac containing four copper atoms in the active site has been studied from *T. versicolor* and *C. maxima* [111, 112]. Bourbonnais et al. [106] reported that the white-rot fungus *T. versicolor* produces two laccase isozymes (I and II).

For effective biological pretreatment of lignocellulose, various white-rot fungi can be used in addition to copper ions in order to induce the secretion of Lac enzymes. In some special cases, Lac can also be induced by addition of aromatic compounds like VA and 2-5 xylidine [32]. Although Lac generally oxidizes phenolic residues of lignin, it also oxidizes non-phenolic compounds of lignin with addition of ABTS as discussed earlier. Therefore, Lac action can be induced further by addition of some special catalyst in the biological pretreatment. For some fungi such as *C. subvermispora* and *Ganoderma lucidum* the Lac production could be increased in the presence of lignocellulosic materials. Recently, some bacterial Lacs have also been characterized from *Azospirillum lipoferum*, *Bacillus subtilis*, *Streptomyces lavendulae*, *Streptomyces cyaneus*, and *Marinomonas mediterranea* [113].

1.3.4 Versatile-Peroxidase (VP)

Versatile-peroxides (VP) are found in various *Bjerkandera* species and *Pleurotus* species [114]. VA can oxidize both phenolic and non-phenolic compounds of lignin as well as Mn^{2+} [32]. The catalytic mechanism is similar to LiP [101]. For example, VP oxidize nonphenolic model compounds such as veratrylglycerol β-guaiacyl ether and results to the formation of veratraldehyde. VP also oxidize Mn^{2+} to Mn^{3+}, VA to veratraldehyde and *p*-dimethoxybenzene to *p*-benzoquinone [32].

1.3.5 Peroxide-producing Enzymes

In lignin biodegradation, hydrogen peroxide (H_2O_2) plays an important role and the rate of oxidation of lignocellulose entirely depends upon the availability of H_2O_2 [98]. Therefore, white-rot fungi have to produce some accessory enzymes for H_2O_2 production in order to support the ligninolytic oxidative reaction of LiPs and MnPs. Such enzymes are glyoxal oxidase (GLOX) found in *P. chrysosporium* and many other white-rot fungi and AAO. Naturally, fungi secrete the GLOX substrates, which are reduced into H_2O_2. For example, *P. chrysosporium* produces glyoxal and methylglyoxal as natural extracellular metabolites. In some cases, the product of lignin oxidation reaction may also undergo a reduction reaction by GLOX, for example, arylglycerol β-aryl ether structure of lignin is oxidized by LiPs to glycolaldehyde and this cleavage product acts as a substrate for GLOX [115].

On the other hand, AAOs produce H_2O_2 through another route in some white-rot fungi. Chlorinated anisyl alcohols are secreted as extracellular metabolites in LiPs producing strains *Bejrkandera* species and secreted metabolites are further reduced to H_2O_2 by specific AAO. Many alkoxybenzyl alcohols are LiP substrates, but not chloroanisyl alcohol. Therefore, this enzymatic mechanism of H_2O_2 production clearly elucidates that fungus separates its lignolytic and H_2O_2-generating pathways [32].

1.3.6 Cellobiose Dehydrogenase (CDH) in Ligninolysis

Cellobiose dehydrogenase (CDH; EC 1.1.99.18; cellobiose: [acceptor] 1-oxidoreductase) is an extracellular flavocytochrome secreted by several wood-degrading fungi (white-rot and brown-rot fungi) under cellulolytic culture conditions. It oxidizes soluble cellodextrins, mannodextrins, and lactose efficiently to their corresponding lactones by a ping-pong mechanism using a wide spectrum of electron acceptors including quinones, phenoxyradicals, Fe^{3+}, Cu^{2+}, and tri-iodide ion [116]. CDH activity was first discovered by Ulla Westermark and Karl-Erik Eriksson as a cellobiose-dependent reduction of quinones in the two white-rot fungi *T. versicolor* and *P. chrysosporium*. This enzyme has been isolated from the white-rot fungi *P. chrysosporium*, *T. versicolor*, *P. cinnabarinus*, *Schizophyllum commune*; the brown-rot fungus *Coneophora puteana*; and the soft-rot fungi *Humicola insolens* and *Myceliophtore thermophila* (*Sporotrichum thermophile*) [117]. Interestingly, no CDH activity has been reported so far from cultures of *C. subvermispora,* even though it is a selective delignifier [118].

Recently, it was found that CDH has shown to participate in the ligninolytic metabolism of white-rot fungi in the presence of H_2O_2 [118]. Henriksson et al. [119] have summarized the findings of various researchers regarding the CDH activity in ligninolysis that it reduce Fe^{3+} to Fe^{2+} and cellobiose or cello-oligosaccharides to H_2O_2. In the presence H_2O_2, the reaction favors the formation of Fenton's reagent that trigger the production of hydroxyl radicals. This hydroxyl radical is highly reactive and known to attack lignin and cellulose. Further, Henriksson et al. [117] have discussed the following hypothesis/theory about the CDH activity:

- CDH supports the lignin degradation by reducing the aromatic radicals, which is produced from lignin oxidation reaction by LiP and Lac. Enzymatic reaction is a reversible reaction; therefore lignin degraders may favor the polymerization of the radicals in vitro condition. CDH may inhibit polymerization by reducing the radicals created by LiP and Lac.
- CDH supports MnP.
- CDH reduces toxic quinones to phenols that can be used as redox mediators by lignolytic enzymes.
- CDH reduces compound II of lignolytic peroxidases and thus, complete the catalytic cycle in the absence of peroxidase substrate.
- CDH degrades and modifies cellulose, hemicelluloses, and lignin by generating hydroxyl radicals in a Fenton type reaction.

All the above theory/hypothesis is not yet proved practically and still unclear concepts. Although, the hypothesis is unclear, the last point about generation of hydroxyl radicals gives plausible explanations for many of the characteristic properties of CDH and it may be the most attractive suggestion for the function of CDH [117]. Further, Dumonceauxa et al. [120] suggested that CDH is not important in lignin degradation, at least for *T. versicolor* delignifying and concluded that it is possible that some other enzyme masked the effect of the lack of CDH by performing reductive reactions. Hence, the CDH-deficient mutant can still degrade or modify the lignin in a similar manner as the wild type but does not degrade cellulose [121].

1.3.7 Low-Molecular Weight Compounds Involved in Lignin Degradation (Mediators)

As per theoretical and practical views, it is elucidated that enzymes Lac and peroxidase (LiPs and MnPs) are larger than the pore size of the cell wall, and they cannot have direct contact with the lignin. Various low-molecular weight compounds are found in white-rot fungi, which play an important role in ligninolytic enzyme system of white-rot fungi. It has been studied the role of mediators or co-factors in various in vitro studies, which revealed optimum concentration of H_2O_2, lignin, O_2, and suitable mediators [122]. Therefore, for effective biological removal of lignin components from lignocellulosic biomass, the fungi and/or bacteria should not fail to produce these mediators or co-factors.

1.3.7.1 Veratyl Alcohol

As discussed earlier, the role and importance of VA is very essential in bio-ligninolysis. It is generally synthesized de novo from glucose via shikimate pathway at the early stage of secondary metabolism in parallel with the LiPs production. The biosynthetic pathway for VA was performed with ^{14}C isotope trapping experiments in the ligninolytic fungus *P. chrysosporium* (ATCC 34541); and concluded that the pathway proceeds as follows: Phenylalanine → cinnamic acid → Benzoate/Benzaldehyde → VA [115, 123]. In *P. chrysosporium*, VA production is induced by nitrogen-limitation, whereas in *Bjerkandera* sp., the nitrogen element does not have any significant regulatory effect on the VA biosynthesis [124]. Furthermore, LiPs action on non-phenolic residue of lignin can be enhanced by addition of VA in biological pretreatment of lignocellulose. As per Hammel et al. [115], VA protects LiPs against H_2O_2-mediated inactivation reaction (rate limiting step) in the LiPs catalytic cycle reaction (Fig. 1.2) and it has been proposed that VA act, in vivo as a stabilizer for the enzymes.

1.3.7.2 Manganese

Naturally, all wood materials and residues contain manganese elements, which are present sometime in high concentration depending upon the type of the wood materials, varying from 10 mg/kg to100 mg/kg of dry wood. The importance of Mn^{2+} can clearly be found during the fungal decay on woody materials as it accumulates in the form of MnO_2 precipitates. Indeed, the insoluble Mn^{4+} species deposits at the tip of new fungal hyphae in the early stages of infestation and growth [125]. As mentioned earlier, Mn^{2+} stimulates the production of MnP and enhances the degradation of lignin components during oxidation reaction, where Mn^{3+} is generated by MnP and acts as a mediator for the oxidation of various phenolic compounds. Therefore, addition of Mn^{2+} increases the biological oxidation rate in biological pretreatment of lignocellulose. On the other hand, addition of Mn^{2+} inhibits the action of LiPs and its production [98, 125]. Hence, it is very essential to optimize the concentration of Mn^{2+} in order to achieve better biological pretreatment. Indeed, in decaying wood, naturally a manganese concentration gradient is established, allowing soluble forms of manganese (Mn(I1) and Mn(III)) to diffuse into regions of low manganese concentration [126].

1.3.7.3 Oxalate

Two enzymes, oxaloacetase and glyoxylate oxidase that catalyze the hydrolysis of oxaloacetate and the oxidation of glyoxylate, respectively, are responsible for the biosynthesis of oxalate. An important aspect is that LiPs and MnPs are capable of decomposing oxalate in the presence of VA or Mn^{2+} [98,125]. The breakage of oxalate results in the formation of carbon dioxide and formate anion radical (R-CO_2^{+-}), which is further oxidized by O_2 to give CO_2 and superoxide (O_2^{+-} or HOO^{+-}) under aerobic conditions. The active oxygen species are suggested to directly participate in the oxidation of lignin. This reaction can be observed in oxidation of phenol red and kojic acid by MnP in the presence of Mn^{2+} and oxalate without exogenous addition of H_2O_2. This suggests that oxalate may be regarded as a passive sink for H_2O_2 production [98]. If the oxalate reduces the VA^{+-} and Mn^{3+} ions, the mineralization rate of lignin will be affected adversely (Figs. 1.2 and 1.3). As mentioned earlier, VA^{+-} and Mn^{3+} both should be reduced by phenolic and/or non-phenolic compounds of lignin for the effective degradation of lignin. For better biological treatment, it is important to conquer the excessive action of oxalate on VA^{+-} and Mn^{3+}.

1.3.7.4 2-Chloro-1,4-dimethoxybenzene

White-rot fungi produce a wide range of organohalogen metabolites. The most commonly produced halogens are chlorinated anisyl metabolites (CAM) and chlorinated hydroquinone metabolites (CHM). CAM has an important physiological function in lignin degradation, contributing as substrates for AAO involved in extracellular H_2O_2 production. Among CHM metabolites, chlorinated 1,4-dimethoxybenzene such as 2-chloro-1,4-dimethoxybenzene, 2,6-dichloro-1,4-dimethoxybenzene, tetrachloro-1,4-dimethoxybenzene, and tetrachloro-4-methoxyphenol are identified. 2-Chloro-1,4-dimethoxybenzene (2-Cl-1,4-DMB) is another substrate for LiP, indicating a possible active function in the wood decomposition process. Like VA, it can also act as a redox mediator [98, 127].

1.4 Effect of Biological Treatment on Lignocelluloses

Biological pretreatment of lignocellulosic biomass changes the physico-chemical characteristic of biomass. Among the changes, lignin degradation is the most attractive and most studied. For example, lignin loss in wheat straw was found 25 % after 1 week [128]; lignin loss in corn straw was up to 54.6 % after 30 days pretreatment with *T. vericolor* [129]; lignin loss increased from 75.67 % to 80 % when corn stalk treated with *Irpex lacteus* [130]; lignin extractability and glucose yield could be improved in canola straw with fungus strain *T. vericolor* and cellobiose dehydrogenase-deficient strain (m4D) [44]. Degradation of lignin by microbes is mainly due to a non-specific oxidative reaction, which leads to complete oxidation of lignin. Among bio-delignifier, white-rot fungus is one of the mostly studied microbes, as discussed earlier, which has unique capability to cleave carbon–carbon linkages of lignin and oxidizes with the help of various lignolytic enzymes. The changes in terms of the ratio between *p*-hydroxyphenyl (H), guaiacyl (G), and syringyl (S) units of lignin were analyzed using pyrolysis-gas chromatography–mass spectrometry (Py-GC–MS) and concluded that the susceptibility of lignin units are in the following order: S > G > H. This order indicates that the biomass with S-rich lignin is more susceptible to fungal degradation than the biomass with other lignin units [32].

During fungal attack on biomass, hemicellulose and cellulose are also consumed and among biomass components, hemicellulose is easier to degrade. White-rot fungi such as *P. chrysosporium* [131], *P. citrinopileatus* and *P. florida* [132], *Trametes ochracea* and *E. taxodii* 2538 [77], *C. subvermispora* [133] have been found to degrade hemicellulose along with lignin loss (Table 1.1) and showed the multiple endoxylanase activity. This effect results in reduction of recalcitrance of lignocellulose but increases the risk of loss of cellulose or lowering the all sugar recovery in bioconversion process [130].

White-rot fungi also secrete cellulase enzyme with different specificity and synergistic characteristics during biological treatment of lignocellulose. Cellulase hydrolyzes β-1,4-linkage of cellulose to glucose and the hydrolyzed products are utilized by same fungi or other microbes. As mentioned earlier, non-selective white-rot fungi mineralize all lignocellulosic components equally. Selective white-rot fungi generally degrade negligible amount of cellulose and have promising role in the biological pretreatment of lignocellulose. Cellulose loss can be analyzed by X-ray diffraction (XRD) method in terms of crystallinity index (CrI). When the corn stover was treated with brown-rot fungi *Fomitopsis* sp. IMER2, the crystallinity degree of treated biomass could be increased from 33.22 % to 46.06 % and crystalline portion from 59.96 % to 94.96 % [134]; and it was found that the crystalline change of the treated biomass is due to *Fomitopsis* sp. IMER2 preferential degradation of the amorphous region of cellulose. In contrast, crystallinity decreased from 68.4 % to 64–65.9 % after the biological pretreatment of Japanese red pine (*Pinus densiflora*) with three white-rot fungi [5].

Further, Xu et al. [31] investigated the surface morphological changes during white-rot fungus *I. lacteus* CD2 attack on corn stover by scanning electron microscopy (SEM). SEM images showed some physical changes after biological treatment and resulted in irregular holes in the corn stover. The functional group changes and bond arrangement in the treated corn stover were analyzed by Fourier transform infrared (FTIR) spectroscopy [31], wheat straw biodegradation by *P. chrysosporium* [131] and bamboo culms (*Phyllostachys pubescence*), which was treated by *E. taxodii* 2538 and *T. versicolor* G20 [135]. The various characterization results, obtained by distinguished researchers, indicate that biological treatment increases the pore volume, pore size and remarkably enhance the surface area of the lignocellulose. A more-defined surface area obtained from wheat straw treated by *P. chrysosporium* supplemented with Tween 80 inorganic salts, indicating removal of lignin and making more accessible the surface of hemicellulose and cellulose [128]. Xu et al. [31] also indicated that biological treatment of corn stover with *I. lacteus* CD2 enhanced the pore size and pore volume of corn stover, resulted more accessible surface area for enzymatic saccharification.

Table 1.1 Effect of biological treatment on lignocellulosic components

Raw material	Strain name	Weight loss (%)	Lignin loss (%)	Cellulose loss (%)	Hemicellulose loss (%)	Reference
Softwood, *Pinus densiflora*	*Ceriporia lacerata*lacerate	9.5 ± 0.5	13.1 ± 0.4	8.0 ± 0.5	–	[5]
	Polyporus brumalis	9.9 ± 0.4	11.6 ± 0.3	10.6 ± 0.3	–	
	Stereum hirsutum	10.7 ± 0.7	14.5 ± 0.4	7.8 ± 0.3	–	
Sugarcane trashes	*Cellulomonas cartae*	15.5 ± 3.83	5.5 ± 0.26	25.4 ± 0.66	–	[13]
	Cellulomonas uda	24.3 ± 2.06	5.5 ± 0.25	21.8 ± 1.25	–	
	Bacillus macerans	17.5 ± 0.49	5.5 ± 0.22	30.4 ± 0.51	–	
	Zymomonas mobilis	17.9 ± 0.54	8 ± 0.51	26.8 ± 0.63	–	
Prosopis juliflora wood	*Pycnoporus cinnabarinus*	18.87 ± 1.11	8.87 ± 0.22	4.06 ± 0.18	–	[136]
Lantana camara wood	*Pycnoporus cinnabarinus*	15.4 ± 1.88	13.13 ± 1.32	2.34 ± 0.54[a]	–	
Chinese willow (*Salix babylonica*, hardwood)	*Echinodontium taxodii*	32.5 ± 1.7	45.6 ± 2.0	26.7 ± 0.2	50.8 ± 1.8	[77]
China-fir (*Cunninghamia lanceolata*, softwood)	*Echinodontium taxodii*	24.1 ± 0.9	39.8 ± 1.2	12.6 ± 0.1	31.4 ± 2.7	
Corn stover	*Ceriporiopsis subvermispora*	18.8	39.2	4.8 ± 0.25	28 ± 0.5	[137]
	Ceriporiopsis subvermispora	14.59 ± 0.28	31.33 ± 1.01	4.49 ± 1.29	22.45 ± 0.54	[138]
Bamboo clums	*Echinodontium taxodii*	10.58	24.28	1.64	28.46	[135]
	Flammulina velutipes	2.27	3.14	3.88	4.82	
	Ganoderma lucidum	12.1	10.56	12.83	15.16	
	Trametes ochracea	15.21	18.63	10.79	29.22	
	Trichaptum biforme	11.04	12.54	8.48	32.7	
Water hyacinth	*Pleurotus citrinopileatus*	31.9 ± 0.2	19.1 ± 0.4	30.1 ± 0.5	37.5 ± 0.1	[132]
	Pleurotus plorida	28.8 ± 0.4	19.7 ± 0.3	28.5 ± 0.8	30.5 ± 0.7	
Moso Bamboo	*Irpex lacteus*	–	17.87 ± 0.83	48.20 ± 0.92	18.50 ± 0.97	[139]
Wheat straw	*Fomes fomentarius*	–	35 ± 1	45 ± 1	51 ± 27	[140]
Corn stover	*Auricularia polytricha, Irpex lacteus*	–	17.8 ± 1.0	31.5 ± 0.8	16.8 ± 0.9	[141]

The effect of biological pretreatment of lignocellulose in terms of weight loss, cellulose, hemicellulose, and lignin losses is summarized in Table 1.1.

1.5 Combined Biological Treatment with Other Pretreatment Methods

In view of achieving the effective biological pretreatment, the process can be combined with physical and chemical treatment methods as the main drawback of biological pretreatment is loss of polysaccharide (cellulose/hemicellulose) and the longer pretreatment duration than chemical and physical pretreatment. Combination of biological pretreatment with chemical/physical pretreatment can enhance the fermentable sugar conversion from biomass and can improve the performance of pretreatment as compared to sole pretreatment. It is obvious that chemical/physical pretreatment prior to biological pretreatment allows the substrate more assessable for microbes to degrade lignin. Therefore, optimization is required in order to minimize the overall cost of the pretreatment, time, and energy and maximize the fermentable sugar yield after the enzymatic treatment. This combination can be carried out by two ways (i) chemical/physical treatment prior to biological pretreatment, (ii) chemical/physical treatment after biological pretreatment. The combined biological pretreatment with chemical/physical treatment and pretreatment process conditions are summarized in Tables 1.2 and 1.3, respectively.

Taniguchi et al. [142] treated rice straw with steam explosion prior to biological pretreatment using *P. ostreatu* and found that the pretreatment duration could be reduced from 60 days to 36 days required for obtaining 33 % net glucose yield. Yu et al. [129] reported that the treatment time could be reduced from 60 days to 18 days with considerable sugar yield, when rice straw was pretreated with H_2O_2 (2 %, 48 h).Itoh et al. [143] reported that ethanol yield could be increased by 1.16 times when biological pretreatment was carried out prior to organosolv treatment by using *C. subvermispora* and saved 15 % electrical energy. Indeed, biological treatment can also be used in lignin-based oil production. For example, *Fomitopsis* sp. IMER2 was used in removal of amorphous region of cellulose from corn stover and resulted a significant increase in the oil yield from 32.7 % to 50.8 % in pyrolysis process. Therefore, it can be concluded that biopretreatment favors thermal decomposition of corn stover [134].

1.6 Challenges in Biological Pretreatment

The fermentable sugar loss and relatively long time of the pretreatment compared to physical/chemical pretreatment are major challenges in biological pretreatment process. As discussed earlier, brown-rot fungi are the major consumers of fermentable sugars in the biological pretreatment. Furthermore, biological pretreatment requires

Table 1.2 Combined biological pretreatment of lignocellulose with chemical/physical treatment

Raw materials	Chemical/physical pretreatment	Biological treatment	Achievement	Reference
Rice straw	Steam explosion prior to biological pretreatment	*Pleurotus ostreatus*	Reduction in pretreatment duration from 60 daysto 36 days for obtaining 33 % glucose yield	[142]
Rice straw	Pretreated with H_2O_2 (2 %, 48 h) before biological pretreatment	*Echinodontium taxodii*	Reduction in pretreatment duration from 60 daysto 18 days	[129]
Water hyacinth	After the biological pretreatment, 0.25 % H_2SO_4 acid treatment	*Echinodontium taxodii*, *Eichhorina crassipes*	Sugar yield increased by a factor of 1.13 to 2.11	[47]
Beech wood chips	Biological pretreatment prior to organosolv treatment	*Ceriporiopsis subvermispora*	Ethanol yield increased by 1.16 times and saved 15 % electrical energy	[143]
Pinus radiata	Biological pretreatment carried out prior to ethanolysis	*Gloephyllum trabeum*	Increased solvent accessibility and decreased H factor from 6,000 to 1,156 for obtaining 161 g ethanol/kg of wood	[144]
Corn stover	Thermochemical decomposition after the biological pretreatment	*Fomitopsis* sp. IMER2	Oil yield increased from 32.7 % to 50.8 %	[134]
Corn stalks	Alkaline treatment after the biological pretreatment	*Irpex lacteus*	Lignin loss increased from 75.67 % to 80 %	[130]
Wheat straw	Thermal decomposition after the biological pretreatment	*Phanerochaete chrysosporium*	Significant reduction in the thermal degradation temperature	[131]
Corn straw	Biological pretreatment for 15 days followed by alkali/oxidative pretreatment	*Echinodontium taxodii*	Sugar yield increased by 50.7 %	[129]

more space and longer time; hence the probability of risk of contamination increases. Consequently, these factors increase the process cost. In order to overcome the above problems and making the process more cost effective and beneficial, a dedicative microorganism must be used in the process, where it could decrease the lignocelluloses recalcitrance with a minimum loss of sugar and a short time for incubation. The effective biological pretreatment process is influenced by many factors, such as (i) strain selection: The strain must have a high affinity to lignin rather than the other part of

Table 1.3 Biological pretreatment conditions for various applications

Strain	Raw material	Pretreatment condition	Result(s) achieved	Reference
Trametes hirsuta	Paddy straw	Solid state fermentation at 30 °C for 10 days	Enhanced carbohydrate content by 11.1 %	[146]
Irpex lacteus	Corn stover	In 250 ml Erlenmeyer flasks at 28 °C for 25 days	Highest saccharification ratio reached 66.4 %	[114]
Stereum hirsutum	Japanese red pine chips	cultivated at 30 °C for 8 weeks in cultivation bottle	Sugar yield increased up to 21.01 %	[5]
Phanerochaete chrysosporium	Wheat straw	Solid state fermentation at 37 °C for one week supplemented with Tween 80	Highest lignin loss (25 %) and approx. 250 % higher efficiency for the total sugar release	[128]
Epitrimerus taxodii	Bamboo culms	Cultures maintained at 25 °C for 120 days in 250 ml Erlenmeyer flasks	Sugar yield increased 8.7 fold and caused high lignin loss (>20 %)	[135]
P. chrysosporium	Wheat straw	solid substrate fermentation at 30 °C in 500 ml Erlenmeyer flasks for 3 weeks	30 % loss of total lignin	[131]
Ceriporiopsis subvermispora	Japanese cedar wood	300 ml Erlenmeyer flask at 28 °C with 70 % relative humidity for 4–8 weeks supplemented with wheat bran	74–76 % of β-O-4 aryl ether linkages in the lignin and methane yield reached 35 %	[147]
C. subvermispora	Corn stover	solid-state fermentation at 28 °C for 42 days	57–67 % overall glucose yield increased	[137]
C. subvermispora	Corn stover	pretreated at 28 °C with 75 % moisture content for 35 days	Lignin degradation up to 31.59 % and glucose yields of 66.61 %	[138]

lignocelluloses; (ii) high degradation rate of lignin; (iii) simple nitrogen source requirement; (iv) simple micronutrient requirements. These factors have already been optimized and implemented by many researchers in their biological pretreatment process for various applications.

In view of reducing the capital cost, incubation time and effective biological pretreatment with minimum fermentable sugar loss, the following approaches can be implemented in near future:

1. Combined biological and chemical/physical treatment may be effective for treatment of lignocelluloses.

2. Using some advance tools like bioinformatic tools, metagenomic tools, and high throughput screening, the process can be implemented effectively. For example, as discussed earlier, altering the pathway of lignolytic enzyme or removing cellulase/hemicellulase enzymes may provide the alternative solution.
3. Novel strains or novel enzymes can be isolated with the help of metagenomic tools for the better degradation or conversion of lignocelluloses.
4. To inhibit the action of cellulolytic enzyme or to increase the lignolytic enzyme action during the process, a specific enzyme inhibitor or mediator can be used.

Acknowledgments We thank Dr. A.K. Jain, Director, SSS-NIRE, Kapurthala, for his valuable suggestions and encouragement to write this chapter. One of the authors (Nandhagopal N.) gratefully acknowledges the Senior Research Fellowship under 'Bio-energy Promotion Fellowship' awarded by the SSS-NIRE, Kapurthala and MNRE, Govt. of India, New Delhi.

References

1. Botha T, Blottnitz HV (2006) A comparison of the environmental benefits of bagasse-derived electricity and fuel ethanol on a life-cycle basis. Energy Policy 34:2654–2661
2. Patrick MF, Champagne P, Cunningham MF, Whitney RA (2010) Biorefinery processing perspective: treatment of lignocellulosic materials for the production of value-added products. Bioresour Technol 101:8915–8922
3. Hendriks ATWM, Zeeman G (2009) Pretreatment to enhance the digestibility of lignocellulosic biomass. Bioresour Technol 100:10–18
4. Rajkumar Singh S, Singh OV (2008) Bioconversion of lignocellulosic biomass: biochemical and molecular perspectives. J Ind Microbiol Biotechnol 35:377–391
5. Lee JW, Gwak K, Park JY, Park MJ, Choi DH, Kwon M, Choi IG (2007) Biological pretreatment of softwood *Pinus densiflora* by three white-rot fungi. J Microbiol 45:485–491
6. Fengel D, Wegener G (1984) Wood-chemistry, ultrastructure, reactions. J Polym Sci, Part C: Polym Lett 23:601–602
7. Mielenz JR (2001) Ethanol production from biomass: technology and commercialization status. Curr Opin Microbiol 4:324–329
8. Kim KH, Tucker MP, Nguyen QA (2002) Effects of pressing lignocellulosic biomass on sugar yield in two-stage dilute-acid hydrolysis process. Biotechnol Prog 18:489–494
9. Kadam KL, McMillan JD (2003) Availability of corn stover as a sustainable feedstock for bioethanol production. Bioresour Technol 88:17–25
10. Knauf M, Moniruzzaman M (2004) Lignocellulosic biomass processing: a perspective. Int Sugar J 106:147–150
11. Demirbas A (2008) Products from lignocellulosic materials via degradation processes. Energy Sources Part A 30:27–37
12. Agbor VB, Cicek N, Sparling R, Berlin A, Levin DB (2011) Biomass pretreatment: fundamentals toward application. Biotechnol Adv 29:675–685
13. Singh P, Suman A, Tiwari P, Arya N, Gaur A, Shrivastava AK (2008) Biological pretreatment of sugarcane trash for its conversion to fermentable sugars. World J Microbiol Biotechnol 24:667–673
14. Ohkuma M, Maeda Y, Johjima T, Kudo T (2001) Lignin degradation and roles of white-rot fungi: study on an efficient symbiotic system in fungus-growing termites and its application to bioremediation. Riken Rev 42:39–42
15. Kumar P, Barrett DM, Delwiche MJ, Stroeve P (2009) Methods for pretreatment of lignocellulosic biomass for efficient hydrolysis and biofuel production. Ind Eng Chem Res 48:3713–3729

16. Lee J (1997) Biological conversion of lignocellulosic biomass to ethanol. J Biotechnol 56:1–24
17. Lin Y, Tanaka S (2006) Ethanol fermentation from biomass resources: current state and prospects. Appl Microbiol Biotechnol 69:627–642
18. Champagne P (2008) Bioethanol from agricultural waste residues. Environ Prog 27:51–57
19. Kumar S (2010) Biethanol production from biomass by using biomass. PhD Thesis, Indian institute of Technology, Roorkee
20. Galbe M, Zacchi G (2002) A review of the production of ethanol from softwood. Appl Microbiol Biotechnol 59:618–628
21. Demirbaş A (2005) Bioethanol from cellulosic materials: a renewable motor fuel from biomass. Energy Sources Part A 27:327–337
22. Van Maris AJA, Abbott DA, Bellissimi E, Van Den Brink J, Kuyper M, Luttik MAH, Wisselink HW, Scheffers WA, Van Dijken JP, Pronk JT (2006) Alcoholic fermentation of carbon sources in biomass hydrolysates by *Saccharomyces cerevisiae*: current status. Antonie van Leeuwenhoek 90:391–418
23. Taherzadeh MJ, Niklasson C, Lidén G, Eklund R, Gustafsson L (1997) Characterization and fermentation of dilute-acid hydrolyzates from wood. Ind Eng Chem Res 36:4659–4665
24. Zaldivar J, Nielsen J, Olsson L (2001) Fuel ethanol production from lignocellulose: a challenge for metabolic engineering and process integration. Appl Microbiol Biotechnol 56:17–34
25. Li W, Xu J, Wang J, Yan Y, Zhu X, Chen M, Tan Z (2008) Studies of monosaccharide production through lignocellulosic waste hydrolysis using double acids. Energy Fuels *22:*2015–2021
26. Hamelinck CN, van Hooijdonk G, Faaij A (2005) Ethanol from lignocellulosic biomass: techno-economic performance in short-, middle- and long-term. Biomass Bioenergy 28:384–410
27. Karimi K, Kheradmandini S, Taherzadeh MJ (2006) Conversion of rice straw to sugars by dilute-acid hydrolysis. Biomass Bioenergy 30:247–253
28. Zhu L, O'Dwyer JP, Chang VS, Granda CB, Holtzapple M (2008) Structural features affecting biomass enzymatic digestibility. Bioresour Technol 99:3817–3828
29. O'Dwyer JP, Zhu L, Granda CB, Holtzapple MT (2007) Enzymatic hydrolysis of lime-pretreated corn stover and investigation of the HCH-1 Model: inhibition pattern, degree of inhibition, validity of simplified HCH-1 Model. Bioresour Technol 98:2969–2977
30. Dadi AP, Varanasi S, Schall CA (2006) Enhancement of cellulose saccharification kinetics using an ionic liquid pretreatment step. Biotechnol Bioeng 95:904–910
31. Xu C, Ma F, Zhang X, Chen S (2010) Biological pretreatment of corn stover by *Irpex lacteus* for enzymatic hydrolysis. J Agric Food Chem 58:10893–10898
32. Isroi, Millati R, Syamsiah S, Niklasson C, Cahyanto MN, Lundquist K, Taherzadeh MJ (2011) Biological pretreatment of lignocellulose with white-rot fungi and its application: review. BioResources 6:1–36
33. Ghosh P, Ghose TK (2003) Bioethanol in India: recent past and emerging future. In: Scheper T (ed) Advanced biochemical engineering/biotechnology, biotechnology in India II. Springer, New York 20:1–27
34. Kumar S, Singh SP, Mishra IM, Adhikari DK (2009) Recent advances in production of bioethanol from lignocellulosic biomass. Chem Eng Technol 32:517–526
35. Badger PC (2002) Ethanol from cellulose: a general review. In: Janick J, Whipkey A (eds) Trends in new crops and new uses. ASHS Press, Alexandria
36. Mosier N, Wyman C, Dale B, Elander R, Lee YY, Holtzapple M, Ladisch M (2005) Features of promising technologies for pretreatment of lignocellulosic biomass. Bioresour Technol 96:673–686
37. Campo I D, Alegría I, Zazpe M, Echeverría M, Echeverría I (2006) Diluted acid hydrolysis pretreatment of agri-food wastes for bioethanol production. Ind Crops Prod 24:214–221
38. Cardona CA, Sánchez ÓJ (2007) Fuel ethanol production: process design trends and integration opportunities. Bioresour Technol 98:2415–2457
39. Saritha M, Arora A, Lata (2011) Biological pretreatment of lignocellulosic substrates for enhanced delignification and enzymatic digestibility. Ind J Microbiol. doi: 10.1007/s12088-011-0199-x

40. Claassen PAM, van Lier JB, Lopez-Cóntreras AM, van Niel EWJ, Sijtsma L, Stams AJM, deVries SS, Weusthuis RA (1999) Utilisation of biomass for the supply of energy carriers. Appl Microbiol Biotechnol 52:741–755
41. Sun Y, Cheng J (2002) Hydrolysis of lignocellulosic materials for ethanol production: a review. Bioresour Technol 83:1–11
42. Yang B, Wyman CE (2008) Pretreatment: the key to unlocking low-cost cellulosic ethanol. Biofuels Bioprod Biorefin 2:26–40
43. Zheng Y, Pan Z, Zhang R (2009) Overview of biomass pretreatment for cellulosic ethanol production. Int J Agric Biol Eng 2:51–68
44. Canam T, Town JR, Tsang A, McAllister TA, Dumonceaux TJ (2011) Biological pretreatment with a cellobiose dehydrogenase-deficient strain of *T. versicolor* enhances the biofuel potential of canola straw. Bioresour Technol 102:10020–10027
45. Keller FA, Hamilton JE, Nguyen QA (2003) Microbial pretreatment of biomass. Appl Biochem Biotechnol 105:27–41
46. Monrroy M, Ibãnez J, Melin V, Baeza J, Mendonc RT, Contreras D, Freer J (2010) Bioorganosolv pretreatments of *P radiata* by a brown-rot fungus (*Gloephyllum trabeum*) and ethanolysis. Enzyme Microb Technol 47:11–16
47. Ma F, Yang N, Xu C, Yu H, Wu J, Zhang X (2010) Combination of biological pretreatment with mild acid pretreatment for enzymatic hydrolysis and ethanol production from water hyacinth. Bioresour Technol 101:9600–9604
48. Falkowski P, Scholes RJ, Boyle E, Canadell J, Canfield D et al (2000) The global carbon cycle: a test of our knowledge of earth as a system. Science (New York) 290:291–296
49. Hatakka A (2001) Biodegradation of lignin. In: Hofrichter M, Steinbuchel A (eds) Biopolymer. Biology, chemistry, biotechnology, application, vol 1 lignin, humic substance and coal. Wiley, New York
50. Waksman SA, Cordon TC, Umbreit WW (1939) Thermophilic actinomycetes and fungi in soils and composts. Soil Sci 47:37–61
51. Kirk TK (1971) Effects of microorganisms on lignin. Ann Rev Phytopathol 9:185–210
52. Hatakka A (1983) Pretreatment of wheat straw by white-rot fungi for enzymatic saccharification of cellulose. Appl Microbiol Biotechnol 18: 350–357
53. Russell JD, Henderson Moira EK, Farmer VC (1961) Metabolism of lignin model compounds by *Polystictus versicolor*. Biochim Biophys Acta 52:565–570
54. Ishikawa H, Schubert WJ, Nord FF (1963) Investigations on lignins and lignification XXVIII. The degradation by *Polyporus versicolor* and *Fomes fomentarius* of aromatic compounds structurally related to soft wood lignin. Arch Biochem Biophys 100:140–149
55. Kirk T K, Harkin JM, Cowling EB (1968) Oxidation of guaiacyl and veratryl-glycerol, 8-guaiacyl ether by *Polyporous versicolor* and *Stereum frustulaturn*. Biochem Biophys Acta 165:135–144
56. Fukuzumi T, Takatuka H, Minami K (1969) Enzymic degradation of lignin V The effects of NADH on the enzymic cleavage of the arylalkyl ether bond in veratrylglycerol-β-guaiacyl ether as lignin model compound. Arch Biochem Biophys 129:396–403
57. Haider K, Trojanowski J (1975) Decomposition of 14 C-labelled phenols and dehydropolymers of coniferyl alcohol as models for lignin degradation by soft and white-rot fungi. Arch Microbiol 105:33–41
58. Kirk TK, Connors WJ, Bleam RD, Hackett WF, Zeikus JG (1975) Preparation and Microbial Decomposition of Synthetic [14C]lignins. Proc Natl Acad Sci 72:2515–2519
59. Ander P, Eriksson KE (1976) The importance of phenol oxidase activity in lignin degradation by the white-rot fungus *Sporotrichum pulverulentum*. Arch Microbiol 109:1–8
60. Ander P, Hatakka A, Eriksson KE (1980) Vanillic acid metabolism by the white-rot fungus *Sporotrichum pulverulentum*. Arch Microbiol 125:189–202
61. Cowling EB (1961) Comparative biochemistry of the decay of sweetgum sapwood by white-rot fungi and brown-rot fungi. US Department of Agriculture Technical Bulletin 1258 Washington

62. Hatakka A, Uusi-Rauva A (1983) Degradation of 14C-labeled poplar wood lignin by selected white-rot fungi. Eur J Appl Microbiol Biotechnol 17:235–242
63. Tien M, Kirk TK (1983) Lignin-degrading enzyme from the hymenomycete *Phanerochaete chrysosporium* burds. Science 221:661–663
64. Glenn JK, Meredith A, Morgan MA, Mayfield MB, Kuwahara M, Gold MH (1983) An extracellular H2O2 requiring enzyme preparation involved in lignin biodegradation by the white-rot basidiomycete *Phanerochaete chrysosporium*. Biochem Biophys Res Commun 114:1077–1083
65. Kuwahara M, Glenn JK, Morgan MA, Gold MH (1984) Separation and characterization of two extracellular H2O2-dependent oxidases from ligninolytic cultures of *Phanerochaete chrysosporium*. FEBS Lett 69:247–250
66. Kersten PJ, Tien M, Kalyanaraman B, Kirk TK (1985) The Iigninase of *Phanerochaete chrysosporium* generates cation radicals from methoxybenzenes. J Biol Chem 260:2609–2612
67. Kirk TK, Tien M, Johnsrud SC, Eriksson KE (1986) Lignin-degrading activity of *Phanerochaete chrysosporium* burds: comparisons of cellulase-negative and other strain. Enzyme Microb Technol 8:75–80
68. Kirk TK, Farrell RL (1987) Enzymatic combustion: the microbial degradation of lignin. Ann Rev Microbiol 41:465–505
69. Reid ID (1989) Optimization of solid-state fermentation for selective delignification of aspen wood with *Phlebia tremellosa*. Enzyme Microb Technol 11:786–803
70. Tien M, Tu C (1987) Cloning and sequencing of a cDNA for ligninase from *Phanerochaete*. Nature 326:520–523
71. Pribnow D, Mayfield MB, Nipper VJ, Brown JA, Gold MH (1989) Characterization of a CDNA encoding a manganese peroxidase from the lignin-degrading basidomycete *Phanerochaete chrysosporium*. J Biol Chem 264:5036–5040
72. Mayfield SP, Cohen A, Danon A, Yohn CB (1994) Translation of the psbA mRNA of *Chlamydomonas reinhardtii* requires a structured RNA element contained within the 5′ untranslated region. J Cell Biol 127:1537–1545
73. Saloheimo M, Niku-Paavola ML, Knowles JKC (1991) Isolation and structural analysis of the laccase gene from the lignin-degrading fungus *Phlebia radiate*. J Gen Microbiol 137:1537–1544
74. [74] Sollewijn Gelpke MD, Mayfield-Gambill M, Lin Cereghino GP, Gold MH (1999) Homologous expression of recombinant lignin peroxidase in *Phanerochaete chrysosporium*. Appl Environ Microbiol 65:1670–1674
75. Ducros V, Brzozowski AM, Wilson KS, Brown SH, Ostergaard P, Schneider P, Yaver DS, Pedersen AH, Davies GJ (1998) Crystal structure of the type-2 Copper depleted laccase from *Coprinus cinereus* at 22 Å resolution. Nat Struct Biol 5:310–316
76. Hofrichter M, Vares T, Kalsi M, Galkin S, Scheibner K, Fritsche W, Hatakka A (1999) Production of manganese peroxidase and organic acids and mineralization of 14 C-labelled lignin (14 C-DHP) during solid-state fermentation of wheat straw with the white-rot fungus *Nematoloma frowardii*. Appl Environ Microbiol 65:1864–1870
77. Yu H, Guo G, Zhang X, Yan, Xu C (2009) The effect of biological pretreatment with the selective white-rot fungus *Echinodontium taxodii* on enzymatic hydrolysis of softwoods and hardwoods. Bioresour Technol 100:5170–5175
78. Pérez J, Munoz-Dorado J, de la Rubia T, Martinez J (2002) Biodegradation and biological treatments of cellulose, hemicellulose and lignin: an overview. Int Microbiol 5:53–63
79. Shaw AJ, Jenney Jr FE, Adams MWW, Lynd LR (2008) End-product pathways in the xylose fermenting bacterium, *Thermoanaerobacterium saccharolyticum*. Enzyme Microb Technol 42:453–458
80. Muller HW, Trosch W (1986) Screening of white-rot fungi for biological pretreatment wheat straw for biogas production. Appl Microbiol Biotechnol 24:180–185
81. Tuyen VD, Cone JW, Baars JJP, Sonnenberg ASM, Hendriks WH (2012) Fungal strain and incubation period affect chemical composition and nutrient availability of wheat straw for rumen fermentation. Bioresour Technol 111:336–342

82. Gilbertson RI (1980) Wood-rotting fungi of North American polypores wood-rotting fungi of North America. Mycologia 72:1–49
83. Burlat V, Ruel K, Martinez AT, Camarero S, Hatakka A, Vares T, Joseleau JP (1998) The nature of lignin and its distribution in wheat straw affect the patterns of degradation by filamentous fungi. In: Proceedings of the 7th international conference on BioTechnol in pulp and paper industry, Canadian Pulp and Paper Association, Montreal, Canada
84. Hatakka A, Hammel KE (2010) Fungal biodegradation of lignocelluloses. In: Esser K (series ed) The Mycota, a comphrehensive treatise on fungi as experimental system for basic and applied research, Hofrichter M (vol ed) Industrial application, 2nd edn. Springer, Berlin
85. Samsuri M, Gozan M, Hermansyah H, Prasetya B, Nasikin M, Watanabe T (2008) Ethnanol production from bagasse with combination of cellulasecellubiase in simultaneous saccharification and fermentation (SSF) using white-rot fungi pre-treatment. J Chem Nat Resour Eng 3:20–32
86. Blanchette RA (1995) Degradation of the lignocellulose complex in wood. Can J Bot 73:999–1010
87. Tian X, Fang Z, Guo F (2012) Impact and prospective of fungal pre-treatment of lignocellulosic biomass for enzymatic hydrolysis. Biofuels Bioprod Biorefin 6:335–350
88. Taniguchi M, Suzuki H, Watanabe D, Sakai K, Hoshino K, Tanaka T (2005) Evaluation of Pretreatment with *Pleurotus ostreatus* for enzymatic hydrolysis of rice straw. J Biosci Bioeng 100:637–643
89. Fernandez-Fueyoa E, Ruiz-Dueasa FJ, Ferreirab P, Floudasc D, Hibbettc DS, Canessad P et al (2012) Comparative genomics of *Ceriporiopsis subvermispora* and *Phanerochaete chrysosporium* provide insight into selective ligninolysis. Proc Natl Acad Sci 109:5458–5463
90. Hatakka A (1983) Pretreatment of wheat straw by white-rot fungi for enzymatic saccharification of cellulose. Appl Microbiol Biotechnol 18: 350–357
91. Monrroy M, Ortega I, Ramírez M, Baeza J, Freer J (2011) Structural change in wood by brown-rot fungi and effect on enzymatic hydrolysis. Enzyme Microb Technol 49:472–477
92. Rasmussen ML, Shrestha P, Khanal SK, Pometto AL, Leeuwen AHL (2010) Sequential saccharification of corn fiber and ethanol production by the brown-rot fungus *Gloeophyllum trabeum*. Bioresour Technol 101:3526–3533
93. Ray MJ, Leak DJ, Spanu P D, Murphy RJ (2010) Brown-rot fungal early stage decay mechanism as a biological pretreatment for soft wood biomass in biofuel production. Biomass Bioenergy 34:1257–1262
94. Nilsson T, Daniel G, Kirk TK, Obst JR (1989) Chemistry and microscopy of wood decay by some higher ascomycetes. Holzforschung 43:11–18
95. Crawford DL, Pometto III AL, Crawford RL (1983) Lignin degradation by *Streptomyces viridosporus*: isolation and characterization of a new polymeric lignin degradation in termediate. Appl Environ Microbiol 45:898–904
96. Godden B, Ball AS, Helvenstein P, McCarthy AJ, Penninckx MJ (1992) Towards elucidation of the lignin degradation pathway in actinomycetes. J Gen Microbiol 138:2441–2448
97. Tuomela M, Vikman M, Hatakka A, Itavaara M (2000) Biodegradation of lignin in a compost environment: a review. Bioresour Technol 72:169–183
98. Have RT, Teunissen PJM (2001) Oxidative mechanisms involved in lignin degradation by white-rot fungi. Chem Rev 101:3397–3413
99. Archibald FS (1992) A new assay for lignin-type peroxidases employing the dye azure B. Appl Environ Microbiol 58:3110–3116
100. Cullen D, Kersten PJ (2004) Enzymology and molecular biology of lignin degradation. In: Bramblr R, Marzluf GA (eds) The mycota III, biochemistry and molecular biology, 2nd edn. Springer, Berlin
101. Wong DWS (2009) Structure and action mechanism of ligninolytic enzymes. Appl Biochem Biotechnol 157:174–209
102. Hofrichter M (2002) Review: lignin conversion by manganese peroxidase (MnP). Enzyme Microb Technol 30:454–4662

103. Hammel KE, Cullen D (2008) Role of fungal peroxidases in biological ligninolysis. Curr Opi Plant Biol 11:1–7
104. Baldrian P, Snajdr J (2006) Production of ligninolytic enzymes by litter-decomposing fungi and their ability to decolorize synthetic dyes. Enzyme Microb Technol 39:1023–1029
105. Heinzkill M, Bech L, Halkier T, Schneider P, Anke T (1998) Characterization of laccases and peroxidases from wood-rotting fungi (Family *Coprinaceae*). Appl Environ Microbiol 64:1601–1606
106. Bourbonnais R, Paice MG, Reid ID, Lanthier P, Yaguchi M (1995) Lignin oxidation by laccase isozymes from *T. versicolor* and role of the mediator 2,29-azinobis(3-ethylbenzthiazoline-6-sulfonate) in kraft lignin depolymerization. Appl Environ Microbiol 61:1876–1880
107. Li KC, Helm RF, Eriksson KEL (1998) Mechanistic studies of the oxidation of a non-phenolic lignin model compound by the laccase/1-hydroxybenzotriazole redox system. Biotechnol Appl Biochem 27:239–243
108. Xu F, Kulys JJ, Duke K, Li KC, Krikstopaitis K, Deussen HJW, Abbate E, Galinyte V, Schneider P (2000) Redox chemistry in laccase-catalyzed oxidation of N-hydroxy compounds. Appl Environ Microbiol 66:2052–2056
109. Johannes C, Majcherczyk A (2000) Natural mediator in the oxidation of polycyclic aromatic hydrocarbon by laccase mediator system. Appl Environ Microbiol 66:524–528
110. Eggert C, Temp U, Dean JFD, Eriksson KEL (1996) A fungal metabolite mediates degradation of non-lignin structures and synthetic lignin. FEBS Lett 391:144–148
111. Bertrand T, Jolivalt C, Briozzo P, Caminade E, Joly N, Madazak C, Mougin C (2002) Crystal structure of a four-copper laccase complexed with an arylamine: insight into substrate recognition and correlation with kinetics. Biochemistry 41:7325–7333
112. Piontek K, Antorini M, Choinowski T (2002) Crystal structure of a laccase from the fungus *T. versicolor* at 190 Å resolution containing a full complement of coppers. J Biol Chem 277:37663–37669
113. Kunamneni A, Ballesteros A, Plou FJ, Alcalde M (2007) Fungal laccase—a versatile enzyme for biotechnological applications. In: Méndez-Vilas A (ed) Communicating in current research and educational topics and trends in applied microbiology. Formex, Spain
114. Chen M, Yao S, Zhang H, Liang X (2010) Purification and characterization of a versatile peroxidase from edible mushroom *Pleurotus eryngii*. Chin J Chem Eng 18:824–829
115. Hammel KE, Mozuch MD, Jensen KA, Kersten PJ (1994) H2O2 recycling during oxidation of the arylglycerol β-aryl ether lignin structure by lignin peroxidase and glyoxal oxidase. Biochemistry 33:13349–13354
116. Harreither W, Sygmund C, Augustin M, Narciso M, Rabinovich ML, Gorton L, Haltrich D, Ludwig R (2011) Catalytic properties and classification of cellobiose dehydrogenases from ascomycetes. Appl Environ Microbiol 77:1804–1815
117. Henriksson G, Johansson G, Pettersson G (2000) A critical review of cellobiose dehydrogenases. J Biotechnol 78:93–113
118. Harreither W, Sygmund C, Dünhofen E, Vicuna R, Haltrich D, Ludwig R (2009) Cellobiose dehydrogenase from the Ligninolytic Basidiomycete *Ceriporiopsis subvermispora*. Appl Environ Microbiol 75:2750–2757
119. Henriksson G, Ander P, Pettersson B, Pettersson G (1995) Cellobiose dehydrogenase (cellbiose oxidase) from *Phanerochaete chrysosporium* as a wood degrading enzyme, studies on cellulose, xylan and synthetic lignin. Appl Microbiol Biotechnol 42:790–796
120. Dumonceauxa T, Bartholomewa K, Valeanua L, Charlesd T, Archibalda F (2001) Cellobiose dehydrogenase is essential for wood invasion and nonessential for kraft pulp delignification by *Trametes versicolor*. Enzyme Microb Technol 29:478–489
121. Phillips CM, Beeson WT, Cate JH, Marletta MA (2011) Cellobiose dehydrogenase and a copper-dependent polysaccharide monooxygenase potentiate cellulose by *Neurospora crassa*. ACS Chem Biol 6:1339–1406
122. Hammel KE, Jensen AR Jr, Mozuch MD, Landucci L, Tien M, Pease EA (1993) Ligninolysis by purified lignin peroxidase. J Biol Chem 268:12274–12281

123. Shimada MF, Nakatsubo, Kirk TK, Higuchi T (1981) Biosynthesis of the secondary metabolite veratryl alcohol in relation to lignin degradation in *Phanerochaete chrysosporium*. Arch Microbiol 129:321–324
124. Mester T, Jong E, Field JA (1995) Manganese regulation of veratryl alcohol in white-rot fungi and its indirect effect on Lignin Peroxidase. Appl Environ Microbiol 61:1881–1887
125. Hames BR, Kurek B, Pollet B, Lapierre C, Monties B (1998) Interaction between MnO2 and oxalate: formation of a natural and abiotic lignin oxidizing system. J Agric Food Chem 46:5362–5367
126. Roy BP, Paice MG, Archibald FS (1993) Creation of metal-complexing agents, reduction of manganese dioxide, and promotion of manganese peroxidase-mediated Mn(III) production by cel l obiose: quinone oxidoreductase from *Trametes versicolor*. J Biological Chem 269:19745–19750
127. Teunissen PJM, Field JA (1998) 2-Chloro-1,4-Dimethoxybenzene as a novel catalytic cofactor for oxidation of anisyl alcohol by lignin peroxidase. Appl Environ Microbiol 64:830–835
128. Zeng J, Singh D, Chen S (2011) Biological pretreatment of wheat straw by *Phanerochaete chrysosporium* supplemented with inorganic salts. Bioresour Technol 102:3206–3214
129. Yu H, Zhang X, Song L, Ke J, Xu C, Du W, Zhang J (2010) Evolution of white-rot fungi-assisted alkaline/oxidative pretreatment of corn straw undergoing enzymatic hydrolysis by cellulose. J Biosci Bioeng 110:660–664
130. Yu H, Du W, Zhang J, Ma F, Zhang X, Zhong W (2010) Fungal treatment of cornstalks enhances the delignification and xylan loss during mild alkaline pretreatment and enzymatic digestibility of glucan. Bioresour Technol 101:6728–6734
131. Singh D, Zeng J, Laskar DD, Deobald L, Hiscox WC, Chen S (2001) Investigation of wheat straw biodegradation by *Phanerochaete chrysosporium*. Biomass Bioenergy 35:1030–1040
132. Mukherjee R, Nandi B (2004) Improvement of in vitro digestibility through biological treatment of water hyacinth biomass by two *Pleurotus* species. Int Biodeterior Biodegrad 53:7–12
133. Mendonca RT, Jara JF, Gonzalez V, Elissectche JP, Freer J (2008) Evalution of white-rot fungi *Ganoderma austral* and *Ceriporiopsis subvermispora* in biotechnological application. J Ind Microbiol Biotechnol 35:1323–1330
134. Zeng Y, Yang X, Yu H, Zhang X, Ma F (2011) Comparative studies on thermochemical characterization of corn stover pretreated by white-rot and brown-rot fungi. J Agric Food Chem 59:9965–9971
135. Zhang X, Yu H, Huang H, Liu Y (2007) Evaluation of biological pretreatment with white-rot fungi for the enzymatic hydrolysis of bamboo culms. Int Biodeterior Biodegrad 60:159–164
136. Gupta R, Mehta G, Khasa YP, Kuhad RC (2011) Fungal delignification of lignocellulosic biomass improves the saccharification of cellulosics. Biodegradation 22:797–804
137. Wan C, Li Y (2010) Microbial delignification of corn stover by *Ceriporiopsis subvermispora* for improving cellulose digestibility. Enzyme Microb Technol 47:31–36
138. Wan C, Li Y (2010) Microbial pretreatment of corn stover with *Ceriporiopsis subvermispora* for enzymatic hydrolysis and ethanol production. Bioresour Technol 101:6398–6403
139. Zeng Y, Yang X, Yu H, Zhang X, Ma F (2011) The delignification effects of white-rot fungal pretreatment on thermal characteristics of moso bamboo. Bioresour Technol. doi: 10.1016/j.biortech.2011.10.036
140. Salvachúa D, Prieto A, López-Abelairas M, Lu-Chau T, Martínez AT, Martínez MJ (2011) Fungal pretreatment: an alternative in second-generation ethanol from wheat straw. Bioresour Technol 102:7500–7506
141. Ma F, Wang J, Zeng Y, Yu H, Yang Y, Zhang X (2011) Influence of the co-fungal treatment with two white rot fungi on the lignocellulosic degradation and thermogravimetry of corn stover. Process Biochem 46:1767–1773
142. Taniguchi M, Takahashi D, Watanabe D, Sakai K, Hoshino K, Kouya T, Tanaka T (2010) Effect of steam explosion pretreatment on treatment with *Pleurotus ostreatus* for the enzymatic hydrolysis of rice straw. J Biosci Bioeng 110:449–452

143. Itoh H, Wada M, Honda Y, Kuwahara M, Watanabe T (2003) Bioorganosolve pretreatments for simultaneous saccharification and fermentation of beech wood by ethanolysis and white-rot fungi. J Biotechnol 103:273–280
144. Monrroy M, Ibanez J, Melin V, Baeza J, Mendon RT, Contreras D, Freer J (2010) Bioorganosolv pretreatments of *P radiata* by a brown-rot fungus (*Gloephyllum trabeum*) and ethanolysis. Enzyme Microb Technol 47:11–16
145. Zeng J, Singh D, Chen S (2011) Thermal decomposition kinetics of wheat straw treated by *Phanerochaete chrysosporium*. Int Biodeterior Biodegrad 65:410–414
146. Saritha M, Arora A, Nain L (2012) Pretreatment of paddy straw with *Trametes hirsuta* for improved enzymatic saccharification. Bioresour Technol 104:459–465
147. Amirta R, Tanabe T, Watanabe T, Hondaa Y, Kuwahara M, Watanabe T (2006) Methane fermentation of Japanese cedar wood pretreated with a white rot fungus, *Ceriporiopsis subvermispora*. J Biotechnol 123:71–77

Chapter 2
Microbial Production of Extracellular Polysaccharides from Biomass

Ebru Toksoy Öner

Abstract The interest in polysaccharides has increased considerably in recent years, as they are candidates for many commercial applications in different industrial sectors like food, petroleum, and pharmaceuticals. Because of their costly production processes, industrial microbial polysaccharides like xanthan, dextran, curdlan, gellan, and pullulan constitute only a minor fraction of the current polymer market. Therefore, much effort has been devoted to the development of cost-effective and environmentally friendly production processes by switching to cheaper fermentation substrates. In this chapter, various microbial polysaccharide production processes utilizing cheap biomass resources like syrups and molasses, olive mill wastewater, cheese whey, various vegetable and fruit pomace, pulp and kernels as well as carbon dioxide and lignocellulosic biomass like rice hull and bran, sawdust, and fibers are discussed with a special focus on the employed pretreatment methods.

Keywords EPS · Microbial exopolysaccharides · Polysaccharides · Biomass · Fermentation

2.1 Introduction

Since the beginning of twentieth century, technologies related to microbial production of biomolecules like enzymes, antibiotics, metabolites, and polymers have matured to a great extend. Currently, microbes are used for commercial production of a wide variety of products such as pesticides, fertilizers, and feed additives in agrochemical sector, biopharmaceuticals and therapeutics in the healthcare sector, biopolymers and biofuels in the energy and environment sectors. According to recent market reports, growing environmental concerns and increasing demands from end-use sectors are expected to increase the global market for microbial products to about 250 billion US dollars by 2016 [1].

E. T. Öner (✉)
IBSB—Industrial Biotechnology and Systems Biology Research Group,
Bioengineering Department, Marmara University,
Goztepe Campus, 34722 Istanbul, Turkey
e-mail: ebru.toksoy@marmara.edu.tr

Z. Fang (ed.), *Pretreatment Techniques for Biofuels and Biorefineries*,
Green Energy and Technology,
DOI 10.1007/978-3-642-32735-3_2,

Polysaccharides are natural, non-toxic, and biodegradable polymers that cover the surface of most cells and play important roles in various biological mechanisms such as immune response, adhesion, infection, and signal transduction [2, 3]. Investigations on the alternative treatments applied by different cultures throughout the history revealed the fact that the utilized plants and fungi were rich in bioactive polysaccharides with proven immunomodulatory activity and health promoting effects in the treatment of inflammatory diseases and cancer. Hence considerable research has been directed on elucidating the biological activity mechanism of these polysaccharides by structure-function analysis [4].

Besides the interest on their applications in the health and bionanotechnology sectors, polysaccharides are also used as thickeners, bioadhesives, stabilizers, probiotic, and gelling agents in food and cosmetic industries [5–7] and as emulsifier, biosorbent, and bioflocculant in the environmental sector [8].

Polysaccharides are either extracted from biomass resources like algae and higher-order plants or recovered from the fermentation broth of bacterial or fungal cultures. For sustainable and economical production of bioactive polysaccharides at industrial scale, rather than plants and algae, microbial sources are preferred since they enable fast and high yielding production processes under fully controlled fermentation conditions. Microbial production is achieved within days and weeks as opposed to plants where production takes 3–6 months and highly suffers from geographical or seasonal variations and ever increasing concerns about the sustainable use of agricultural lands. Moreover, production is not only independent of solar energy which is indispensible for production from microalgae but also suitable for utilizing different organic resources as fermentation substrates [5].

According to recent reports, the global hydrocolloid market dominated by algal and plant polysaccharides like starch, galactomannans, pectin, carrageenan, and alginate is expected to reach 3.9 billion US dollars by 2012 [9]. Superseding these traditionally used plant and algal gums by their microbial counterparts requires innovative approaches and considerable progress has been made in discovering and developing new microbial extracellular polysaccharides (exopolysaccharides, EPSs) that possess novel industrial significance [6, 7]. A recent review pointed out to four EPSs, namely, xanthan, pullulan, curdlan, and levan, as biopolymers with outstanding potential for various industrial sectors [5]. However, when compared with the synthetic polymers, natural origin polymers still represent only a small fraction of the current polymer market, mostly due to their costly production processes. Therefore, much effort has been devoted to the development of cost-effective and environmentally friendly production processes such as investigating the potential use of cheaper fermentation substrates.

In this chapter, after a brief description of microbial polysaccharides, various microbial production processes utilizing cheap biomass resources as fermentation substrates are discussed with a special focus on the employed pretreatment methods.

2.2 Microbial Polysaccharides

In nature, biopolymers often play important roles in maintaining cell viability by conserving genetic information, by storing carbon-based macromolecules, by producing either energy or reducing power, and by defending an organism against attack from hazardous environmental factors [10]. Microbial polysaccharides are high molecular weight carbohydrate polymers present either at the outer membrane as lipopolysaccharides (LPS) that mainly determine the immunogenic properties or secreted as capsular polysaccharides (CPS) forming a discrete surface layer (capsule) associated with the cell surface or excreted as EPS that are only loosely connected with the cell surface [11]. Whereas CPSs are assigned with functions directly related with pathogenicity like resistance to specific and nonspecific host immunity and adherence [12], EPSs fulfill a variety of diverse functions including adhesion, cell to-cell interactions, biofilm formation [13], and cell protection against environmental extremes [2].

Polysaccharides show considerable diversity in their composition and structure. They are generally classified as homopolysaccharides and heteropolysaccharides based on their monomeric composition [14]. Homopolysaccharides are composed of one type of monosaccharide repeating unit where sugar monomers are bound to form either linear chains (pullulan, levan, curdlan or bacterial cellulose) or ramified chains (dextran). Heteropolysaccharides are composed of two or more types of monosaccharides and are usually present as multiple copies of oligosaccharides, containing three to eight residues (gellan or xanthan) [15, 16]. Table 2.1 summarizes the chemical characteristics of major bacterial and fungal polysaccharides.

The microorganisms used as industrial or technical producers of EPS are chiefly the bacteria. Species of *Xanthomonas*, *Leuconostoc*, *Sphingomonas*, and *Alcaligenes* which produce xanthan, dextran, gellan, and curdlan are the best known and most industrially used (Table 2.1).

Dextran (synthesized by certain lactic-acid bacteria such as *L. mesenteroides*) was the first microbial polysaccharide to be commercialized and to receive approval for food use [22]. Xanthan gum is the EPS from the plant pathogen *X. campestris* bacterium and due to its exceptional rheological properties; xanthan has a considerable market [28]. Gellan produced by *S. paucimobilis* is gaining increasing attention due to its novel property of forming thermo-reversible gels and hold great commercial potential for food, pharmaceuticals, and predominantly environmental bioremediation [23]. On the other hand, because of its very high immunocompatibility and water-binding and retention capacity, hyaluronan is widely used in regenerative medicine and cosmetic applications [16]. Curdlan produced by the alkaline tolerant mesophilic pathogen *Alcaligenes faecalis* [20] can form aqueous suspensions which can form high-set gels upon heating and curdlan is also produced by *Cellulomonas flauigena* as an extracellular storage polymer [21]. Species of *Pseudomonas* and *Azotobacter* are the only microbial sources for alginate that is widely used as a thickening, stabilizing, and gelifying agent in food, textile, paper, and pharmaceutical industries [17, 18]. Contrary to plant cellulose, bacterial cellulose is produced and excreted to

Table 2.1 Principal bacterial and fungal polysaccharides

EPS	Monomers	Charge	Characteristics of chemical structure	Organism	Reference
Bacteria					
Alginate	Guluronic acid Mannuronic acid	Anionic	Blocks of β-1,4-linked D-mannuronic residues, blocks of α-1,4-linked L-guluronic acid residues, and blocks with these uronic acids in random or alternating order	*Pseudomonas aeruginosa* *Azotobacter vinelandii*	[17] [18]
Cellulose	Glucose	Neutral	β-1,4-D-glucan	*Gluconacetobacter xylinus*	[19]
Curdlan	Glucose	Neutral	β-1,3-D-glucan	*Alcaligenes faecalis* *Cellulomonas flavigena*	[20] [21]
Dextran	Glucose	Neutral	α-D-glucan linked by α-1,6-glycosidic bonds; some 1,2-, 1,3-, or 1,4-bonds are also present in some dextrans	*Leuconostoc mesenteroides*	[22]
Gellan	Glucose Rhamnose Glucuronic acid	Anionic	Partially O-acetylated polymer of D-glucose-1,4-β-D-glucuronic acid-1,4-β-D-glucose-1,4-β-L-rhamnose tetrasacharide units connected by α-1,3-glycosidic bonds	*Sphingomonas paucimobilis*	[23]
Hyaluronan	Glucuronic acid Acetylglucosmine	Anionic	Repeating units of β-1,4-linked disaccharides of β-D-*N*-Acetylglucosmine-β-1,3-D-Glucuronic acid	*Pseudomonas aeruginosa* *Pasteurella multocida*	[24] [25]
Levan	Fructose	Neutral	β-2,6-D-fructan	*Bacillus subtilis* *Zymomonas mobilis* *Halomonas sp.*	[26] [27]
Xanthan	Glucose Mannose Glucuronic acid	Anionic	β-1,4-D-glucan with β-D-mannose-1,4-β-D-glucuronic acid-1,2-α-D-mannose sidechain. Approximately 50 % of terminal mannose residues are pyruvated and the internal mannose residue is acetylated at C-6	*Xanthomonas campestris*	[28]
Fungi					
Pullulan	Glucose	Neutral	α-1,6-linked α-1,4-D-triglucoside maltotriose units	*Aureobasidium pullulans*	[29] [30]
Scleroglucan	Glucose	Neutral	β-1,3-D-glucan with β-1,6-D-glucose linked to every third unit	*Sclerotium glutanicum*	[31]

the fermentation medium and directly recovered as a highly pure polymer free from lignin and other noncellulosic materials. It is now a high-value, specialty chemical with highly specific applications like skin regeneration [16]. Levan is a water soluble, strongly adhesive, and film-forming biopolymer with many potential uses as emulsifier, stabilizer and thickener, encapsulating agent, osmoregulator, and cryoprotector in food, cosmetics, pharmaceutical or chemical industries [26]. Fungal polysaccharides are still somewhat limited, with pullulan from *A. pullulans* [30] and scleroglucan produced by *Sclerotium glucanicum* [32] being the most known and already obtained at technical scales. Whereas the major market for pullulan is still in food sector, there are numerous reports for its potential applications in pharmaceutical, medical, and environmental remediation areas [30]. Similarly, due to its exceptionally high stability, the first application of scleroglucan was in the oil recovery; however, other applications in pharmaceutical, cosmetic, and agriculture sectors have also been proposed [31, 32].

2.3 Microbial EPS Production Processes

Fermentation is a very versatile process technology for producing value added products such as microbial biopolymers and since fermentation parameters have a high impact upon the viability and economics of the bioprocess, their optimization holds great importance for process development. Especially, microbial polysaccharide production is greatly influenced by fermentation conditions such as pH, temperature, oxygen concentration and agitation as well as by the composition of the culture medium [7, 10, 16]. Moreover, besides the fermentation conditions, the chemical structure, monomer composition, and physicochemical and rheological properties of the final product also change with the type of strain. This in turn allows the industrial production of polysaccharides with desired specifications via controlling the fermentation conditions, choosing feasible feedstocks, and using high-level producer strains.

EPS synthesis is a tightly regulated carbon and energy-intensive process resulting in a wide range of nutritional and environmental requirements of the EPS producer strains. Consequently, dependency of the production on microbial growth, nutrient availability, and fermentation conditions are subject to significant controversy in literature and hence generalizations should be avoided [7].

Fermentations for EPS production are batch, fed-batch or continuous processes depending on the microbial system used. In most cases, optimum values of temperature and pH for biomass formation and EPS production differ considerably so that typical fermentations start with the growth phase followed by the production phase. Moreover, considerable changes in the rheological properties occur during the course of fermentation due to EPS production. This results in a highly viscous and non-Newtonian broth which in turn may not only cause serious problems of mixing, heat transfer, and oxygen supply but also give rise to instabilities in the quality of the end product. Whereas this is a common technical difficulty in commercial xanthan and pullulan production processes [30], it is not encountered in levan production due to the exceptionally low intrinsic viscosity of the polymer [33] as well as in microbial processes utilizing thermophilic microorganisms where production is realized at high temperatures [7]. Whether the production is small laboratory scale or large at industrial scale, the fermentation media are almost always designed to have high carbon to nitrogen ratio where nitrogen serves as the growth limiting nutrient [34]. Under conditions employed for industrial production of microbial polysaccharides, the same principle of high carbon to nitrogen ratios is used, but the substrates utilized are the cheapest available form.

2.3.1 Low-Cost Biomass Resources

Fermentation medium can represent almost 30 % of the cost for a microbial fermentation. Complex media commonly employed for growth and production are not economically attractive due to their high amount of expensive nutrients such as yeast extract, peptone, and salts. In order to reach high production titers at reasonable costs, fermentation medium should be carefully designed to make the end product compatible with its synthetic petrochemical counterpart.

Fermentation feedstock has been the most expensive constituent in microbial biopolymer production. Till the1990s, studies were generally focused on using defined culture conditions in order to recover ultra pure biopolymers with minimum batch-to-batch variation and free of impurities that would interfere with their chemical and biological characterization. However, to maximize the cost effectiveness of the process, recent work has shifted to use multi-component feedstock systems and the synthetic media were replaced by cheaper alternatives such as olive mill wastewater (OMW), syrups, and molasses [7, 16, 35]. Currently, a wide range of industrial and agricultural by-products and waste materials are used as nutrients for industrial fermentations. In Table 2.2, various biomass resources and the applied pretreatment methods have been listed for some microbial EPS producers together with the EPS yields obtained after a certain fermentation period.

2.3.1.1 Syrups and Molasses

Syrups and molasses have long been used as substrates for fermentative production of commercial polysaccharides such as pullulan [46–49], xanthan [57], dextran [38], scleroglucan [32], levan [33, 41, 42], and gellan [39] due to their many advantages like high sucrose and other nutrient contents, low cost and ready availability, and ease of storage. Using molasses in crude form resulted in low pullulan [46] and scleroglucan [32] yields which in turn pointed out the need for pretreatments. Pretreatment of sugar beet molasses with sulfuric acid has been reported for pullulan [47] and levan [33] production but, when acid treatment was combined with activated carbon treatment, significant improvements in pullulan [48, 49] and levan [33] yields

Table 2.2 Biomass resources and applied pretreatments for some microbial EPSs

EPS	Microorganism	Biomass	Pretreatment	Yield (Time)	Reference
Curdlan	*Agrobacterium* sp. ATCC 31749	CCS	Clarification by filtration	7.72 g/L (120 h)	[36]
Dextran	*L. mesenteroides* NRRL B512	Carob extract	Milling	8.56 g/L (12 h)	[37]
		Aqueous extraction			
Dextran	*L. mesenteroides* NRRL B512	Carob extract and cheese whey	Deproteinization of whey	7.23 g/L (12 h)	[37]
Dextran	*L. mesenteroides* V-2317D	Sugar beet M	No treatment	50 g/L (9 days)	[38]
Gellan	*S. paucimobilis* ATCC-31461	Sugarcane M	Dilution	13.81 g/L (48 h)	[39]
Gellan	*S. paucimobilis* ATCC 31461	Cheese whey	Neutralization	7.9 g/L (100 h)	[40]
		Heat treatment			
Levan	*Halomonas* sp. AAD6	Starch M	Clarification by centrifugation	12.4 g/L (210 h)	[33]
	Sugar beet M	pH adjustment			
		Acid hydrolysis			
		TCP treatment			
		AC treatment			
Levan	*Paenibacillus polymyxa* NRRL B-18475	Sugar beet M	Dilution	38.0 g/L (5 days)	[41]
		Gel filtration chromatography			
		Anion exchange chromatography			
Levan	*P. polymyxa* NRRL B-18475	Sugarcane syrup	Clarification by filtration	19.6 g/L (5 days)	[41]
Levan	*Zymomonas mobilis* ATCC 31821	Sugarcane M	Clarification by centrifugation and filtration	2.53 g/L (24 h)	[42]
Levan	*Z. mobilis* ATCC 31821	Sugarcane syrup	Clarification by centrifugation	15.5 g/L (24 h)	[42]
		Filtration			
Pullulan	*Aureobasidium* sp. NRRL Y	CCS	Clarification by centrifugation	4.5 g/L (9 days)	[43]
Pullulan	*A. pullulans* SU-M18	Carob extracts	Aqueous extraction	6.5 g/L (3 days)	[44]
Pullulan	*A. pullulans*	OMW	Clarification by filtration	8 g/L	[45]
Pullulan	*A. pullulans* NRRLY-6220	OMW	No treatment	10.7 g/L (7 days)	[46]
Pullulan	*A. pullulans* NRRLY-6220	Grape pomace	Aqueous extraction	22.3 g/L (7 days)	[46]
Pullulan	*A. pullulans* NRRLY-6220	Sugar beet M	Dilution	6.0 g/L (7 days)	[46]
Pullulan	*A. pullulans*	Sugar beet M	Acid hydrolysis	32.0 g/L	[47]

Table 2.2 (continued)

EPS	Microorganism	Biomass	Pretreatment	Yield (Time)	Reference
Pullulan	*A. pullulans* P 56	Sugar beet M AC treatment	Acid hydrolysis	24 g/L (144 h)	[48]
Pullulan	*A. pullulans* P56	Sugar beet M $K_3[Fe(CN)_6]$ treatment AC treatment	Acid hydrolysis	35 g/L (96 h)	[49]
Scleroglucan	*Sclerotium rolfsii* MTCC 2156	Sugarcane juice	Dilution	23.87 g/L (72 h)	[32]
Scleroglucan	*S. rolfsii* MTCC 2156	Sugarcane M	Dilution	19.21 g/L (72 h)	[32]
Scleroglucan	*S. rolfsii* MTCC 2156	Coconut water	Dilution	12.58 g/L (72 h)	[32]
Scleroglucan	*S. rolfsii* MT-6	Waste loquat kernel Acid Hydrolysis Detoxification	Milling	12.08 g/L (72 h)	[50]
Scleroglucan	*S. glucanicum* NRRL 3006	CCS	Dilution	14.8 g/L (144 h)	[51]
Xanthan	*X. campestris*	Carob extracts Pressing Heat treatment	Aqueous extraction	0.126 g/L/h	[52]
Xanthan	*X. campestris* PD 656	Apple pomace Alkaline treatment	Drying and crushing	52.1 g/L (6 days)	[53]
Xanthan	*X. campestris*	Grape pomace Alkaline treatment	Drying and crushing	10 g/L (6 days)	[53]
Xanthan	*X. campestris* PD 656	Tangerine peels	Alkaline treatment	32.9 g/L (6 days)	[53]
Xanthan	*X. campestris* NRRL B-1459	Sugar beet pulp	No pretreatment	1.19 g/L (4 days)	[54]
Xanthan	*X. campestris* NRRL B-1459	OMW	Clarification	4 g/L (5 days)	[55]
Xanthan	*X. campestris* T646	OMW	Clarification	7.7 g/L (5 days)	[56]
Xanthan	*X. campestris* ATCC 1395	Sugar beet M	No pretreatment	53 g/L (24 h)	[57]
Xanthan	*X. campestris* EBK-4	Ram horn hydrolysate Heat treatment Clarification by filtration	Acid hydrolysis	25.6 g/L (48h)	[58]
Xanthan	*X. campestris* 1182	Cheese whey	No pretreatment	26.35 g/L (72 h)	[59]
β-Glucan	*Botryosphaeria rhodina*	OMW	Clarification by centrifugation	17.2 g/L (120 h)	[60]
EPS	*Paenibacillus jamilae* CECT 5266	OMW	Clarification by filtration	2.5 g/L (100 h)	[61]
EPS	*P. jamilae* CP-38	OMW	Clarification by filtration	5 g/L (72 h)	[62]
EPS	*Halomonas* sp. AAD6	Sugar beet pulp	Drying and milling	2.22 g/L (3 days)	[63]

AC: activated carbon, *CCS*: condensed corn solubles, *M*: molasses, *OMW*: Olive mill wastewater, *TCP*: tricalcium phosphate were obtained, most probably due to the removal of heavy metals and colored substances. Activated carbon is particularly known for its efficiency in removing heavy metal pollutants [64]. However, after a systematic study on the effect of different pretreatments on the heavy metal distribution of starch and beet molasses samples, Küçükaşık et al. [33] reported a drastic increase in the dissolved iron (Fe^{2+}) content after the activated carbon treatment. This has been attributed to the reduction of iron from its impregnated Fe^{3+} form to its soluble form since this increase in soluble iron was more profound when acid treated samples were subjected to activated carbon treatment [33]. Same authors suggested tricalcium phosphate ($Ca_3(PO_4)_2$,TCP) treatment as an effective method for selective removal of iron and zinc from molasses or other mixtures of comparable composition. Heavy metals like iron, zinc, and nickel are known to enter the apatite crystal structure of TCP by replacing the Ca atom [65]. Göksungur et al. [49] applied potassium ferrocyanide ($K_3[Fe(CN)_6]$) treatment to precipitate heavy metals. For the microbial levan production with *Paenibacillus polymyxa* NRRL B-18475, clarified sugar cane syrup and crude sugar beet molasses resulted in very low yields. Therefore, peptone was added to the cane syrup and beet molasses was subjected to various expensive pretreatments like passing it through gel filtration and anion exchange columns in order to increase the levan yields to levels comparable with sucrose [41]. To produce levan from *Zymomonas mobilis*, both sugarcane molasses and sugarcane syrup were clarified by centrifugation followed by filtration and then used at 250 g/L carbohydrate concentration [42]. For levan production by halophilic *Halomonas* sp. cultures, sugar beet molasses, and starch molasses were subjected to five different physical and chemical pretreatment methods and their combinations, that is, clarification, pH adjustment, sulfuric acid, TCP and activated carbon treatment [33]. In both molasses types, pretreatments like clarification, pH adjustment were not adequate as also reflected by the low EPS production yields due to the retained undesirable constituents (e.g., heavy metals, impurities) which influence the growth of microorganism and associated polysaccharide production [47, 66, 67]. Highest levan yields were obtained with sugar beet molasses pretreated with TCP followed by acidification with sulfuric acid and then subjected to activated carbon pretreatment [33]. On the other hand, Kalogiannis et al. [57] applied various treatment methods to sugar beet molasses including aeration, acid, activated carbon, $K_3[Fe(CN)_6]$ treatments, and ion exchange chromatography, however, none of the pretreatments improved the xanthan yield of *X. campestris* ATCC 1395 cultures and the highest production was obtained with the untreated crude molasses. Crude beet molasses is also used to produce dextran by *L. mesenteroides* bacterial cultures and the yields were comparable to those of media containing pure sucrose [38]. Banik et al. [39] used Response surface methodology to optimize the production of gellan gum by *S. paucimobilis* ATCC-31461 using crude sugarcane molasses and reported a maximum yield of 13.81 g/L gellan. Survase et al. [32] used various dilutions of coconut water, sugarcane molasses, and sugarcane juice for scleroglucan production by filamentous fungi *S. rolfsii* MTCC 2156 and obtained the highest yields (23.87 g/L in 72 h) from sugarcane juice that was obtained from a local market and hence did not require any pretreatments before use. Coconut water and sugarcane juice were also used for EPS production by *Lactobacillus confusus* cultures [68].

2.3.1.2 Sugar Beet Pulp

Sugar beet pulp (SBP), a by-product of the sugar beet industry, is the fibrous material left over after the sugar is extracted from sugar beets and is mainly composed of cellulose, hemicellulose, and pectin. Beet pulp is used in countries with an intensive cattle raising industry, as livestock feed. In other countries, it is dumped in landfills. However, SBP can be an important renewable resource and its bioconversion appears to be of great biotechnological importance [69].

There are several reports on the pretreatments applied to SBP. Autoclaving of SBP at 122 °C and 136 °C for 1 h was reported to change its composition and physicochemical properties causing increased swelling and improved solubility of pectins and arabinans [70]. Ammonia pressurization depressurization (APD) pretreatment where SBP is exploded by the sudden evaporation of ammonia, was found to substantially increase hydrolysis efficiency of the cellulose component [71]. Recently, Kühnel et al. [72] examined the influence of six mild sulfuric acid or hydrothermal pretreatments at different temperatures on the enzymatic degradability of SBP and they found that optimal pretreatment at 140 °C of 15 min in water was able to solubilize 60 % w/w of the total carbohydrates present, mainly pectins.

Taking into consideration that SBP is carbohydrate-rich with a high carbon-to-nitrogen ratio (C/N, 35–40), that sugar beet farming is a widespread and already mature industry, and that beet pulp is abundant and cheap, this coproduct has potential for use as a renewable biomass feedstock for microbial fermentations for biopolymer production [71]. On the other hand, there are very limited numbers of reports on the use of SBP as a resource for microbial polysaccharide production. Yoo and Harcum investigated the feasibility of using autoclaved SBP as a supplemental substrate for xanthan gum production from *X. campestris* and they reported a production yield of 0.89 g xanthan per gram of SBP in 4 days of fermentation time [54]. Söğütçü et al. [63] investigated the effects of autoclaving, reducing the particle size by milling and accessibility of SBP for EPS production by halophilic *Halomonas* sp. AAD6 cultures. In this study, milling of dried SBP in a mortar grinder, supplying SBP in dialysis tubes rather than directly in culture media and autoclaving SBP separately, and then adding to the fermentation media were all found to increase the EPS yields.

2.3.1.3 Olive Mill Wastewater

The manufacture of olive oil produces large amounts of a dark colored juice called OMW that consists of a mixture of water from the olive, machinery cooling waters, fruit washings, and remainder of the fruit. Typically, OMW comprises about 15 % organic material that is composed of carbohydrates, proteins, and lipids as well as a number of other organic compounds including monoaromatic and polyaromatic molecules [62] and the toxic effects are mainly derived from its extremely high organic load and the presence of recalcitrant organic compounds such as polyphenols with strong antimicrobial properties. Hence valorization of OMW produced by the olive oil industry has long been an environmental concern in Mediterranean countries [61]. Beside various conventional technological treatment methods applied, biovalorization of OMW to value-added chemicals is considered as the most cost-effective and environmentally compatible option [73]. Due to its composition with high carbon-to-nitrogen ratio, its use as a suitable substrate for microbial polymer production has been proposed [45] and applied to produce pullulan [45] and xanthan gum [74]. By studying the resource variability factors [55] and then by use of a high-producer strain and medium optimization, Lopez et al. [56] reported significant improvements in xanthan yields, reaching 7.7 g/L in 5 days. In all these studies, in order to reduce the inhibitory effect of phenols, the OMW obtained from the industry was clarified by filtration, diluted with distilled water or saline, neutralized, sterilized by autoclaving, and then used in microbial fermentation. OMW pretreated by this approach has also been used for the microbial production of a metal-binding EPS by *Paenibacillus jamilae* bioreactor cultures. Due to the high phenol biodegradation ability of *Paenibacillus* genus, these cultures were not only proposed for the production of an EPS that could be used as a biofilter but also for the bioremediation of OMW [62]. The main constraint associated with the use of OMW is the need for dilution in order to lower the amount of phenols which in turn limits the concentration of the used waste as culture medium [62]. On the other hand, for the β-glucan production from the fungus *Botryosphaeria rhodina* DABAC-P82, OMW was only clarified by centrifugation and then after steam sterilization, directly applied as substrate without dilution. Due to the lack of oxidase activity, high biopolymer yields and decreases in phenol content of the culture were attributed to the adsorption action of the fungal biofilm [60]. Undiluted OMW was found to be a poor substrate for pullulan production by *A. pullulans* [46]. Besides EPSs, OMW has also been used as a fermentation substrate for the microbial production of other biopolymers including polyhydroxyalkanoates (PHAs) [75].

2.3.1.4 Cheese Whey

Whey is the major by-product obtained during the preparation of dairy products such as cheese. The nutrient composition of whey is based on the nutrient composition of milk from which it is derived, which in turn is affected by many factors including how the milk was processed. Lactose is the major component comprising about 70 % of the total solids of whey. Whey also contains a pool of nutrients and growth factors that have the potential to stimulate the growth of microorganisms but the suitability of whey for EPS production highly depends on the ability of the microorganism to utilize lactose. Cheese whey has been used as carbon and nitrogen source for xanthan [59] and gellan [40] production. Mozzarella cheese whey has been used for xanthan production by two different *X. campestris* strains and although both strains reached comparable yields, the polymers were found to differ in their chemical characteristics [59]. The low yields were attributed to the low capacity of the *X. campestris* strains to utilize lactose. On the other hand, Fialho et al. [40] evaluated the gellan gum production by the *S. paucimobilis* ATCC 31461 strain in media containing lactose, glucose, and sweet cheese whey as substrates. The strain was known to grow on lactose and to produce highly viscous gellan directly from lactose [76]. Sweet cheese whey obtained from the industry was neutralized and disinfected by three cycles of heat treatment at 80 °C for 30 min. A maximum gellan yield of 7.9 g/L could be recovered from the flask cultures after 100 h of fermentation period [40]. Cheese whey has also been investigated as a potential substrate for dextran production by *L. mesenteroides* NRRL B512 cultures [37]. For this, proteins were removed from whey by precipitation through autoclaving and then centrifugation. Though lactose in the supernatant was found to repress the dextransucrase activity, 7.23 g/L dextran could be produced when carob extract was also present in the medium [37].

2.3.1.5 Pomace

Only few researchers have published work on grape pomace, which today is a very significant waste product in agriculture industries. Grape pomace is the residue left after juice extraction by pressing grapes in the wine industry. Globally, about 10 million tons of grape pomace (seeds, skin, and stem) is produced each year. In Spain alone, over 250 million kg of this by-product are used every year either as animal feed (with low nutritional value) or for ethanol production by fermentation and distillation (low level benefit). This material is underexploited and most of it is generally disposed in open areas, leading to serious environmental problems [77]. Israilides et al. [46] extracted the sugars in the grape pulp by using hot water at 65–70 °C and then clarified the solution and used for pullulan production by *A. pullulans* NRRLY-6220 cultures. Since the grape pomace extract mainly contained sugars and very low amounts of protein, the polymer produced was very similar in its amount as well as molecular weight to the pullulan produced in defined medium. Moreover, the pullulan yields were high reaching 22.3 g/L after 7 days of fermentation period. Stredansky and Conti [53] tested grape pomace, tangerine peels, and apple pomace as substrate for xanthan production by solid state fermentation (SSF) with *X. campestris* NRRL B-1459 cultures. These substrates were soaked in alkaline solution to neutralize the organic acids and then added to the fermentation media. Performances of these feedstocks were also evaluated in the presence and absence of spent malt grains as inert support and apple pomace proved to be a superior substrate yielding high amounts of xanthan under both conditions. Low xanthan yields with grape pomace were attributed to the low sugar content used and the low absorption capacity of the solid material.

2.3.1.6 Carbon Dioxide

CO_2 is a nontoxic, nonflammable, abundant, and renewable feedstock and its biotransformation into industrially important chemicals can not only have a positive impact on the global carbon balance but also provide novel routes for the green biotechnology. As one of the oldest life forms on earth, microalgae have very high CO_2 biofixation capacity, grow fast, and accumulate large quantities of lipids and carbohydrates and hence became the most promising feedstock for production of next generation biofuels like biodiesel and bioethanol [78]. Considering the fact that CO_2 is a very cheap carbon source, microalgal systems should also be considered as potential resources for EPS production. However, in the literature, there are very few reports on microalgal polysaccharide production. In general, for production of value added products, the biggest advantage in using open microalgae culture is the direct use of solar energy which in turn is highly energy efficient and cheap [79]. Actually, these systems applied to phototropic and mixotrophic cultures are considered to be the most technically and economically feasible methods at commercial scale [78]. On the other hand, this advantage does not hold for EPS production where use of monocultures, closed, and controlled cultivation systems are required to reach high levels of productivity [80]. Although photobioreactors and fermenters are advantageous in providing optimum conditions for biomass growth and EPS production, these systems are expensive and energy intensive when compared with open systems [81].

There are various types of bioreactors that can be used for EPS production such as airlift flat plate photobioreactors (well reviewed by Zhang et al. [82]). Generally, culture conditions for lipid-rich biomass production and EPS production are remarkably different. A systematic study conducted with the green colonial fresh water microalgae *Botryococcus braunii* strains on the effect of culture conditions on their growth, hydrocarbon and EPS production also revealed two distinct culture conditions so that cultivation in 16:8 h light dark cycle yielded higher hydrocarbons whereas continuous illumination with agitation yielded higher amounts of EPSs with 1.6 g/L—maximum yield obtained from *B. braunii* LB 572 strain [83]. In a study on the effect of salinity with the same strain, EPS yields of 2–3 g/L were also reported [84]. The difference in cultivation conditions could also be used for the high-level EPS production by use of a two-stage culture as reported for spirulan production by *Spirulina platensis* [85]. In this method, whereas the first stage focuses on rapidly increasing microalgal biomass, culture conditions in the second stage are modified to maximize the polysaccharide yield. *Rhodella violacea* [86] and *Porphyridium cruentum* [87] are well known as producers for viscous bioactive EPS [88] and the highest yield of 543.1 mg/L EPS production was reported for *P. cruentum* after optimization of initial pH, light intensity, inoculation ratio, and liquid volume of shaking batch cultures [89]. By culturing *P. cruentum* semi-continuously in flat plate photobioreactors, a production rate of 68.64 mg/L per day could be reached by Sun et al. [90]. Very low EPS concentrations (less than 30 mg/L) were reported for planktonic diatoms like *Amphora holsatica*, *Navicula directa*, and *Melosira nummuloïdes* [5, 91]. However, these yields can be improved by further studies on optimizing the bioreactor conditions in favor of EPS production.

Another important issue for microalgal cultivation is the need for using high concentrations of chemical fertilizers as a source for nitrogen and phosphorus. Whereas high nitrogen concentrations in the cultivation medium favors polysaccharide synthesis pathways and biomass formation, lipid accumulation is favored under nitrogen limited conditions where polysaccharide pathways are blocked and the photosynthetically fixed carbon is directed towards fatty acid synthesis [92]. Microalgae could become a favorable source for EPS production if the high expenses associated with fertilizers could be reduced by replacing them with their low cost alternatives. Besides the use of wastewater as an inexpensive source, the literature is very limited in such studies.

2.3.1.7 Lignocellulosic Biomass

Lignocellulosic biomass is also a cheap and abundant alternative for microbial biopolymer production, especially for microbial systems with hydrolytic capability via endoglucanases or cellobiose. Otherwise, it is utilized to a limited extend during the fermentation and hence requires pretreatments beforehand. The filamentous fungi *S. rolfsii* and other medicinal mushrooms (*Basidiomycetes*) can naturally metabolize different five carbon sugars like xylose and arabinose and hence are especially well suited for EPS production from lignocellulosic substrates. Some examples include the simultaneous production of schizophyllan and arabinoxylan by *Schizophyllum commune* strain ATCC 38548 cultures grown on alkaline H_2O_2-pretreated corn fiber as a sole carbon source [93]. Same strain was also used for schizophyllan production from activated charcoal detoxificated rice hull hydrolysate [94]. Influences of individual or combined inhibitors as well as the importance of detoxification step in EPS production were systematically investigated in this study. Under SSF conditions, a high temperature tolerant white rot fungus *Lentinus squarrosulus* MBFBL 201 was reported to degrade cornstalks very fast and up to 5 g/L EPS could be recovered from the fermentation media [95].

On the other hand, there are only very few reports on the bacterial EPS production using cellulose-rich biomass. Acid hydrolaysates of wood were used for succinoglycan production by *Pseudomonas* sp. ATCC 31260 cultures. The produced EPS was found to be rheologically comparable with commercially available xanthan [96]. When cultured on acid-hydrolyzed sawdust, *Brevundimonas vesicularis* LMG P-23615 and *Sphingopyxis macrogoltabida* LMG 17324 bacterial strains were found to accumulate high amounts of PHA with yields ranging from 64 to 72 % of the dry cell weight [97]. In another study, lignocellulosic fibers with 58–63 % cellulose content were used as a low cost natural complex carbon source for EPS production by *Bacillus megaterium* RB-05 cells with known cellulase activity. The fibers immersed in production medium were pretreated by autoclaving for 15 min, however, once inoculated with cells, EPS production was found to be driven solely by the bacterial cellulase activity. Moreover, recovery of the EPS from the culture required several steps due to the biofilm formed along the fibers [98]. In another recent study, rice bran was subjected to serial enzymatic treatment using amylase, amyloglucosidase, alcalase, and lipase enzymes and then the hydrolysate was used for the co-production of intacellular and extracellular polymers by nitrogen-fixing *Sinorhizobium meliloti* MTCC 100 shaking bacterial cultures. Supplementation of the medium with 20 % rice bran hydrolysate resulted in maximum yields of 11.8 and 3.6 g/L EPS and PHA, respectively [99].

2.3.1.8 Others

Carob (*Ceratonia siliqua* L.), which has long been regarded as just a nitrogen-fixing tree grown in the Mediterranean region, has recently found its place in the food industry as a biomass substrate due to its very high sugar content [100]. Moreover, it has been established as a viable biomass resource for bioethanol production [101]. Carob extracts have also been used for microbial production of xanthan [52] and pullulan [44] polysaccharides. Roseiro et al. [52] developed a multistep pretreatment process for carob-based feedstocks that involves aqueous extraction of carob pulp followed by pressing. By recycling of press liquor, the final sugar content of the carob extract was improved however, as a result of esterasic activities, the syrup was found to contain increasing concentrations of isobutyric acid with time in a pH-dependent manner. Though accumulation of isobutyric acid could be controlled by an additional heat treatment step, its presence was found to inhibit the growth of *X. campestris* cells [52]. When carob extracts with 25 g/L initial sugar content were used for pullulan production by a pigmented strain of *A. pullulans* (SU-M18), a pullulan productivity of 2.16 g/L/day could be reached at pH 6.5 and 25 °C [44]. For dextran production, carob pod residues obtained from the galactomannan industry were milled and the sugars were extracted at 70 °C by use of an acetate buffer. A dextran yield of 8.56 g/L could be reached by *L. mesenteroides* NRRL B512 cultures within 12 h of fermentation period [37].

Condensed corn solubles (CCS) is a by-product of bioethanol industry. While ethanol is separated from the fermentation broth via distillation, the remaining solids are first recovered by centrifugation and then concentrated using evaporators. The final product CCS contains changing levels of carbohydrates, proteins, vitamins, and nutrients [102]. CCS obtained from a dry-mill ethanol plant has been diluted and used for the cost-effective production of scleroglucan by *S. glucanicum* [51, 103]; however, the yields were lower than those of *S. rolfsii* cultures grown on sugarcane juice or molasses [32], coconut water [32], and waste loquat kernel [50]. In another study, CCS was diluted, neutralized, clarified by centrifugation and filtration, and then used for the poly-(3-hydroxybutyrate-*co*-3-hydroxyvalerate) production by *Rhodospirillum rubrum* cultures [102]. CCS from a wet-mill ethanol production plant has been clarified by centrifugation and then used as substrate for pullulan production by *Aureobasidium* sp. strain NRRL Y-12974 cultures and the yields (4.5 g/L in 9 days) were found to be comparable with those of soluble starch (5.4 g/L in 9 days), however, much lower than glucose containing medium (10.1 g/L in 9 days) [43]. CCS has also been used for curdlan production by *Agrobacterium* sp. ATCC 31749 shake flask cultures. A maximum curdlan yield of 7.72 g/L was recovered after 120 h of fermentation in media containing 400 g/L CSS [36].

Ram horn hydrolysates were also reported to be a suitable enhancer for xanthan production by *X. campestris* EBK-4 because of their high amino acid and mineral content. To obtain the hydrolysates, ram horns, which are usually discharged as waste in slaughterhouses, were subjected to acid hydrolysis with sulfuric acid followed by heat treatment at 130 °C. After neutralization, the hydrolysates were clarified by filtration and then added to the fermentation medium [58].

Waste loquat kernel is another potential biomass resource for EPS production due to its high protein and carbohydrate content. Waste kernels are dried and milled and then subjected to acid hydrolysis with 2 M HCl using an autoclave. Then the hydrolysates were detoxified with $Ca(OH)_2$, neutralized, and then used for scleroglucan production by *S. rolfsii* MT-6 [50] and EPS production by *Morchella esculenta* [104].

Other biomass residues used for microbial EPS production include corn-steep, spent grain and spent sulfite liquors, hydrolyzed potato starch, peach pulp, and peat hydrolysate [28, 30].

2.4 Summary

Synthesis of value added biochemicals from biomass using microorganisms serves as a promising alternative to harsh chemical synthesis processes that employ expensive, hazardous, and non-renewable raw materials. Though functional characteristics of a biopolymer establish its market potential, a useful biopolymer cannot find its proper place in the polymer market unless it can be produced economically. In microbial polysaccharide production, the shift in feedstock utilization requires intensive research activities for the application of innovative concepts on a large scale. These concepts could involve novel resources and pretreatments as well as fermentation and downstream processing techniques.

Suitability of the feedstock is largely determined by the metabolic needs of the microorganism for EPS production. If the production relies on glucose or sucrose, then syrups and molasses of varying origin are established substitutes. However, other biomass resources rich in pectin, sucrose or glucose like ram horn, pomace, and pulps from food industry could also be used along with an appropriate pretreatment to extract the precursors for polysaccharide synthesis. Similarly, suitability of cheese whey for gellan rather than xanthan and dextran production is a direct outcome of the lactose metabolism of the producer strain. Cellulose-rich biomass is an already established resource for fungal bioactive EPSs; however, more studies on the isolation of novel EPS producing cellulolytic strains, development of pretreatment methods, and optimization of fermentation conditions are needed for feasible production of bacterial EPSs from lignocellulosic biomass resources.

Another very important issue in microbial EPS production is that the chemical and physical properties of the polymers are largely determined by the cultivation conditions and this variation in polymer properties is more pronounced when biomass residues are used as feedstock. This point becomes more important when the product is launched to the market. Hence precise characterization of the EPS as well as mechanism of its biosynthesis is of utmost importance in the search for suitable biomass resources.

For biopolymers with high-value applications, rather than the production yields, consistency in both product quality and yield are important which in turn can be ensured to a high extend by preventing the carryover of impurities and metabolic by-products. In such cases, chemically defined medium conditions are usually preferred for cultivation over complex medium with varying composition. Microalgae enable the use of CO_2 as a cheap, simple, and abundant carbon source and hence cell-free microalgal cultivation media could be a good source for the recovery of bioactive polysaccharides. With accumulating knowledge and awareness on the biological importance of microalgal polysaccharides, a growing number of studies are now focused on optimizing the fermentation conditions for their production. Problems associated with the diversity of culture conditions for lipid-rich biomass production, and EPS production could be overcome by further studies on the development of multi-stage process strategies. Such studies on integration of value added chemical production to microalgal biofuel production based on biorefinery approach hold great importance for the long-term sustainability of the whole process.

Acknowledgments Financial support provided by The Scientific and Technological Research Council of Turkey (TUBITAK) through project 111M232 is gratefully acknowledged.

References

1. McWilliams A (2011) Microbial products: technologies, applications and global markets. BCC Research. http://www.giiresearch.com/report/bc180728-glob-microbial-prod.html. Accessed 16 Feb 2012
2. Kumar AS, Mody K, Jha B (2007) Bacterial exopolysaccharides—a perception. J Basic Microb 47:103–117
3. Sutherland IW (1998) Novel and established applications of microbial polysaccharides. Trends Biotechnol 16:41–46
4. Ramberg JE, Nelson ED, Sinnott RA (2010) Immunomodulatory dietary polysaccharides: a systematic review of the literature. Nutr J 9:1–60
5. Donot F, Fontana A, Baccou JC, Schorr-Galindo S (2012) Microbial exopolysaccharides: main examples of synthesis, excretion, genetics and extraction. Carbohyd Polym 87:951–962
6. Freitas F, Alves VD, Reis MAM (2011) Advances in bacterial exopolysaccharides: from production to biotechnological applications. Trends Biotechnol 29:388–398
7. Nicolaus B, Kambourova M, Toksoy Öner E (2010) Exopolysaccharides from extremophiles: from fundamentals to biotechnology. Environ Technol 31:1145–1158
8. Sam S, Küçükaşık F, Yenigun O, Nicolaus B, Toksoy Öner E, Yükselen MA (2011) Flocculating performances of Exopolysaccharides produced by a Halophilic bacterial strain cultivated on agro-industrial waste. Bioresource Technol 102:1788–1794
9. Phillips K (2007) Demand for healthier foods stirs up growth in hydrocolloids. Chemical Week. http://www.accessmylibrary.com/article-1G1-167346654/demand-healthier-foods-stirs.html. Accessed 20 Feb 2012
10. Kazak H, Toksoy Öner E, Dekker RFH (2010) Extremophiles as sources of exopolysaccharides. In: Ito R, Matsuo Y (eds) Handbook of Carbohydrate Polymers: development, properties and applications. Nova Science Publishers, New York
11. Cuthbertson L, Mainprize IL, Naismith JH, Whitfield C (2009) Pivotal roles of the outer membrane polysaccharide export and polysaccharide copolymerase protein families in export of extracellular polysaccharides in gram-negative bacteria. Microbiol Mol Biol R 155–177

12. Taylor CM, Roberts IS (2005) Capsular polysaccharides and their role in virulence. In: Russell W, Herwald H (eds) Concepts in bacterial virulence. Contrib Microbiol. Karger, Basel
13. Mann EE, Wozniak DJ (2012) Pseudomonas biofilm matrix composition and niche biology. FEMS Microbiol Rev. doi:10.1111/j.1574–6976.2011.00322.x
14. Sutherland IW (1982) Biosynthesis of microbial exopolysaccharides. Adv Microb Physiol 23:79–150
15. Purama RK, Goswami P, Khan AT, Goyal A (2009) Structural analysis and properties of dextran produced by Leuconostoc mesenteroides NRRL B-640. Carbohyd Polym 76:30–35
16. Sutherland IW (2007) Bacterial exopolysaccharides. In: Kamerling JP (ed) Comprehensive glycoscience. Elsevier, Amsterdam
17. Hay ID, Gatland K, Campisano A, Jordens JZ, Rehm BHA (2009) Impact of alginate overproduction on attachment and biofilm architecture of a supermucoid Pseudomonas aeruginosa strain. Appl Environ Microb 75:6022–6025
18. Gaona G, Nunez C, Goldberg JB, Linford AS, Najera R, Castaneda M, Guzman J, Espin G, Soberon-Chavez, G (2004) Characterization of the Azotobacter vinelandii algC gene involved in alginate and lipopolysaccharide production. FEMS Microbiol Lett 238:199–206
19. Nguyen VT, Flanagan B, Gidley MJ, Dykes GA (2008) Characterization of cellulose production by a Gluconacetobacter xylinus strain from Kombucha. Curr Microbiol 57:449–453
20. Matsushita M (1990) Curdlan, a (1–3)-beta-D-glucan from Alcaligenes faecalis var. myxogenes IFO13140, activates the alternative complement pathway by heat treatment. Immunol Lett 26:95–97
21. Kenyon WJ, Buller CS (2002) Structural analysis of the curdlan-like exopolysaccharide produced by Cellulomonas flavigena KU. J Ind Microbiol Biotechnol 29:200–203
22. Sarwat F, Ul Qader SA, Aman A, Ahmed N (2008) Production and characterization of a unique dextran from an indigenous Leuconostoc mesenteroides CMG713. Int J Biol Sci 4:379–386
23. Bajaj IB, Survase SA, Saudagar PS, Singhal RS (2007) Gellan gum: fermentative production, downstream processing and applications. Food Technol Biotech 45:341–354
24. Bonde GJ, Carlsen FE, Jensen CE (1957) Production of hyaluronic acid by Pseudomonas aeruginosa. Acta Pharmacol Tox 13:205–212
25. deAngelis PL, Jing W, Drake RR, Achyuthan AM (1998) Identification and molecular cloning of a unique hyaluronan synthase from Pasteurella multocida. J Biol Chem 273:8454–8458
26. Kang SA, Jang K-H, Seo J-W, Kim KH, Kim YH, Rairakhwada D, Seo MY, Lee JO, Ha SD, Kim C-H, Rhee S-K (2009) Levan: applications and perspectives. In: Rehm BHA (ed) Microbial production of biopolymers and polymer precursors. Academic Press, Caister
27. Poli A, Kazak H, Gürleyendağ B, Tommonaro G, Pieretti G, Toksoy Öner E, Nicolaus B (2009) High level synthesis of levan by a novel Halomonas species growing on defined media. Carbohyd Polym 78:651–657
28. Palaniraj A, Jayaraman V (2011) Production, recovery and applications of xanthan gum by Xanthomonas campestris. J Food Eng 106:1–12
29. Singh RS, Saini GK, Kennedy JF (2008) Pullulan: microbial sources, production and applications. Carbohyd Polym 73:515–531
30. Cheng KC, Demirci A, Catchmark JM (2011) Pullulan: biosynthesis, production, and applications. Appl Microbiol Biotechnol 92:29–44
31. Schmid J, Meyer V, Sieber V (2011) Scleroglucan: biosynthesis, production and application of a versatile hydrocolloid. Appl Microbiol Biotechnol 91:937–947
32. Survase SA, Saudagar PS, Singhal RS (2007) Use of complex media for the production of scleroglucan by Sclerotium rolfsii MTCC 2156. Bioresource Technol 98:1509–1512
33. Küçükaşık F, Kazak H, Güney D, Finore I, Poli A, Yenigün O, Nicolaus B, Toksoy Öner E (2011) Molasses as fermentation substrate for levan production by Halomonas sp. Appl Microbiol Biot 89:1729–1740
34. Sutherland IW (1996) Microbial biopolymers from agricultural products: production and potential. Int Biodeter Biodegr 38:249–261
35. Salehizadeh H, van Loosdrecht MCM (2004) Production of polyhydroxyalkanoates by mixed culture: recent trends and biotechnological importance. Biotechnol Adv 22:261–279

36. West TP, Nemmers B (2008) Curdlan production by Agrobacterium sp. ATCC 31749 on an ethanol fermentation coproduct. J Basic Microb 48:65–68
37. Santos M, Rodrigues A, Teixeira JA (2005) Production of dextran and fructose from carob pod extract and cheese whey by Leuconostoc mesenteroides NRRL B512(f). Biochem Eng J 25:1–6
38. Vedyashkina TA, Revin VV, Gogotov IN (2005) Optimizing the conditions of dextran synthesis by the bacterium Leuconostoc mesenteroides grown in a molasses-containing medium. Appl Biochem Microbiol 41:361–364
39. Banik RM, Santhiagu A, Upadhyay SN (2007) Optimization of nutrients for gellan gum production by Sphingomonas paucimobilis ATCC-31461 in molasses based medium using response surface methodology. Bioresource Technol 98:792–797
40. Fialho AM, Martins LO, Donval ML, Leitao JH, Ridout MJ, Jay AJ, Morris VJ, Corria I (1999) Structures and properties of gellan polymers produced by Sphingomonas paucimobilis ATCC 31461 from lactose compared with those produced from glucose and from cheese whey. Appl Environ Microb 65:2485–2491
41. Han YW, Watson MA (1992) Production of microbial levan from sucrose, sugarcane juice and beet molasses. J Ind Microbiol 9:257–260
42. Oliveira MR, da Silva RSSF, Buzato JB, Celligoi MAPC (2007) Study of levan production by Zymomonas mobilis using regional low-cost carbohydrate sources. Biochem Eng J 37:177–183
43. Leathers TD, Gupta SC (1994) Production of pullulan from fuel ethanol byproducts by Aureobasidium sp. strain NRRLY-12974. Biotechnol Lett 16:1163–1166
44. Roukas T, Biliaderis CG (1995) Evaluation of carob pod as a substrate for pullulan production by Aureobasidium pullulans. Appl Biochem Biotechnol 55:27–44
45. Ramos-Cormenzana A, Monteoliva-Sánchez M, López MJ (1995) Bioremediation of alpechin. Int Biodeter Biodegr 35:249–268
46. Israilides CJ, Smith A, Harthill JE, Barnett C, Bambalov G, Scanlon B (1998) Pullulan content of the ethanol precipitate from fermented agro-industrial wastes. Appl Microbiol Biotechnol 49:613–617
47. Roukas T (1998) Pretreatment of beet molasses to increase pullulan production. Process Biochem 33:805–810
48. Lazaridou A, Biliaderis CG, Roukas T, Izydorczyk M (2002) Production and characterization of pullulan from beet molasses using a nonpigmented strain of Aureobasidium pullulans in batch culture. Appl Biochem Biotechnol 97:1–22
49. Göksungur Y, Uçan A, Güvenç U (2004) Production of pullulan from beet molasses and synthetic medium by Aureobasidium pullulans. Turk J Biol 28:23–30
50. Taşkın M, Erdal S, Canlı O (2010) Utilization of waste loquat (Eriobotrya japonica) kernels as substrate for scleroglucan production by locally isolated Sclerotium rolfsii. Food Sci Biotechnol 19:1069–1075
51. Fosmer A, Gibbons W (2011) Separation of scleroglucan and cell biomass from Sclerotium glucanicum grown in an inexpensive, by-product based medium. Int J Agric Biol Eng 4:52–60
52. Roseiro JC, Costa DC, Collaco MTA (1992) Batch and fed-cultivation of Xanthomonas campestris in carob extracts. Food Sci Technol-Lebensm-Wiss Technol 25:289–293
53. Stredansky M, Conti E (1999) Xanthan production by solid state fermentation. Process Biochem 34:581–587
54. Yoo SD, Harcum SW (1999) Xanthan gum production from waste sugar beet pulp. Bioresource Technol 70:105–109
55. Lopez MJ, Moreno J, Ramos-Cormenzana A (2001a) The effect of olive mill wastewaters variability on xanthan production. J Appl Microbiol 90:829–835
56. Lopez MJ, Moreno J, Ramos-Cormenzana A (2001b) Xanthomonas campestris strain selection for xanthan production from olive mill wastewaters. Water Res 35:1828–1830
57. Kalogiannis S, Iakovidou G, Liakopoulou-Kyriakides M, Kyriakidis DA, Skaracis GN (2003) Optimization of xanthan gum production by Xanthomonas campestris grown in molasses. Process Biochem 39:249–256

58. Kurbanoglu EB, Kurbanoglu NI (2007) Ram horn hydrolysate as enhancer of xanthan production in batch culture of Xanthomonas campestris EBK-4 isolate. Process Biochem 42:1146–1149
59. Silva MF, Fornari RCG, Mazutti MA, Oliveira D, Padilha FF, Cichoski AJ, Cansian RL, Luccio MD, Treichel H (2009) Production and characterization of xanthan gum by Xanthomonas campestris using cheese whey as sole carbon source. J Food Eng 90:119–123
60. Crognale S, Federici F, Petruccioli M (2003) Beta-Glucan production by Botryosphaeria rhodina on undiluted olive-mill wastewaters. Biotechnol Lett 25:2013–2015
61. Morillo JA, Antizar-Ladislao B, Monteoliva-Sánchez M, Ramos-Cormenzana A, Russell NJ (2009) Bioremediation and biovalorisation of olive-mill wastes. Appl Microbiol Biotechnol 82:25–39
62. Aguilera M, Quesada MT, Aguila VG, Morillo JA, Rivadeneyra MA, Romos-Cormenzana A, Monteoliva-Sanchez M (2008) Characterization of Paenibacillus jamilae strains that produce exopolysaccharide during growth on and detoxification of olive mill wastewaters. Bioresource Technol 99:5640–5644
63. Söğütçü E, Akyıldız SM, Toksoy Öner E (2011) Exopolysaccharide production from waste sugar beet pulp by Halomonas sp. In: The 3rd international conference on biodegradable and biobased polymers (BIOPOL-2011), Strasbourg, 29–31 August 2011
64. Mohan D, Chander S (2001) Single component and multi-component adsorption of metal ions by activated carbons. Colloids Surf A 177:183–196
65. Yin X, Calderin L, Stott MJ, Sayer M (2002) Density functional study of structural, electronic and vibrational properties of Mg- and Zn-doped tricalcium phosphate biomaterials. Biomaterials 23:4155–4163
66. Jiang L, Wang J, Liang S, Wang X, Cen P, Xu Z (2009) Butyric acid fermentation in a fibrous bed bioreactor with immobilized Clostridium tyrobutyricum from cane molasses. Bioresource Technol 100:3403–3409
67. Kundu S, Panda T, Majumdar SK, Guha B, Bandyopadhyay KK (1984) Pretreatment of Indian cane molasses for increased production of citric acid. Biotechnol Bioeng 26:1114–1121
68. Seesuriyachan P, Techapun C, Shinkawa H, Sasaki K (2010) Solid state fermentation for extracellular polysaccharide production by Lactobacillus confusus with coconut water and sugar cane juice as renewable wastes. Biosci Biotechnol Biochem 74:423–426
69. Hutnan M, Drtil M, Mrafkova L (2000) Anaerobic biodegradation of sugar beet pulp. Biodegradation 11:203–211
70. Guillon F, Barry J, Thibault J (1992) Effect of autoclaving sugar-beet fibre on its physicochemical properties and its in-vitro degradation by human faecal bacteria. J Sci Food Agr 60:69–79
71. Foster BL, Dale BE, Doran-Peterson JB (2001) Enzymatic hydrolysis of ammonia-treated sugar beet pulp. Appl Biochem Biotechnol 91–93:269–282
72. Kühnel S, Schols HA, Gruppen H (2011) Aiming for the complete utilization of sugar-beet pulp: examination of the effects of mild acid and hydrothermal pretreatment followed by enzymatic digestion. Biotechnol Biofuels 4:14
73. Mantzavinos D, Kalogerakis N (2005) Treatment of olive mill effluents: Part I. Organic matter degradation by chemical and biological processes—an overview. Environ Int 31:289–295
74. Lopez MJ, Ramos-Cormenzana A (1996) Xanthan production from olive mill wastewaters. Int Biodeter Biodegr 59:263–270
75. Cerrone F, Sánchez-Peinado MM, Juárez-Jimenez B, González-López J, Pozo C (2010) Biological treatment of two-phase olive mill wastewater (TPOMW, alpeorujo): polyhydroxy alkanoates (PHAs) production by Azotobacter strains. J Microbiol Biotechnol 3:594–601
76. Pollock TJ (1993) Gellan-related polysaccharides and the genus Sphingomonas. J Gen Microbiol 139:1939–1945
77. Botella C, de Ory I, Webb C, Cantero D, Blandino A (2005) Hydrolytic enzyme production by Aspergillus awamori on grape pomace. Biochem Eng J 26:100–106
78. Lam MK, Lee KT (2012) Microalgae biofuels: a critical review of issues, problems and the way forward. Biotechnol Adv 30:673–690

79. Mata TM, Martins AA, Caetano NS (2010) Microalgae for biodiesel production and other applications: a review. Renew Sust Energ Rev 14:217–232
80. Freitas AC, Rodrigues D, Rocha-Santos TAP, Gomes AMP, Duarte AC (2012) Marine biotechnology advances towards applications in new functional foods. Biotechnol Adv. doi:10.1016/j.biotechadv.2012.03.006
81. Harun R, Singh M, Forde GM, Danquah MK (2010) Bioprocess engineering of microalgae to produce a variety of consumer products. Renew Sust Energ Rev 14:1037–1047
82. Zhang Y, Arends JBA, Wiele TV, Boon N (2011) Bioreactor technology in marine microbiology: from design to future application. Biotechnol Adv 29:312–321
83. Dayananda C, Sarada R, Usha Rani M, Shamala TR, Ravishankar GA (2007) Autotrophic cultivation of Botryococcus braunii for the production of hydrocarbons and exopolysaccharides in various media. Biomass Bioenerg 31:87–93
84. Rao AR, Dayananda C, Sarada R, Shamala TR, Ravishankar GA (2007) Effect of salinity on growth of green alga Botryococcus braunii and its constituents. Bioresource Technol 98:560–564
85. Lee MC, Chen YC, Peng TC (2012) Two-stage culture method for optimized polysaccharide production in Spirulina platensis. J Sci Food Agric 92:1562–1569
86. Evans LV, Callow ME, Percival E, Fareed V (1974) Studies on the synthesis and composition of extracellular mucilage in the unicellular red alga Rhodella. J Cell Sci 16:1–21
87. Fábregas J, García D, Morales ED, Lamela T, Otero A (1999) Mixotrophic production of phycoerythrin and exopolysaccharide by the microalga Porphyridium cruentum. Cryptogamie Algol 20(2):89–94
88. Badel S, Callet F, Laroche C, Gardarin C, Petit E, El Alaoui H, Bernardi T, Michaud P (2011) A new tool to detect high viscous exopolymers from microalgae. J Ind Microbiol Biot 38:319–326
89. Wang J, Chen B, Rao X (2007) Optimization of culturing conditions of Porphyridium cruentum using uniform design. World J Microbiol Biotechnol 23:1345–1350
90. Sun L, Wang C, Ma C, Shi L (2010) Optimization of renewal regime for improvement of polysaccharides production from Porphyridium cruentum by uniform design. Bioproc Biosyst Eng 33:309–315
91. Leandro SM, Gil MC, Delgadillo I (2003) Partial characterization of exopolysaccharides exudated by planktonic diatoms maintained in batch cultures. Acta Oecologic 24:S49-S55
92. Li Y, Han D, Hu G, Sommerfeld M, Hu Q (2010) Inhibition of starch synthesis results in overproduction of lipids in Chlamydomonas reinhardtii. Biotechnol Bioeng 107:258–268
93. Leathers TD, Nunnally MS, Price NP (2006) Co-production of Schizophyllan and Arabinoxylan from Corn Fiber Biotechnol Lett 28:623–626
94. Shu CH, Hsu HJ (2011) Production of schizophyllan glucan by Schizophyllum commune ATCC 38548 from detoxificated hydrolysate of rice hull. J Taiwan Inst Chem E 42(3):387–393
95. Isikhuemhen OS, Mikiashvili NA, Adenipekun CO, Ohimain EI, Shahbazi G (2012) The tropical white rot fungus, Lentinus squarrosulus Mont.: lignocellulolytic enzymes activities and sugar release from cornstalks under solid state fermentation. World J Microbiol Biotechnol 28:1961–1966
96. Meade MJ, Tanenbaum SW, Nakas JP (1995) Production and rheological properties of a succinoglycan from Pseudomonas sp. 31260 grown on wood hydrolysates. Can J Microbiol 41(12):1147–1152
97. Silva JA, Tobella LM, Becerra J, Godoy F, Martinez MA (2007) Biosynthesis of poly-β-hydroxyalkanoate by Brevundimonas vesicularis LMG P-23615 and Sphingopyxis macrogoltabida LMG 17324 using acid-hydrolyzed sawdust as carbon source. J Biosci Bioeng 103:542–546
98. Roy Chowdhury S, Basak RK, Sen R, Adhikari B (2012) Utilization of lignocellulosic natural fiber (jute) components during a microbial polymer production. Mater Lett 66:216–218
99. Saranya Devi E, Vijayendra SVN, Shamala TR (2012) Exploration of rice bran, an agro-industry residue, for the production of intra- and extra-cellular polymers by Sinorhizobium meliloti MTCC 100. Biocatal Agric Biotechnol 1:80–84

100. Turhan I, Bialka KL, Demirci A, Karhan M (2010) Enhanced lactic acid production from carob extract by Lactobacillus casei using Invertase pretreatment. Food Biotechnol 24: 364–374
101. Sánchez-Segado S, Lozano LJ, de los Ríos AP, Hernández-Fernández FJ, Godínez C, Juan D (2012) Process design and economic analysis of a hypothetical bioethanol production plant using carob pod as feedstock. Bioresource Technol 104:324–328
102. Smith RL, West TP, Gibbons WR (2008) Rhodospirillum rubrum: utilization of condensed corn solubles for poly-(3-hydroxybutyrate-co-3-hydroxyvalerate) production. J Appl Microbiol 104:1488–1494
103. Fosmer A, Gibbons WR, Heisel NJ (2010) Reducing the cost of scleroglucan production by use of a condensed corn soluble medium. J Biotech Res 2:131–143
104. Taşkın M, Erdal S, Genisel M (2011) Biomass and exopolysaccharide production by morchella esculenta in submerged culture using the extract from waste loquat (Eriobotrya japonica) kernels. J Food Process Pres 35:623–630

Part II
Thermal Pretreatment

Chapter 3
Lignocellulosic Biomass—Thermal Pre-treatment with Steam

Saqib Sohail Toor, Lasse Rosendahl, Jessica Hoffmann, Jens Bo Holm-Nielsen and Ehiaze Augustine Ehimen

Abstract With the ever rising demand for more energy and the limited availability of depleted world resources, many are beginning to look for alternatives to fossil fuels. Liquid biofuel, in particular, is of key interest to decrease our dependency on fuels produced from imported petroleum. Biomass pre-treatment remains one of the most pressing challenges in terms of cost-effective production of biofuels. The digestibility of lignocellulosic biomass is limited by different factors such as the lignin content, the crystallinity of cellulose and the available cellulose accessibility to hydrolytic enzymes. A number of different pre-treatment methods are known to enhance the digestibility of lignocellulosic biomass by affecting these limiting factors. Some of them are: milling, thermal pre-treatment with steam or hot water, acid pre-treatment, and alkaline pre-treatment. This chapter will focus on one of the more promising technologies; thermal pre-treatment with steam.

The Norwegian company Cambi developed a process for treatment of sludge from waste water treatment plants, and the idea was based on the experience that cooking sludge under pressure at temperature from 150 °C to 180 °C improved the digestibility and at the same time increased the dewaterability of the sludge. If Cambi's process is to be used for treatment of biomass, it will have to compete with other processes on market. The strongest competitor at present is the integrated biomass utilisation system process of DONG Inbicon which is used for pre-treatment of straw. Both processes are being described and discussed in this chapter.

Keywords Biomass · Steam pre-treatment · Lignocellulose · Hydrolysis

3.1 Introduction

With the ever rising demand for more energy and the limited availability of depleted world resources, it is crucial to look for alternatives to fossil fuels. Liquid biofuel, in particular, is of key interest to decrease our dependency on fuels produced from imported petroleum. Hereby lignocellulosic biomass is an important bioresource for

L. Rosendahl (✉) · S. S. Toor · J. Hoffmann · J. B. Holm-Nielsen · E. A. Ehimen
Department of Energy Technology, Aalborg University, Pontoppidanstræde 101, DK-9220, Aalborg Ø, Denmark
e-mail: lar@et.aau.dk

Z. Fang (ed.), *Pretreatment Techniques for Biofuels and Biorefineries*, Green Energy and Technology,
DOI 10.1007/978-3-642-32735-3_3,

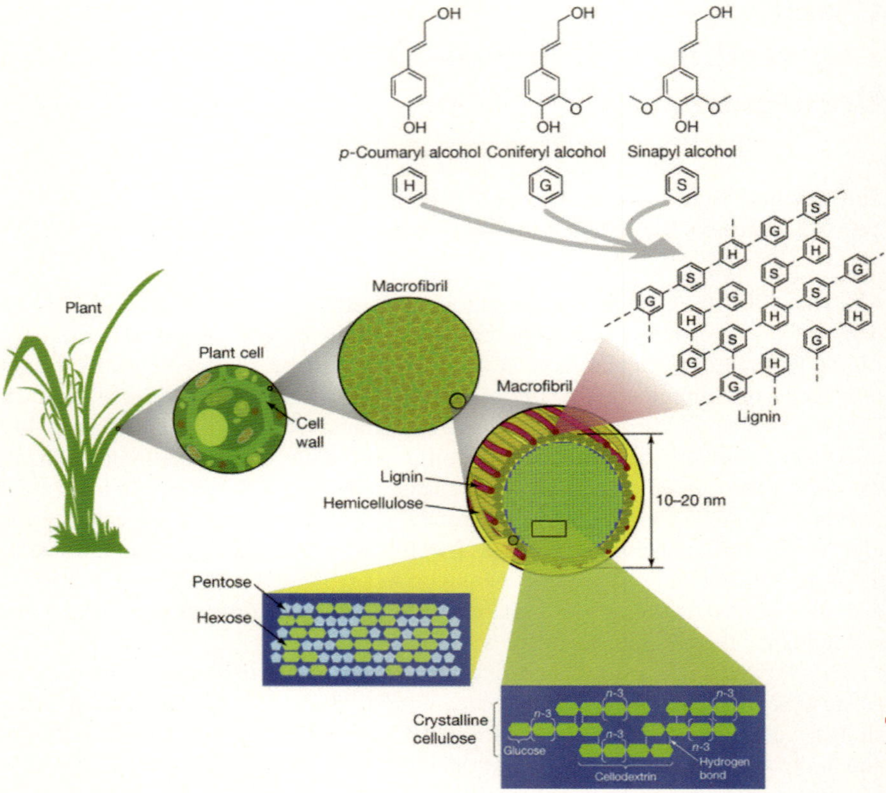

Fig. 3.1 Structure of lingo-cellulose. (Adapted from Macmillan Publishers Ltd: Ref. [2], copyright 2008)

producing second generation biofuels. Through either thermochemical or biochemical conversion processes lignocellulosic biomass can be converted into energy or energy carrier. Biochemical conversion processes deliver biofuel, that is, bioethanol or biogas through fermentation of the biomass. It has been seen that pre-treatment of lignocellulosic biomass prior to conversion enhances the product yield and is therefore of great importance also for the economical viability of the process. Lignocellulose forms the cell wall of plant material and has a complex structure, but its main constituents are cellulose, hemicellulose and lignin, with 40–50 %, 25–30 % and 15–20 %, respectively [1].

Cellulose fibers are embedded in a matrix of other structural biopolymers, primarily hemicellulose and lignin. Cellulose is a linear polymer of glucose. Hemicellulose, a branched heteropolymer consisting of various 5–6 carbon sugars and lignin, is composed of three major phenolic components [2]. Cellulose, hemicellulose and lignin form structures so-called macrofibril which are responsible for the structural stability of the plant cell wall. Those macrofibrils are sub-divided into microfibrili, where cellulose is packed inside lignin and hemicellulose as can be seen in Fig. 3.1. Hydrogen bonds between different layers of the cellulose lead to polymerisation. Crystalline cellulose has a high degree of polymerisation and contributes to the resistance to biodegradation due to low cellulolytic enzyme accessibility, whereby high accessibility amorphous cellulose could have a lower degree of polymerisation and will be more susceptible to enzymes [1]. For a most effective biomass to ethanol conversion process all major lignocellulosic components should be utilised; therefore, pre-treatment to increase the amenability for hydrolytic enzymes is of great importance. Different kind of pre-treatment technologies exist. They can be separated into physical pre-treatment methods, chemical pre-treatment methods, biological methods and physico-chemical pre-treatment methods. During physical pre-treatment the improved hydrolysis results from decreased crystallinity and improved mass transfer characteristics from reduction in particle size. Through chemical pre-treatment, for example, by adding acids, alkali, pH-controlled hot water or ionic liquids, lignin and hemicellulose were partially removed and the degree of polymerisation of cellulose is lowered [3]. Physico-chemical processes, through, for example, steam pre-treatment or explosion, combine both physical and chemical methods and lead to hemicellulose degradation and lignin transformation due to high temperatures, thus increasing the potential of cellulose hydrolysis [1]. During steam pre-treatment and steam explosion which is part of physico-chemical method, the biomass is treated with high pressure steam. Steam pre-treatment and explosion which is being discussed in this chapter leads to a low by-product generation, has a high sugar yield, is applicable to different feedstock, has low investment costs and is already being used commercially [1].

3.2 The Steam Pre-Treatment of Lignocellulosic Biomass

The application of steam techniques for pre-treatment of lignocellulosic biomass can be broadly differentiated into two schemes (a) the uncatalysed steam pre-treatment and (b) catalysed steam explosion methods. With this chapter dealing specifically with the use of steam as a pre-treatment tool, liquid hot water pre-treatment systems which are operated at high pressures using elevated temperatures (i.e. as demonstrated in [4–6]) will not be covered here.

3.2.1 Un-Catalysed Steam Pre-Treatment

Also referred to as auto-hydrolysis, the use of steam for the pre-treatment of lignocellulosic biomass is one of the most widely demonstrated and implemented methods in research and commercial facilities [7, 8]. This method employs the use of high-pressured saturated steam for the treatment of the lignocellulosic biomass. Steam pre-treatment is usually conducted at reaction temperatures of 160–260 °C using high pressures of 0.69–4.83 MPa with the period in which the biomass is exposed to this conditions ranging from several seconds to a few minutes [9]. A further classification of the steam pre-treatment can be made depending on if the use of high pressured steam in the process is followed by a sudden reduction of the process pressures. The process is thus regarded as 'Steam explosion' where such abrupt depressurisation and cooling of the lignocellulosic biomass after a specified reaction period is implemented. This is since the abrupt pressure reduction results in the explosive decompression of the biomass materials, thus facilitating a disruption of the lignocellulosic biomass cell walls and enhancing the accessibility of the biomass macromolecular contents. The conditions facilitated by the high-pressured steam (*for both explosive and non-explosive methods*) facilitates a disruption of the lignin sheath and enhances a solubilisation of the biomass hemicellulosic component (via hydrolysis), thus aiding the accessibility of cellulose to further conversion methods, that is, enzymatic hydrolysis [10]. For example, a 90 % enzymatic hydrolysis of *Poplar* biomass chips was achieved in 24 h after it was subjected to a steam explosion pre-treatment compared to only a 15 % hydrolysis obtained for the untreated chips [11].

During the steam pre-treatment, some of the biomass lignin is solubilised (*which however re-polymerises on cooling and forms a part of the acid-soluble lignin fraction*), with the biomass hemicellulose solubilised after the application of the steam pre-treatment and is subsequently recovered in the aqueous fraction (*or is further degraded to other compounds, i.e. furfural*) [12]. Most of the cellulose content of the biomass is preserved in the solid fraction, however, its hydrolysis to glucose could be obtained under high steam pre-treatment temperature conditions (i.e. $>$200 °C) [12]. The hydrolysis of the hemicelluloses is proposed to be brought about mainly by the action of acetic acid formed from the acetyl groups released during the steam pre-treatment [9]. In addition to this, other acids, that is, formic, levulinic and pyromucic acids, produced during the steam pre-treatment process as described [13], may also play an important role in the acid catalysed breakdown of the hemicellulosic glycosidic bonds [13]. Furthermore, water at high temperatures has been demonstrated to possess some acid properties, which could also enhance the hemicelluloses hydrolysis [14]. The acidic conditions provided by the steam pre-treatment could thus also lead to the degradation of available sugars in the biomass materials [15].

The potential benefits accruable with the use of steam pre-treatment over other pre-treatment systems have been widely demonstrated in the literature. Compared to the use of commonly applied chemical pre-treatment techniques, that is, sodium hydroxide, calcium hydroxide (lime) and dilute sulphuric acid hydrolysis, higher treated products recovery as well as improved substrate availability for further processing has been obtained with the application of auto-hydrolysis [16–19]. The implementation of mechanical biomass pre-treatment routes such as milling and novel methods like microwave radiation have also been demonstrated to less effective than the use of high-pressure steam for the pre-treatment of lignocellulosic biomass [18, 19]. Furthermore, the application of non catalytic steam pre-treatment methods was discussed in to have a lower process energy requirements compared to mechanical comminution methods (70 % less, to achieve similar size reductions), with no or a low recycling or environmental costs attached [9].

The non-catalysed steam pre-treatment method (including steam explosion techniques) has been widely demonstrated both in the literature and in practice to be utilisable for a variety of lignocellulosic biomass, that is, from forest products and residues (including the use of short rotation woody biomass) [17, 18, 20–22], purpose grown energy crops [9], as well as from agricultural residues [16, 19]. Regarding the use of woody biomass, it was seen that the use of younger biomass materials were easier to fractionate during the steam pre-treatment process and thus better substrates production for subsequent enzymatic hydrolysis, when compared to the use of older materials [13, 23]. This is, therefore, promising for the integration of steam pre-treatment with ongoing and proposed large-scale short rotation forest and fast growing purpose grown lignocellulosic biomass projects. The use of this method is however less effective for the pre-treatment of softwood (i.e. pine) where the acid catalysed route as described in Sect. 3.2.2.1 is better suited.

3.2.1.1 Factors Influencing the Auto-Hydrolysis Process

The major factors influencing the un-catalysed steam explosion processes are residence time, temperature, moisture content and particle size [20]. The process conditions which result in the best substrates for hydrolysis and the least amount of soluble sugars lost to side reactions (i.e. sugar dehydration) usually considered to be the optimum conditions [13]. The use of steam temperatures ranging from 140 to 240 °C have been investigated in the literature and its influence on the overall process efficiencies is usually associated with the period in which the biomass is exposed to the steam pre-treatment. An optimum solubilisation (and hydrolysis) of the hemicellulosic component of lignocellulosic biomass was reported to be realised either by using a combination of high temperatures and short residence times (i.e. 270 °C, 1 min) or lower temperatures with longer residence times (i.e. 190 °C, 10 min) [24]. This is due to the fact that at lower steam pre-treatment temperatures (i.e. 190 °C) the recovery of the obtained hemisellulosic sugars in the lignocellulosic hydrolysate is maximised, with the acid-labile biomass polysaccharides partially converted to water soluble sugars [25]. On the other hand, the drastic conditions provided by the use of increased steam temperatures would most likely facilitate an enhanced accessibility of the macromolecules of the biomass material, but with inevitable sugar losses [13]. The employment of high steam temperatures has also been demonstrated to lead to an increase in the relative amount of acid-insoluble lignin in the post-treated materials [23]. With the use of steam temperatures (i.e. 220–240 °C) and residence times, condensation reactions involving the by-products are derived from the biomass lignin and hemicelluloses. The acid-soluble lignin was observed to increasingly occur, resulting in the formation and accumulation of acid-insoluble polymeric materials [26]. This knowledge of the extensive condensation driven modification of the biomass lignin during steam pre-treatment has two important implications: That the formed polymeric materials could cause an apparent increase in the overall lignin yields (*even potentially higher than the theoretical lignin yield, based on the content of the starting biomass*), and the fact that part of these by-products are likely to remain in the steam pre-treated material even after employing additional washing steps (i.e. alkaline washing) [13]. A careful compromise must therefore be considered with the use of these two important conditions with opposite trends. The selection of the best temperature and time conditions could therefore be dependent on the other parameters such as the subsequent processing and conversion steps to be applied after the biomass pre-treatment and the targeted fuel or chemical which is aimed to be produced. In general, an increase in the steam temperatures would correspond to a decrease in the carbohydrate yields, while longer reaction times have been seen to favour an increased lignin condensation and a degradation of pentosan, with acid hydrolysis observed to predominate over degradation reactions with the use of shorter exposure times [13]. Furthermore, the use of an energy–material assessment would prove useful to ascertain the benefits (if any) of compromising high-energy inputs with the potential substrate conversion efficiencies.

The extent of biomass drying or water content is an important economic and technical parameter with the processing of lignocellulosic biomass. This is since the biomass costs could be substantially increased depending on the methods used in achieving a reduction in the biomass moisture content. To minimise processing costs, single fuel and chemical production systems as well as biorefineries would therefore ideally prefer the utilisation of cheap unprocessed biomass (containing high moisture content) or naturally dry biomass (containing ~5–15 % moisture). Investigations have been carried out on the influence of the use of lignocellulosic biomass with varying moisture contents on the steam pre-treatment process efficiencies. Using 'green' freshly harvested and air-dried *Aspen* wood chips (3.2 mm, *in the direction of the fibres*), with moisture contents (oven dried basis) of 108.2 and 7.16 %, respectively, investigations carried out in [21] showed that statistically comparable reducing sugar yields (after enzymatic hydrolysis) were obtainable with the use of the different levels of water in the biomass materials. However, with an increase in the chip size, the samples with lower moisture contents (i.e. the air died chips) were observed to attain the steam temperatures within a shorter time, thus facilitating improved and quicker pre-treatment conversions. The steam requirements for the auto-hydrolysis process as examined by that study were therefore observed to increase with an increase in the lignocellulosic biomass size and moisture content.

Regarding the influence of particle sizes on the biomass steam pre-treatment, the application of mechanical size reduction schemes, that is, chipping and milling is usually carried out before the pre-treatment stage mainly to improve handling and to ease the biomass transportation from the acquisition site. The biomass size has also been described to be a critical parameter to be considered for the pre-treatment and conversion reactor designs [21]. The use of small biomass sizes were discussed to be preferable for most batch and continuous process operations due to the enhanced heat transfer facilitated with the use of such sizes for the biomass treatment [13]. However, much finer materials with smaller particle sizes (i.e. sawdust) have been seen to be difficult to utilise in batch units, with the use of plug flow reactors employed to increase the pre-treatment conversion efficiencies [13]. An investigation on the influence of a range of lignocellulosic biomass (*Brassica sp.*) with particle sizes of 2–5, 5–8 and 8–12 mm respectively using process steam temperatures of 190 and 270 °C and a residence time of 4 and 8 min as carried out in [16] showed that an extensive size reduction was not desirable for optimum pre-treatment of the lignocellulosic biomass using steam for all the different temperature and residence time conditions examined. This was demonstrated by the higher cellulose concentrations and subsequent enzymatic digestibilities exhibited by the largest particle sizes (i.e. 8–12 mm) studied.

Another important aspect with the application of the non-catalytic steam pre-treatment route is the consideration of the need to employ the explosive decompression step or not. Studies carried out in [22] using *Aspen* chips concluded that the explosive decompression step of the steam explosion process was not important and contributed little or nothing to hydrolysis of the lignocellulosic biomass and eventual accessibility of the biomass cellulose contents. Similarly, the use of steam explosion for the pre-treatment of green Eucalyptus chips was observed not to yield any considerable hydrolytic improvements; however, the use of high-pressured steam without an abrupt decompression step for the pre-treatment of air-dried Eucalyptus chips was seen to result in the production of poor substrates for further hydrolysis, thus suggesting that the use of the explosive decompression scheme is most suited for hardwood chips with a low moisture content [27]. The results of that study were however in somewhat in contrast to that presented in the patent proposed in [28], which discussed that the explosion step was essential for the production of hydrolytic substrates with improved macromolecular accessibility.

3.2.1.2 Limitations with the Use of the Auto-Hydrolysis Process

The major problems associated with the application of high-pressure steam for the pre-treatment of lignocellulosic materials has been the observed destruction of the xylan polysaccharide content, an incomplete disruption of the biomass lignin–carbohydrate matrix and the formation of by-products after the treatment process which inhibit any microbial and enzymatic activities utilised in subsequent downstream conversion schemes [29]. The production of microbial inhibitors has been mainly attributed to the formation of compounds such as fufural and hydroxymethylfurfural from the biomass pentoses and hexoses respectively and release acetate from acetyl-group from hemicelluloses during the steam pre-treatment process [13]. This sugar degradation usually occurs as a result of dehydration processes take place with the use of high steam temperatures for the biomass pre-treatment [13]. With these by-products potentially hindering the application of potential biological conversion, detoxification schemes might therefore be required to improve the use of lignocellulosic hydrolysates to fuels and chemicals, especially where enzymatic methods are to be applied. The use of water washing as a cheap method to aid the removal of potential inhibitory substances, as well as water soluble hemicelluloses has been discussed in the literature [30]. The use of such a method has however been reported to lead to a decrease in the overall saccharification yields obtained from the steam pre-treatment method, since the soluble sugars (i.e. generated via hemicelluloses hydrolysis) are also removed by the water washing step [9]. Other detoxification techniques, that is, the use of neutralisation and fungal treatments to aid the removal of such potential process inhibitors has also been investigated [31]. Depending on the availability of oxygen in the pre-treatment process, sugar degradation could also occur via pyrolysis (absence of oxygen) and oxidation processes resulting in the thermal decomposition of the organic matter and a partial conversion of the biomass pentoses to carboxylic acids and other by-products [13].

3.2.2 Catalysed Steam Pre-Treatment

The mode of action of the catalysed steam pre-treatment method is similar to the auto-hydrolytic process described in the preceding section with the major difference being the impregnation of the lignocellulosic biomass with acidic gases or liquids (i.e. sulphur dioxide (SO_2), sulphuric acid (H_2SO_4), nitric acid (HNO_3) and hydrochloric acid (HCl)) which act as process catalysts prior to applying the steam pre-treatment.

The implementation of acid catalysis has been demonstrated to exhibit a higher pre-treatment efficiency than the use of the steam pre-treatment method alone. Compared with the use of the steam pre-treatment alone, the catalysed route has been observed to lead to a greater hemicellulose (near complete) removal with a reduced generation of inhibitory compounds [32]. This in turn results in an increased conversion potential of the post-treated fractions via an improvement of the biomass digestibility using enzymatic methods. An improved sugar recovery, especially as pentoses in the post treated aqueous phases has been shown by [13].

As highlighted in Sect. 3.2.1, due to the characteristic low reactivity and permeability of the fibres of softwoods to the penetrating steam used in the auto-hydrolytic process [13], acidic conditions could thus be employed with the use of this particular type of lignocellulosic biomass to enhance the conversion efficiencies obtainable with the use of the steam pre-treatment method. In addition, the use of the acid catalysts on such biomass materials have been seen to facilitate the use of lower steam temperatures and shorter exposure times [33]. As demonstrated in [31], more than 85 % of the original hemicellulose derived sugars was recovered in the aqueous phase after an acid catalysed steam explosion of Douglas fir wood chips at relatively lower steam severity conditions (175 °C, 7.5 min.) with SO_2 (4.5 %, w/w) used as the process catalyst. In that study, it was also observed that all of the hexose components obtained from the biomass using low- and medium severities were readily converted to ethanol [31].

3.2.2.1 Factors Influencing the Catalysed Pre-Treatment Route

The reaction parameters (i.e. steam temperature, reaction time and particle size) which have been previously highlighted to influence the auto-hydrolytic process similarly have been demonstrated to affect the catalysed steam pre-treatment method [21, 12]. In addition to those factors, the selection of the type and concentration of acid catalyst employed, as well as the biomass type and its moisture content are also major considerations which must be addressed with the use of this method.

Although a variety of acid catalysts prior to steam pre-treatment have been demonstrated in the literature (i.e. the use of aqueous phosphoric acid (H_3PO_4) for the pre-treatment of sugarcane baggase [34]), the use of H_2SO_4 and SO_2 as the process acid catalysts currently predominates acid catalysts investigated and in practice. This section will, thus, concentrate on the use of these acid catalysts. A study carried out in [35] compared the use and influence of the choice of H_2SO_4 (0–3 %, w/w) and SO_2 (1 %, w/w) for the pre-treatment of *Salix caprea* (Willow) wood chips with a steam temperature range of 160–230 °C investigated with a fixed reaction time of 10 min (*with a 15 min treatment of the biomass samples with saturated steam at 1 bar prior to the acid catalyst addition*). With an increase in the H_2SO_4 acid concentrations (and steam temperatures), a reduction in the fibrous material yields were observed. This was attributable to the improved solubilisation of the extractives and hemicelluloses under these conditions [35]. The xylose yields (based on the calculated biomass xylan content) were also seen to increase with an increase in the H_2SO_4 concentrations. With a steam temperature of 190 °C and using the highest study H_2SO_4 concentration of 3 %, a xylose yield of 80 % was obtained, the glucose yields (*after enzymatic hydrolysis*) were however seen to be reduced under these conditions [35]. This was considerably higher than the <15 % xylose yields observed when no acid catalysts were used for the steam pre-treatment [35]. With the use of SO_2 as the process catalyst, a maximum xylose yield of 62 % was observed using a steam temperature of 200 °C [35]. Correspondingly, a glucose yield of 95 % (on the basis of the of the biomass glucan availability) was obtained after enzymatic hydrolysis with the use of the same reaction parameters [35]. With glucans being the main constituents of willow biomass and with higher glucose yields upon hydrolysis considered to be more beneficial than xylose yields in the water soluble fractions, the results of that study led to the conclusion that the SO_2 acid catalysts were preferable for use as the acid catalyst candidate for use in the catalysed steam pre-treatment method.

The use of SO_2 concentrations (1–4 %, w/w) as the process catalyst for the impregnation of wood prior to the steam pre-treatment was seen to significantly reduce the overall process temperature and time requirements to achieve optimum biomass solubilisation, recovery and hydrolysis of the post-treated substrates [13].

The use of SO_2 as the process catalyst over H_2SO_4 has also been discussed to be preferred in industrial catalysed steam pre-treatment systems since the former was reported to require a comparatively less expensive reactor materials, leads to the generation of less gypsum as a process by-product, and less process steam requirements than the former [13, 36]. The ease in which SO_2 is incorporated evenly within lignocellulosic materials has also been a factor favouring its use over H_2SO_4, especially with regards to the handling inconveniences encountered with the soaking process necessary with the use of aqueous H_2SO_4 [13].

Regarding the influence of the type of lignocellulosic biomass employed in the catalysed steam pre-treatment process, softwoods have been demonstrated to require more harsh pre-treatment conditions for the production of suitable substrates aimed at effecting high product yields upon hydrolysis [23]. Based on this, the use of SO_2 as an acid catalyst was shown to be beneficial when used for the pre-treatment of hardwood, but highly essential for use for the impregnation of softwood, that is, spruce and fir prior to their steam pre-treatment [23, 31]. Furthermore, with an increasing moisture content of the wood chips, it was observed that a higher efficiency of the acid catalyst (expressed in relation to the dry weight of the wood chips) was retained within the chips, subsequently resulting in an enhanced biomass pre-treatment [23].

3.2.2.2 Limitations with the Acid Catalysed Method

A major problem encountered with the use of gaseous acid catalysts for the steam pre-treatment is the fact that most of these gases, for example SO_2, are regarded as a major air pollutant. Potential SO_2 escapes could be toxic, having significant negative impacts on human health and safety, as well as the environmental impacts (i.e. as a precursor to acid rain) associated with such emissions [36]. The use of acid catalysts (especially using high pre-treatment severities) was seen to lead to an increase in the formation of toxic degradation products from the biomass sugars which in turn negatively affect further hydrolysis [33]. Other operating issues associated with the use of acid catalysts such as acid corrosion and the need to implement an extensive downstream effluent treatment processes further limit the use of this method [13].

3.3 Steam Pre-Treatment Technologies: Status and Commercial Applications

3.3.1 Batch Steam Pre-Treatment Processes

Large scale steam pre-treatment of lignocellulosic biomass has mainly been carried out using batch processes, where quantities of the biomass samples are radically modified from the high pressure steam conditions resulting in an alteration of the cell wall structures. This is usually visualised by a change of the original biomass colour to a near dark brown material after the treatment step, from which the partially hydrolysed hemicellulose can be obtained by water-washing, leaving a water-insoluble hydrolysate fraction made up of cellulose, residual hemicellulose and modified lignin components which can be further extracted using mild alkali, ethanol or oxidative methods [13, 23].

Regarding industrial batch pre-treatment processes, the Cambi and Inbicon processes have enjoyed successful applications for improving the digestibility of a wide variety of biomass especially aimed for subsequently biological conversion processes, that is, for methane and ethanol production. These processes will be highlighted below as useful industrial batch process case studies.

3.3.1.1 The Cambi Thermal Hydrolysis (TH) Process

The thermal steam hydrolysis process developed by Cambi A/S, Norway was originally designed for treatment of sludge from waste-water treatment plants, on the experience that under high pressure and temperatures (i.e. 150–180 °C), an improvement in the sludge digestibilities was observed. The large-scale steam pre-treatment applications developed by Cambi has thus been mainly applied for the prior degradation of organic waste sludge, aimed for methane production in anaerobic digestion plants, but can also be utilised for the treatment of a wide variety of feedstocks, including lignocellulosic biomass. A brief description of this steam technology is outlined in Fig. 3.2.

Initially, the feedstocks to be pre-treated is dewatered using centrifugation methods or via belt press to a dry solids content of 16–17 %. Before the biomass is fed into the thermal hydrolysis reactor, it is initially preheated to a temperature of ≈97 °C in a pulper (for a period of ≈1.5 h), where it is mixed with recycled steam from the thermal hydrolysis process. This potentially reduces the overall external process energy consumption and ensures a maximum homogeneity of the substrates [38]. After the biomass has been heated in the pulper, it is transferred to the thermal hydrolysis reactor, where it is mixed with hot steam at around 165–170 °C at an operational pressure of 5–6 bar [39]. The treatment is usually conducted in batches with a reaction time of 20–30 min [39]. However, with the use of multiple reactors to a system operating on a staggered basis, the effect will be that of a continuous flow. A gradual release of the steam in the pulper is then carried out until the pressure has fallen to ≈2 bars [39]. The released steam is then recycled for re-use in the pulper as previously described, with the residual pressure used transfer the sludge into a flash tank. Here the rest steam is released to the pulper, with the post treated samples attaining a temperature of 105 °C. Using heat exchangers and water, the treated substrates are then cooled to the temperature levels suited for the subsequent enzymatic conversion methods which they would be subjected to, that is, fermentation or anaerobic digestion.

3.3.1.2 Inbicon Integrated Biomass Utilisation System (IBUS)

An operational large-scale steam pre-treatment platform similar to the Cambi process is available with the integrated biomass utilisation system (IBUS) developed by Inbicon A/S (a subsidiary company of Dong Energy, Denmark), using straw as the principal process input. The Inbicon IBUS process is mainly focused on the production of ethanol from lignocellulosic biomass. Besides this main product consideration, to improve the process economics, other residuals from the process are reused in different process applications. The IBUS process has been tested since 2003 in a pilot plant and was realised in 2009 with the establishment of a demonstration plant in Kalundborg, Denmark. Although the entire IBUS technology incorporates other processing steps to meet the desired product goals, with the emphasis on steam pre-treatment, only the steps involving the auto-hydrolysis process will be highlighted in this chapter.

Prior to the steam pre-treatment step of the IBUS process a mechanical crushing step is initially applied to facilitate a reduction of the biomass particle size of the biomass. The particles are then treated hydrothermally with steam under pressure. In the thermal hydrolysis, the biomass is continuously mixed with water to a dry matter

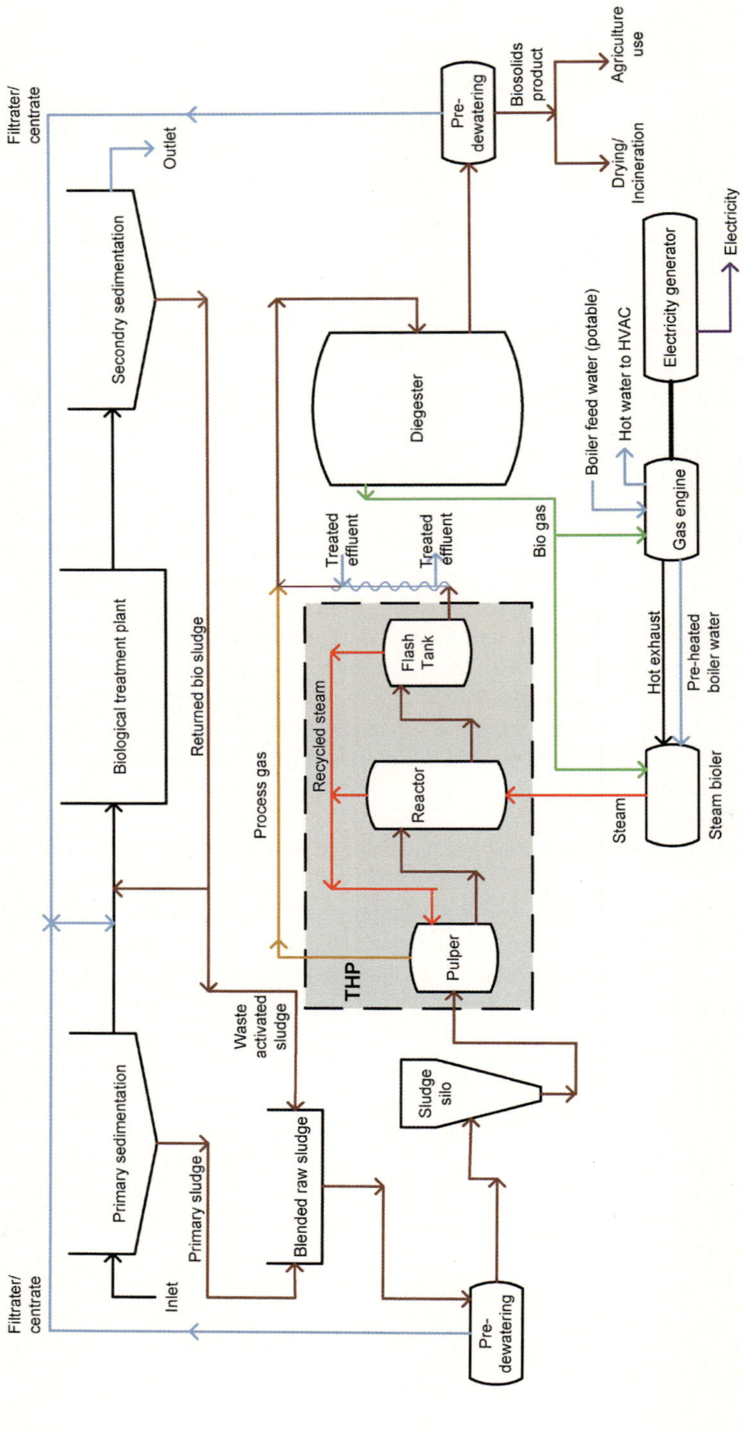

Fig. 3.2 Cambi process. (Adapted from [37])

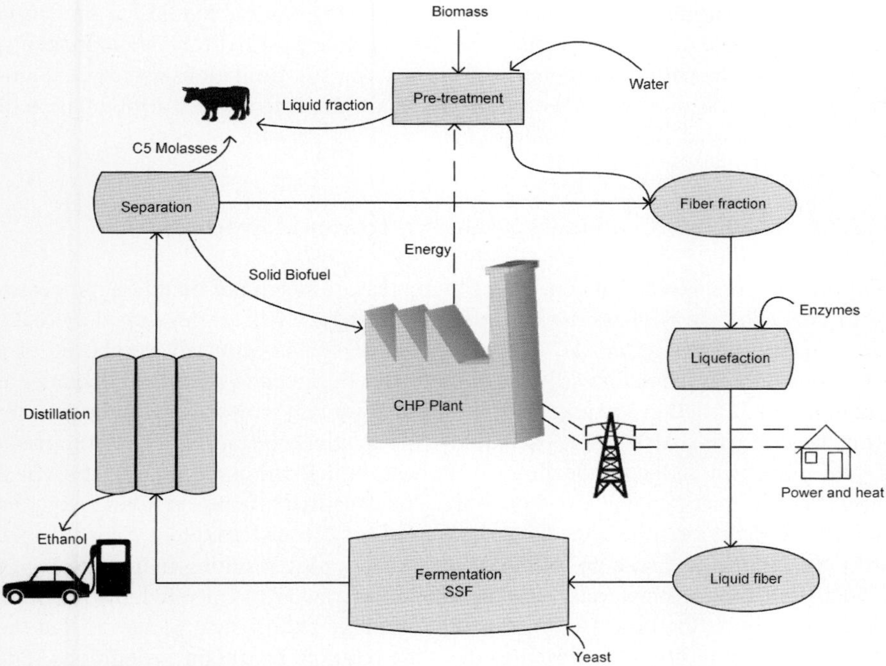

Fig. 3.3 The overall process flow in the IBUS process [40]

content of 30–40 % and then treated at 180–200 °C for 5–15 min. Results of the optimisation of the IBUS process has shown that a 100 % of the cellulose and ≈70 % of the hemicellulosic content of the biomass can be recovered in practice [40]. Following the steam pre-treatment, the easier to degrade polysaccharide content of the post treated lignocellulosic biomass is then subjected to enzymatic fermentation processes, which coupled with thermal processes, are used for the production of ethanol, other chemicals and energy. With the IBUS process capable of handling high dry matter contents, higher concentrations of sugar can thus be available as a feedstock for the subsequent fermentation step, resulting in higher ethanol yields [40]. This also leads to a potential reduction in the handling and upgrading costs, since minimal water is used in processing and with the output stream containing less water, hence higher ethanol concentrations. An overview of the IBUS process can be seen in Fig. 3.3.

3.3.2 *Continuous Steam Pre-Treatment Processes*

When compared to the use of batch reactors, the development and application of continuous high pressure steam reactors however could provide a better control of the pre-treatment variables which are critical to obtaining optimal biomass degradation and conversions, and a higher purity of the valuable post-hydrolysis components. This is due to the partial overcoming of the heat transfer limitations usually encountered in batch systems which would subsequently result in a lower accumulation of undesirable degradation by-products [13].

3.3.2.1 StakeTech Continuous Steam Pre-Treatment Reactors

One of the more successful continuous high pressure steam pre-treatment processes for lignocellulosic biomass use has been the StakeTech reactor developed by Stake Technology, Ontario, Canada. The StakeTech reactor is primarily composed of a stainless steel horizontal pressure vessel which is designed to withstand operational pressures of 31 bar (i.e. 450 psig) [13]. Using an upstream screw conveyor the digester (*i.e. the part of the vessel where the main stream treatment occurs*) is fed continuously by the movement of the lignocellulosic biomass through the compression tube which also serves to build up the vessel pressure. The densified biomass resulting from the feeder compression tube upon entering the reactor is transferred to a conical choke which facilitates a 'break up' of the dense biomass plug resulting in a scattering of undensified biomass materials onto the retention screw [13]. Once fed, the retention screw transports the biomass towards the discharge end in such a manner that the precise retention time is achieved to meet the required processing conditions. The post treated material is then conveyed to the discharge valve (*which is made up of a rotary ball valve with its opening timed according to the desired production rate*), using the discharge screw at the end of the digester [13]. The discharge valve further aids in providing an explosion effect on emptying the pre-treated samples while readying it for the subsequent processing stages. The StakeTech continuous technology has been widely used in research and has been reported to reach full commercialisation [13].

Apart from the batch and continuous pre-treatment processes previously outlined, various patents, that is [38–41] on the use and optimisation of the catalysed and auto-hydrolytic steam pre-treatment technologies, as well as improving the purity and accessibility of the desired process outputs from a wide variety of lignocellulosic biomass inputs have been granted and are continuously being researched.

3.4 Conclusion

In this chapter, different types of steam pre-treatment have been described, and examples of commercial implementations of steam pre-treatment have been given. It is clear that for biomass to play the future role of major supplier of carbon for energy and products, full utilisation of all parts of the biomass is necessary. This necessitates technologies that go beyond high quality, cellulosic feedstocks, targeting the more abundant yet significantly more complex lignocellulosic fractions. For this purpose, steam pre-treatment offers an efficient yet environmentally benign approach to opening the biomolecular structure, and preparing the substrate for subsequent separation or conversion. Uniqueness for steam pre-treatment is that it is able to address most of the limiting factors for subsequent digestion (conversion) and at the same time turn the feed into a pumpable slurry or paste, even without the addition of catalysts or other additives, which call for subsequent separation or effluent treatment. Being a thermal process, it can be possible to design the entire plant with a significant amount of heat integration, reducing the need for external supply of heat for steam generation.

References

1. Menon V, Rao M (2012) Trends in bioconversion of lignocellulose: biofuels, platform chemicals & biorefinery concept. Prog Energ Combust Sci 38:522–550
2. Rubin EM (2008) Genomics of cellulosic biofuels. Nature 454(14):841–845
3. Arantes V, Saddler JN (2010) Access to cellulose limits the efficiency of enzymatic hydrolysis: the role of amorphogenesis. Biotechnol Biofuels 3–4:1e–11
4. Brandon SK, Eiteman MA, Patel K, Richbourg MM, Miller DJ, Anderson WF, Peterson JD (2008) Hydrolysis of Tifton 85 bermuda grass in a pressurized batch hot water reactor. J Chem Technol Biotechnol 83:505–512
5. Negro MJ, Manzanares P, Ballesteros I, Oliva JM, Cabanas A, Ballesteros M (2003) Hydrothermal pretreatment conditions to enhance ethanol production from poplar biomass. Appl Biochem Biotechnol 105:87–100
6. Rogalinski T, Ingram T, Brunner GJ (2008) Hydrolysis of lignocellulosic biomass in water under elevated temperatures and pressures. J Supercrit Fluids 47:54–63
7. Hsu T-A (1996) Pretreatment of biomass. In: Wyman C E (ed) Handbook on bioethanol, production and utilization. Taylor & Francis, Washington, DC, pp 179–212
8. Chandra RP, Bura R, Mabee WE, Berlin A, Pan X, Saddler JN (2007) Substrate pretreatment: the key to effective enzymatic hydrolysis of lignocellulosics? Adv Biochem Eng Biotechnol 108:67–93
9. Sun Y, Cheng JY (2002) Hydrolysis of lignocellulosic materials for ethanol production: a review. Bioresour Technol 83:1–11
10. Blaschek HP, Ezeji T, Scheffran J (2010) Biofuels from agricultural wastes.Blackwell, Ames, p 26
11. Grous WR, Converse AO, Grethlein HE (1986) Effectof steam explosion pretreatment on pore size and enzymatic hydrolysis of poplar.Enzyme Microb Technol 8:274–280
12. Sassner P, Galbe M, Zacchi G (2005) Steam pretreatment of Salix with and without SO2 impregnation for production of bioethanol. Appl Biochem Biotech 21:1101–1117
13. Ramos LP (2003) The chemistry involved in the steam pre-treatment of lignocellulosic materials. Quim Nova 26:863–871
14. Weil JR, Sarikaya A, Rau S-L, Goetz J, Ladisch CM, Brewer M, Hendrickson P, Ladisch MR (1997) Pretreatment of yellow poplar sawdust by pressure cooking in water. Appl Biochem Biotechnol 68:21–40
15. Garcia-Aparicio MP, Ballesteros I, Gonzalez A, Oliva JM, Ballesteros M, Negro MJ (2006) Effect of inhibitors released during steam-explosion pretreatment of barley straw on enzymatic hydrolysis. Appl Biochem Biotechnol 129:278–288
16. Ballesteros I, Oliva JM, Negro MJ, Manzanares P, Ballesteros M (2002) Enzymatic hydrolysis of steam exploded herbaceous agricultural waste (*Brassica carinata*) at different particle sizes. Process Biochem 38:187–192

17. Playne MJ (1984) Increased digestibility of bagasses by pretreatment with alkalis and steam explosion. Biotechnol Bioeng 26;426–433
18. Saddler JN, Brownell HH, Clermont LP, Levitin N (1982) Enzymatic hydrolysis of cellulose and various pretreated wood fractions. Biotechnol Bioeng 24:1389–1402
19. Chang VS, Nagwani M, Holtzapple M (1998) Lime pretreatment of crop residues bagasse and wheat straw. Appl Microbiol Technol 74:135–159
20. Negro MJ, Manzanares P, Oliva JM, Ballesteros I, Ballesteros M (2003) Changes in various physical/chemicals parameters of *Pinus pinaster* wood after steam explosion pretreatment. Biomass Bioenergy 25:301–308
21. Brownell HH, Yu EKC, Saddler JW (1986) Steam explosion pretreatment of wood: effect of chip size, acid, moisture content and pressure drop. Biotechnol Bioeng 28:792–801
22. Brownell HH, Saddler JN (1987) Steam pre-treatment of lignocellulosic material for enhanced enzymatic hydrolysis. Biotechnol Bioeng 29:228–235
23. Ramos LP, Breuil C, Saddler JN (1992) Comparision of steam pretreatment of *eucalyptus*, *aspen* and spruce wood chips and their enzymatic hydrolysis. Appl Biochem Biotechnol 34/35:37–48
24. Duff SJB, Murray WD (1996) Bioconversion of forest products industry waste celluloses to fuel ethanol: a review. Bioresour Technol 55:1–33
25. Wright JD (1988) Ethanol from biomass by enzymatic hydrolysis. Chem Eng Prog 84:62–74
26. Ramos LP, Saddler JN (1994) Enzyme recycling during fed-batch hydrolysis of cellulose derived from steam exploded *Eucalyptus viminalis*. Appl Biochem Biotechnol 45–46:193–207
27. Ramos LP, Breuil C, Kushner DJ, Saddler JN (1992) Steam pretreatment conditions for effective enzymatic hydrolysis and recovery yields of *Eucalyptus viminalis* wood chips. Holzforschung 46:149–154
28. DeLong EA (1983) Method of rendering lignin separable from cellulose and hemicelluloses and the product so produced. Can Patent No 1441376, 1983
29. Kumar P, Barrett DM, Delwiche MJ, Stroeve P (2009) Methods for pretreatment of lignocellulosic biomass for efficient hydrolysis and biofuels production. Ind Eng Chem Res 48:3713–3729
30. McMillan JD (1994) Pretreatment of lignocellulosic biomass. In: Himmel ME, Baker JO, Overend RP (eds) Enzymatic conversion of biomass for fuels production. American Chemical Society, Washington, DC, pp 292–324
31. Boussaid A, Robinson J, Cai J-Y, Gregg DJ, Saddler JN (1998) Fermentability of the hemicellulose-derived sugars from steam-exploded softwood (Douglas Fir). Biotechnol Bioeng 64:284–289
32. Morjanoff PJ, Gray PP (1987) Optimization of steam explosion as method for increasing susceptibility of sugarcane bagasse to enzymatic saccharification. Biotechnol Bioeng 29:733–741
33. Grethlein HE, Converse AO (1991) Common aspects of acid prehydrolysis and steam explosion for pretreating wood. Bioresour Technol 36:77–82
34. Deschamps FC, Ramos LP, Fontana JD (1996) Pretreatment of sugar cane bagasse for enhanced ruminal digestion. Appl Biochem Biotechnol 57/58:171–182
35. Eklund R, Galbe M, Zacchi G (1995) The influence of SO2 and H2SO4 impregnation of willow prior to steam pretreatment. Bioresour Technol 52:225–229
36. Zheng Y, Pan Z, Zhang R (2009) Overview of biomass pretreatment for cellulosic ethanol production. Int J Agric Biol Eng 2:51–68
37. Camacho P, Ewert W, Kopp J, Panter K, Perez-Elvira SI, Piat E (2008) Combined experiences of thermal hydrolysis and anaerobic digestion—latest thinking on thermal hydrolysis of secondary sludge only for optimum dewatering and digestion. In: Proceedings of the water environment federation, pp 1964–1978
38. Snekkenes V, Almqvist L, Trolin D, Gustavsson L (2011) System and a method for the pretreatment of chips in association with the production of chemical cellulose pulp. US Patent No US 8052843 B2, 2011
39. Norli M (2006) The Cambi thermal hydrolysis process (THP)—Dramatically improving sludge digestion and dewatering. In: Proceedings of the Gdansk conference, Gdansk, 24–26 October 2006

40. Larsen J, Petersen MØ, Thirup l, Li HW, Iversen FK (2008) The IBUS process—lignocellulosic bioethanol close to a commercial reality. Chem Eng Technol 31:765–772
41. Wingerson RC (2003) Cellulose production from lignocellulosic biomass. US Patent No US 6620292 B2, 2003
42. Foody B (2007) Pre-treatment of bales of feedstock. US Patent No US 7198925 B2, 2007
43. Baker JT (2002) Apparatus for the pretreatment of wood chips. US Patent No US 6497791 B1, 2002

Chapter 4
Stalk Inhomogeneity and Steam Explosion Integrated Fractional Refining Technology System

Hongzhang Chen and Junying Zhao

Abstract This chapter presents several integrated refining and fractionation technologies for multiple products platform based on the understanding of the heterogeneous property of corn stalk. This heterogeneous property was found at the levels of tissue, cell, and chemical composition. This property can be advantageous when proper fractionation technologies are adopted to pro-duce different products. Based on low pressure steam explosion technology, steam explosion integrated me-chanical carding, steam explosion integrated super grinding and steam explosion integrated washing and alkali extraction are designed to realize stalk fractiona-tion at different levels. Industry implementation of these technologies was also presented.

Keywords Lignocelluloses · Inhomogeneity · Steam explosion · Integrated · Fractional refining · Multi-production

4.1 Introduction

At present, stalk-based products from stalk refining has been regarded as perspective both in research and industrialization [1, 2]. In that, the world is rich in stalk [3, 4] without application completely. However, stalk refining is hard to be industrialized for lack of cost-competitiveness.

The main reason for lack of cost-competitive is that stalk is converted into one product as a whole with single linear technology. For example, corn stalk is converted into ethanol with dilute acid pretreatment and simultaneous saccharification and fermentation technology [5]. The whole corn stalk is regarded as raw material with single conversion property. As a result, only about 30 % cellulose and a little hemicellulose are converted into ethanol. Other components including lignin are discharged as wastes. Therefore, this increases cost and causes pollution. Lignin from stalk could also be converted into various products [6]. However, if lignin in stalk

H. Chen (✉) · J. Zhao
National Key Laboratory of Biochemical Engineering, Institute of Process Engineering, Chinese Academy of Sciences, Beijing 100190, China
e-mail: hzchen@home.ipe.ac.cn

Z. Fang (ed.), *Pretreatment Techniques for Biofuels and Biorefineries*, Green Energy and Technology,
DOI 10.1007/978-3-642-32735-3_4, © Springer-Verlag Berlin Heidelberg 2013

is converted into products as the only effective component, there would be the same problems including high cost and pollution. Hemicellulose in corn cob is usually converted into furfural [7]. There is report that microorganism could also convert pentose from hemicellulose [8]. Therefore, cellulose, hemicelluloses, and lignin in stalk are all potential resources. If only one component of stalk is converted with single linear technique, high cost and pollution would be unavoidable.

Lignocellulose materials including stalk are not only heterogeneous in the level of component but also in the levels of tissue and cell [9, 10]. Therefore, the conversion properties of stalk are also different. If the whole stalk is converted into one product, the product quality is hard to be controlled.

So it is necessary to fractionate stalk according to its heterogeneous characteristics in levels of component, cell, and tissue; then different fractions are converted into optimum products. The whole process from stalk to products includes pretreatment, saccharification, and fermentation. Fractionation should be realized in pretreatment process. Pretreatment methods reported focus on separation of one component including steam explosion [11], dilute acid or alkali [12], and hot water [13], and the whole stalk could not be fractionated completely into different fractions with only one of these technologies. Though there is report about integrated pretreatment technology, it is also amid at one product [14]. The intrinsic property of heterogeneity is ignored.

For example, steam explosion method is integrated with many other pretreatment technologies to improve ethanol yield including methanol, hydrogen dioxide, sodium hydroxide [15], ammonia, and sulfur dioxide. Sodium hydroxide and 1,4-dihydroxy-anthraquinone pretreatment are integrated for corn stalk to improve the methane yield [16]. Chemical and ultrasonic techniques are integrated to remove lignin and hemicelluloses in wood and get 5–10 nm fiber [17]. Therefore, integrated technology system should be set up to realize fractionation of stalk.

The author team have found heterogeneous characteristics of stalk and its different conversion properties [1, 9]. Moreover, steam explosion integrated pretreatment technology is set up to fractionate stalk into fractions and converted into multiple products. Different fractionation technologies are designed for different levels of fractionations.

This paper analyzes the heterogeneous characteristic of stalk and its requirement for conversion technology, and then steam explosion integrated pretreatment technologies are introduced as a multiple products model. Finally, industrial demonstration is given and advantages of fractionation refining technology oriented by multiple products are presented.

4.2 Heterogeneous Characteristics and Conversion Property of Stalk

For vascular plant [18], different organs are all composed of vascular tissue embedded in parenchyma tissue and epidermis tissue cover. For vascular tissue, the wall of vessel cell and fiber cell wall (bundle sheath cell) is rich in lignin because secondary wall is lignified. While for parenchyma tissue, parenchyma cell wall is rich in cellulose because there is only primary cell wall.

4.2.1 Tissue Level

According to Sachs' convenient classification [18], the body of a vascular plant is composed of three tissue systems, the dermal, the vascular, and the fundamental (or ground). The dermal tissue system comprises the epidermis and the periderm. The vascular tissue system contains two kinds of conducting tissues, the phloem (food conduction) and the xylem (water conduction). The fundamental tissue system (or ground tissue system) includes parenchyma tissue, secretory tissue, collenchymas tissue, and sclerenchyma tissue.

There is cutin layer out of the dermal tissue system [19]. Cutin layer is composed of cutin and wax. Cutin is mostly polymers composed of C_{16} and C_{18} monomer. Parenchyma cell under dermal is hardened, leading to high lignin content. For vascular tissue system, vessel cell in xylem is thickened so that lignin content is high too. There are mainly parenchyma tissue cell in ground tissue system. For parenchyma tissue cell, there is only primary cell wall which is composed of cellulose. In terms of mass content, vascular plant is mainly composed of ground tissue system and vascular tissue system. Therefore, vascular plant could be simply fractioned into vascular tissue fraction which is rich in lignin and ground tissue fraction which is rich in cellulose.

Corn stalk could be fractionated into vascular tissue fraction and parenchyma tissue fraction manually. Vascular tissue fraction includes fiber cell around vascular because fiber cell is connected tightly to xylem and phloem, and both fractions are hydrolyzed with enzyme for 48 h. It is found that glucose content from parenchyma tissue is higher than that from vascular tissue by 22.5 %. Corn stalk could be fractionated into vascular tissue fraction and parenchyma tissue fraction with steam explosion integrated mechanical carding. For rind and leaf, the enzyme hydrolysis rates of parenchyma tissue fraction are 1.77 and 1.37 times higher than vascular tissue fraction, respectively.

Vascular tissue fraction from corn stalk is tested by pulping with ethanol self catalyzing. It reveals that pulp yield could reach to 57.6 % while pulp whiteness reached 65.9 % when catalyzing time is 2.0 h at 160 °C with ethanol concentration 50 % and solid-to-liquid ratio 1:10. The whiteness meets the requirement of printing paper [20].

Corn stalk has not been applied in pulping in that there is much non-fiber cell. Non-fiber cells are mainly parenchyma cell and dermal cell, and the fiber cell content in corn stalk is low. Moreover, the ratio of length to width for corn stalk fiber cell is smaller than other pulping materials. Therefore, if the whole corn stalk is applied for pulping, pulp yield is low and the quality of paper could hardly meet the requirement. If non-fiber cell could be removed and fiber cell content is high, corn stalk would become another resource for pulping.

4.2.2 Cell Level

Cells in stalk could be classed into two categories according to the thickness of cell wall. Category one is the cell that only has a primary cell wall, such as parenchyma tissue cell, sieve cell, companion cell. Category two is the cell that has both primary cell wall and secondary cell walls, such as sclerenchyma cell (including fiber and hardened cell), vessel cell, tracheid cell, and collenchyma cell. Though there is cellulose both in primary cell wall and secondary cell wall. Lignin content in the secondary cell wall is high. Lignin is regarded as recalcitrance of stalk hydrolysis [21]. For bioconversion of stalk, the property for two kinds of cells would be different.

Besides, there are special cells with a different cell wall structure [19]. For a silicon cell, the cell wall is often silicified, and for dermal cell, the wall is usually hornificated. There are also secretory tissue cells including secretory cell, glandular hair, nectar, secretory sac. Secretory cell belongs to parenchyma cell, so their cell walls are rich in cellulose. Though there are few special cells in vascular plant, they affect bioconversion process. For example, hornification dermal cell is regarded as recalcitrance for enzyme hydrolysis [21].

Therefore, stalk is heterogeneous in the level of cell because of different components for a cell wall. So it is necessary to research their different conversion properties.

When corn stalk is pretreated with steam explosion integrated Bauer screening, two fractions could be obtained. The fraction bigger than 28 mesh contains more than 89 % fiber cell and the fraction smaller than 200 mesh contains 64 % parenchyma cell. Therefore, these two fractions are chosen to analyze the hydrolysis property of fiber cell and parenchyma cell. The fraction bigger than 28 mesh is crushed first to smaller than 200 mesh to remove the effect of a particle size. When two fractions are hydrolyzed for 48 h, glucose concentration is 5.15 g/L for parenchyma cell, which is two times higher than that of the fiber cell, and the hydrolysis rate of parenchyma cell reaches 70 %.

If corn stalk is fractionated with steam explosion integrated super grinding, powder fraction and residues fraction could be obtained. Parenchyma cell in powder fraction is 26.6 % higher than that of residues (area percentage), and fiber cell content in residue fraction is 26.4 % higher than that of powder fraction [10]. Dermal cell content is also different in two fractions. Therefore, two fractions obtained from corn stalk pretreated by steam explosion integrated super grinding could be used to analyze the hydrolysis property of different cells. After hydrolysis for 24 h, reducing sugar content is 61.4 % for powder fraction, which is 3.8 times higher than that of residues [10].

The cell content before and after 48 h hydrolysis for powder fraction is analyzed. It reveals that fiber cell reduces by 22.8 % after hydrolysis. Parenchyma cell percentage reduces from 54.2 % to 7.3 %, while dermal cell percentage increases from 10.4 % to 80.9 %. These changes demonstrate that different cells have different hydrolysis properties. The enzyme hydrolysis property could be arranged as parenchyma cell > fiber cell > dermal cell [10].

In the process of pretreatment with steam explosion integrated super grinding, moisture content could affect fractionation results. If moisture content of steam exploded materials is 40 %, the fiber cell content in residue fraction would be more than 60 %. Therefore, residue fraction could be applied to analyze fiber cell conversion properties.

Ethanol self-catalyzing method is applied for pulping. Pulping process is carried out at 180 °C for 2 h with 50 % ethanol concentration and solid-to-liquid ratio 0.8/10 (g/mL). It reveals that crud pulp yield of residue fraction could reach 61.4 %. However, pulp yields of steam exploded rice straw and rice straw are 35.5 % and 32.1 %, respectively. So it demonstrates that there is positive correlation between fiber cell content and pulping property. Therefore, it would be an effective way to fractionate different cells and convert them, respectively.

4.2.3 Component Level

Stalk is composed of cellulose, hemicellulose, and lignin. Cellulose connects with hemicellulose and lignin with hydrogen bond, while hemicelluloses and lignin connects with each other by covalent bond. Therefore, the structure of cell wall is tight and complex. Cellulose, lignin, and hemicelluloses are composed of glucose, phenyl propane units, and pentose, respectively. It is obvious that three main component are different from each other. Cellulose and hemicellulose are carbohydrate, and lignin is aromatic.

Heterogeneous bioconversion property of stalk in component level presents in two processes: hydrolysis and fermentation.

In the hydrolysis process, hemicelluloses and cellulose could be hydrolyzed by cellulase. However, lignin is regarded as recalcitrance for cellulose enzymatic hydrolysis [21]. On one hand, lignin prevents cellulase contacting to substrate. On the other hand, lignin absorbs cellulase non-productively.

In the fermentation process, glucose from cellulose could be used as the main carbon source. Pentose from hemicellulose could also be used by several microorganisms [8]. However, the hydrolysate of lignin, especially small molecular [22] is proved to be inhibitors for fermentation.

Therefore, if stalk is converted as a whole, cellulose conversion rate would be low because of lignin effect. It would be necessary to fractionate stalk into different components and then convert them, respectively.

For pulping, cellulose is extracted in pulping process with different chemicals pretreatment. However, lignin is removed to improve the property of pulp. Chemical structures of cellulose, hemicellulose, and lignin change differently during pulping.

4.3 Requirement of Stalk Heterogeneous Characteristics for Conversion Technology

4.3.1 Fractionation Technology

As per the analysis in Sects. 4.1 and 4.2, heterogeneous characteristics of stalk lead to different conversion properties of different fractions. If corn stalk is converted into single products as a whole with linear technology, different conversion properties of each fraction could hardly be applied [1]. As a result, conversion yield is reduced and waste treatment cost increased. Therefore, it is necessary to integrate various technologies according to the intrinsic characteristics of stalk for multiple products.

Petroleum refining provided a good example for raw materials conversion into universal products. Heterogeneous raw materials are converted into homogeneous fraction at first, and then various fractions are converted into final products according to market requirement. It is an effective way for nature resource to fulfill human requirement. There are hundreds of hydrocarbon in petroleum. It would lead to high cost if only one component is applied ignoring others. The very reason for petroleum to play an important role in life and industry is the refining process invitation. Petroleum is split into different fractions according to a different boiling point. Pure fractions made it possible to explore technologies for further conversion.

Therefore, it would play an important role for stalk-based products to fractionate stalk and integrate various technologies.

4.3.2 Multiple Products Technology

Heterogeneous raw materials demonstrate heterogeneous conversion property as mentioned above. Therefore, if only one product is prepared from stalk, other components would be wasted or even become pollution. What is worse, yield is low if only one fraction is applied and all cost is assumed to one product. On the contrary, stalk could be fractionated into different fractions according to intrinsic characteristics, and different fractions could be converted into different products, respectively. With this refining way, the whole stalk is converted into multiple products without pollution. Moreover, the cost of each product is reduced because of cost apportion.

To solve the problems of energy, environment, and poverty, ethanol production from stalk cellulose has been researched for years [1, 23]. However, it has not been industrialized till now. The essential reason is that the yield is low for single product, leading to high cost. Therefore, it would be an effective model to convert stalk into multiple products.

4.3.3 Integrated Technology

To realize multiple products conversion from stalk with fractionation technique, a single technique is unavailable. At first, the single technique reported could not fractionate stalk into different parts. Dilute acid or alkali pretreatment could only remove lignin and hemicelluloses [24, 25]. Steam cooking and steam explosion just make hemicellulose hydrolyzed and whole stalk loosen [26]. Ionic liquid dissolves cellulose component [27]. Pyrolysis degrades each component into small molecule first [28], and then, small molecules are fractionated with petroleum refining equipment. Secondly, advanced biological, chemical, and physical processes would be integrated to convert different fractions into different products.

Therefore, it is necessary to integrate various refining and conversion technologies in stalk conversion process to realize fractionation oriented by multiple products.

4.4 Research Progress of Steam Explosion Integrated Fractionation Refining Technology

Based on the heterogeneous characteristics of stalk, the authors invent low pressure and non-pollution steam explosion technology. Moreover, steam explosion integrated technology system has been set up. As a result, fractionation refining is realized to obtain multiple products.

Steam explosion has been regarded as one of potential technologies [29, 30] since 1986 for stalk conversion. Nowadays, to solve problems of energy, environment, and poverty, steam explosion is proved to be an effective pretreatment technology [31–34]. The advantages of steam explosion technology have been summarized as follows.

1. At the instant of steam explosion, steam in and among the cell spurts out, destroying cell wall and the connection of tissues and cells. As a result, tissues and cells separate with each other.
2. After steam explosion, more than 50 % hemicellulose is degraded. Therefore, components separated from each other to some extent.
3. Much space is formed because tissues and cells separate. Plus the hydrolysis of hemicellulose, tight structure among three components is destroyed by steam explosion. Therefore, multiple pore structure with wide pore diameter arrangement is formed and specific surface area is increased. So, the mass transfer rate could be enhanced and accessibility is improved when solvent, cellulase, and other chemicals are applied afterwards.
4. Purity of the conversion products is enhanced after separating hydrolysate of hemicellulose and lignin, as well as other water soluble components.
5. Steam explosion pretreatment is more flexible. Steam explosion tank and conditions could be adjusted and various medium could be applied.
6. Steam explosion pretreatment is clear, effective, efficient, and none chemicals addition, and steam explosion is easy to operate and spread.

Loose and porous structure of steam exploded materials is suitable for other pretreatment. Therefore, steam explosion pretreatment technology became the core to integrate many other pretreatment technologies for stalk fractionation refining.

4.4.1 Steam Explosion Integrated Mechanical Carding Technology

As mentioned in Sect. 4.2, vascular plant is mainly including lignin-rich vascular tissue and cellulose-rich parenchyma tissue. Lignin is hypothesized as recalcitrance

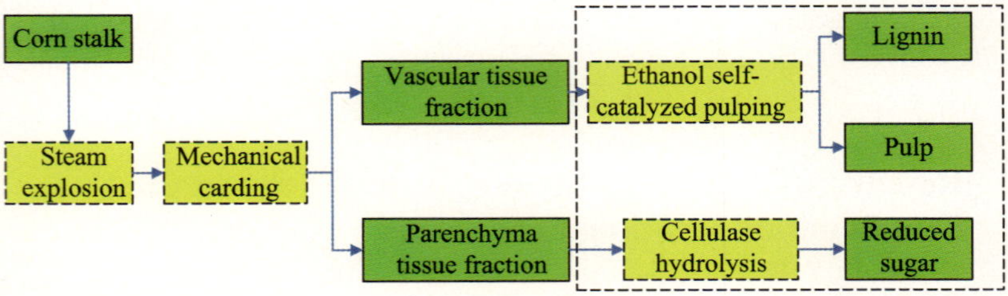

Fig. 4.1 Steam explosion integrated mechanical carding chart

for enzyme hydrolysis process. After steam explosion vascular tissue is obviously longer than parenchyma tissue. According to the morpha characteristics of steam exploded materials, mechanical carding equipment is designed (Application Number, 201110233853.6). With steam explosion integrated mechanical carding technology, it is expected to fractionate stalk into vascular tissue fraction and parenchyma tissue fraction. It is composed of feed inlet, central axis, variable frequency motor, connecting conveyer, short fraction sliding door, fan, short fraction outlet, long fraction outlet. Materials pretreated with steam explosion would be fractionated in this equipment without drying procedure or adding a large amount of water. Steam exploded materials could be loosened firstly, and then fractionated. So crashing process is avoided. Rotating speed and carding time could be regulated to fractionate different materials into vascular tissue fraction and parenchyma tissue fraction. This equipment is simple and easy to maintain.

In the operating process (Fig. 4.1), steam exploded feed is put into inlet and is loosened by manipulator. In this process, parenchyma tissue and the minute fiber cell are separated from vascular tissue and flown out by air flow through short fraction outlet. At the same time, vascular tissue and a small amount of epidermis tissue are flown out through long fraction outlet. Therefore, parenchyma tissue is effectively separated from vascular tissue.

In designing mechanical carding equipment, two parameters could be considered: manipulator density and manipulator length.

In terms of manipulator density, the higher the density is, the shorter the vascular tissue fraction is. Therefore, the density of manipulator is decided by raw material property and the final aim of fractionation.

In terms of manipulator length, the longer the manipulator and axis are, the higher the throughput is, and then, the parameter about other parts could be decided by the length of the manipulator and axis including variable frequency motor, connecting conveyer, inlet, and outlet.

Prophase research revealed that, if corn stalk is steam exploded and mechanical carded as a whole, different organs demonstrate different fractionation results because of heterogeneity in structure and component [9]. Therefore, different organs are steam exploded and fractionated, respectively. As a result, vascular tissue fraction and parenchyma tissue fraction are effectively fractionated when rind, leaf, and pith are steam exploded at 1.5 MPa for 7 min, 5 min, and 2 min, respectively.

If rice straw is taken as an example, with steam explosion integrated mechanical carding fractionation, cellulose content in parenchyma tissue fraction and vascular tissue fraction are 57.8 % and 46.2 %, respectively. The separation degree is 1.25 (separation degree = cellulose percentage in parenchyma tissue fraction/cellulose percentage in vascular tissue fraction). However, the separation degree is just 1.08 when rice straw is fractionated with the steam explosion integrated air flow method [10, 35]. Therefore, steam explosion integrated mechanical carding is an effective technique to enrich cellulose by fractionating raw material into vascular tissue fraction and parenchyma tissue fraction. As a result, the homogeneity of each fraction is improved in structure and component. Compared with the fiber carding method by water flow applied in paper making, mechanical carding could fractionate vascular tissue and parenchyma tissue by manipulator instead of water that would lead to pollution.

Therefore, steam explosion integrated mechanical carding is an effective technique for corn stalk tissue fractionation and is clean for industrialization.

4.4.2 Steam Explosion Integrated Super Grinding Technology

As analyzed in Sect. 4.4.1, cells of materials pretreated with steam explosion are separated to some extent. However, a microscope observation demonstrates that cells from different tissues separate differently. That is, fiber cells and parenchyma cells separate into a single cell which is light. Several dermal cells connected with each other because there is cutin out of them. Connected dermal cells are heavier than a single cell. According to the different weight of cell, air flow super grinding technology is applied to materials steam exploded to realize fractionation in the cell level.

A fluidized bed opposed jet mill [10], FJM-200 super grinding equipment (Beijing, Jinxin Technology Ltd., China) is applied. Main parameters are as follows: working pressure 0.6–1.0 MPa, air consumption rate 1.10–1.73 m^3/min, particle size smaller than 60 mesh, throughput 2–10 kg/h, motor power 13–15 kW.

In the crushing process, air becomes high-pressure gas through high-pressure nozzle. Steam-exploded rice particles are accelerated to high velocity and collided with each other to become powder. Powder fraction is collected from the grinding cavity, and residue fraction is collected from the discharge opening. Heat produced in the process of grinding is taken away by air. Therefore, heating effect is avoided.

For rice straw pretreated with steam explosion severity $\log_{10} R_0$ [min] = 3.1, the powder yield could reach 78 % when operating parameter is as follows: feed load 15 kg/h, rotational speed of classifier wheel 4,544 rpm, moisture content lower than 5 %, grinding time 25 min. With the same conditions, powder yield is just 28 % for raw rice straw.

Fiber cell is easy to be crushed by high-pressure air. To avoid deconstruction of fiber cell, steam explosion integrated wet grinding technology is researched. By this way, fiber cell yield could be enhanced.

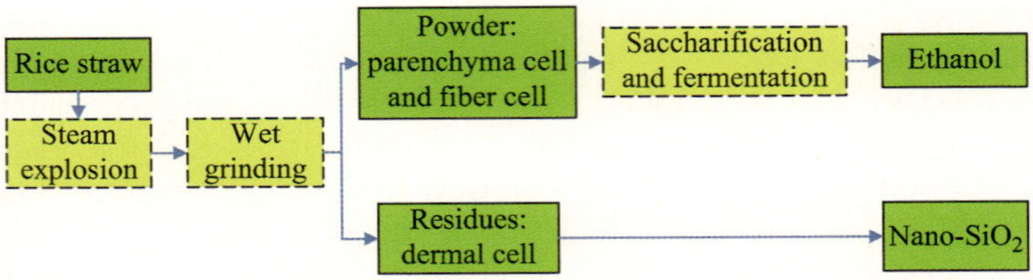

Fig. 4.2 Steam explosion integrated wet grinding chart

The degree of fibrous tissue separation is defined as the ratio of fiber cell content to non-fiber cell content in powder divided by the ratio of fiber cell content to non-fiber cell content in rice straw pre-fractionated.

$$\text{Degree of fraction} = \frac{C'_{\text{fiber cell}}/C'_{\text{non fiber cell}}}{C^0_{\text{fiber cell}}/C^0_{\text{non fiber cell}}}$$

where C' is the cell content of post-fractionation, C^0 is the cell content of pre-fractionation.

The conditions to fractionate rice straw is as follows [35]: steam explosion severity $\log_{10}R_0$ [min] = 3.5, raw material size 3–6 cm, feed load (wet weight) 350–450 g, moisture content of raw materials 40–50 %, grinding time 30 min. Degree of fraction reaches 2.04 (degree of fraction for raw material is 1.00). Powder fraction yield is 70.4 %. Fiber cell content in powder fraction is 63.1 %, and parenchyma cell is 33.5 %. Fiber cell content in powder fraction is higher than raw rice straw by 37.8 %. Cellulose content in powder is 65.6 %, which is higher than that of raw rice straw by 74.9 %. For residues fraction, there is mainly dermal cell and silicon cell. For silicon cell, it could be applied for nano-SiO_2 preparation. It reveals that steam explosion integrated wet super grinding technology is an effective way to fractionate rice straw in cell level (Fig. 4.2).

4.4.3 Steam Explosion–Washing–Alkali Extraction Integrated Technology

In the process of steam explosion, more than 50 % hemicellulose degrades into pentose. Therefore, it is possible to fractionate hemicellulose by washing materials pretreated with steam explosion. As per the analysis in Sect. 4.1, materials pretreated with steam explosion formed porous structure, which enhance the accessibility of a substrate for solvent or other chemicals. So, steam exploded materials are suitable for component fractionation.

Research reveals that lignin could be separated by alkali oxygenation [36]. Therefore, stalk is pretreated with steam explosion-washing-alkali extraction technology to realize fractionation in the component level.

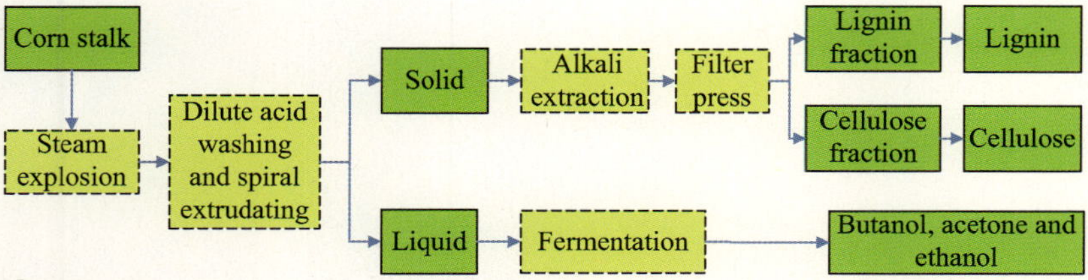

Fig. 4.3 Steam explosion–washing–alkali integrated technology chart

Organic solvent is applied in the process of clean fractionation technology from National Renewable Energy Laboratory [37]. By this way, cellulose, hemicellulose, and lignin in lignocelluloses are also separated. However, compared with organic solvent steam and alkali are cost-competitive. Moreover, alkali solvent could be recycled by nanofiltration.

To improve the yield of hemicellulose [38], corn stalk exploded with steam at severity of $\log_{10}R_0$ [min] = 3.05 is pretreated with 0.3–0.5 % dilute sulfuric acid at 110–120 °C for 0.5–1 h. The ratio of dilute acid to solid is 1:5–1:7, and then liquid and solid are separated with spiral extruder. As a result, hemicellulose yield reaches to 30.7 %. Lignin in solid is fractionated with 0.5–2 % NaOH for 2–3 h at 150–160 °C. The ratio of solid to liquid is 1:5–1:7 (based on the hydrolysis residues). The extract liquid is filtrated sequentially with precise filter, ultrafiltration membrane, and nanofiltration membrane. Filtrate from the nanofiltration membrane is a NaOH solution that could be reused. Concentrated solution from the ultrafiltration membrane is a lignin solution. The lignin solution is sent to neutralization tank to get lignin precipitation that is washed to get lignin product. The whole chart is as Fig. 4.3. As a result, lignin yields are 16.80 %, and cellulose content in cellulose fraction reaches 63 % from 25.4 %.

4.5 Commercial Implementation of Steam Explosion Integrated Fractionation Refining Technology

Stalk is heterogeneous, the same as its conversion property. According to the heterogeneous characteristics of stalk, three platforms have been set up including low pressure, non-pollution steam explosion platform, solid-state fermentation platform and solid enzyme hydrolysis, fermentation, and separation coupling platform. For each platform, pilot test apparatus are designed, as well as industrial equipment. By process integration, multi-layered stalk refining has been established and industrialized in plant about ethanol, sheet, spinning. Especially, in September, 2010, Laihe Chemcial Co., Ltd., located in Jilin Province, China introduced this integrated process (Fig. 4.4 and 4.5). Steam explosion–alkali extraction–mechanical carding integrated pretreatment is applied. Its capacity is 300 thousand tons corn stalk annually. The products from this technology are 50 thousand tons butanol, acetone, and

Fig. 4.4 A 6.5 m^3 steam explosion tank [39]

Fig. 4.5 Butanol acetone distillation tower [39]

ethanol, 30 thousand tons pure lignin which could be converted into 20 thousand tons phenol formaldehyde resin adhesive, 120 thousand tons cellulose which could be converted into 50 thousand tons biological polyether polyol. The cost of solvent products reduce more than 50 % after cost apportion by lignin and cellulose. Many other stalk refining plants are also on the way.

Table 4.1 Economic comparison between multiple products oriented fractionation technology and single product oriented technology [39]

Cost apportion		Cost	Multiple products oriented model					Single product oriented model		
			Butanol 1 t	Acetone 0.28 t	Ethanol 0.14 t	Lignin 0.96 t	Cellulose 3.6 t	Butanol 1 t	Acetone 0.28 t	Ethanol 0.14 t
Raw material	Corn stalk (RMB)	4,715	788	221	110	757	2,838	3,320	929	465
	Corn (RMB)	2,400	1,690	473	237			1,690	473	237
	Industrial sulfuric acid (RMB)	177	124	35	17			124	35	17
	Alkali (RMB)	650				650				
	Activated carbon (RMB)	1,200					1,200			
Kinetic energy consumption	Steam (RMB)	2,800	468	131	66	450	1,686	1,972	552	276
	Water (RMB)	470	79	22	11	76	283	331	93	46
	Electricity (RMB)	750	125	35	18	120	452	528	148	74
Labour cost (RMB)		6,000	1,500	650	50	2,800	1,000	4,091	1,773	136
Total cost (RMB)		19,162	5,447	1,755	603	5,498	9,878	14,887	4,795	1,649
Cost price (RMB/t)			5,447	6,268	4,306	5,727	2,744	14,887	17,126	11,776
Market price (RMB/t)			8,000	6,000	4,000	10,000	2,200	8,000	6,000	4,000
Total value (RMB)		27,7760	8,000	1,680	560	9,600	7,920	8,000	1,680	560
Profit and tax (RMB)		8,598	3,225	113	51	4,748	462	−6,887	−3,115	−1,089

4.6 Advantages of Steam Explosion Integrated Fractionation Refining Technology

In the terms of technological economics, the more by-products are, the higher cost is apportioned, leading to low cost of the main product. That is to say, multiple products are an effective way to improve economic rationality.

The economic comparison between fractionation refining technology with multiple products and single technology with one product is presented in Table 4.1.

It demonstrates that if only the hydrolyzed carbohydrate of corn stalk is converted into acetone, butanol, and ethanol, products lack market competitiveness. What is worse, waste disposal would increase cost. On the contrary, if the whole stalk is fractionally converted into products, each product is market competitive. Moreover, there are no wastes left, which fulfills green industry requirement and realizes resource total utilization.

Compared with single pretreatment technology, the features of integrated pretreatment technology for stalk are as follows.

1. Integrated pretreatment is multiple products oriented, making best of each component for different products. Therefore, by-products increase the economic value of stalk and reduce discharge.
2. Integrated pretreatment is set up according to the heterogeneity of stalk in different levels, such as tissue level, cell level, and component level.
3. Integrated pretreatment is of wide adaptability because there are more adjustable parameters. Therefore, integrated pretreatment is also flexible and fit for various raw materials from different regions and planting manners.
4. For integrated technology, any idea and technology could be integrated including those from paper making, pinning and sheet industry, especially petroleum refining which is rational, economic, effective, clear, and operable.
5. Each technology involved in integrated pretreatment could be complement in advantages to reduce the whole pretreatment cost.

4.7 Conclusion

Stalk heterogeneous property in the level of tissue, cell, and component is analyzed in terms of stalk morphology and anatomy. It is proved that different tissues, cells, and components have different conversion properties because of heterogeneous characteristics. Steam explosion integrated fractional refining technology is set up according to stalk heterogeneous intrinsic characteristics. The whole stalk is fractionated in the level of tissue, cell, and component and then converted into different products, respectively. Industrial demonstration proves that steam explosion integrated fractional refining technology is an effective way to improve the economic benefit of stalk utilization.

Acknowledgments Financial support to this study is provided by the National Basic Research Program of China (973 Project, No. 2011CB707401), the National High Technology Research and Development Program of China (863 Program, SS2012AA022502), the National Key Project of Scientific and Technical Supporting Program of China (No. 2011BAD22B02).

References

1. Chen HZ, Qiu WH (2010) Key technologies for bioethanol production from lignocellulose. Biotechnol Adv 28(5):556–562
2. FitzPatrick M, Champagne P, Cunningham MF et al (2010) A biorefinery processing perspective: treatment of lignocellulosic materials for the production of value-added products. Bioresour Technol 101(23):8915–8922
3. Tim Beringer WLaSS (2011) Bioenergy production potential of global biomass plantations under environmental and agricultural constraints. GCB Bioenergy 3:299–312
4. Erb KH, Krausmann F, Lucht W et al (2009) Embodied HANPP: mapping the spatial disconnect between global biomass production and consumption. Ecol Econ 69(2):328–334
5. Ballesteros M, Oliva J, Negro M et al (2004) Ethanol from lignocellulosic materials by a simultaneous saccharification and fermentation process (SFS) with *Kluyveromyces marxianus* CECT 10875. Process Biochem 39(12):1843–1848
6. Gargulak J, Lebo S (2000) Commercial use of lignin-based materials. ACS Publications, Washinton, DC, pp 304–320
7. Zhang MJ, Wang F, Su RX et al (2010) Ethanol production from high dry matter corncob using fed-batch simultaneous saccharification and fermentation after combined pretreatment. Bioresour Technol 101(13):4959–4964
8. Hahn-Hägerdal B, Karhumaa K, Fonseca C et al (2007) Towards industrial pentose-fermenting yeast strains. Appl Microbiol Biotechnol 74(5):937–953
9. Chen HZ, Li HQ, Liu LY (2011) The inhomogeneity of corn stover and its effects on bioconversion. Biomass Bioenergy 35(5):1940–1945
10. Jin SY, Chen HZ (2006) Superfine grinding of steam-exploded rice straw and its enzymatic hydrolysis. Biochem Eng J 30(3):225–230
11. Chen HZ, Liu LY (2007) Unpolluted fractionation of wheat straw by steam explosion and ethanol extraction. Bioresour Technol 98(3):666–676. doi:10.1016/j.biortech.2006.02.029
12. Saha BC, Iten LB, Cotta MA et al (2005) Dilute acid pretreatment, enzymatic saccharification and fermentation of wheat straw to ethanol. Process Biochem 40(12):3693–3700
13. Pérez J, Ballesteros I, Ballesteros M et al (2008) Optimizing liquid hot water pretreatment conditions to enhance sugar recovery from wheat straw for fuel-ethanol production. Fuel 87(17):3640–3647
14. Nguyen TAD, Kim KR, Han SJ et al (2010) Pretreatment of rice straw with ammonia and ionic liquid for lignocellulose conversion to fermentable sugars. Bioresour Technol 101(19):7432–7438
15. Maekawa E (1996) On an available pretreatment for the enzymatic saccharification of lignocellulosic materials. Wood Sci Technol 30(2):133–139
16. Cui QJ, Zhu HG, Wang DY et al (2011) Effect on biogas yield of straw with twin-screw extruder physical-chemical combination pretreatment. Trans Chin Soc Agric Eng 27(1):280–285
17. Chen W, Yu H, Liu Y et al (2011) Individualization of cellulose nanofibers from wood using high-intensity ultrasonication combined with chemical pretreatments. Carbohydr Polym 83(4):1804–1811
18. Sachs J (1875) Text-book of botany, morphological and physiological. Clarendon Press, Oxford
19. Evert RF, Esau K (2006) Esau's plant anatomy: meristems, cells, and tissues of the plant body: their structure, function, and development. Wiley, New York

20. Liu LY, Chen HZ (2006) The fractionation of straw and its high value conversion. Dissertation, Institute of Process Engineering, Chinese Academy of Sciences
21. Himmel ME, Ding SY, Johnson DK et al (2007) Biomass recalcitrance: engineering plants and enzymes for biofuels production. Science 315(5813):804–807
22. Larsson S, Cassland P, Jönsson LJ (2001) Development of a *Saccharomyces cerevisiae* strain with enhanced resistance to phenolic fermentation inhibitors in lignocellulose hydrolysates by heterologous expression of laccase. Appl Environ Microbiol 67(3):1163–1170
23. Wyman CE (1994) Ethanol from lignocellulosic biomass: technology, economics, and opportunities. Bioresour Technol 50(1):3–15
24. Guo GL, Hsu DC, Chen WH et al (2009) Characterization of enzymatic saccharification for acid-pretreated lignocellulosic materials with different lignin composition. Enzyme Microb Technol 45(2):80–87
25. Persson P, Andersson J, Gorton L et al (2002) Effect of different forms of alkali treatment on specific fermentation inhibitors and on the fermentability of lignocellulose hydrolysates for production of fuel ethanol. J Agric Food Chem 50(19):5318–5325
26. Sun FB, Chen HZ (2008) Comparison of atmospheric aqueous glycerol and steam explosion pretreatments of wheat straw for enhanced enzymatic hydrolysis. J Chem Technol Biotechnol 83(5):707–714
27. Liu LY, Chen HZ (2006) Enzymatic hydrolysis of cellulose materials treated with ionic liquid [BMIM] Cl. Chin Sci Bull 51(20):2432–2436
28. Meier D, Faix O (1999) State of the art of applied fast pyrolysis of lignocellulosic materials—a review. Bioresour Technol 68(1):71–77
29. Guo P, Mochidzuki K, Zhang D et al (2011) Effects of different pretreatment strategies on corn stalk acidogenic fermentation using a microbial consortium. Bioresour Technol 102(16):7526–7531
30. Grous WR, Converse AO, Grethlein HE (1986) Effect of steam explosion pretreatment on pore size and enzymatic hydrolysis of poplar. Enzyme Microb Technol 8(5):274–280
31. Jacquet N, Quievy N, Vanderghem C et al (2011) Influence of steam explosion on the thermal stability of cellulose fibres. Polym Degrad Stab 96(9):1582–1588
32. Rocha G, Gonçalves A, Oliveira B et al (2012) Steam explosion pretreatment reproduction and alkaline delignification reactions performed on a pilot scale with sugarcane bagasse for bioethanol production. Ind Crops Prod 35(1):274–279
33. Kumar Biswas A, Umeki K, Weihong Y et al (2011) Change of pyrolysis characteristics and structure of woody biomass due to steam explosion pretreatment. Fuel Process Technol 92(10):1849–1854
34. Brugnago RJ, Satyanarayana KG, Wypych F et al (2011) The effect of steam explosion on the production of sugarcane bagasse/polyester composites. Composites Part A 42(4):364–370
35. Jin SY, Chen HZ (2007) Fractionation of fibrous fraction from steam-exploded rice straw. Process Biochem 42(2):188–192. doi:10.1016/j.procbio.2006.07.030
36. Chen HZ, Han YJ, Xu J (2008) Simultaneous saccharification and fermentation of steam exploded wheat straw pretreated with alkaline peroxide. Process Biochem 43(12):1462–1466
37. Laborotary NRE (2008) Clean fractionation. http://www.nrel.gov/biomass/publications.html. Accessed Oct 2008
38. Chen HZ, Wang L (2011) Method for fully utilizing biobased products prepared by semicellulose of straws as well as components thereof. PCT/CN2011/000142
39. Wang L (2010) Key process of butanol production from steam-exploded corn stover. Dissertation, Institute of Process Engineering, Chinese Academy of Sciences

Chapter 5
Pretreatment and Pelletization of Woody Biomass

Pak Sui Lam, Zahra Tooyserkani, Ladan Jafari Naimi and Shahab Sokhansanj

Abstract Pretreatment is a first crucial step to modify the structure of wood via physical, chemical, and biological treatment for cost effective and sustainable fuels and chemicals production. Different pretreatments would be selected to upgrade the characteristics of wood with respect to different applications and process efficiencies. High-temperature pretreatment (e.g., torrefaction) at the temperature range greater than 250 °C led to higher degradation rate of sugars and extractives, which is not preferable for fuel and chemicals production from ligno-cellulosic biomass. Instead, high-temperature pretreat-ment was used to upgrade the solid fuel for thermo-chemical conversion (e.g., combustion and gasification). It can remove the moisture and volatiles with a low-heating value of the native biomass, which favors for the ease of fuel combustion compared to the raw wood. In addition, it can increase the hydrophobicity of the biomass which improves their handling and storage performance. In this chapter, the production chain of the wood pellet production with incorporating recent novel pretreatment technologies (torrefaction, steam explosion, and hydrothermal carbonization) were discussed. The resulted pellets are a uniform feedstock for producing chemicals, heat, and energy via biochem-ical and thermochemical conversion, respectively.

Keywords Pretreatment · Wood pellet · Drying · Grinding · Biomass preprocessing · Torrefaction · Steam explosion · Hydrothermal carbonization · Pellet quality

---

P. S. Lam (✉) · Z. Tooyserkani · L. J. Naimi · S. Sokhansanj
Biomass and Bioenergy Research Group, Clean Energy Research Center,
Department of Chemical and Biological Engineering,
University of British Columbia,
2360 East Mall, Vancouver, B.C., V6T 1Z3, Canada
e-mail: wilsonlam82@yahoo.com

Z. Fang (ed.), *Pretreatment Techniques for Biofuels and Biorefineries*,
Green Energy and Technology,
DOI 10.1007/978-3-642-32735-3_5,

5.1 Introduction

Renewable energy has been targeted as a strategic important area for many countries for both environmental and economic reasons [1]. The establishment of a clean energy supply can provide greater energy independence and security, has notable environmental benefits due to reduced CO_2 emissions, as well as promoting positive economic growth for the local area. Among various types of renewable energy, bioenergy is attractive as biomass is considered to be carbon neutral that absorbs CO_2 from the atmosphere during production [2]. Besides, bioenergy systems can create the highest job creation effect, particularly in the rural areas with high unemployment rate, and resulting in the stimulation of economic growth [2, 3]. Since biomass is dispatchable, it is economically preferable to deploy when required. The biomass feedstock supply logistic cost contributes around 30–50 % of the total bioenergy production cost [4]. An optimized pre-processing of the biomass into densified pellets is essential to achieve a cost-effective production process for bioenergy.

Wood pellets are a type of solid fuel made from sawdust with uniform shape and dimensions. Pellets are made by densifying the ground particles of woody biomass. Wood residues usually come as sawdust from saw mills. Their bulk densities are around 40–60 kg/m^3 (wet basis) depending on species and moisture content (MC) [4]. Drying is required to ensure that the size-reduced feedstock is good for pelletization (densification) to produce durable wood pellets. The bulk densities of the biomass pellets are around 550–700 kg/m^3 depending on the size of the pellets [4]. The volume reduction reduces the space required for storage and transportation. For storage, biomass densification helps to reduce the space required to store the materials. Biomass pellets improve the heating efficiency and have lower emissions during combustion than using the low bulk density and fluffy biomass. The typical example is the co-firing plant using wood pellets and coal as feedstock where the difference between these two materials' densities cause difficulties in feeding due to the uneven, fluffy, and low bulk density of the biomass feedstock [4].

The typical production process of biofuel pellets is collecting the residues from saw mill and following by drying using a rotary drum dryer, further size reduction to the granular form by hammer mill, and finally pelletizing into fuel pellets using a pellet mill (Fig. 5.1). The fuel pellets are then cooled, screened, and transported to an export port by trains. They are usually transported on conveyor belts and dropped from the height of 10–15 m above the storage silos for temporary storage. They are stored under a well-monitored environment to prevent self-heating and off-gas accumulation. Pellets are then loaded into the ocean vessel. The details of each unit operation will be discussed in the following sections. The pellet quality needs to be maintained during transportation in order to meet the import specification of the European standard [5].

Biomass preprocessing is aimed at enhancing the energy density of a bulk biomass. Further optimization of the process can be achieved by enhancing the production yield and reducing the energy required for the preprocessing process. Two major technical problems during the preprocessing process need to be addressed. Poor mechanical strength of biomass pellets contributes to disintegration of pellets into fines during

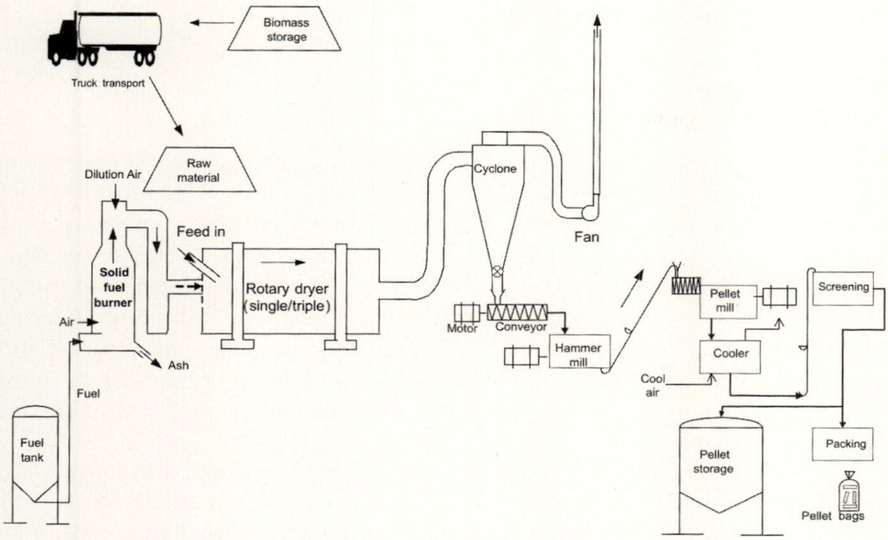

Fig. 5.1 Schematic layout of a typical biomass pelleting plant. (Reprinted with permission from [5]. Copyright 2006 American Society of Agricultural and Biological Engineers)

transportation. This usually happens for the pellets transporting on the conveyor and loading from the top of the silo to form piles and pellets break into fines due to impact. The fines cause blockage of the conveyor or hopper during processing and also lead to an occupational health problem to the workers inhaling the fines [6]. Moreover, the fines also lead to dust explosion which causes severe fire damage to the expensive handling facilities. This is related to the lack of natural binding between the fibers of the pellets, and most biomass species including straws and stover are difficult to densify without any expensive binders [7–10]. Only wood pellet can be formed with good durability due to their binderless characteristics.

Pellets easily adsorb moisture and disintegrate into fines under high-humidity conditions. The high surface area of the small fines favors the susceptibility of the attack by the micro-organisms during storage [11]. Anaerobic conditions lead to local heat generation and generation of toxic off-gassing that may include terpenes [12]. Local heat generated may ignite the volatiles in the pellets to cause fire, and the off-gas accumulations inside the storage silos are toxic to the workers. High MC reduces pellets combustion efficiency at the power plant.

In the following, we will focus on discussing the development and optimization of the entire biomass pellet production chain by different pretreatment techniques. This not only aims to produce fuel pellets for ease of chemical conversion and energy production, but also to reduce the preprocessing cost and improve the safe handling of pellets during transport and storage.

5.2 Biomass Preprocessing

5.2.1 Drying

The biomass, in the forms of wood chips, sawdust, bagasse, grass, and agricultural residues is bio-origin material that initially contains moisture from 50 % to over 150 % (dry basis) in the fresh form. In order to increase energy efficiency, improve energy product quality, and reduce emissions in its thermochemical energy conversion, drying of the biomass to the required MC is important in the development of energy production systems [13, 14]. In addition, it was found that uniformity of drying also significantly affects the energy efficiency in a combined heat and power (CHP) plant [15]. Other issues needed to be considered in biomass drying include forms of feedstock, energy conversion technology, environmental impact, risk of fire and explosion, available energy source, and drying costs [13].

Three common types of dryers widely used in industry are packed moving bed (PMB) dryer [16, 17], rotary dryers [18–20], and pneumatic or flash dryers [19, 21]. For the development of biomass drying technologies, multistage drying, exhaust air recycling, heat recovery, and the optimization of the drying conditions have been explored [13, 19, 22]. In particular, superheated steam drying has attracted great interest in order to prevent the risk of fire, reduction of emissions and increase the energy efficiency in drying [13, 23, 24].

5.2.1.1 Rotary Dryer

The rotary dryer is one of the most commonly used technologies for drying wood. It is effective to handle both sawdust and chips [18, 20, 25]. The most commonly used rotary dryers in industry are direct contact type, which consists of a hollow, rotational metal cylinder providing space for direct contact between the material to be dried and the drying medium, usually hot air. The heat and mass transfer between these two streams is high with a direct contact and can be further enhanced by installing a series of flights on the inner wall of the cylinder to promote contact of the two streams. In addition, uniform MC distribution of the dried product can be achieved because every piece of the solid material has an equal chance to contact with the hot air.

For the co-current rotary dryer, the wet material and the hot air enter from one end, and the dried material and the humid air exit from the other end. The wet material and the hot air enter the drum from the opposite ends and move inside the dryer in the opposite directions. A typical drying temperature used in commercial drying of woody biomass is up to 500 °C that ensures high drying rate and energy efficiency of drying. The co-current mode is usually employed for industrial application as this ensures the dried biomass will not overheat and cause self-ignition. The operation conditions have to be well monitored to prevent fire hazard. This requires reliable measurement techniques of dryer temperature and MC in order to develop a prediction model for biomass drying.

5.2.1.2 Packed Moving Bed Dryer

For packed moving bed dryers, the wet biomass is fed from one side to a moving bed that has openings on it, allowing the drying medium (hot gas) to flow through. In order to increase energy efficiency of the drying, the drying medium is recycled by flowing back through the biomass bed in the second half of the dryer. The overall air flow rate is only half of that in the arrangement where the drying air always flows upwards. Because with the drying air reversal in the second half of the dryer, the exhaust gas has low temperature and high humidity, the corresponding equilibrium moisture content (EMC) is relatively high for the bio-originated material. This indicates that if low final MC is required, the exhaust gas temperature must be kept higher than a certain value to achieve the required dryness. On the other hand, higher exhaust air temperature results in greater heat losses [13, 16, 17].

During drying, the MC varies across the bed thickness [26]. In a co-current arrangement, the bottom layers of biomass in the first half of the dryer dry faster than that of top layers. Although the reverse flow of the drying air reduces the uneven moisture distribution, the MC gradient may not be totally eliminated because the recycling drying gas in the second half of the dryer has a lower drying temperature and higher humidity; thus, the drying rate is lower compared to the first half of the dryer.

In the case that the low final MC is required for the dried biomass, a countercurrent arrangement can be used where the hot gas is fed from the second half of the dryer, flowing upwards, and is then reversed to flow through the biomass bed in the first half of the dryer. In this improvement, the drying efficiency is higher and the required final MC can be achieved [13].

5.2.1.3 Pneumatic Dryer

Pneumatic dryers are gas–solid transport systems with continuous convective heat and mass transfer process [21]. This type of dryers can achieve rapid drying with short residence time (5–10 s) by fully entraining the material with a high velocity gas flow [19, 27]. The high-velocity gas transports the solid particles along the pipeline and mixing takes place between them. The gas stream also serves as a drying medium to supply heat to remove moisture away from the biomass particles. Since pneumatic dryers operate with a low solid content, it is easy to control and allow the materials to be dried to desired equilibrium MC rapidly [13].

5.2.2 Size Reduction

Size reduction is an important step in preparing biomass for pelletization. It is energy intensive. Depending on the original form of feedstock, there are one or two steps of size reduction necessary prior to pelletization. Woody biomass can be transported from the forest to the end user (e.g., power plant) in the chips form. They were processed from logs into chips. Logs have to be debarked, and the clean wood logs are size reduced to chip shape using a chipper. Wood chips with 25–50 mm length and width are usually further size reduced to 3.0–6.4 mm ground particles before compacting the material into pellets using either a hammer mill or a knife mill. Often chipping is carried out in the forest, and the chips are transported to the pellet plant where they are ground.

Size reduction is a sensitive step for controlling the ground particles size, distribution is done by installing screen with different sizes in the mill. For the wood pellet industry, a hammer mill with 3.0–6.4 mm screen is mostly preferred for pellet production. A larger particle size (greater than 1 mm) will also act as predetermined breaking points in the pellet, and therefore the optimum particle size range between 0.5 and 0.7 mm is sometimes suggested [28]. In reality, a mixture of different particle size gives the optimum durability of the pellet as they have a better inter-particle bonding with less interspaces [29–31]. Therefore, there is a need to study the particle size distribution on the durability of the wood pellets.

5.2.2.1 Size Reduction Mechanisms

The dominant grinding mechanisms of size reduction are impact and shear. Different feedstock requires different size reduction mechanisms. The most common devices used for size reduction for preparing feedstock for pelletization are chippers (shear) and hammer mills (impact). Figure 5.2 shows the images of typical size reduction equipment that are used for biomass grinding [32].

5.2.2.2 Chippers

Chippers are used when big logs and short rotational trees are used as feedstock for pelletization. They are classified as drum chippers and disk chippers. The mechanism of grinding in chippers is shear. The material is size reduced by cutting with a sharp knife. Either the knives are mounted on a disc (disc chipper) or a cylinder (drum chippers). The chippers knives are sensitive to impurities in the feedstock. A high maintenance cost is resulted when dirty feedstock are used. The ground materials from chippers are usually uniform in size and shape.

5.2.2.3 Hammer Mill

Hammer mills can be used for a wide range of feedstocks. The mechanism of size reduction in hammer mills is impact. There are fixed or swing hammers mounted on the rotor. Hammers may be blunt or sharp. Sharp hammers allow a combination of both mechanisms of shear and impact for size reduction. The ground particles have various shapes and a wider range of size distributions in comparison to chippers. Maximum particle size can also be controlled by installing screens with different sizes.

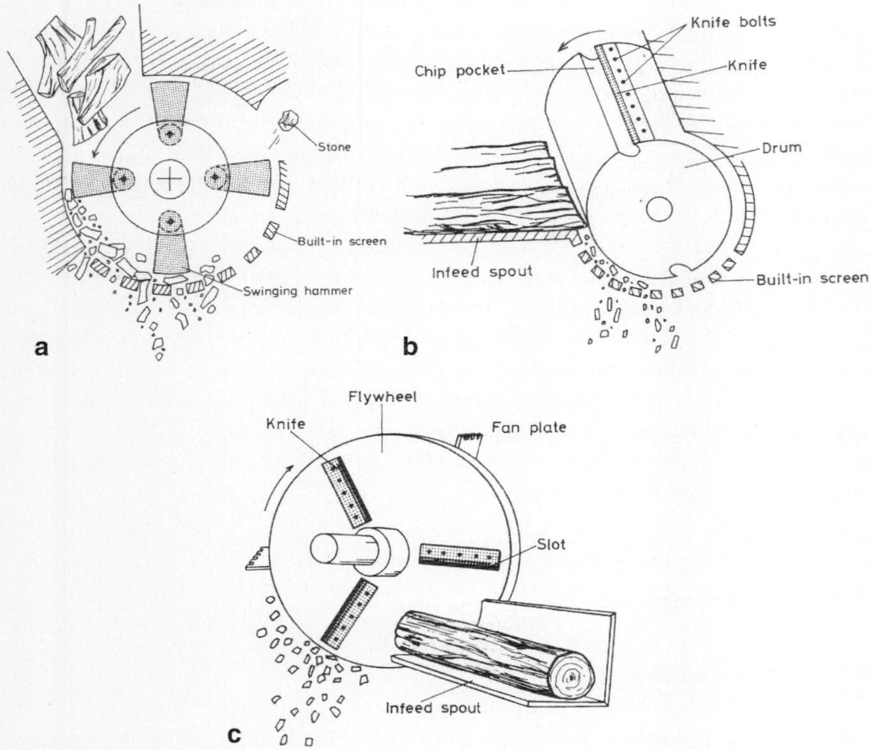

Fig. 5.2 Size reduction equipment. *Top left*: a hammer mill, *top right*: a drum chipper, *bottom*: a disc chipper. (Reprinted from [32], copyright 1989, with permission from Springer)

5.2.2.4 Size Reduction and Energy Consumption

The trade-off of producing fine grinds is to consume increased energy to produce finer particles. An optimum particle size should be defined that fulfills the need of pelletization process and downstream process (e.g., combustion, ethanol production, and gasification).

In general, manufacturing 1 t of dried pellets may use 100–180 MJ for grinding [5]. It is the third highest energy consumption unit operations of the overall pellet production after drying and pelletization. It is reported that for a pellet mill that uses sawdust as feedstock and has a production rate of 5 t (w.b) pellet per hour, a 110-kW hammer mill is needed [33]. The energy expense is around 18.7 kW h or 0.38 % of the energy content of the pellets, based on an assumed energy content of the pellets 4900 kW h/t (w.b) [33].

5.2.2.5 Size Reduction Cost

Grinding unit is the second largest electricity consumer in pellet production (pellet mill is the maximum electricity consumer) [34]. The whole process of pellet production is divided into: raw material, general investment, drying, grinding, pelletization, cooling, storage, peripheral equipment, and personnel. If the raw material is wet, grinding cost is 3 % of the whole cost, and if the raw material is dry, the grinding cost is 2 % of the whole cost. It should be considered that sawdust is the base raw material [34]. A hammer mill cost ranges between 62,000 € (with a capacity of about 2.5–3 t (d.b.)/h) and 168,000 € (two large hammer mills with a total capacity of about 9 t(d.b.)/h) depending on the plant size and the equipment used. The maintenance cost is 18 % of the investment costs per year. Scale of operation can impact overall costs, for example the overall cost of pellet production is lower in Sweden because of their larger capacity plants and lower electricity price.

One step of grinding is sufficient for material preparation for the traditional pellet production using sawdust and shaving as feedstock. When the feedstock sources change from sawdust and shaving to other mill residues (e.g., logging residues, thinning materials, agricultural residues, and short rotational trees), two or more steps of size reduction is needed to prepare the material for pelletization. Two stages of size reduction using a coarse grinder and a hammer mill were used for the pellet production of short rotation trees [35].

5.2.3 *Pretreatment*

5.2.3.1 Torrefaction

Torrefaction is used as a pretreatment prior to pelletization for upgrading the woody biomass primarily for energy production [36]. Torrefaction is a thermochemical treatment that subjects the biomass under heat at the reaction temperature between 200 and 300 °C in an inert medium (e.g., nitrogen) for a certain period of time (ranging between several seconds and an hour) depending on the particle size.

Reaction Chemistry

A detailed review of torrefaction chemistry and its process conditions is summarized in the literature [37, 38]. The reactions that take place for torrefaction are mainly decomposition and which can further divide into (1) drying, (2) depolymerisation and recondensation, (3) limited devolatilization and carbonisation, (4) extensive devolatilization and carbonisation throughout the reaction temperature regime (Fig. 5.3). Hemicelluloses soften at temperatures between 150 and 200 °C and undergo dehydration, deacetylization, and depolymerization reactions at processing temperatures in the 200–300 °C [39, 40]. Xylan is the predominant form of hemicellulose for hardwood while glucomannan is the predominant form of hemicellulose of softwood. Xylan tends to break down more quickly than glucomannan at lower temperatures. Therefore, hardwood has a higher breakdown of hemicellulose (or higher mass loss) than that of softwood when treated at the same temperature. This suggests

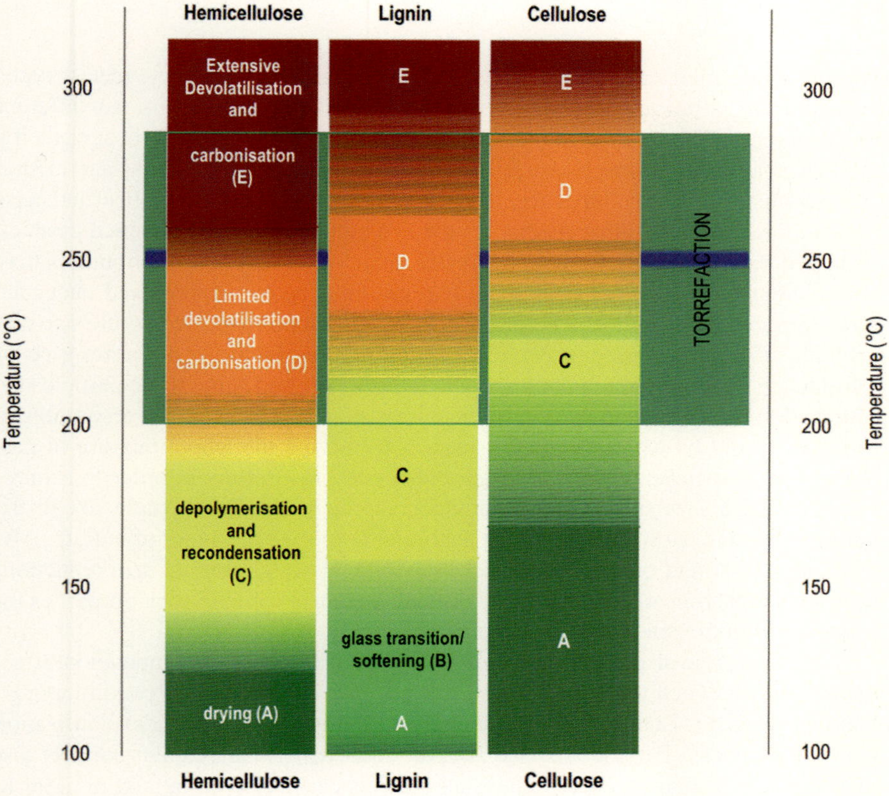

Fig. 5.3 Main physico-chemical phenomena during heating of lignocellulosic materials at pre-pyrolytic conditions (torrefaction). (Reprinted with permission from [42], copyright 2005, Energy Research Centre of the Netherlands)

that different species of wood have different torrefaction kinetics which is worth for further in-depth investigation. A simple one-step (single stage) kinetic model with the first-order reaction of chemical reactions of cellulose, hemicellulose, and lignin during torrefaction of British Columbia (BC) softwoods was reported for slow residence time reaction while a two-component and one-step first-order reaction yielded a better prediction of chemical components for the torrefaction with short residence time [41].

A small degree of degradation of cellulose and/or lignin also occurs during torrefaction [43]. When the reaction temperature is high (>270 °C), a greater proportion of cellulose degradation was reported. In contrast, lignin is relatively stable and does not undergo significant chemical changes during torrefaction even at high temperature.

Particle Size and Residence Time

Particles with small size distribution would be recommended for some reactor types to achieve an optimized torrefaction efficiency and product quality. Note that different biomass species have different physical properties (porosity, specific heat capacity, thermal conductivity, particle size distribution of grinds under the same size reduction process, etc.) that further introduces a non-homogenous reaction due to heat and mass transfer limitation and subsequently results in a non-homogenous torrefied product.

In general, large particles within a wider range of particle size distributions may not be completely torrefied, and small particles may be over-carbonized under the same torrefaction condition. Therefore, a narrow or mono-disperse particle size distribution of a certain specific wood particles may be ideal for a certain type of torrefaction reactor operating at a specific temperature and time. If the particle size distribution of raw materials is a bi-modal distribution, sieving and separation of mixtures would be recommended instead of torrefying the whole mixture of particles. Large particles may require more than one pass milling in order to achieve smaller particle size suitable for homogenous torrefaction. The particle size distribution of ground particles depends on the biomass species, the types of mill, the MC of the biomass, and other factors, etc. Pretreatments (e.g., drying, size reduction) prior to torrefaction are necessary and critical to control the feedstock properties for producing a homogenous product quality.

Residence time also affects the mechanism of torrefaction decomposition. Residence time is related to the reactor design (i.e., size) and operation condition (e.g., feeding velocity). Long residence time allows a greater degree of devolatilization from the biomass. Particle size and MC of wood particle also affect the heat and mass transfer. Small particles require less residence time to be treated in order to achieve the same degree of chemical reaction as the longer residence time is required for the large particles to allow nitrogen diffusion to initiate the decomposition reaction. A slow heating rate is critical for product homogeneity. Slow heating facilitates a uniform temperature gradient across the particle during torrefaction. Less volatile with low heating value is vaporized and thus results in a higher yield of solid residue with high heating value for pellet production.

Reactor Types

Commercial torrefaction technology suppliers require a strict/narrow particle size distribution as feedstock for producing fuels with a homogenous quality. Many reactor types require a maximum chip size of 2.5 cm (1″). In particular, the reactors with a shorter residence time (in the order of seconds or a few minutes) require a more stringent particle size limitation (less than 0.5 cm), for example, Torbed and moving bed reactor. If the feedstock particles are with a larger proportion of fines, fines will block the gas to limit the heat transfer and also become difficult to fluidize or transport pneumatically.

Some reactors allow using large particles as feedstock, for example, multiple hearth reactor and rotary drum reactor. The rotary drum reactor can accommodate a mixture of large and small particles. Both groups of particle will be treated at the same temperature and reside within the same chamber at the same time. Meanwhile, the attrition of wood chips or particles occurs and generates lots of fines due to the rotational motion. However, large particles would undergo with non-homogenous reactions within a single particle and result in non-homogenous product quality. Feedstock with a wider particle size range can be used if the torrefaction temperature range is within a mild region (i.e., 260–290 °C). Besides, a pack bed of large particles are usually with lower bulk density. This limits the throughput as well as introducing a limited load capability for some reactor types including rotary drum reactor, screw reactor, and belt reactor.

A moving bed scale reactor setup for studying complex gas–solid reactions has been designed in order to obtain kinetic data for scale-up purpose. In the bench scale reactor setup, gas and solid reactants can be contacted in a co-current and counter-current manner at high temperatures. Gas and solid sampling can be performed through the reactor bed with their composition profiles determined at steady state. The reactor setup can be used to evaluate and corroborate model parameters accounting for intrinsic reaction rates in both simple and complex gas–solid reaction systems. The moving bed design allows experimentation over a variety of gas and solid compositions in a single experiment unlike differential bed reactors where the gas composition is usually fixed. The data obtained from the reactor can also be used for direct scale-up of designs for moving bed reactors. The recommended particle size range: 5 mm < diameter < 20 mm; length <70 mm.

5.2.3.2 Steam Explosion

Steam explosion is one of the hydrothermal treatments that subject the biomass to high-pressure saturated steam at the reaction temperature between 180 and 240 °C for several minutes followed by a rapid decompression. The residence time of steam treatment depends on the reactor type and the desired degree of treatment. The degree of steam treatment can be described by a severity equation developed by previous researchers [44]. This equation was developed based on modeling complex reaction systems by assuming that each reaction is homogenous, and Arrhenius dependence rate law and the temperature function were linearized by Taylor series [45, 46]. This equation is developed based on the data from batch reactor.

$$R_0 = \int_0^t \exp\left(\frac{(T - 100)}{14.75}\right) dt' \tag{5.1}$$

where R_0 = is the reaction severity, T is the reaction temperature (°C), t is the reaction time (min).

Another equation is also developed for scaling up from a batch process to a continuous process [47],

$$\log R_{0,\text{Batch}} = 1.50 \times \left(\log R_{0,\text{Continuous}} - 1\right) \tag{5.2}$$

where $R_{,0,\mathrm{Batch}}$ is the reaction severity studied based on batch reactor, $R_{,0,\mathrm{Continuous}}$is the reaction severity applied to the continuous large scale reactor.

Chemistry

Under high pressure of saturated steam, the initial reaction that takes place inside the chemical components in presence of woody biomass is hydrolysis. Biomass is a mixture of polymer composite mainly made up of carbohydrates (cellulose, hemicelluloses) and lignin. In particular, hemicelluloses and lignin hydrolyze in the presence of acetic acid to release low molecular weight components: mono-sugars and acid-soluble lignin. Some mono-sugars will further degrade into other chemicals at high temperature by dehydration reaction. One typical example is the formation of furfural by dehydration reaction of xylose (mono-sugar of xylan, a type of hemicelluloses). For lignin, repolymerization or condensation reaction of low molecular weight lignin takes place as reaction coordinate proceed. This changes the structure and morphology of lignin, which is important for improving the binding ability of the wood fibers during pellet production. Detailed of chemistry has been reported previously [48].

Particle Size and Reaction Time

From the mass transfer point of view, the diffusion resistance of the gas phase is larger for larger particles (e.g., wood chips). This implies that less heat is brought by saturated steam to diffuse into the inner core of a single biomass particle. A longer time is required for saturated steam to diffuse into the core of the large particle to initiate the hydrolysis. Hydrolysis usually starts to take place when the reaction temperature is higher than 160 °C, excluding the mass transfer limitation.

Saturated steam is a better heat transfer medium thanNitrogen. Its high-energy state (a higher latent heat of vaporization) allows them to reach the core of the particle with a smaller diffusion resistance. Hydrolysis is preferentially taking place inside the polymeric system instead of pyrolysis reaction using high-pressure saturated steam. In general, the development of chemical reaction kinetics models usually work with the particles with a <8 mm diameter by assuming no diffusion limitation [49, 50]. Since sawdust has a smaller particle size and a higher specific surface area than wood chip, more glycosidic bonds are accessible for cleavage by hydronium ions (H_3O^+) resulting in higher degree of hemicelluloses solubilisation. Hydronium ions have a typical size of 0.4 nm in wood chips [49]. Its small size allows them to penetrate into the wood pores to hydrolyze the glycosidic bonds of xylan molecules (hemicelluloses) as well as to promote lignin solubilization and restructuring. Restructuring of lignin helps to activate and enhance the accessible lignin as a self-binding agent for durable pellet production.

Reactor Types

Steam explosion pretreatment is a commercial available process. It has been used as a pretreatment for bioethanol production. In Europe, Andritz Sprout developed a screw type horizontal continuous scale steam explosion unit. This reactor is similar to the one used by Mascoma Corporation. The feed material can be in a wide particle range in ground particles (diameter: 6 mm) or chips (diameter: 25–75 mm) form depending on the size of the hopper and the blow valve. There is an opening on the top of the horizontal chamber that can be connected to a hopper for material feeding. A steam chamber in horizontal direction is used for steam treatment of biomass. Saturated steam is supplied from the other steam generator via the nozzles from the side wall of the chamber. At the end of the chamber, there is a rapid opening blow valve controlled by a controller to release the high-pressure steam-treated biomass to the ambient condition and further conveying to the other downstream process. Discharged steam is separated by a flow discharger and recovered steam can be recycled for the pretreatment process. Both horizontal reactor and vertical reactor can be equipped with a dry discharger. The resulted material can be ready for pellet production without post-drying.

5.2.3.3 Hydrothermal Carbonization

Hydrothermal carbonization (HTC), also termed wet torrefaction, is a pretreatment process for woody biomass where the biomass is treated with hot compression water at the temperature between 180 and 250 °C for 1–12 h depending on the system design [51, 52]. This is different from steam explosion as the woody biomass is treated at highly pressurized liquid water instead of saturated steam. The reaction chemistry is similar to the steam explosion for which hydrolysis is the preliminary reaction to take place during treatment [52]. HTC is used to carbonize the biomass, making products with higher carbon content. The product characteristics, their relative proportions in the gas/liquid/solid phases, and the process energy requirements depend upon the input material and the process conditions. The advantage of HTC is that it can convert wet input material (MC $>$ 50 % (wb) into carbonaceous solids at relatively high yields without the need for an energy-intensive drying before or during the process. Moreover, some mono-sugars can be recovered from the liquid water, and the dried solid char can be used to produce pellets for energy production. However, there are limited published literatures evaluating the quality of the HTC wood pellet. There is lots of room for future R&D to evaluate the quality of HTC wood pellet.

5.2.4 Pelletization

Pellets quality is controlled by the pelletizing conditions, system design of the pellet mill, and feedstock parameters. Pelletization conditions include die temperature, pretreatment conditioning, biomass preheating, pressure, feeding speed, and retention time/relaxation time. The design parameters of densification units including types of densification unit, die shape, die specifications, and material to make the die determine the pellet quality.

5.2.4.1 Type of Pellet Mill and Die Specification

Pellet mill has two types of pellet presses: ring die and flat die. Ring die is a more common type of pellet mill used in commercial unit. For the ring die pellet mill, the die remains stationary and the rollers rotate. Some models have the dies to rotate and the rollers remain stationary during pellet production. The die of a pelletizer is made of hardened steel that is perforated allowing the ground wood particles to be forced through by the rotating die or rollers. As the die revolves, the friction driven rollers force the feed through holes in the die (steps A and B, Fig. 5.4). A layer of biomass (step C, Fig. 5.4) is needed to develop between the roller and the flat die in order to start flowing the particles into the die channels (step D, Fig. 5.4) to push the materials to extrude through the flat die. Cut-off knives mounted on the swing cover cut the pellets when they extruded from the die (step E, Fig. 5.4). The advantage of the ring die type pellet mill allows a higher production throughput compared to other types of presses (e.g., piston and screw presses), while maintaining the power consumption in the range of 15–40 kW h/t [53].

Flat die pellet mill is widely used for producing animal feed. Similar to the ring die pellet mill, the ground particles are loaded into the densification area. The rollers keep rotating in a clockwise direction in a vertical plane while the flat die is rotating anticlockwise in a horizontal plane. A layer of biomass powder is needed to develop between the roller and the flat die to start flowing into the die channels. Pressure is built up due to friction in the press channel and exponential correlation between press channel length and pressure is found [55, 57]. Finally, the materials are extruded through the flat die. A cut-off knife is installed at the exit of the flat die to cut the pellets into a certain specific length.

According to Tumuluru's review [54], die geometry refers to the size and shape of the die. The die geometry determines the pellet dimensions and the resulted density and durability. The aspect ratio (length to diameter of the pellet) can be a metric for the degree of compression during pelletization. An increase in pelletizing pressure increases the length of the pellet, whereas an increase in pellets diameter decreases the pelletizing pressure. Hence, the geometry of the die has a significant determination on the pressure applied onto the resulted pellet. A mathematical model for hardwood and softwood was developed to show how the variation in the model parameters (sliding friction coefficient, the ratio of compression, and the material-specific parameters, such as elastic modulus and Poisson ratio) significantly changes the necessary pelletizing pressure along the press channels of the matrix [55, 56]. Later, the co-relations of processing conditions between single die and commercial pellet mill was attempted [57].

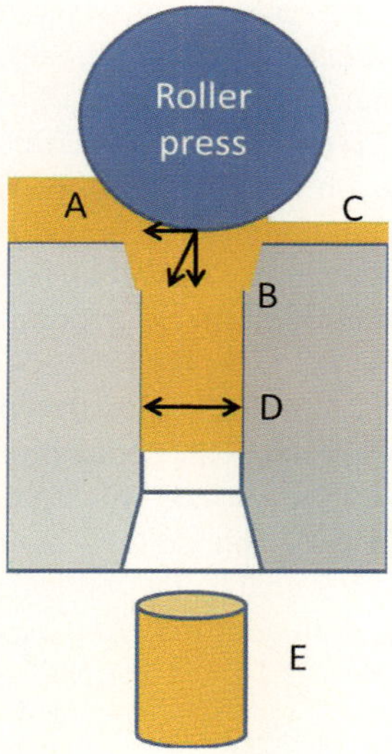

Fig. 5.4 Mechanisms of pelletization by a ring die

The inner diameter (*D*) and the effective length (*L*) of the die determine the pellet density. The effective length is the die thickness that actually performs work on the feed. *L*/*D* ratio is the effective length divided by the inner diameter of the die. High *L*/*D* ratios provide high pellet die resistance as feed moves through the die. Low *L*/*D* ratios provide less resistance. Each material has an *L*/*D* ratio requirement to form the material into a pellet. The durability of the pellets improves when a smaller die with higher *L*/*D* ratios is used [58]. However, if a longer die length is used, which will introduce a higher friction with no more improvement in the pellet durability. This will consume excess energy for production. Therefore, there should have optimum *L*/*D* ratios of the die for different types of biomass to produce durable pellets.

5.2.4.2 Die Temperature

The die temperature is determined by the frictional heat generated during the extrusion/pelletization. It is not a controlled parameter in the real commercial process, although it may be indirectly controlled by the following factors: species, particle size, flowability, MC, and feeding rate.

The durability and density of the densified products is highly influenced by the die temperature. The relationship between the durability of pellets and the die temperature of the pellet mill was reviewed [54] and found that the durability of pellets increased with die temperature.

The use of glass transition temperature (T_g) to determine the processing temperature is a key to produce the durable pellets [8, 10]. The die temperature is required to be higher than the glass transition temperature (T_g) in order to produce durable pellets. However, the die temperature should not be too high as this will dry the feedstock and cause frequent blocking of materials inside the die (i.e., reducing the throughput capacity or even unable to produce).

Note that different species of wood have different types of lignin, in which they exhibit different T_g. The T_g of different species of wood ranges between 50 and 100 °C depending on the MC. Hardwood lignins have fewer phenolic hydroxyl groups, and substantially more methoxyl groups [59], resulting in a significantly lower softening temperature of hardwood lignin than softwood lignin [60]. This helps to explain that different wood species requires different die temperature to produce high-quality pellets. For example, the strength of hardwood pellets is stronger than that of softwood pellets, regardless of the low lignin content of hardwood [10]. The higher strength of beech pellets compared to spruce can be explained by the different type of lignin on the surface, since the T_g of lignin of hardwoods are lower than softwood. A lower T_g means that the wood fibers are easier to deform to facilitate strong bonding among each other under high-pressure pelletization at a lower compression temperature. In addition, T_g is also highly dependent on the MC of the wood. The T_g of wood is 75 °C at 10 % MC while the T_g of wood can be as high as 110 °C at 0 % MC [61].

5.2.4.3 Steam Conditioning and Preheating

Preheating the biomass prior to densification helps to improve a better quality pellet. It influences pellet quality more than the effect of die specifications. This helps to reduce the time to bring the material to reach the die temperature around 80–140 °C. When the biomass reaches the temperature above the glass transition temperature of lignin, it can deform and rearrange easily to form a densely packed structure [9, 10]. Preheating biomass could significantly increase the throughput of the pelletizing machine and reduce the energy requirement per kilogram of pellets formed [62]. In some cases, steam conditioning of biomass is applied to preheat the materials. Saturated steam is used because of its higher heat transfer capacity (indicated by heat transfer coefficient) compared to air. It facilitates a faster heat transfer to biomass.

5.2.4.4 Pressure Retention and Relaxation Time

Pellets usually expand significantly right after the extrusion and their density decrease with time until reaching a constant volume; this is termed a spring-back effect. The relaxed density of the pellet is always lower than the initial pellet density (i.e., the pellet density measured right after the extrusion). This is due to the rheological behavior of the polymer components of the lignocellulosic biomass fiber. When the pellets extruded from the die, the individual biomass particles are free from the high compression force. The resulted pellets suddenly undergo dilation. The elasticity of polymers (i.e., cellulose, hemi-cellulose, and lignin) allows them to have a tendency to restore to their original un-deformed structure prior to compression. If the binding of the individual particles within a single pellet is not strong enough, the pellets will have inertia to expand and the resulted density decreases.

Pressure affects the density and durability of the pellets. High pressures and temperatures during densification may develop solid bridges by a diffusion of molecules from one particle to another at the points of contact, thereby increasing the pellet density and durability. However, the pellet density initially increases significantly with the pressure and reaches a maximum point, and beyond that it does not increase with pressure. Fractures may occur in the pellets due to sudden dilation when the pressure exceeds the optimum level.

Retention/relaxation time refers to the hold times of the biomass inside the die. It is usually around 5–30 s. During this time, it allows enough time for the biomass particles to build up a certain pressure and forms a dense structure without a significant spring-back effect. The relaxation time has a significant effect on the final density of the pellet during low-pressure compaction. At high-pressure compaction, the relaxation time did not show a significant effect on pellet density.

5.3 Effect of Different Pretreatment on Pellet Quality

5.3.1 Drying

Drying is a crucial step in the densification process of moist material. The optimum MC for pelletization was reported between 9 % and 15 % depending on feedstock species. However, the final quality of the feedstock can be affected by different factors during drying, and these include the type of dryer, drying conditions, drying medium, and the biomass characteristics. The degree to which the material is agitated and broken up in the drying process, the residence time and temperature of the material in the dryer influence the downstream densification process. In rotary dryers, the biomass is agitated to some extent and thus a significant amount of fines is usually produced in these dryers. The prolong drying of the fine particles caused by lag time of rotary dryer may degrade the biomass that affects the final quality of the resulted pellet. In contrast, PMB dryers are designed to circulate the drying medium through biomass layers, and the biomass is relatively stable on the bed, which can act as a filtration bed to trap any fines from the exhaust, resulting in low particulate emission.

Different components and products may be emitted in gas phase during drying. These include entrained fine particulates, volatile organic components (VOCs), and products of thermal degradation of the biomass [13]. The vaporized components can be further categorized into those that remain volatile at ambient conditions and those that condense after drying the stack. The most volatile components consist of mono-terpenes, which are naturally emitted from wood at ambient temperatures and the emission rate increases with temperature, particularly above 100 °C.

The condensable category consists of extractable components such as fatty acids, resin acids, di-terpenes, and tri-terpenes. Although these have high boiling points, they have sufficient vapor pressure at high drying temperatures (180–220 °C) to be released from wood. They are responsible for the formation of blue haze, a blue-gray discoloration of the exhaust gas from a wood dryer [16].

Thermal degradation products, such as formic and acetic acids, alcohols, aldehydes, furfurals, and carbon dioxide, are released at higher drying temperatures (200 °C or higher) when pre-pyrolysis occurs. Increasing the wood temperature rapidly increases the amount of thermal degradation products that have a strong smell. The formation of these degradation products (pseudo-lignin) can influence the quality of produced wood pellets and improve the binding characteristics of woody pellets [63].

The wood color change (darkening) during drying is another issue to produce high-value pellets. Chromophoric groups (carboxylates and phenol) may be produced within the lignin or extractive molecules at high temperatures and humidity. During drying furfural and some polysaccharides with low molecular weights are created from hemicelluloses degradation. As these components are dark in color (blue green), they lead to the darker color in wood appearance [64].

5.3.2 Size Reduction

Particle size of 3 mm was recommended by Swedish Exergy AB, in their energy-integrated production line. A chipper and a milling stage are used for size reduction of available biomass which is fresh wood or energy crop. Both size reduction equipment are installed before drying step in the production line.

The impact of particle size on pellets made from a mixture of white-wood sawdust with 18.3 % of bark was reported [65]. The mixture was sieved through three sieves with sizes of 0.63, 1, and 2 mm. They found particle in these range sizes is not a significant factor on breakability of pellets.

5.3.3 Torrefaction

The effect of torrefaction conditions (temperature and time) on grinding performance and pellets quality has been reported [66–70].

The fuel characteristics and grindability of pine chips and logging residues torrefied under nitrogen at temperatures ranging from 225 to 300 °C and 30 min residence time were investigated [66]. It was found that torrefied pine chips had an improved grindability with lower energy consumption as torrefaction temperature increased. Another research group reported an improvement in the grindability of the eucalyptus brought by dry torrefaction treated at the reaction temperature between 240, 260 and 280 °C [67]. Others studied two torrefaction temperatures of 240 and 275 °C for 1 h on the properties of bamboo, willow, coconut shell, and wood using a thermogravimetry analyzer (TGA) [68]. They pointed out that light torrefaction at 245 °C only has a drastic impact on hemicelluose while cellulose and lignin have a

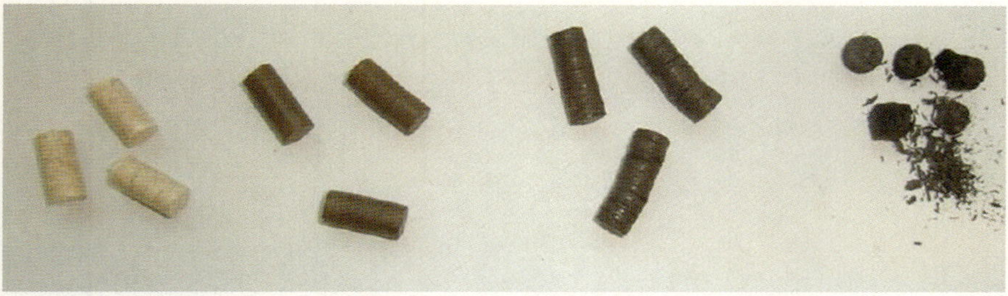

Spruce **T250** **T275** **T300**

Fig. 5.5 Pellets made from spruce and torrefied spruce. From *left* to *right*: spruce and torrefied spruce at 250, 275, and 300 °C. (Reprinted with permission from [69], copyright 2011, Elsevier)

slight impact. Light torrefaction is a proper operation to pretreat biomass for producing fuels. Severe torrefaction at 275 °C is not recommended to pretreat biomass as some cellulose and lignin in the biomass were further reacted. This is in line with a prediction that lignin degradation takes place at temperature higher than 260 °C will lose its binding role for pelletization after severe treatment. The degradation temperature of lignin depends on species and its content.

The pelletizing properties of the torrefied spruce at 250, 275 and 300 °C were also investigated and reported [69]. The torrefied spruce particles were compressed at 100 °C die temperature and at 200 MPa maximum pressure. They found that no pellets can be produced for the spruce torrefied at 300 °C and lots of defects were found in the pellets made from torrefied spruce treated at 275 °C (Fig. 5.5). The pellets made from torrefied spruce treated at 250 °C are the best, but still with only one-third of the compressive strength compared to the untreated spruce pellet.

The pellets made from torrefied sawdust from a fluidized bed reactor were reported with lower densities and lower hardness with increasing treatment severity [70]. The decrease in hardness was due to the loss of hydroxyl groups of hemicelluloses and inhibition of the availability of hydrogen bonding formation by the depositions of decomposition products from hemicelluloses and lignin, resulting in the decrease of particle plasticity. To compensate for this effect, adding external binder or to lower the torrefaction temperature and shorten torrefaction time, or different torrefaction media (e.g., wet torrefaction) or surface increase (e.g., steam explosion conditions) to minimize the loss of hemicelluloses to improve the binding ability of the particles to produce durable pellets.

5.3.4 *Steam Explosion*

The effect of steam explosion conditions (temperature and time) and feedstock characteristics (feed particle size and MC) on wood pellet quality (hardness, hydrophobicity, combustion) were studied by a few researchers [71–74]. The severity treatment levels being studied are covered from the range of $R_0 = 3.64 - 4.84$.

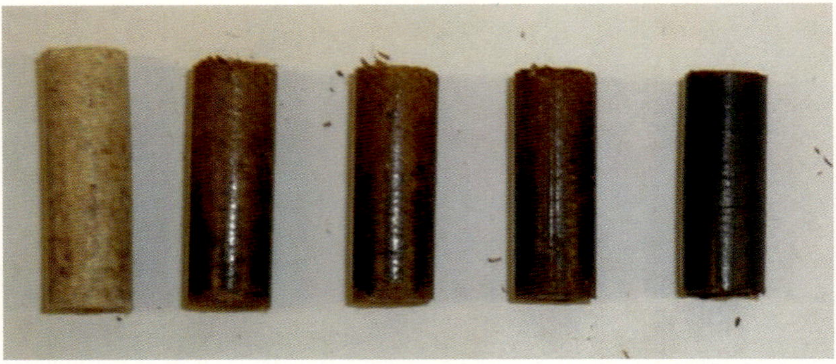

Fig. 5.6 Physical appearance of wood pellets treated at different steam explosion conditions. From *left* to *right*: untreated, 200 °C—5 min, 200 °C—10 min, 220 °C—5 min, and 220 °C—10 min. (Reprinted with permission from [71], copyright 2011, American Chemical Society)

The effect of steam explosion processing condition on the quality of the steam treated pellets was studied [71]. The studied temperatures are 200 and 220 °C and the reaction time are 5 and 10 min. The severity treatment levels being studied are in the range of $R_0 = 3.64$, 3.94, 4.23, and 4.53. The steam exploded pellets (Fig. 5.6) were found with breaking strength 1.4–3.3 times stronger than the untreated pellets. The steam exploded pellets also had a reduced equilibrium MC of 2– 4 % that helped to reduce the expansion of the pellets over the storage period. The only significant factor for improving the hydrophobicity of pellet is the reaction temperature instead of time. The heating value of steam exploded pellets had a slight increase from 18.94 to 20.09 MJ/kg. This may be contributed to the mild torrefaction effect, as some volatiles with low heating value were lost in gas phase during steam explosion treatment.

Steam explosion pretreatment of short-rotation wood crop (Salix wood chips) at two different temperatures (220 and 228 °C) and two different times (6 and 12 min) were investigated [72]. The severity treatment levels being studied are in the range of $R_0 = 4.31, 4.54, 4.61$, and 4.84. Pellets produced from steam treated residue showed increment in physical properties such as higher density, impact resistance, and abrasive resistance. Improvement in strength is attributed to melting of low-molecular weight lignin on the surface of the pellet during pelletization. The heating value of the steam treated Salix increased from 21.4 to 21.9 MJ/kg when the treatment severities increase from 4.3 to 4.84. However, no heating value of pellets made from untreated willow was reported for comparison. Moreover, there were no results reported on the hydrophobicity. Steam treatment was promising in reducing the alkali and heavy metal components of biomass. This was attributed by the water leaching and disrupted cell structure to lower the ash content in steam-treated residue. Degradation in ash fusibility was found for steam treated willow from the ash fusibility test.

In general, the mechanical strength of the steam-exploded pellet increased with treatment severity. However, it increases up to a maximum point and decreases with further increasing the treatment severity. It was found that steam exploded Fir wood pellets made from the most severely treated wood at 220 °C and 10 min show a decrease in mechanical strength [71]. It was due to the cellulose decomposition. This introduces the breakup of crystalline structure of cellulose, which causes the decrease in the hardness of the pellet.

5.3.5 Hydrothermal Carbonization

There is limited literature reporting the pellet quality of HTC wood pellet [51, 75]. Loblolly pine pellets made from HTC treated at 200, 230, and 260 were reported with higher durability of 99.5 %, 99.7 %, and 99.8 %, respectively, compared to the untreated pellets with 98.1 % [51]. The abrasion index (i.e., fines content after tumbling) was decreased with increasing treatment temperature, which indicated the treated pellets are mechanically more stable than the untreated pellets. In addition, the hydrophobicity of HTC pellets was higher than the untreated pellets and increased with the treatment temperature at the same reaction time. The equilibrium MC of pretreated HTC pellets were between 4.7 % and 12.4 % at the relative humidity of 83.6 %, which were higher than that of untreated pellet with 15.6 %. The researchers also tested the mechanical strength of pellets by immersing them in the water. They reported that the untreated pellet swelled and disintegrated after 15 s, while the pellet made from HTC biochar at 260 °C took 2 weeks to create a fracture without disintegration. This indicates that the binding ability between fibers is improved by the HTC treatment. This is critical for the safe handling and storage of biomass.

5.4 Summary

Wood pellets must be handled and stored safely in order to encourage its use as a feedstock for chemicals and energy production. The current major barriers of developing high-quality wood pellet are its low durability and high hygroscopicity. Several pretreatment techniques (torrefaction, steam explosion, and HTC) have been studied for optimizing the wood pellet quality to enhance the mechanical strength and hydrophobicity, which is crucial for safe handling, storage, and ease of conversion.

For future work, the optimal torrefaction conditions have to be investigated with respect to different species of wood for producing stable pellets. The use of nanotechnology to synthesize chemical binders may help to modify the surface chemistry of wood fiber and overcome the low binding of torrefied wood fiber. Small amount of binders may be required to produce durable torrefied pellet. Pellets made with a mixture of steam-treated wood with raw wood or a mixture of steam treated wood and torrefied wood can reduce the overall cost of production of high-quality wood pellet.

References

1. Wei M, Patadia S, Kammen DM (2010) Putting renewables and energy efficiency to work: how many jobs can the clean energy industry generate in the US? Energy Policy 38:919–931
2. Domac J, Richards K, Risovic S (2004) Socio-economic drivers in implementing bioenergy projects. Biomass Bioenergy 28:97–106
3. Thornley P, Rogers J, Huang Y (2008) Quantification of employment from biomass power plants. Renewable Energy 33:1922–1927
4. Sokhansanj S (2006) Cost benefit of biomass supply and preprocessing. BIOCAP research integration program synthesis paper
5. Mani S, Sokhansanj S, Bi XT et al (2006) Economics of producing fuel pellets from biomass. Appl Eng Agric 22:421–426
6. Aleksandra S, Madsen AM, Martensson L et al (2006) Assessment of microbial exposure risks from handling of biofuel wood chips and straw—effect of outdoor storage. Ann Agric Environ Med 13:139–145
7. Sokhansanj S, Mani S, Bi X et al (2005) Binderless pelletization of biomass, Paper No. 056061, Portland, Oregon, July 9–12. : ASABE, St. Joseph
8. Kaliyan N, Morey RV (2009) Factors affecting strength and durability of densified biomass products. Biomass Bioenergy 33:337–359
9. Kaliyan N, Morey RV (2010) Natural binders and solid bridge type binding mechanisms in briquettes and pellets made from corn stover and switchgrass. Bioresour Technol 101:1082–1090
10. Stelte W, Holm JK, Sanadi AR et al (2011) A study of bonding and failure mechanisms in fuel pellets from different biomass resources. Biomass Bioenergy 35:910–918
11. Lehtikangas P (2000) Storage effects on pelletised sawdust, logging residues and bark. Biomass Bioenergy 19:287–293
12. Rupar K, Sanati M (2005) The release of terpenes during storage of biomass. Biomass Bioenergy 28:29–34
13. Class DL (1998) Biomass for renewable energy, fuels and chemicals. Academic Press, San Diego
14. Pang S, Mujumdar AS (2010) Drying of woody biomass for bioenergy: drying technologies and optimization for an integrated bioenergy plant. Drying Technol 28:690–701
15. Holmberg H (2007) Biofuel drying as a concept to improve the energy efficiency and industrial CHP plant. Dissertation, Helsinki University of Technology
16. Pang S, Xu Q, Mahdum MM (2008) Modeling and optimization of packed moving bed drying of woody biomass for bioenergy. Proc ISD B:1122–1129
17. Poirier D (2007) Conveyor dryers. In: Mujumdar AS (ed) Handbook of industrial drying, 3rd edn. CRC Press, Boca Raton
18. Xu Q, Pang S (2008) Modelling of rotary drying of woody biomass. Drying Technol 26:1344–1350
19. Pang S (2001) Improving MDF fibre drying operation by application of a mathematical model. Drying Technol 19:1789–1805
20. Krokida M, Marinos KD, Mujumdar AS (2007) Rotary drying. In: Mujumdar AS (ed) Handbook of industrial drying, 3rd edn. CRC Press, Boca Raton
21. Borde I, Levy A (2007) Pneumatic and flash drying. In: Mujumdar AS (ed) Handbook of industrial drying, 3rd edn. CRC Press, Boca Raton
22. Spets JP, Ahtila P (2004) Reduction of organic emissions by using a multistage drying system for wood based biomasses. Drying Technol 22:541–561
23. Mujumdar AS (2007) Superheated steam drying. In: Mujumdar AS (ed) Handbook of industrial drying, 3rd edn. CRC Press, Boca Raton
24. Hulkkonen S, Heinonen O, Tijhonen J et al (1994) Drying of wood biomass at high pressure steam atmosphere: experimental research and application. Drying Technol 12:869–887
25. Zabaniotou AA (2000) Simulation of forestry biomass drying in a rotary dryer. Drying Technol 18:1415–1431

26. Bengtsson P (2008) Experimental analysis of low-temperature bed drying of wooden biomass particles. Drying Technol. 26:602–610
27. Brammer JG, Bridgwater AV (1999) Drying technologies for an integrated gasification bio-energy plant. Renewable Sustainable Energy Rev 3:243–289
28. FrankeM, Rey A (2006) Pelleting quality, World Grain, pp. 78–79, University of Georgia
29. MacBain R (1966) Pelleting animal feed. American Feed Manufacturing Association, Chicago
30. Payne JD (1978) Improving quality of pellet feeds. Mill Feed Fert 162:34–41
31. Grover PD, Mishra SK (1996) Biomass Briquetting: technology and practices. Regional Wood Energy Development Program in Asia, Field Document No. 46. Food and Agriculture Organization of the United Nations, Bangkok, Thailand
32. Hakkila P (1989) Utilization of residual forest biomass. Springer, Berlin
33. Obernberger I, Thek G (2010) The pellet handbook: the production and thermal utilization of biomass pellets. Earthscan, London
34. Obernberger I, Thek G (2004) Physical characterization and chemical composition of densified biomass fuels with regard to their combustion behavior. Biomass Bioenergy 27:653–669
35. Nilsson D, Bemesson S, Hansson PA (2011) Pellet production from agricultural raw materials—a systems study. Biomass Bioenergy 35:679–689
36. Uslu A, Faaij APC, Bergman PCA (2008) Pre-treatment technologies, and their effect on international bioenergy supply chain logistics. Techno-economic evaluation of torrefaction, fast pyrolysis and pelletisation. Energy 33:1206–1223
37. Ciolkosz D, Wallace R. (2011) A review of torrefaction for bioenergy feedstock production. Biofuels, Bioprod Biorefin 5:317–329
38. Tumuluru JS, Sokhansanj S, Hess JR, Wright CT, Boardman RD (2011) A review on biomass torrefaction process and product properties for energy applications. Ind Biotechnol 7:384–401
39. Fisher T, Hajallgol M, Waymack B et al (2002) Pyrolysis behavior and kinetics of biomass derived materials. J Anal Appl Pyrolysis 62:331–349
40. Deng J, Wang G, Kuang J et al (2009) Pre-treatment of agricultural residues for co-gasification via torrefaction. J Anal Appl Pyrolysis 86:331–337
41. Peng J, Bi XT, Lim J et al (2012) Development of torrefaction kinetics for British Columbia softwoods. Int J Chem Reactor Eng 10:1–37
42. Bergman PCA, Boersma AR, Zwart RWH et al (2005) Torrefaction of biomass co-firing in existing coal-fired power stations. Report ECN-C-05-013, ECN, Petten
43. Nimlos M, Brooking E, Looker M et al (2003) Biomass torrefaction studies with a molecular beam mass spectrometer. Am Chem Soc, Div Polym Chem 48:590–591
44. Overend RP, Chornet E (1987) Fractionation of lignocellulosics by steam aqueous pretreatments. Philos Trans R Soc London, Ser A 321:523–536
45. Abatzoglou N, Chornet E, Belkacemi K et al (1992) Phenomenological kinetic of complex systems: the development of a generalized severity parameter and its application to lignocellulosics fractionation. Chem Eng Sci 47:1109–1122
46. Montane D, Overend RP, Chornet E (1998) Kinetic models for non-homogenous complex systems with a time dependent rate constant. Can J Chem Eng 76:58–68
47. Zimbardi F, Viggiano D, Nanna F et al (1999) Steam explosion of straw in batch and continuous systems. Appl Biochem Biotechnol 77:117–125
48. Ramos LP (2003) The chemistry involved in the steam treatment of lignocellulosic materials. Quim Nova 26:863–871
49. Converse AO, Kwarteng IK, Grethlein HE et al (1989) Kinetics of thermochemical pretreatment of lignocellulosic materials. Appl Biochem Biotechnol 20–21:63–78
50. Mittal A, Chatterjee SG, Scott GM et al (2009) Modeling xylan solubilisation during autohydrolysis of sugar maple and aspen wood chips: reaction kinetics and mass transfer. Chem Eng Sci 64:3031–3041
51. Reza MT (2011) Hydrothermal carbonization of lignocellulosic biomass. Dissertation, University of Nevada, Reno
52. Llbra JA, Ro KS, Kammann C et al (2011) Hydrothermal carbonization of biomass residuals: a comparative review of the chemistry, process and applications of wet and dry pyrolysis. Biofuels 2:89–124

53. Grover PD, Mishra SK (1996) Biomass briquetting: technology and practices, Food and Agriculture organization of the United Nations Bangkok, Field Document No.46
54. Tumuluru JS, Wright CT, Kenney KL et al (2010) A technical review on biomass processing, densification, preprocessing, modeling and optimization. In Proceedings of 2010 ASABE annual international meeting
55. Holm JK, Henriksen UB, Hustad JE et al (2006) Toward an understanding of controlling parameters in softwood and hardwood pellets production. Energy Fuels 20:2686–2694
56. Holm JK, Henriksen UB, Wand K et al (2007) Experimental verification of novel pellet model using a single pellet unit. Energy Fuels 21:2446–2449
57. Holm JK, Stelte W, Posselt D et al (2011) Optimization of a multi-parameter model for biomass pelletization to investigate temperature dependence and to facilitate fast testing of pelletization behavior. Energy Fuels 25:3706–3711
58. Tabil LG, Sokhansanj S (1996) Process conditions affecting the physical quality of alfalfa pellets. Appl Eng Agric 12:345–350
59. Sjostrom E (1983) Wood chemistry: fundamentals and applications. Academic Press, London
60. Olsson AM, Salmen L (1992) Viscoelasticity of insitu lignin as affected by structure—softwood vs. hardwood. In: Glasser W (ed) Viscoelasticity of biomaterials. ACS Symposium series 489. American Chemical Society, Washington
61. Kelley SS, Rials TG, Glasser WG (1987) Relaxation behavior of the amorphous components of wood. J Mater Sci 22:617–624
62. Aqa S, Bhattacharya SC (1992) Densification of preheated sawdust for energy conservation. Energy 17:575–578
63. Lam PS (2011) Steam explosion of biomass to produce durable wood pellets. Dissertation, The University of British Columbia, Canada
64. Zhang Y, Cai L (2006) Effects of steam explosion on wood appearance and structure of sub-alpine fir. Wood Sci Technol 40:427–436
65. Relova I, Vignote S, Leon MA et al (2009) Optimisation of the manufacturing variables of sawdust pellets from the bark of Pinus caribaea Morelet: particle size, moisture and pressure. Biomass Bioenergy 33:1351–1357
66. Phanphanich M, Mani S. (2011) Impact of torrefaction on the grindability and fuel characteristics of forest biomass. Bioresour Technol 102:1246–1253
67. Arias B, Pevida C, Fermoso J et al (2008) Influence of torrefaction on the grindability and reactivity of woody biomass. Fuel Process Technol 89:169–175
68. Chen WH, Kuo PC (2010) A study on torrefaction of various biomass materials and its impact on lignocellulosic structure simulated by a thermogravimetry. Energy 35:2580–2586
69. Stelte W, Clemons C, Holm JK et al (2011) Pelletizing properties of torrefied spruce. Biomass Bioenergy 35:4690–4698.
70. Li H, Liu X, Legros R et al (2012) Pelletization of torrefied sawdust and properties of torrefied pellets. Appl Energy 93:680–685.
71. Lam PS, Sokhansanj S, Bi XT et al (2011) Energy input and quality of pellets made from steam-exploded Douglas Fir (Pseudotsuga menziesii). Energy Fuel 25:1521–1528
72. Biswas A, Yang W, Blasiak W (2011) Steam pretreatment of Salix to upgrade biomass fuel for wood pellet production. Fuel Process Technol 92:1711–1717
73. Lam PS, Sokhansanj S, Bi X et al (2012a) Colorimetry applied to steam treated biomass and pellets made from Western Douglas Fir (Pseudotsuga menziesii, L.). T. ASABE.
74. Lam PS, Sokhansanj S, Bi X et al (2012b) Drying characteristics and equilibrium moisture content of steam-treated Douglas-fir (Pseudotsuga menziesii, L.). Bioresour Technol. doi:10.1016/j.biortech.2012.03.093
75. Erlach B, Wirth B, Tsatsaronis G (2011) Co-production of electricity, heat and biocoal pellets from biomass: a techno-economic comparison with wood pelletizing. Proceedings of World Renewable Energy Congress (Bioenergy Technology), Sweden

Chapter 6
Microwave-Based Pretreatment for Efficient Biomass-to-Biofuel Conversion

Armando T. Quitain, Mitsuru Sasaki and Motonobu Goto

Abstract Pretreatment has been considered an important step for efficient and effective biomass-to-biofuel conversion. One of many promising methods of pretreatment includes the use of microwave (MW). MW-based pre-treatment approach utilizes both thermal and non-thermal effects generated by an extensive intermolecular collision as a result of realignment of polar molecules with MW oscillations. Compared to conventional heating, electromagnetic field generated by MW has the ability to directly interact with the material to produce heat, thereby accelerating chemical, physical, and biological processes. The advantages of employing MW rather than the conventional heating include reduction of process energy requirements, selective processing, and capability for instantaneous start and ceasing of a process. This also offers enormous benefits such as energy efficiency due to rapid and selective heating, and the possibility for developing a compact process. This chapter reviews recent advances on the utilization of MW irradiation for pretreatment of biomass for more efficient biofuel (bioethanol, biogas (methane), and biodiesel) production.

Keywords Microwave · Biomass · Sludge · Bioethanol · Biogas · Methane · Biodiesel · Free · fatty acids

6.1 Introduction

Pretreatment is an extremely important step in the synthesis of biofuels from lignocellulosic biomass or plant oils containing high fraction of free fatty acids (FFAs). Thorough knowledge of the fundamentals underlying various processes is necessary

---

M. Goto (✉)
Bioelectrics Research Center, Faculty of Engineering, Kumamoto University, 2-39-1 Kurokami, 860-8555 Kumamoto, Japan
e-mail: mgoto@kumamoto-u.ac.jp

A. T. Quitain · M. Sasaki
Graduate School of Science and Technology, Faculty of Engineering, Kumamoto University, 2-39-1 Kurokami, 860-8555 Kumamoto, Japan

Z. Fang (ed.), *Pretreatment Techniques for Biofuels and Biorefineries*, Green Energy and Technology,
DOI 10.1007/978-3-642-32735-3_6,

in choosing a suitable pretreatment method appropriate for a particular structure of the biomass substrate and the hydrolysis agent [1].

The pretreatment process for the lignocellulosic biomass targets the breakdown of the lignin structure and the disruption of the crystalline structure of cellulose for an easy access of acids or enzymes into the matrix resulting into more efficient hydrolysis of the cellulose [2]. Comprehensive review of several methods of pretreatment including physical, pyrolysis, physicochemical, and biological has been reported by Kumar et al. [1], while pretreatment technologies based on the use of enzymes have been extensively summarized by Alvira et al. [3]. Mtui [4] also highlighted recent advances in the treatment of lignocellulosic wastes focusing on domestic and agro-industrial wastes. Mechanical, physical, and biological treatment systems were discussed, and among the systems mentioned, physicochemical and biological treatments seem to be most favored options for production of biofuel or bio-based products. In these published reviews, limited information has been reported regarding the use of microwave (MW), which is considered to be a very promising pretreatment technique.

Research on the development of pretreatment methods utilizing MW irradiation has reached a significant level during the past few years. Literature search using Scopus resulted into heavy turn outs. This dramatic increase in the number of published research literatures is due to the introduction of scientific MW equipment to the market in the 1980s, the search for a more energy efficient methods for biofuel production, and the increasing interests of the academe, research institutes, and industry on its utilization for biomass-to-biofuel conversion.

This chapter summarizes recent advances on the utilization of MW irradiation for pretreatment of biomass for more efficient biofuel (bioethanol, biogas (or methane), and biodiesel) production. The fundamentals of MW technology and its benefits are also discussed, while introducing some challenges of the technique, prospects, and future outlook.

6.2 Fundamentals of Microwave Technology

MW technology relies on the use of electromagnetic waves to generate heat by the oscillation of molecules upon MW absorption. The electromagnetic spectrum for MWs is in between infrared radiation and radiofrequencies of 30 GHz to 300 MHz, respectively, corresponding to wavelengths of 1 cm to 1 m. Domestic and industrial MW systems are required to operate at either 12.2 cm (2.45 GHz) or 33.3 cm (900 MHz) in order not to interfere with the wavelength ranges being utilized for RADAR transmissions and telecommunications [5].

In MW-assisted heating, unlike the conventional methods, the heat is generated within the material, thus rapid heating occurs. As a result of this rapid heating, many MW-assisted organic reactions are accelerated, incomparable with those obtained using the conventional methods. Thus, higher yields and selectivity of target compounds can be obtained at shorter reaction times. In addition, many reactions not

Table 6.1 Brief summary of advantages of microwave heating compared to conventional

	Advantages	Comments
1	Super heating	Heating starts at any local point where microwave is absorbed
2	Internal heating	Heat flows from inside out
3	Selective heating	Polar compounds are heated selectively, thus increasing reaction rate
4	Rapid heating	Quick start/quick stop

possible using the conventional heating methods, had been reported to occur under MW heating. Some very useful information on the fundamentals of MW-enhanced chemistry, its sample preparation, and applications are well presented in the book edited by Kingston and Haswell [6]. The advantages of MW heating are briefly summarized in Table 6.1.

Other than the above-mentioned advantages of rapid, internal, and selective heating, MW non-thermal effects on reaction likely occur, obtaining dramatic increase in the yield even at milder conditions. The MW non-thermal effect is defined as the system response to electromagnetic energy not attributed to temperature variation [7]. Although doubts are cast on the true existence of non-thermal effects, some evidences had been reported and postulates had also been made by several researchers. These were summarized in a review article published by de la Hoz et al. [8] comparing them with the thermal effects. The review of Jacob et al. [9] on thermal and non-thermal interaction of MWs with materials attributed some interesting results on specific MW effects. Evidences on reaction rate enhancement due to some reasons other than the thermal effects such as "hotspots" or localized heating, molecular agitation, improved transport properties were discussed. They suggested that due to the interaction of MW with the materials, heating cannot be simply treated as that similar to the conventional methods as there are a lot of possible mechanisms of activation of materials that might possibly occur.

The above-mentioned thermal and non-thermal effects of MW irradiation offer enormous benefits to the pretreatment of biomass for synthesis of biofuels including energy efficiency, development of a compact process, rapid heating, and instant on-off process (instant heating–cooling process), among many other possible advantages.

6.3 Benefits of Microwave Pretreatment

MW-based pretreatment approach utilizes both thermal and non-thermal effects generated by an extensive intermolecular collision as a result of realignment of polar molecules with MW oscillations. Compared to conventional heating, electromagnetic field generated by MW has the ability to directly interact with the material to produce heat, thereby accelerating chemical, physical, and biological processes. The advantages of employing MW rather than the conventional heating include reduction of process energy requirements, selective processing and capability for instantaneous starting and ceasing of the process. This also offers enormous benefits such as energy efficiency due to rapid and selective heating and the possibility for developing a compact process.

When MW is applied to pretreatment of lignocellulosic biomass, the unique feature of selectively heating the more polar part will result in an improved disruption of the recalcitrant (treatment-resistant) structures of lignocellulose. With the nonthermal effects, electromagnetic field enhances the destruction of crystalline structures and changes the super molecular structure of lignocellulosic material thereby improving its reactivity.

MW pretreatment is also an energy-efficient and environmentally benign technology that aids in the transport of chemicals into the substrates. The project team from the US Department of Energy in partnerships with research institutes including the Oak Ridge National Laboratory [10] has showed that by opening the cellular microstructures of wood, for example, MW pretreatment could permit pumping chemicals for easy access of even large sections (10 cm long x 10 cm diameter) of hardwoods. The project team has demonstrated that, for both hardwood and softwood chips, MW pretreatment can decrease both H-factor and chemicals required to pulp hardwoods and softwoods by greater than 40 % with acceptable quality. The steam pressure generated inside the wood breaks the pit membranes and vessel cell walls, thereby enhancing the woods permeability to chemicals and process liquors.

Other than the lignocellulosic biomass, the use of MW for pretreatment of samples for a more efficient oil extraction and pretreatment of FFAs for biodiesel conversion has also been proposed [11, 12].

6.4 Typical Microwave Experimental Apparatus

A typical scientific MW apparatus normally works at a frequency of 2.45 GHz. It consists of a magnetron that generates the MW and introduces it into the cavity, where samples to be treated are placed. Figure 6.1 shows the typical apparatus for MW experiments (Ethos, Milestone Co. Ltd.). The MW power is programmable from 0 to 1000 W. In this apparatus, the temperature, time, and MW power output can be monitored and automatically controlled.

The reactors being used in the experiments are usually made of MW transparent materials such as glass or Teflon. However, for high-temperature experiments, high-pressure build-up inside the reactor is expected and specially designed reactor should be used. For this purpose, tough material such as those made from poly-ether ether ketone (PEEK) could serve as a casing for the reactor to resist high pressure. Till date, reactors made for use up to a maximum temperature of 250 °C and maximum pressure of 2.5 MPa are readily available but demand high costs.

Another commercially available laboratory scale MW apparatus for high-temperature and high-pressure experiments is the MW Accelerated Reaction System (MARS-5, CEM Corporation). This apparatus, comparable with the one shown in Fig. 6.1, can be operated at a programmable MW power up to 1250 W at 2.45 GHz

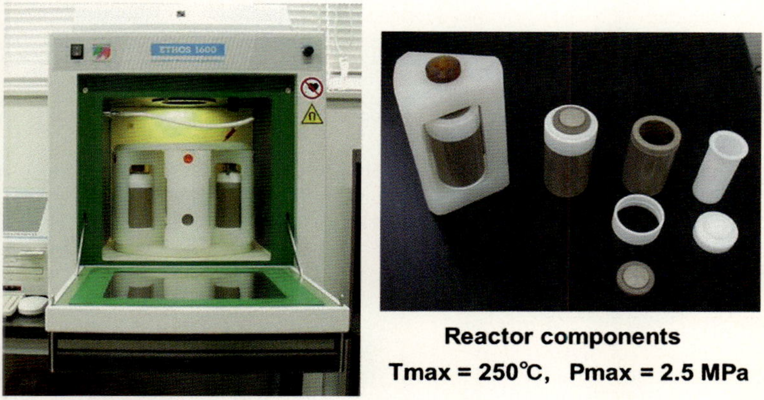

Fig. 6.1 Typical microwave apparatus for high temperature and high pressure experiments (Ethos, Milestone Co. Ltd.)

frequency. This is equipped with fiber optic temperature and pressure probes within the cavity and a turning carousel with a maximum of 14 pressure sealed 100 mL vessels (XP-1500), which can be used up to a maximum temperature of 260 °C and maximum pressure of 3 MPa. This equipment operates with a focused MW irradiating beam and is capable of heating and holding at desired cooking (or temperature ramp) rates and holding times.

6.5 Recent Advances on Microwave-Based Pretreatment

6.5.1 Bioethanol Production

The heterogeneous nature of lignocellulosic biomass feedstock used for production of bioethanol makes the treatment process challenging. An efficient pretreatment to maximize enzymatic hydrolysis efficiency is necessary to reduce economics of the total process. Besides, there is a limitation on the use of acid and alkali in conventional high temperature and high-pressure pretreatment method due to its high-energy input. Alternative heating techniques are sought to reduce energy input at the same time increase the total process efficiency. MW pretreatment could be a good alternative because it can reduce the pretreatment time at higher temperature, and for this reason its use for bioethanol production has been extensively investigated as summarized in Table 6.2.

Binod et al. [13] have reported that MW treatment of sugarcane bagasse with 1 % NaOH at 600 W for 4 min followed by enzymatic hydrolysis gave reducing sugar yield of 0.665 g/g dry biomass. Combining MW-alkali-acid treatments with 1 %

Table 6.2 Summary of recent researches done on the use of microwave for pretreatment of biomass for bioethanol production

Material	Pretreatment conditions	Results	References
Sugarcane bagasse	1 % NaOH, 450 W, 5 min	90 % Lignin removal	Binod et al. [13]
Switchgrass	3 % NaOH, 250 W, 10 min	Highest yield of reducing sugars (30 mg/mL)	Keshwani et al. [14]
Sawdust (Oak, Fir, Hemp)	0.82 % H_2SO_4, 140 °C, 15 min	Yield: 14 mg/mL	Balcu et al. [15]
Corn stalk	Two-stage method: 1) Microwave–alkali 2) Microwave–glycerine	2.48 g Hemicellulose 0.95 g Lignin 3.55 g Sugars	Zou et al. [16]
Rice straw/sugarcane bagasse	Glycerine as medium 240 W, 10 min, atmospheric	Twice the amount of reducing sugars	Intanakul et al. [17]
Green coconut fiber	Alkaline hydrogen peroxide, 250 W, 10 min	Yield of reducing sugar (35.98 mg/g) Ethanol (1.16 g/g)	Jeyanthi and Subramanian [18]
Cotton cellulose	110 °C, Ionic liquids	50-Fold increase in hydrolysis rate	Xu et al. [19]

NaOH followed by 1 % sulfuric acid resulted in an increase in reducing sugar yield to 0.83 g/g dry biomass. MW–alkali treatment at 450 W for 5 min resulted in almost 90 % of lignin removal from the bagasse. From the results, they found that combined MW-alkali-acid treatment for short duration enhanced the fermentable sugar yield.

The positive effects of the synergy between alkali and MW irradiation had also been confirmed by the works of Keshwani et al. [14] on hydrolysis of switchgrass. They reported that pretreatment using MW irradiation at lower power levels resulted in more efficient enzymatic hydrolysis. The application of MW irradiation for 10 min at 250 W to switchgrass immersed in 3 % NaOH (w/v) produced the highest yields of reducing sugars. The finding suggests that combined MW and alkali is a promising pretreatment method to enhance enzymatic hydrolysis of switchgrass.

On the contrary, instead of alkali, Balcu et al. [15] proposed the use of acid in combination with MW pretreatment. They found that elevated temperatures close to 180 °C are not necessary for better conversion of lignocellulosic biomass into sugars. They suggested the use of 0.82 % aqueous solution of sulfuric acid, getting very good yield even at low temperature of 140 °C.

Moreover, a two-stage MW pretreatment method, which includes MW–alkali pretreatment for hemicelluloses extraction and MW–glycerine pretreatment for lignin extraction of corn stalk as proposed by Zou et al. [16], seems to be more promising. They reported that MW–alkali pretreatment is suitable for hemicelluloses extraction with the following suitable treatment conditions: liquid-to-solid ratio of 20 mL/g, alkali consumption of 150 wt%, treatment time of 10 min, MW power of 116 W/g and the particle size of 40–80 mesh. Using MW and pure glycerine, lignin can be extracted at optimal treatment time of 30 min, and MW power of 66.7 W/g.

The use of glycerine as a solvent has also been investigated by Intanakul et al. [17] for an improved enzymatic hydrolysis of lignocellulosic wastes by MW pretreatment under atmospheric pressure. The benefits of using glycerine as a solvent include no high-pressure build-up even if the temperature reaches 200 °C. Their results showed that with the pretreatment, more than twice the amount of reducing sugars could be produced from enzyme saccharification compared with no pretreatment at all. Unlike the steam explosion process which requires high-pressure and subsequent pressure release, this technique provides some advantages regarding high temperature and high pressure handling.

The technique was also applied to pretreatment of green coconut fiber for bioethanol production [18]. Prior to MW irradiation for 10 min at 250 W, pretreatment using alkaline hydrogen peroxide gave higher yield of reducing sugar (35.98 mg/g) and high ethanol yield (1.16 g/g) compared to alkaline sodium peroxide.

Optimization of the method as applied to pretreatment of wheat straw for ethanol production has also been investigated using various techniques such as orthogonal design (L9(34)) [19]. This optimization technique was applied by investigating the effects of four factors including the ratio of biomass to NaOH solution, pretreatment time, MW power, and the concentration of NaOH solution with three different levels on the chemical composition, cellulose/hemicellulose recoveries, and ethanol concentration. Results showed that pretreatment with the ratio of biomass to liquid at 80 g/kg, the NaOH concentration of 10 kg/m^3, and the MW power of 1000 W for 15 min was the optimal condition. They obtained ethanol yield of 148.93 g/kg wheat straw at this optimum condition, much higher than that from the untreated material which was only 26.78 g/kg.

The use of ionic liquids (IL) in combination with MW pretreatment was also reported [20]. In this method, MW irradiation enhances the solubility of cellulose in IL while decreasing the degree of polymerization. This results into improved cellulose hydrolysis. Results showed a 50-fold increase in the rate of enzymatic hydrolysis of cotton cellulose when MW irradiation is used in combination with IL dissolution pretreatment at 110 °C, about four times better than when only IL was used for dissolution.

Development of a continuous process, although a very challenging approach for the use of MW irradiation, could be the most economical and efficient for large-scale commercial production of bioethanol. The group of Mitani et al. [21] from Kyoto University (Japan) attempted to develop a prototype for a continuous MW pretreatment system for bioethanol production from woody biomass as shown in schematic diagram in Fig. 6.2. A ceramic pipe was set in a metal vessel, and a mixture of woody biomass, water, and solvents flows through the pipe. MW propagates in the internal space of the metal vessel, and it is absorbed by the mixture since MW can penetrate through the ceramic pipe. In the present system, the mixture flows through a metal pipe, and it is irradiated with MW at T-junction metal pipe sections. The *gray arrows* and *white arrows* in Fig. 6.2 show the mixture flow and the MW irradiation direction, respectively. MW frequency of the present pretreatment system is 2.45 GHz-band, which is the same as that of a MW oven. The diameter of the metal pipe is 75 mm.

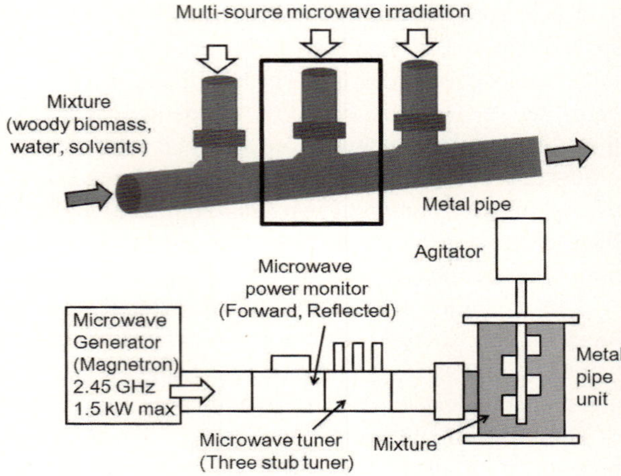

Fig. 6.2 Schematic diagram of continuous microwave pretreatment system of woody biomass for bioethanol production. (Adapted from source [21])

Performing MW pretreatment of a mixture consisting of 70 g of Japanese cedar sapwood chips and 770 g of solvents (ethylene glycol:phosphoric acid = 95:5), about 45.9 % of the total saccharide from woody biomass can be obtained as compared to only 43.6 % with the conventional heating. However, the energy consumption was quite higher at 552 kJ compared to only 498 kJ with the conventional heating. The amount of bioethanol that can be produced from this experiment was estimated to be 14.8 g of bioethanol corresponding to an energy of about 439 kJ.

6.5.2 Biogas (Methane) Production

MW has also been applied to the pretreatment of biomas or sludge for methane production as summarized in Table 6.3. Sólyom et al. [22] investigated the use of MW as a pretreatment of secondary wastewater sludge for biogas production. They found that the highest solubilization was achieved at an absorbed MW energy of 0.54 kJ/mL using MW power of 1000 W for 90 g of sludge. Under this condition, an improvement of 7.1 % in methane production was observed compared to the untreated sample. As expected, the methane production could be further increased using a higher absorbed energy but solubilization decreased.

MW pretreatment was also applied to disintegration and digestion of different types of sludges including waste-activated sludge (WAS), primary sludge (PS), combined (PS + WAS), sequencing batch reactor (SBR) sludge, and anaerobically digested biocake [23]. Results showed that MW pretreatment could increase bioavailability of sludge components under batch anaerobic digestion and enhance the dewaterability of pretreated sludges after digestion. The degree of solubilization and biodegradation depended on the types of waste sludges. The characteristics of

Table 6.3 Summary of recent researches done on the use of microwave for pretreatment of biomass/sludge for methane production

Material	Pretreatment conditions	Results	References
Secondary wastewater sludge	Absorbed MW energy: 0.54 kJ/mL, 1000 W	7.1 % Improvement in methane production	Solyom et al. [22]
Various types of sludges	Various temperatures and MW intensities	Increased bioavailability Enhanced dewaterability	Eskicioglu et al. [23]
Sludge	Combined MW–alkali (600 W,85 °C, 2 min, 1.5 g NaOH/L)	46 % COD solubilization	Chang et al. [24]
Thickened waste activated sludge	Combined MW-alkali	27 % Improvement in methane production	Chi et al. [25]
Anaerobic sludge	Heating rate 9.1 °C/min Final $T = 90\,°C$	Methane production: 2.02 L/L	Park and Ahn [27]
Food industrial waste sludge	Different MW intensities	Biogas production increased from 220 to 600 mL/g	Beszedes et al. [28]
Wheat straw	MW at 150 °C	28 % Improvement on methane production	Jackowiak et al. [29]

the sludge may influence final pretreatment outcomes, and general effect of MW on the pretreatment of sludge cannot be concluded.

In the works of Chang et al. [24] on the effects of MW and alkali on pretreatment of sludge, they reported that the synergistic effects of MW and alkali could enhance sludge solubilization, obtaining 46 % COD solubilization. This is equivalent to almost two-fold increase compared to the combined values obtained with only MW (8.5 %) or alkali pretreatment (18 %) (total = 26.5 %).

The combined MW and alkali method was also investigated on the pretreatment of thickened waste activated sludge (TWAS) to improve thermophilic anaerobic digestion efficiency [25]. The effects of 12 different pretreatment methods were investigated in 28 thermophilic batch reactors by monitoring cumulative methane production. Improvements in methane production in the TWAS were directly related to the MW and alkali pretreatment of the sludge. An improvement in the highest cumulative methane production of about 27 % over the control was obtained.

MW pretreatment was also found to be most effective compared with ultrasonic and chemo-mechanical pretreatments of pulp mill wastewater treatment sludge, increasing specific methane yields of WAS samples by 90 % compared to controls after 21 days of mesophilic digestion [26].

Park and Ahn [27] investigated optimum MW pretreatment conditions for methane production in anaerobic sludge digestion. They found out that both MW heating rate and final temperature significantly affected solubilization of the sludge and methane production as well, obtaining maximum methane production of 2.02 L/L at optimum heating rate of 9.1 °C/min and final temperature of 90 °C. MW pretreatments also showed to enhance efficiency of anaerobic digestion of food industrial sewage sludge [28]. Due to increased solubility (from 9.7 % to more than 40 %), the specific biogas product could be increased three-folds from 220 mL/g to more than 600 mL/g.

Also, with MW pretreatment, the decomposing bacteria could easily access organic compounds resulting to more efficient digestion.

Jackowiak et al. [29] also made an optimization of MW pretreatment for solubilization and anaerobic biodegradability of wheat straw. An improvement of about 28 % on methane production was obtained with MW pretreatment, with the maximum yield obtained at 150 °C.

6.5.3 Microwave-Pretreatment of Biodiesel Feedstock

MW pretreatment has a potential to induce stress reactions in plant systems or oil seeds by the rupture of their cell membranes. This results into high mass transfer coefficients, thus obtaining higher extraction yield. Cheng et al. [30] applied MW to the pretreatment of palm fruits prior to extraction of its oil. Applying MW for 3 min, an extraction yield comparable to the conventional and commercial palm oil milling process with an average of 20 % was achieved. The resulting palm oil also exhibits desirable and very low FFA content of about 0.26 % and moisture content of 0.05 %.

Moreover, most of the oil feedstocks for biodiesel syntheses contain relatively high amount of FFAs especially the waste cooking oil. This has become a big hurdle for industrialization of the proposed process, because the presence of fatty acids significantly affects the solubility of Ca-based or alkali catalysts in the products. Government quality standards for biodiesel require the level of Ca to be below 5 ppm, while the fatty acid content should not exceed 1 wt%. Thus, pretreatment of FFAs in oil is necessary prior to transesterification of the triglyceride components. In this regard, MW irradiation could also be applied to convert FFAs into biodiesel. Our previous studies showed that about 88 % conversion of FFAs in waste cooking oil could be obtained in 1 min of MW irradiation at a power of 700 W using ion exchange resin (Amberlyst 15) as catalysts [12] as shown in Fig. 6.3. With these results, a two-step process is proposed for the conversion of waste oil, or any other types of oil feedstock containing high amount of FFAs, to biodiesel fuel. The process consists of a first step of esterification of fatty acids followed by a second step of transesterification of the triglyceride. While the two-step process seems ideal for the treatment of FFAs in oil, this also minimizes the solubility of Ca-based catalysts as a result of the reduction of fatty acid contents.

6.6 Technological Challenges of Using Microwave for Pretreatment

MW-based pretreatment method offers several great advantages to the synthesis of biofuels, however, there are also some technological challenges associated with its use. MW could not work well with large quantities of materials, and thus could not be easily converted from laboratory to a multikilogram industrial-scale production.

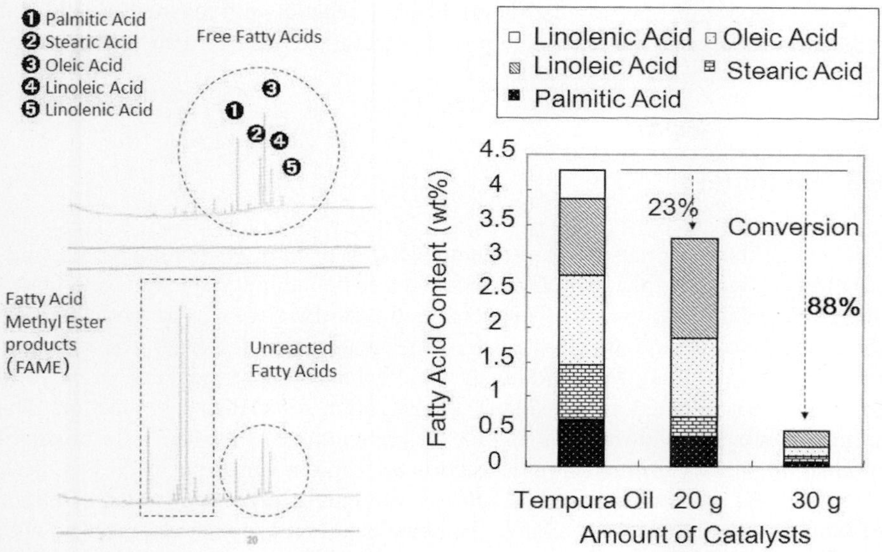

Fig. 6.3 Results of experiments on esterification of free fatty acids in waste oil using microwave (MW Power = 700 W, t = 1 min)

The penetration depth of MW irradiation into the absorbing materials is only a few centimeters, and this significantly limits scale up of the technology.

MW irradiation is non-homogeneous and formation of "hotspots" is likely, thus control of reaction is too difficult. Mixing may improve homogeneity, however, appropriate methods of mixing under MW irradiation remains a challenge.

Safety consideration is another factor for industrial utilization of MW. The use of batch MW reactors, for the processing of comparatively large volumes under pressure may not be safe because any malfunction or rupture of a large pressurized reaction vessel, which are usually made of Teflon or glass materials, may result into massive spillage causing significant operational damages to the working place and the environment.

6.7 Outlook and Future Prospects

The demand for biofuels are expected to increase in the near future, and while the search for an efficient and low-cost production process continues, the global outlook is positive for the use of MW irradiation for the pretreatment of lignocellulosic biomass, sludge or biodiesel feedstock.

To overcome the limitations for scaling up MW-assisted technology for pretreatment, development of a continuous process offers numerous advantages, but still poses several challenges that require detailed investigation especially when working with high temperature and high pressure.

While the use of MW irradiation offers great benefits with regards to rapid and efficient pretreatment approach, safety is a big factor to consider in designing a large scale production plant.

6.8 Summary

Reviews of the recent advances of the application of MW for pretreatment of lignocellulosic biomass had proven the technique to be an important step for efficient and effective biomass-to-biofuel conversion. It was also shown to be more effective than the ultrasonic and chemo-mechanical pretreatments of sludge. The synergistic effect of combining MW and alkali could enhance the fermentable sugar yield for bioethanol production and sludge solubilization for methane production. The technique was also shown promising for the pretreatment of feedstock for biodiesel production, including efficient oil extraction and rapid treatment of FFAs.

MW-based pretreatment method offers several great advantages to the synthesis of biofuels, however, some technological challenges still remain. To meet up with the demands of the foreseen shift to the use of renewable bioenergy, the future should also look at the development of a continuous pretreatment process involving solid wastes while taking serious consideration of the safety in designing a large scale production plant.

References

1. Kumar P, Barrett DM, Delwiche MJ, Pieter S (2009) Methods for pretreatment of lignocellulosic biomass for efficient hydrolysis and biofuel production. Ind Eng Chem Res 48:3713–3729
2. Mosier NS, Wyman C, Dale B, Elander R, Lee YY, Holtzapple M, Ladisch MR (2005) Features of promising technologies for pretreatment of lignocellulosic biomass. Bioresour Technol 96:673–686
3. Alvira P, Tomas-Pejo M, Ballesteros M, Negro MJ (2010) Pretreatment technologies for an efficient bioethanol production process based on enzymatic hydrolysis: a review. Bioresour Technol 101:4851–4861
4. Mtui GYS (2009) Recent advances in pretreatment of lignocellulosic wastes and production of value added products. Afr J of Biotechnol 8(8):1398–1415
5. Mingos DMP, Baghurst DR (1997) Applications of microwave dielectric heating effects to synthetic problems in chemistry. In: Kingston HM (Skip), Haswell SJ (eds) Microwave-enhanced chemistry. American Chemical Society, ISBN 0-8412-3375–6, Washington, DC, pp 3–50
6. Kingston HM (Skip), Haswell SJ (1996) Microwave-enhanced chemistry: fundamentals, sample preparation and applications. American Chemical Society, ISBN 0-8412-3375–6, Washington, DC
7. Banik S, Bandyopadhyay S, Ganguly S (2003) Bioeffects of microwave—-a brief review. Bioresour Technol 87:155–159
8. De la Hoz A, Diaz-Ortiz A, Moreno A (2005) Critical review: microwaves in organic synthesis. Thermal and non-thermal microwave effects. Chem Soc Rev 34:164–178
9. Jacob J, Chia LHL, Boey FYC (1995) Review: thermal and non-thermal interaction of microwave irradiation with materials. J Mater Sci 30:5321–5327

10. Compere A, Parent D et al (2007) Increasing yield and quality of low-temperature, low-alkali kraft cooks with microwave pretreatment, US Department of Energy, Energy Efficiency and Renewable Energy Reports
11. Abu Bakar S, Yusup S, Ahmad MM, Goto M, Quitain A (2011) Production of biodiesel from crude palm oil and jatropha oil using microwave technique. In: Proceedings of international symposium on ecotopia science, Nagoya, Japan, p 82
12. Quitain AT, Katoh S, Goto M (2011) Microwave-assisted synthesis of biofuels. In: Bernardes MA Dos Santos (ed) Biofuel production-recent developments and prospects. ISBN: 978-953-307-478-8. InTech. http://www.intechopen.com/books/biofuel-production-recent-developments-and-prospects/microwave-assisted-synthesis-of-biofuels
13. Binod P, Satyanagalakshmi K, Sindhu R, Janu KU, Sukumaran RK, Pandey A (2012) Short duration microwave assisted pretreatment enhances the enzymatic saccharification and fermentable sugar yield from sugarcane bagasse. Renew Energ 37:109–116
14. Keshwani DR, Cheng JJ, Burns JC, Li L, Chiang V (2007) Microwave pretreatment of switchgrass to enhance enzymatic hydrolysis. An ASABE meeting presentation number 077127. In: ASABE annual international meeting, Minneapolis Convention Center, Minneapolis, 2007
15. Balcu I, Macarie CA, Segneanu A-E, Pop RO (2011) Combined microwave-acid pretreatment of the biomass. In: Shaukat S (ed) Progress in biomass and bioenergy production. ISBN: 978-953-307-491-7. InTech. http://www.intechopen.com/books/progress-in-biomass-and-bioenergy-production/combined-microwave-acid-pretreatment-of-the-biomass
16. Zou A, Shen C, Zhao L, Dai G (2011) Effect of microwave pretreatment on component fractionation and saccharification of corn stalk. Trans Chin Soc of Agric Eng 27(12):269–274
17. Intanakul P, Krairiksh M, Kitchaiya P (2003) Enhancement of enzymatic hydrolysis of lignocellulosic wastes by microwave pretreatment under atmospheric pressure. J Wood Chem Technol 23(2):217–225
18. Jeyanthi GP, Subramanian J (2011) A comparison between microwave assisted alkaline sodium hydroxide and alkaline hydrogen peroxide pretreatments of green coconut fiber for bioethanol production. Asian J Microbiol Biotechnol Environ Sci 13(2):365–369
19. Xu J, Chen H, Kadar Z, Thomsen AB, Schmidt JE, Peng H (2011) Optimization of microwave pretreatment on wheat straw for ethanol production. Biomass Bioenergy 35:3859–3864
20. Ha SH, Mai NL, An G, Koo Y-M (2011) Microwave-assisted pretreatment of cellulose in ionic liquid for accelerated enzymatic hydrolysis. Bioresour Technol 102(2):1214–1219
21. Mitani T, Oyadomari M, Suzuki H, Yano K, Shinohara N, Tsumiya T, Sego H, Watanabe T (2011) A feasibility study on a continuous-flow-type microwave pre-treatment system for bioethanol production from woody biomass. J Jpn Inst Energy 90:881–885
22. Sólyom K, Mato RB, Perez-Elvira SI, Cocero MJ (2011) The influence of the energy absorbed from microwave pretreatment on biogas production from secondary wastewater sludge. Bioresour Technol 102:10849–10854
23. Eskicioglu C, Kennedy KJ, Droste RL (2008) Initial examination of microwave pretreatment on primary, secondary and mixed sludges before and after anaerobic digestion. Water Sci Technol 57(3):311–317
24. Chang CJ, Tyagi VK, Lo SL (2011) Effects of microwave and alkali induced pretreatment on sludge solubilization and subsequent aerobic digestion. Bioresour Technol 102(17):7633–7640
25. Chi Y, Li Y, Fei X, Wang S, Yuan H (2011) Enhancement of thermophilic anaerobic digestion of thickened waste activated sludge by combined microwave and alkaline pre-treatment. J Environ Sci 23(8):1257–1265
26. Saha M, Eskicioglu C, Marin J (2011) Microwave, ultrasonic and chemo-mechanical pretreatments for enhancing methane potential of pulp mill wastewater treatment sludge. Bioresour Technol 102(17):7815–7826
27. Park WJ, Ahn JH (2011) Optimization of microwave pretreatment conditions to maximize methane production and methane yield in mesophilic anaerobic sludge digestion. Environ Technol 32(13):1533–1540
28. Beszédes S, László Z, Szabó G, Hodúr C (2011) Effects of microwave pretreatments on the anaerobic digestion of food industrial sewage sludge. Environ Prog Sustainable Energy 30(3):486–492

29. Jackowiak D, Bassard D, Pauss A, Ribeiro T (2011) Optimisation of a microwave pretreatment of wheat straw for methane production. Bioresour Technol 102(12):6750–6756
30. Cheng SF, Nor LM, Chuah CH (2011) Microwave pretreatment: a clean and dry method for palm oil production. Ind Crops Prod 34(1):967–971

Part III
Chemical Pretreatment

Chapter 7
Converting Lignocellulosic Biomass to Low-Cost Fermentable Sugars

Michael Zviely

Abstract Concentrated hydrochloric acid-driven hydrolysis provides the most powerful and industrially proven technology for converting all cellulosic wastes-wood, solids from city sewage plants, bagasse, grasses, etc.—to sugars that can be fermented to ethanol or other biofuels as well as a large variety of bio-products and food and feed.

Our process begins by steam expansion of debarked chipped wood, which undergoes a pre-extraction stage to remove all extractives, for example, tall oils and ash. The pre-extracted wood continues into hydrolysis stage performed using highly concentrated HCl at low-temperature (10–15 °C), thus affording sugars hydrolyzate (98 % of the theoretically available sugars, composing 65 % of the dry weight of the wood chips for pine wood, are converted into sugars) with minimum degradation products, while simultaneously separating the solid lignin.

A key limitation to any concentrated acid hydrolysis is the difficulty in recovering the acid. In particular, Virdia solution forms an azeotrope at between 21 and 25 % depending on the pressure; simple distillation cannot concentrate a dilute solution beyond the azeotropic point. The efficiency of acid recovery is a key condition to making acid hydrolysis of lignocellulosic materials an economically viable source of fermentable sugars.

Full recovery of HCl at high acid concentration and its reuse yields very minor waste stream, no complicating air emissions, and favorable life cycle analysis.

Keywords Biomass · Lignocellulose · Sugars · Hydrochloric acid · Saccharification · Hydrolysis · Extraction · Lignin · Tall oils

7.1 Introduction

Renewable energy can come from technologies using wind, sun, hydroelectric, and geothermal, as their source, and from technologies using liquid biofuels as source. Liquid fuels mostly originate from biological, thermo chemical, bio thermal, and chemical processes.

M. Zviely (✉)
Research & Development Department, Virdia
(Formerly HCL CleanTech), 46733 Herzlyia, Israel
e-mail: michael.zviely@virdia.com

Z. Fang (ed.), *Pretreatment Techniques for Biofuels and Biorefineries*,
Green Energy and Technology,
DOI 10.1007/978-3-642-32735-3_7,

Biological processes mostly use enzymes to hydrolyze the polysaccharides contained in the biomass, and chemical processes use mostly inorganic acids for the polysaccharides hydrolysis.

Of the inorganic acids used as catalysts for the hydrolysis, hydrochloric acid (HCl) plays a major role due to its unique characteristics. Concentrated (fuming) HCl provides the most powerful and industrially proven technology for converting all cellulosic wastes into sugars. This highly concentrated HCl forms hydrates, and it is assumed these species are responsible to the efficient hydrolysis of cellulose.

In addition to the liquid fuels obtained by these technologies, many additional products, for example, lactic acid, amino acids, or terphthalic acid, can be produced from the intermediate materials, mostly from the sugars, both in biological and chemically catalyzed processes.

Cellulosic residues, considered as a chemical raw material, are a source of hexose and pentose sugars. The major portion, the hexoses, is equivalent to sugar from common sources. The pentoses, on the other hand, are unique, as they may be processed to furfural, a chemical which has not been produced from other sources. The pentosans in corncobs and bagasse are now the main source of furfural. The pentosan content of wood, however, is too low for an economic venture supported by this single product. The brief considerations above lead to the conclusion that while there are indirect benefits to be gained by the development of a technically feasible saccharification process, the important measure is economics. To be of value, the process must be capable of producing sugar at a price competitive with corn or cane sugar or molasses in the locality in which it operates. Wood saccharification is but one aspect of a larger problem—the chemical utilization of cellulosic residues. Sugar is but one of many products obtainable from cellulosic residues, and wood is but one of a variety of potential starting materials [1].

7.2 Wood Composition

Wood is essentially composed of cellulose, hemicelluloses, lignin, and extractives. Table 7.1 presents major chemical compositions of some wood species.

7.3 The Biological Approach for Biomass Hydrolysis into Sugars

The biological approach for hydrolysis of biomass is composed of a pretreatment phase, to make the lignocellulosic material such as wood open to hydrolysis, followed by cellulose and hemicelluloses enzymatic hydrolysis to break them down into sugars; finally, separation of the sugar solution from the residual materials, mostly lignin and also some enzymes adsorbed on the lignin [3].

Table 7.1 Chemical composition of some wood species [2]

Wood part	Constituent	Scots pine (*Pinus sylvestrys*) (%)	Eucalyptus (*Eucalyptus camaldulensis*) (%)	Silver birch (*Betula verrucosa*) (%)
Cellulose	Glucan	40.0	45.0	41.0
Hemi-cellulose	Gluco-mannan (Fig. 7.1)	16.0	3.1	2.3
	Glucorono-xylan (Fig. 7.2)	8.9	14.1	27.5
	Other polysac-charides	3.6	2.0	2.6
Lignin		27.7	31.3	22.0

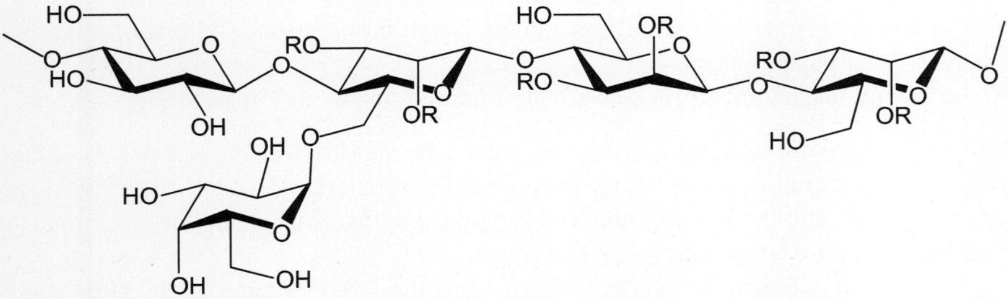

Fig. 7.1 Main structure of galactoglucomannans in softwood hemicellulose ($R = CH_3CO$ or H)

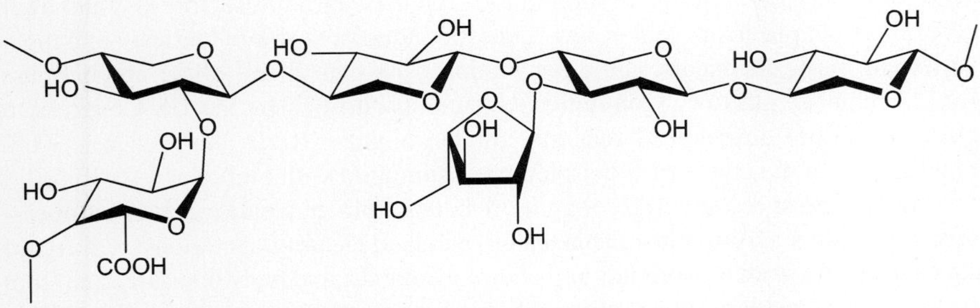

Fig. 7.2 Main structure of arabinoglucuronoxylan in softwood hemicellulose

In addition to this, yields of enzymatic hydrolysis are ca. 80–85 %, the obtained sugar concentration in the medium is low as <12.5 %, and the energy costs for both pretreatment and excess water removal is relatively high, and ca. 20 % of the sugar product is required to make the processing enzymes. Also there is no past experience with such process at industrial scale.

Due to all the above-mentioned withdrawals in using a biological process, which might be the technology of choice in other cases, the use of chemical process seems to be an optimal route for hydrolysis of biomass into sugars. This approach is further presented in this chapter.

7.4 Biomass Availability

Biomass, as a renewable energy source, refers to living or to recently dead biological material that can be used as fuel or for industrial production. Biomass comes in many different types, which may be grouped into five basic categories of material:

- Virgin wood, from forestry, arboricultural activities or from wood processing.
- Energy crops: high yield crops grown specifically for energy applications, for example, energycane, switchgrass, and miscanthus.
- Agricultural residues: residues from agriculture harvesting or processing.
- Food waste, from food and drink manufacture, preparation, and processing, and post-consumer waste.
- Industrial waste and co-products from manufacturing and industrial processes.

Biomass is a feedstock for chemicals, electricity, and natural gas, not just liquid fuels. It is essential to successfully grow energy crops at large scale. Biotechnology improvements and agronomics are both key to improving yields and low cost biomass will be critical to future energy production.

The US Department of Energy (DOE) and the US Department of Agriculture (USDA) are both strongly committed to expanding the role of biomass as an energy source. In particular, they support biomass fuels and products as a way to reduce the need for oil and gas imports; to support the growth of agriculture, forestry, and rural economies; and to foster major new domestic industries—biorefineries—making a variety of fuels, chemicals, and other products. As part of this effort, the biomass R&D Technical Advisory Committee, a panel established by the US Congress to guide the future direction of federally funded biomass R&D, envisioned a 30 % replacement of the current US petroleum consumption with biofuels by 2030 [4].

Primary forest resources are logging residues from conventional harvest operations and residues from forest management and land clearing operations, removal of excess biomass from timberlands and other forestlands, and fuelwood extracted from forestlands. Secondary forest resources are primary wood processing mill residues, secondary wood processing mill residues, and pulping liquors (black liquors). Tertiary resources are urban wood residues—construction and demolition debris, tree trimmings, packaging wastes, and consumer durables.

Primary agricultural resources are crop residues from major crops, for example, corn stover, small grain straw and sugarcane baggase, grains of corn and soybeans used for ethanol, biodiesel, and bioproducts, perennial grasses, and perennial woody crops; secondary agricultural residues are animal manures and food/feed processing residues, and tertiary are municipal solid wastes, post-consumer residues, and landfill gases.

Forest resources (US)	368 million dry tones per year
Agricultural resources (US)	998 million dry tonnes per year
Total (US)	1,366 million dry tones per year

7.5 Biomass Pretreatment

Although cellulose is the most abundant plant material resource, its susceptibility has been curtailed by its rigid structure. As the result, an effective pretreatment is needed to liberate the cellulose from the lignin seal and its crystalline structure so as to render it accessible for a subsequent hydrolysis step [5]. By far, most pretreatments are done through physical or chemical means.

Till date, the available pretreatment techniques include acid hydrolysis, steam explosion, ammonia fiber expansion, alkaline wet oxidation, and ozone pretreatment [6]. Besides effective cellulose liberation, an ideal pretreatment has to minimize the formation of degradation products because of their inhibitory effects on subsequent hydrolysis and fermentation processes [7]. The presence of inhibitors will not only further complicate the ethanol production but also increase the cost of production due to entailed detoxification steps. Even though pretreatment by acid hydrolysis is probably the oldest and most studied pretreatment technique, it produces several potent inhibitors including furfural and hydroxymethyl furfural (HMF) which are by far regarded as the most toxic inhibitors present in lignocellulosic hydrolysate [8]. In fact, ammonia fiber expansion is the sole pretreatment which features promising pretreatment efficiency with no inhibitory effect in resulting hydrolysate.

7.6 Biomass Cell Wall Composition

As already mentioned, the main cell-wall components are cellulose, hemicelluloses, and lignin. Another polysaccharide found in the biomass, in relatively low concentration is pectin. In addition to these, the biomass contains extractives, mainly tall oil and volatile terpenes, ash, and some nitrogen containing molecules.

The main monomeric sugars in the high percentage polysaccharides, that is, cellulose and hemicelluloses, are: glucose, xylose, mannose, galactose, and arabinose.

7.7 Acid Catalyzed Hydrolysis of Biomass

Various methods for the hydrolysis of lignocellulosic materials have recently been described [9]. The dilute acid process is conducted under high temperature and pressure. For example, using a dilute acid process with 1 % sulfuric acid (H_2SO_4) in a continuous flow reactor at a residence time of 0.22 min and a temperature of 237 °C with pure cellulose provided a yield over 50 % sugars. Dilute acids lead to a limited hydrolysis called prehydrolysis. Dilute-acid hydrolysis is carried out using mineral acids such as H_2SO_4 or HCl, at temperatures of 122–202 °C [10]. The chief advantages to using HCl over H_2SO_4 are that HCl permeates the wood more easily than H_2SO_4 and is a volatile compound, which assists in the crucial acid recovery steps. The Udic-Rheinau process was an attempt to make the Bergius-Rheinau process economically advantageous. The latter process will be further discussed. In the improved process, the wood was first prehydrolyzed in 1 % HCI at 130 °C to remove the hemicelluloses. The wood was then dried and subsequently hydrolyzed with 40 % HCl at 12 °C for 10 h. After washing the lignin residue with dilute HCl, the HCl was recovered by vacuum distillation. This process was more economical than the Bergius-Rheinau process [11]. The biggest advantage of dilute acid processes is their fast rate of reaction, which facilitates continuous processing [12].

To summarize, acid catalyzed hydrolysis of biomass can be performed for example by using HCl or H_2SO_4. The H_2SO_4 catalyzed hydrolysis process was developed in the US to yield sugar monomers, however, it seems that the sugar yields are relatively low, due to the formation of relatively high percentage of degradation products. Additional drawbacks are relatively high acid left in the lignin, high energy consumption, and formed gypsum as by-product.

7.8 The Hydrochloric Acid Process

The Bergius-Rheinau wood saccharification process has its origin in Willstatter's [13] discovery that the cellulose of wood is easily hydrolyzed by highly concentrated HCl at low temperature into dextrose. Karl Goldschmidt, one of the leading men in the German chemical industry of that time, who had an open mind for its great future problems, devoted himself to the industrial application of this process. For this the work of Hagglund [14] was decisive, who found that the obtained solution of dextrose in HCl can dissolve fresh cellulose several times over and, besides, that the HCl can be evaporated in a vacuum from the sugar and then recycled. However, from the beginning in the first half of the twentieth century, the technical execution of this process was considered extremely difficult [15].

During the first 20 years, a manufacturing process was developed in the Rheinau pilot plant in Germany under Friedrich Bergius. It represented in these days an excellent technical achievement; especially the problem of the technical manipulation of the highly concentrated HCl was well solved.

In 1927, Bergius was able to conclude his own work on the liquefaction of coal, after the practical possibilities had been proved on a large scale. The I.G. Farbenindustrie and Imperial Chemical Industries then took up the work on an industrial scale. From that time onward Bergius devoted himself to a process of obtaining sugar from cellulose in wood, on which he had already worked during the First World War. He succeeded after 15-year work and an industrial plant was set up, also in the Rheinau works. It is amazing with what intensity Bergius took up the second part of his life's work, namely this hydrolysis of cellulose in wood and similar substances to sugar. It seems as if the well-known difficulties of working with highly concentrated HCl had presented a special challenge to Bergius. Initially the process was taken up only in England and only during the 1930s of the twentieth century did Bergius manage to continue these experiments in Germany; his main concern was to rationalize the process and to ensure complete recovery of the HCl used by constructing intricate devices.

The capacity of the Rheinau plant was raised to 400 tons of raw sugar a month, and at Regensburg a manufacturing plant of 1,600 t of raw sugar a month was erected for food–yeast production. However, a profit with all other wood saccharification processes was only reached as long as the products were protected by the government [16]. At the end of the war, both plants had to be closed down. In Russia, a plant using HCl technology was active for 20 years in Siberia and in the US Dow Chemicals erected a pilot plant in the 1980s. All plants were closed; the main problem was the inability to recycle the HCl efficiently, thus causing the process to be uneconomical.

Concentrated (fuming) HCl-driven hydrolysis provides the most powerful and industrially proven technology for converting all cellulosic wastes—wood, solids from city sewage plants, bagasse, grasses, etc. into sugars that can be fermented to ethanol or other biofuels as well as a large variety of chemicals and bio-products and food and feed. HCl permeates the wood more easily than H_2SO_4. HCl makes the cellulose more susceptible to hydrolysis and it is a volatile compound, which assists in the crucial acid recovery steps.

The Virdia process begins by steam expansion of debarked chipped wood, which undergoes a pre-extraction stage to remove all extractives, for example, tall oil and ash. The pre-extracted wood continues into hydrolysis stage performed using highly concentrated HCl (42 %) at low-temperature (10–15 °C), thus affording sugars hydrolyzate with minimum degradation products (e.g., furfurals), while simultaneously separating the solid lignin. Approximately 98 % of the theoretically available sugars, composing ca. 65 % of the dry weight of the wood chips for pine wood are converted into sugars, which are dissolved in the hydrolyzate. The sugars hydrolyzate is further treated by extracting of the acid for recycling. The soluble oligo-saccharides formed to some extent are converted into the more desired mono-saccharides mixture of glucose, mannose, galactose, xylose, and arabinose, thus removing any impurities that may remain or may have been created during the course of the hydrolysis process.

The hydrolysis catalyst—HCl, forms hydrates, for example: $HCl \cdot 2H_2O$; $HCl \cdot 3H_2O$; $HCl \cdot 4H_2O$ (fuming HCl). It is assumed these species are responsible to the efficient hydrolysis of cellulose. These hydrates are formed mostly at high-HCl concentration in water, that is, 40–42 %; below this concentration, the uniqueness of the HCl hydrates as dispersants of lignocellulose presumably drops sharply.

Another view [17] describes the following structures for hydrated HCl:

- For the dihydrate, $(H_2O{-}H^+{-}OH_2)(Cl^-)$;
- For the trihydrate, $(H_2O{-}H^+{-}OH_2)(H_2O)(Cl^-)$;
- For the hexahydrate, $(H_3O^+)(H_2O)_5(Cl^-)$.

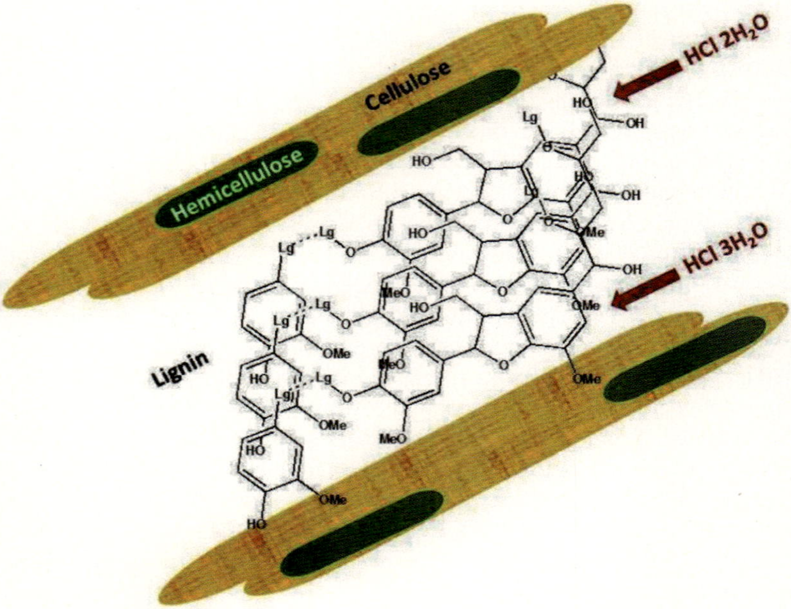

Fig. 7.3 Model of wood penetration by 42 % HCl aq

Or as shown by Botti et al. [18] where these species are so-called Eigen and Zundel-type complexes:

- Eigen $[H_9O_4]^+$
- Zundel $[H_5O_2]^+$

In the above descriptions are not shown any chloride ions that might be in the vicinity of the complex or water molecules that might be bonded to the other end of the Zundel ion. Not shown are also any chloride ions that might be in the vicinity of the Eigen ion complex.

Figure 7.3 shows a model of wood, into which the 42 % HCl succeeds to go through and separate between the lignin and the saccharides and probably penetrates the crystalline cellulose structure.

A key limitation to any concentrated acid hydrolysis is the difficulty in recovering the acid. In particular, HCl solution forms an azeotrope at between 21 and 25 % depending on the pressure; simple distillation cannot concentrate a dilute solution beyond the azeotropic point. The efficiency of acid recovery is a key condition to making acid hydrolysis of lignocellulosic materials an economically viable source of fermentable sugars.

It is important in minimizing the need for make-up HCl, for neutralizing chemicals, and for costly disposal and negative impact on the environment. Full recovery of HCl at high acid concentration and its reuse yields very minor waste stream, no complicating air emissions, and favorable life cycle analysis. As most of the HCl

Fig. 7.4 The extractant roles: removal of HCl and obtaining highly concentrated sugars aqueous solution

recovery happens at relatively low temperatures, this also permits highly efficient energy integration. In addition to the HCl removal from the sugars hydrolyzate, it is also removed from the remaining lignin through a proprietary de-acidification process at low temperatures, thus being almost fully recycled into the process, leaving a very pure lignin. The main innovation in Virdia process is based on the problem the Germans had—recovering the acid, that is, evaporating water from the dilute acid at azeotrope concentration means breaking the bond between the acid and water at ~23 %. This is performed by using a medium, that is, an extractant composition that can perform two contradictory roles at two different circumstances (Fig. 7.4):

1. Taking the HCl out of the water at a relatively low temperature, thus yielding highly concentrated sugars solution;
2. Recovering the acid at high concentration.

Virdia developed several processes for the recovery of HCl from a dilute solution [19]. The following process describes the extraction of the acid by bringing a dilute aqueous HCl solution into contact with a substantially immiscible extractant, thatis, comprising of tris-2-ethylhexyl amine (TEHA; Fig. 7.5), which is substantially water insoluble in both free and salt forms, an oil soluble weak organic acid, for example,

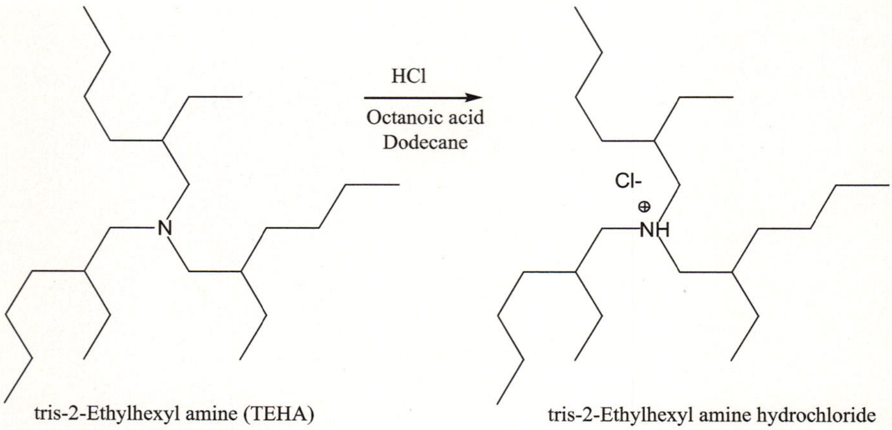

Fig. 7.5 Extraction of HCl using TEHA to form an oil soluble salt

Fig. 7.6 Carboxylic acid enhancer to a secondary amine extractant

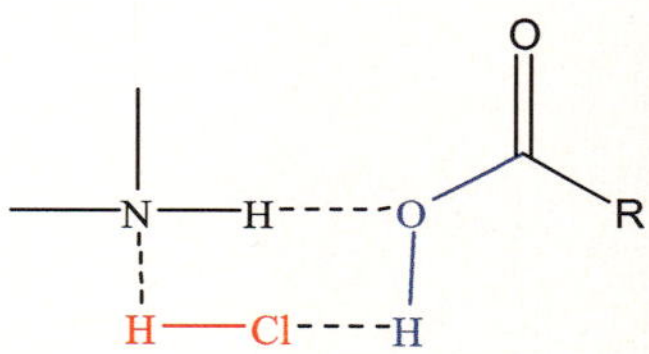

octanoic acid which is substantially water insoluble, in both free and salt forms; and a solvent for the amine and organic acid, for example, dodecane, as a result of which HCl selectively transfers to the extractant to form an HCl-carrying extractant. See the following scheme:

The role of the carboxylic acid, which is used an enhancer, is to form a stable complex with the amine and the HCl, as seen in the following (Fig. 7.6):

Table 7.2 shows equilibrium data of HCI extraction with TEHA/octanoic acid in dodecane extractant:

This HCl-loaded extractant is further treated to obtain gaseous HCl.

Stripping of HCl is performed by passing, for example, xylene vapors stream through the HCl-loaded extractant. The recovered HCl during the stripping is shown in Table 7.3:

The whole extraction/stripping process is presented by the following scheme (Fig. 7.7):

The final product consists of the following sugars (Figs. 7.8 and 7.9):

The HPLC of a typical final sugar product is seen in Fig. 7.10 and a typical final product in Fig. 7.11:

The full Virdia process is presented in the following scheme (Fig. 7.12):

As can be seen in the above scheme, in addition to the sugars produced, two more products are obtained, that is, lignin and tall oil.

Table 7.2 Equilibrium data of HCl extraction with TEHA/octanoic acid 1:0.25 mol/kg in dodecane at 27 °C

HCl in aqueous phase (mol/kg)	HCl in extractant (mol/kg)	HCl in aqueous phase (g/1,000 g H_2O)	HCl in extractant (g/1,000 g extractant)
0.039	0.019	1.4	0.7
0.064	0.050	2.4	1.8
0.154	0.069	5.7	2.5
0.31	0.24	11.4	8.7
0.42	0.35	15.8	13.0
0.62	0.56	23.2	20.9
0.78	0.68	29.5	25.3
1.18	0.87	45.0	32.7

Table 7.3 The HCl recovered by xylene stripping

Time (min)	Recovered HCl (%)
0	0
10	31.0
20	46.7
30	88.0
50	98.1

The lignin (Fig. 7.13) is separated as solid from the process, the HCl is stripped and recycled, and the lignin is dried.

Lignin is used for binders, activated carbon, carbon fibers, fire-retardants, motor fuel, dispersants, sorbents, surfactants, and as starting material for vanillin.

An additional by-product of the Virdia process is tall oil—a generic name for a group of compounds which consist of resin acids, fatty acids, fatty alcohols, some sterols, and other alkyl hydrocarbon derivates. Resin acids occur in pine in a number of isomeric forms having the molecular formula of $C_{20}H_{30}O_2$ and some related structures. The most prevalent are abietic-type acids, such as levopimaric, palustric, abietic, and neoabietic acids; and pimaric-type acids, such as pimaric and isopimaric acids (Fig. 7.14).

The fatty acids include more than 10 different acids: both saturated and unsaturated. The most common are palmitic and stearic acids, which are saturated, and

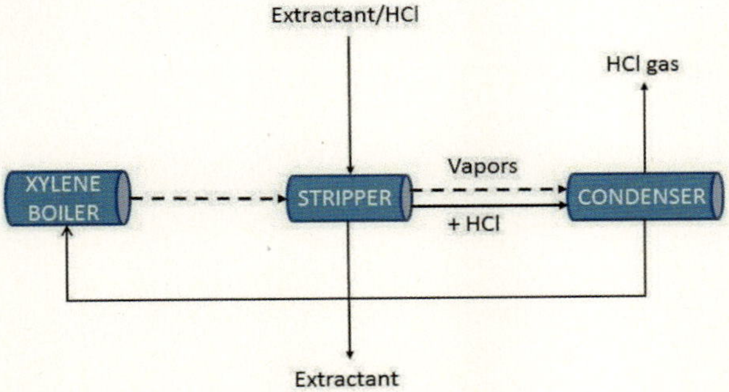

Fig. 7.7 Stripping of HCl gas from loaded extractant

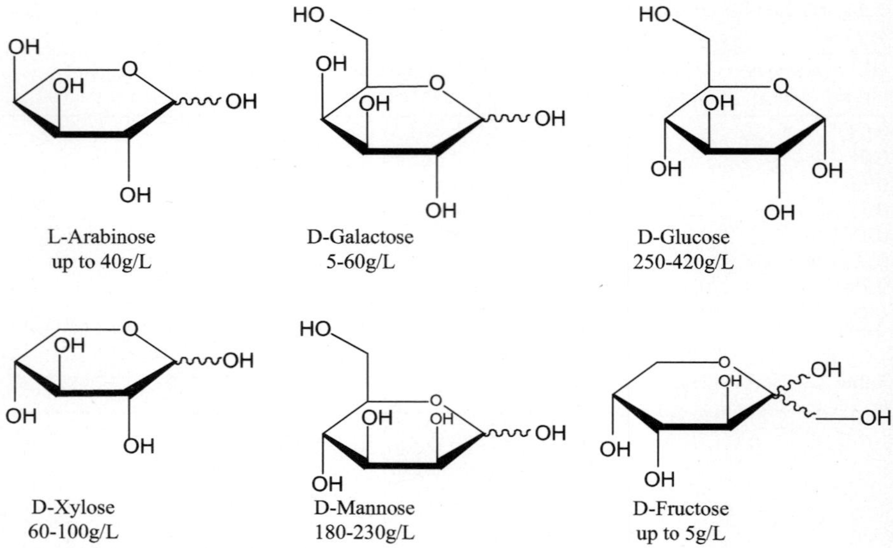

Fig. 7.8 Monosugars from pinewood and their ranges in final product

oleic and linoleic, which are unsaturated. The unsaponifiables present in tall oil include higher fatty alcohols, esters, plant sterols, and some hydrocarbons. The most common sterol present is β-sitosterol (Fig. 7.15) [20].

The resin acids part of the tall oil is used for inks, adhesives, paper-making, road-marking, and tyres. The fatty acids are applied to paints and coatings, bio-lubricants, fuel-additives, and performance polymers. Sterols are used as health-enhancing food additives and for pharmaceuticals.

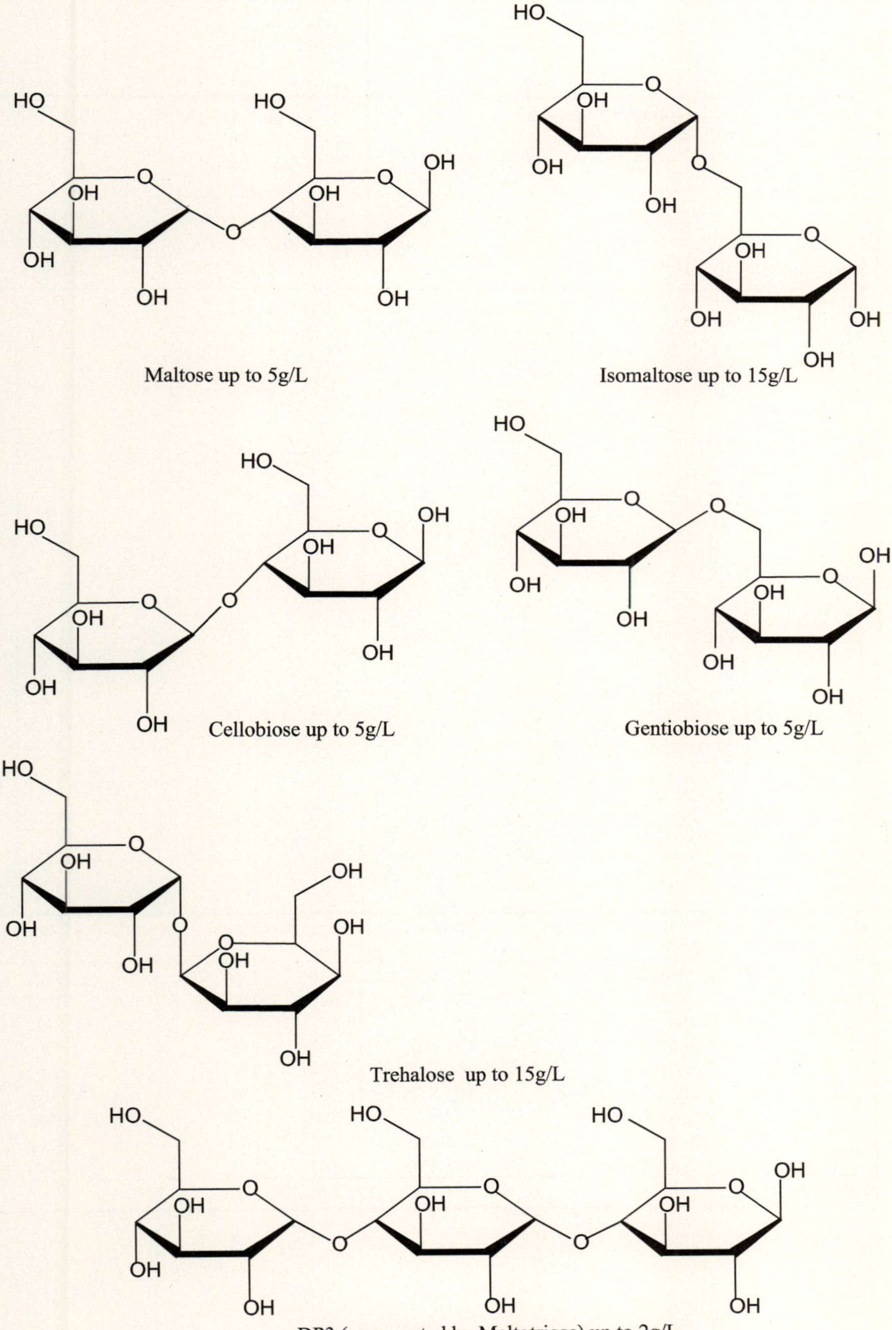

Fig. 7.9 Di- and tri-sugars and their ranges in final product

1 raff060711 filtered

Sample Name:	**raff060711 filtered**	*Injection Volume:*	**10.0**
Vial Number:	**49**	*Channel:*	**ED_1**
Sample Type:	**unknown**	*Wavelength:*	**n.a.**
Control Program:	**gradient_50min**	*Bandwidth:*	**n.a.**
Quantif. Method:	**Quant_1**	*Dilution Factor:*	**1.0000**
Recording Time:	**7/12/2011 9:24**	*Sample Weight:*	**1.0000**
Run Time (min):	**85.10**	*Sample Amount:*	**1.0000**

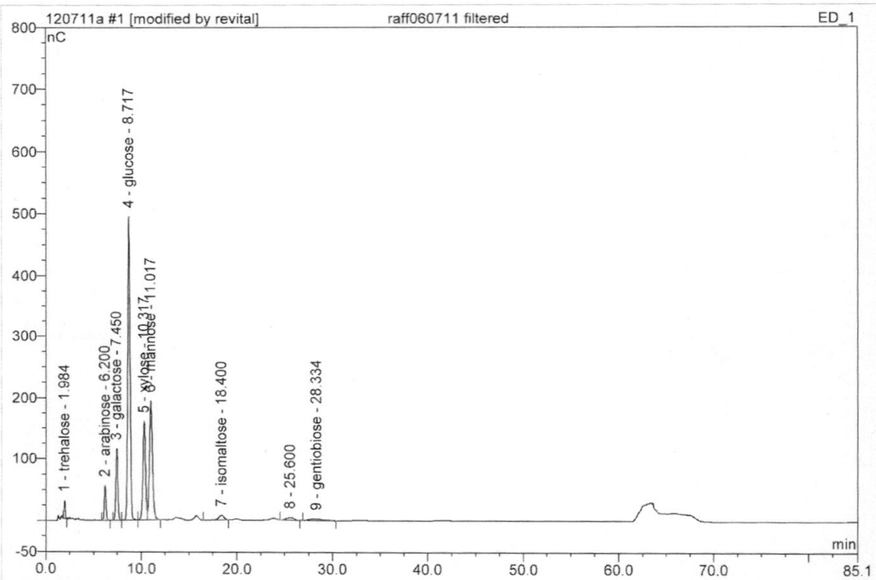

No.	Ret.Time min	Peak Name	Height nC	Area nC*min	Rel.Area %	Amount	Type
1	1.98	trehalose	29.035	4.136	1.27	n.a.	BMB
2	6.20	arabinose	56.447	11.429	3.51	n.a.	BMB
3	7.45	galactose	115.913	28.472	8.75	n.a.	BM
4	8.72	glucose	494.932	143.392	44.06	n.a.	M
5	10.32	xylose	160.011	51.364	15.78	n.a.	M
6	11.02	mannose	193.977	72.827	22.38	n.a.	MB
7	18.40	isomaltose	7.436	4.926	1.51	n.a.	BMB
8	25.60	n.a.	4.476	4.021	1.24	n.a.	BMB
9	28.33	gentiobiose	2.559	4.847	1.49	n.a.	BMB
Total:			1064.786	325.414	100.00	0.000	

Fig. 7.10 HPLC of sugars in a typical product

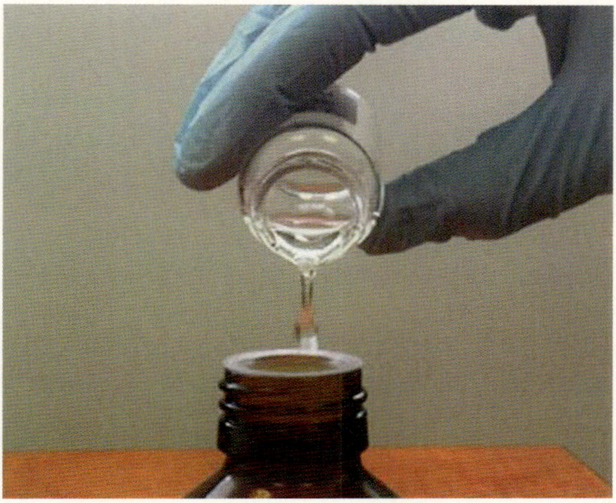

Fig. 7.11 Final sugar product

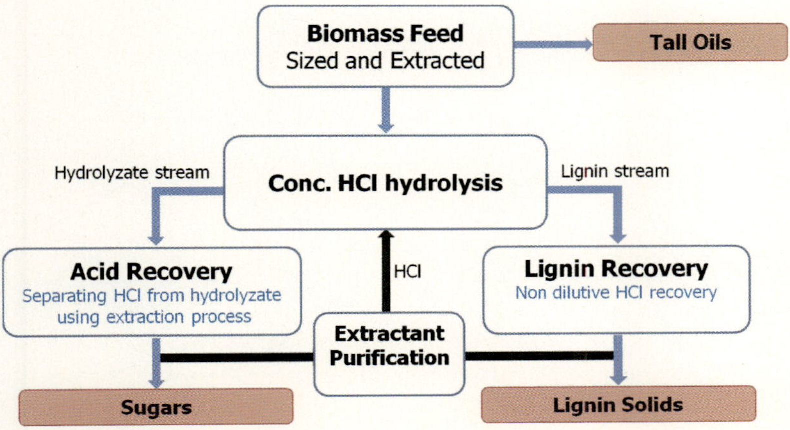

Fig. 7.12 Virdia process for production of sugars, lignin, and tall oil

Fig. 7.13 SEM of lignin obtained in the Virdia process

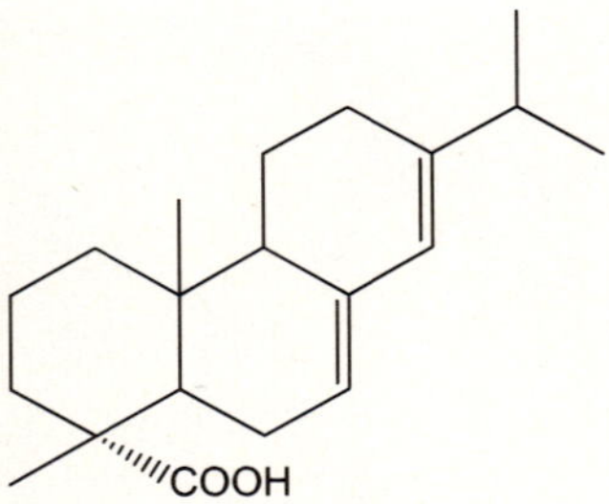

Abietic acid

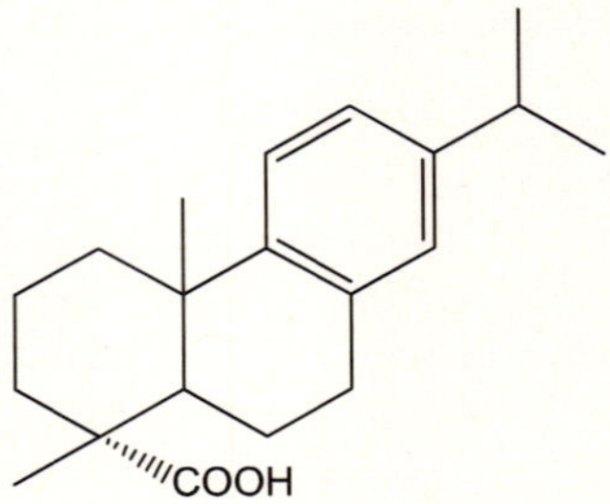

Dehydroabietic acid

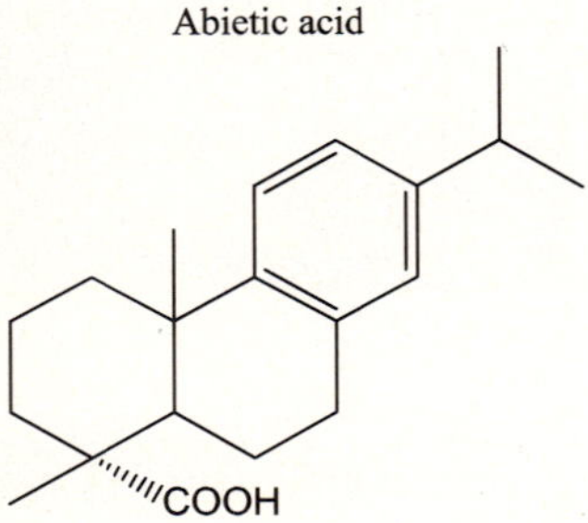

Neoabietic acid

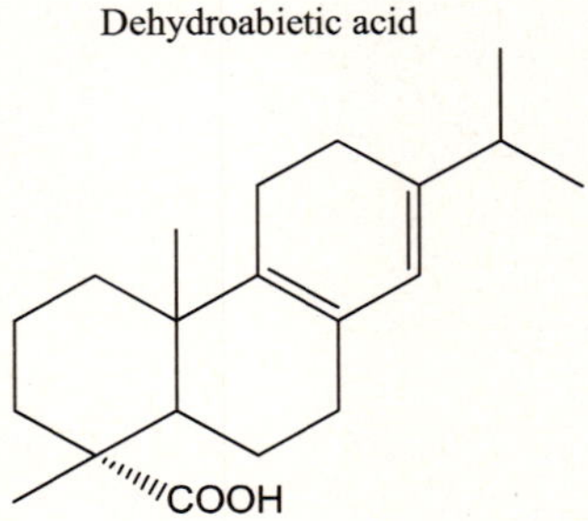

Palustric acid

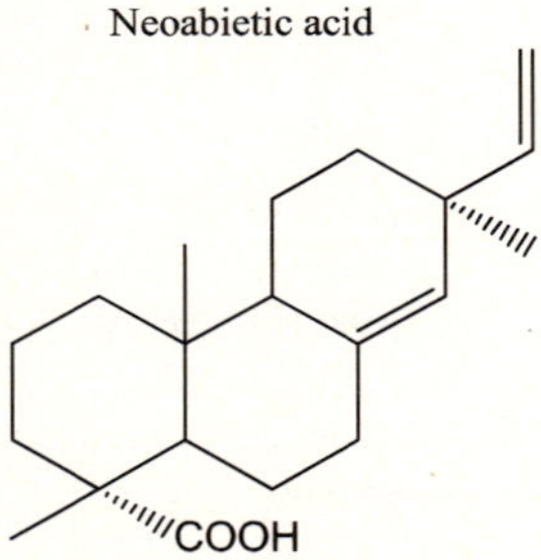

Pimaric acid

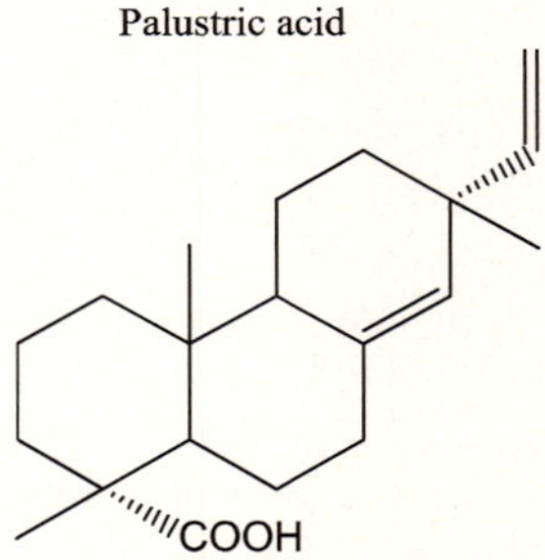

Isopimaric acid

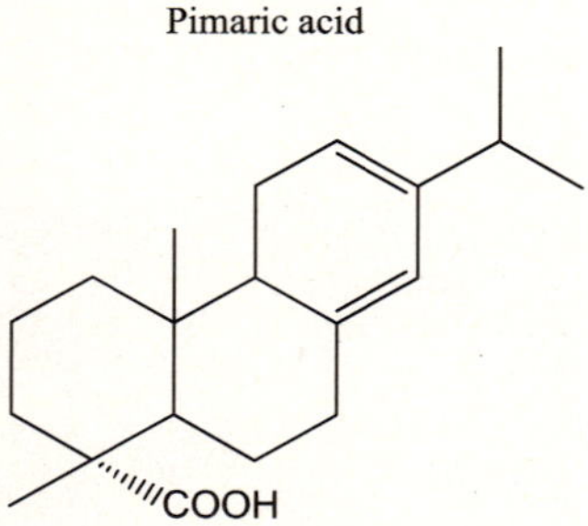

Levopimaric acid

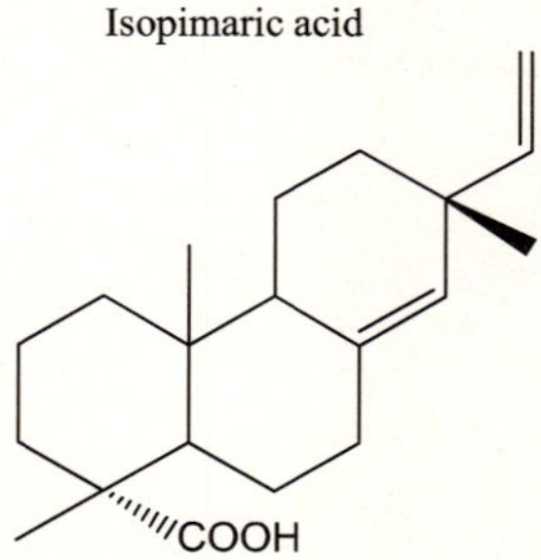

Sandracopimaric acid

Fig. 7.14 Resin acids in pinewood tall oil

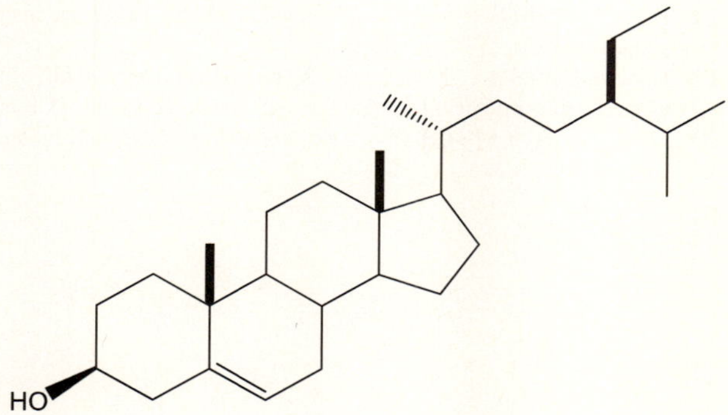

Fig. 7.15 β-Sitosterol structure

References

1. Hall JA, Saeman JF, Harris JF (1956) Wood saccharification. Unasylva—Int Rev For For Prod 10:1
2. Sjostrom E (1993) Wood chemistry. In: Fundamentals and applications, 2nd edn. Academic Press, San Diego, p 292
3. Excoffier G, Toussaint B, Vignont MR (1991) Biotechnol Bioeng 38:11
4. Perlack RD, Wright LL, Turhollow AF, Graham RL, Stokes BJ, Erbach DC (2005) Biomass as feedstock for a bioenergy and bioproducts industry: the technical feasibility of a bil-lion-ton annual supply. Study prepared by the U.S. Department of Energy (DOE) and the U.S. Department of Agriculture (USDA)
5. Mosier N, Wyman C, Dale B, Elander R, Lee YY, Holtzapple M, Ladisch M (2005) Bioresour Technol 96(6):673
6. Klinke HB, Thomsen AB, Ahring BK (2004) Appl Microbiol Biotechnol 66(1):10
7. Olsson L, Hahn-Hagerdal B (1996) Enzyme Microb Technol 18(5):312
8. Palmqvist E, Hahn-Hagerdal B (2000) Bioresour Technol 74(1):17
9. Olsson L, Hahn-Hagerdal B (1996) Fermentation of lignocellulosic hydrolysates for ethanol production. Enzyme Microb Technol 18:312
10. Grethlein HE, Converse AO (1991) Common aspects of acid prehydrolysis and steam explosion for pretreating wood. Bioresour Technol 36:77; Torget R, Hsu TA (1994) Two-temperature dilute acid prehydrolysis of hardwood xylan using a percolation process. Appl Biochem Biotechnol 45–46:23
11. Kucuk MM, Demirbas A (1997) Biomass conversion processes. Energy Convers Manage 38:151
12. Demirbas A (2008) Products from lignocellulosic materials via degradation processes. Energy Sources Part A 30(1):27
13. Willstatter R (1913) Ber Dtsch Chem Ges 46:2405
14. Hagglund E (1939) Holzchemie 2. Aufl., Akad, Verlagsges, Leipzig, S 268 ff
15. Weinberg AV (1919) Emil Fischers Tatigkeit Wahrend des Krieges. Naturw 7:870; and Willstatter R (1949) Aus meinem leben, Verlag Chemie S 223
16. Schoenemann K (1953) Paper presented to the congress of the food and agriculture organisation of the United Nations, Stockholm, 27 July 1953
17. Buch V, Dubrovskiy A, Mohamed F, Parrinello M, Sadlej J, Hammerich AD, Devlin JP (2008) HCl hydrates as model systems for protonated water. J Phys Chem A 112(11):2144

18. Botti A, Bruni F, Ric-ci MA, Soper AK (2006) Eigen versus Zundel complexes in HCl-water mixtures. J Chem Phys 125:1
19. Eyal A, Baniel A (2009) A process for the recovery of HCl from a dilute solution thereof and extractant composition for use therein, Assigned to HCL CleanTech Ltd. WO 2009/125400 A2
20. Drew J, Propst M (1981) Tall oil, compiled and edited by Pulp Chemicals Association. Pulp Chemicals, New-York

Chapter 8
Chemical Pretreatment Techniques for Biofuels and Biorefineries from Softwood

Fang Huang and Arthur J. Ragauskas

Abstract Lignocellulosic materials, such as wood, grass, and agricultural and forest residues, are potential resources for the production of bioethanol. The biochemical process of converting biomass to bioethanol typically consists of three main steps: pretreatment, enzymatic hydrolysis, and fermentation. During the whole process, pretreatment is probably the most crucial step since it has a large impact on the efficiency of the overall bioconversion. The aim of pretreatment is to disrupt recalcitrant structures of cellulosic biomass to make cellulose more accessible to the enzymes that convert carbohydrate polymers into fermentable sugars. Physical, physical-chemical, chemical, and biological processes have been used for pretreatment of lignocellulosic materials. This chapter summarizes the leading technologies in chemical pretreatment on softwood, particularly pine species, which generally show relatively higher recalcitrance than hardwood, grass, and other lignocellulosic materials. Different chemical pretreatment techniques, including dilute acid pretreatment, alkaline hydrolysis, wet oxidation, sulfite pretreatment to overcome recalcitrance of lignocellulose (SPORL), organosolv, ionic liquids pretreatment, and ozonolysis process are intensively introduced and discussed. In this chapter, the key points are focused on the structural changes primarily in cellulose, hemicellulose, and lignin during the above leading pretreatment technologies.

Keywords Chemical pretreatment technology · Biofuel · Biorefinery · Softwood

8.1 Introduction

In order to cope with growing demand for energy, the depletion of fossil fuel resources, and environmental concerns raised by fossil fuel use, countries wishing to limit their energy dependence on petroleum exporting countries are developing

---

A. J. Ragauskas (✉) · F. Huang
School of Chemistry and Biochemistry,
Institute of Paper Science and Technology,
Georgia Institute of Technology, Atlanta, Georgia, 30332-0440, USA
e-mail: arthur.ragauskas@ipst.gatech.edu

Z. Fang (ed.), *Pretreatment Techniques for Biofuels and Biorefineries*,
Green Energy and Technology,
DOI 10.1007/978-3-642-32735-3_8,

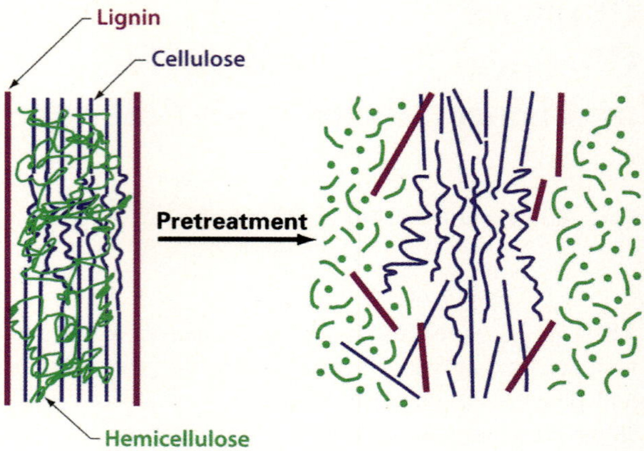

Fig. 8.1 Schematic presentation of effects of pretreatment on lignocellulosic biomass. (Reprint from Ref. [10], U.S. Department of Energy Genomic Science program (http://genomicscience.energy.gov))

alternative energy sources, such as bioethanol produced from renewable biomass [1–4]. Cellulosic bioethanol is regarded as one of the most promising renewable biofuels in the transportation sector for the coming next few decades [5]. Current production of bioethanol relies on sugars that are obtained from starch-based agricultural crops by using first-generation conversion technologies [6]. Nowadays bioethanol produced from lignocellulosic biomass using second-generation technologies has become an interesting alternative, mainly because lignocellulosic raw materials do not compete with food crops or productive agricultural land, and they are also less expensive than conventional agricultural feedstocks [7, 8].

The biological process of converting biomass to bioethanol typically consists of three main steps: pretreatment, enzymatic hydrolysis, and fermentation. During the whole process, pretreatment is the most crucial step since it has a large impact on the efficiency of the overall bioconversion. In lignocellulosic biomass, cellulose and hemicellulose are densely packed together with lignin, which serves several functions including protection against enzymatic hydrolysis [9]. The aim of pretreatment is to disrupt recalcitrant structures of cellulosic biomass to make cellulose more accessible to the enzymes that convert carbohydrate polymers into fermentable sugars (Fig. 8.1). During the pretreatment, the extent of removal of lignin and hemicellulose depends on the pretreatment conditions and severity. For example, acidic chemical pretreatment removes most of hemicellulose. The lignin is condensed when pretreating temperature reaches above 170 °C. On the contrary, the ammonia fiber explosion (AFEX) pretreatment does not significantly remove hemicellulose.

Numerous pretreatment strategies have been developed to enhance the reactivity of cellulose and to increase the yield of fermentable sugars. Typical goals of pretreatment include [11]:

- Production of highly digestible solids that enhances sugar yields during enzyme hydrolysis.
- Avoiding the degradation of sugars (mainly pentoses) including those derived from hemicellulose.
- Minimizing the formation of inhibitors for subsequent fermentation steps.

Table 8.1 Typical lignocellulosic biomass compositions (% dry basis) [3, 4, 5, 13, 14]

	Cellulose	Hemicellulose	Lignin
Pine	43.3	20.5	28.3
Spruce	45.0	22.9	27.9
Douglas fir	45.0	19.2	30.0
Poplar	44.7	18.5	26.4
Eucalyptus	49.5	13.1	27.7
Corn stover	36.8	30.6	23.1
Miscanthus	52.1	25.8	12.6
Wheat straw	44.1	23.8	20.5
Switchgrass	33.5	26.1	17.4

Among the numerous types of biomass, softwoods (SW) are generally recognized as being much more refractory than hardwoods (HW) or agricultural residues in the pretreatment process. This is, in part, due to the fact that SW have a more rigid structure and contains more lignin [12].

The goal of this paper is to review promising chemical pretreatments technologies on softwood, particularly pine species, and to discuss recent developments which have greatly aided the production of bioethanol. For each technology, a brief process description is first given with recent developments, and then the feedstocks on which these technologies are used are highlighted, followed by discussion of the technology's advantages and disadvantages. The key points will be focused on the structural changes primarily in cellulose, hemicellulose, and lignin during the above leading pretreatment technologies.

8.2 Understanding Lignocellulosic Biomass

8.2.1 Composition of Lignocellulosic Biomass

The term "lignocellulosic biomass" is used when referring to higher plants, such as grasses, SW or HW. Understanding lignocellulosic biomass, particularly its chemical composition, is a prerequisite for developing effective pretreatment technologies to deconstruct its rigid structure, designing enzymes to liberate sugars, particularly cellulase to release glucose (Glu), from recalcitrant cellulose, as well as engineering microorganisms to convert sugars into ethanol and other bio-based chemicals. The main components of the lignocellulosic materials are cellulose, hemicellulose, lignin, and a remaining smaller part (extractives and ash). The composition of lignocellulose highly depends on its source. There is a significant variation of the lignin and (hemi)cellulose content of lignocellulosics depending on whether it is derived from hardwood, softwood, or grasses. Table 8.1 summarizes the composition of lignocellulose encountered in some of the most common sources of biomass.

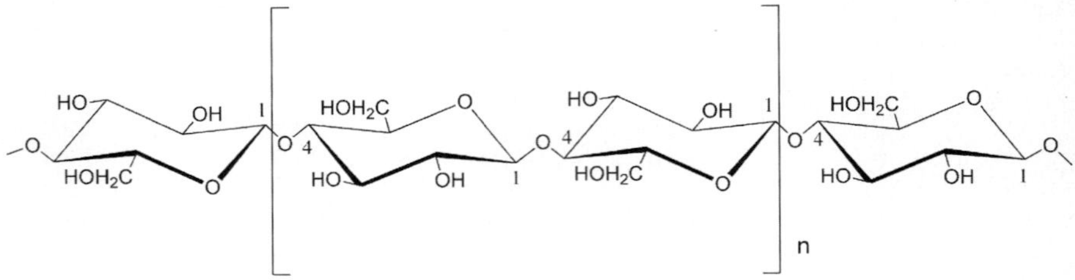

Fig. 8.2 The structure of cellulose [15]

8.2.2 Physical and Chemical Characteristics of Lignocellulosic Biomass

Lignocellulosic biomass has a complex internal structure. The major components of lignocellulosic biomass, that is, cellulose, hemicellulose, and lignin, also have intricate structures. To obtain a clear picture of the material, an analysis of the structure of each main component is made in this section, concluding with the description of the structure of lignocellulose itself. The physical properties of each component and how these components contribute to the behavior of the complex structure are also addressed. The study is oriented toward breaking down the complex of lignocellulose and utilizing the components to produce sugars, as this is one of the main goals of pretreatment.

8.2.2.1 Cellulose

Cellulose is the β-1,4-polyacetal of cellobiose (4-O-β-D-glucopyranosyl-D-Glu). Cellulose is more commonly considered as a polymer of Glu because cellobiose consists of two molecules of Glu. The chemical formula of cellulose is $(C_6H_{10}O_5)_n$ and the structure of one chain of the polymer is presented in Fig. 8.2. Many properties of cellulose depend on its degree of polymerization (DP), that is, the number of Glu units that make up one polymer molecule. The DP of cellulose varies from 5,000 in native wood to approximately 1,000 in bleached wood pulp [15]. Each D-anhydroglucopyranose unit possesses hydroxyl groups at C2, C3, and C6 positions, capable of undergoing the typical reactions known for primary and secondary alcohols. The molecular structure imparts cellulose with its characteristic properties: hydrophylicity, chirality, degradability, and broad chemical variability initiated by the high donor reactivity of hydroxyl groups.

The nature of the bonding between the Glu molecules (β-1,4 glycosidic) allows the polymer to be arranged in long linear chains. The latter arrangement of the molecule, together with the fact that the hydroxyl groups are at C2, C3 and C6 positions, allows for the formation of intra- and inter-molecular hydrogen bonds between the molecules of cellulose [16]. The coalescence of several polymer chains leads to the formation of microfibrils, which in turn are united to form fibers.

The hydrogen bonds in the linear cellulose chains promote aggregation into a crystalline structure and give cellulose a multitude of partially crystalline fiber structures and morphologies [17]. The average degree of crystallinity of native cellulose ranges 50–70 % [18, 19]. The ultrastructure of native cellulose (cellulose I) has been discovered to possess unexpected complexity in the form of two crystal phases: I_{α} and I_{β} [20]. The relative amounts of I_{α} and I_{β} have been found to vary between samples from different origins. The I_{α}-rich specimens have been found in the cell wall of some algae and in bacterial cellulose, whereas I_{β}-rich specimens have been found in cotton, wood, and ramie fibers [21, 22]. Native cellulose also contains para-crystalline and amorphous portion. Para-crystalline cellulose is loosely described as chain segments having more order and less mobility than amorphous chains segments but less-ordered and more mobile than chains within crystals [23, 24]. The presence of crystalline cellulose, with regions of less order, and the size of the elementary fibrils work together to produce interesting combination of contrary properties such as stiffness and rigidity on one hand and flexibility on the other hand [25].

Crystalline cellulose has a very limited accessibility to water and chemicals. Chemical attack can, therefore, be expected to occur primarily on amorphous cellulose and crystalline surface. Cellulose is a relatively hygroscopic material absorbing 8–14 % water under normal atmospheric conditions (20 °C, 60 % relative humidity) [26]. Nevertheless, it is insoluble in water, where it swells. Cellulose is also insoluble in dilute acid solutions at low temperature. The solubility of the polymer is strongly related to the degree of hydrolysis achieved. As a result, factors that affect the hydrolysis rate of cellulose also affect its solubility that takes place. In alkaline solutions extensive swelling of cellulose takes place as well as dissolution of the low molecular weight fractions of the polymer (DP < 200) [27].

8.2.2.2 Hemicellulose

The term hemicellulose is a collective term. It is used to represent a family of polysaccharides that are found in the plant cell wall and have different composition and structure depending on their source and the extraction method. Unlike cellulose, hemicellulose is composed of combinations of pentose (xylose (Xyl) and arabinose (Ara)) and/or hexoses (mannose (Man), galactose (Gal), and Glu); and it is frequently acetylated and has side chain groups such as uronic acid and its 4-O-methyl ester. The chemical nature of hemicellulose varies from species to species. In general, the main hemicelluloses of softwood are galactoglucomannans and arabinoglucuronoxylan, while in hardwood is glucuronoxylan (Fig. 8.3) [28]. Table 8.2 summarizes the main structural features of hemicelluloses appearing in both softwood and hardwood.

Important aspects of the structure and composition of hemicellulose are the lack of crystalline structure, mainly due to the highly branched structure, and the presence of acetyl groups on the polymer chain. Hemicellulose extracted from plants possesses

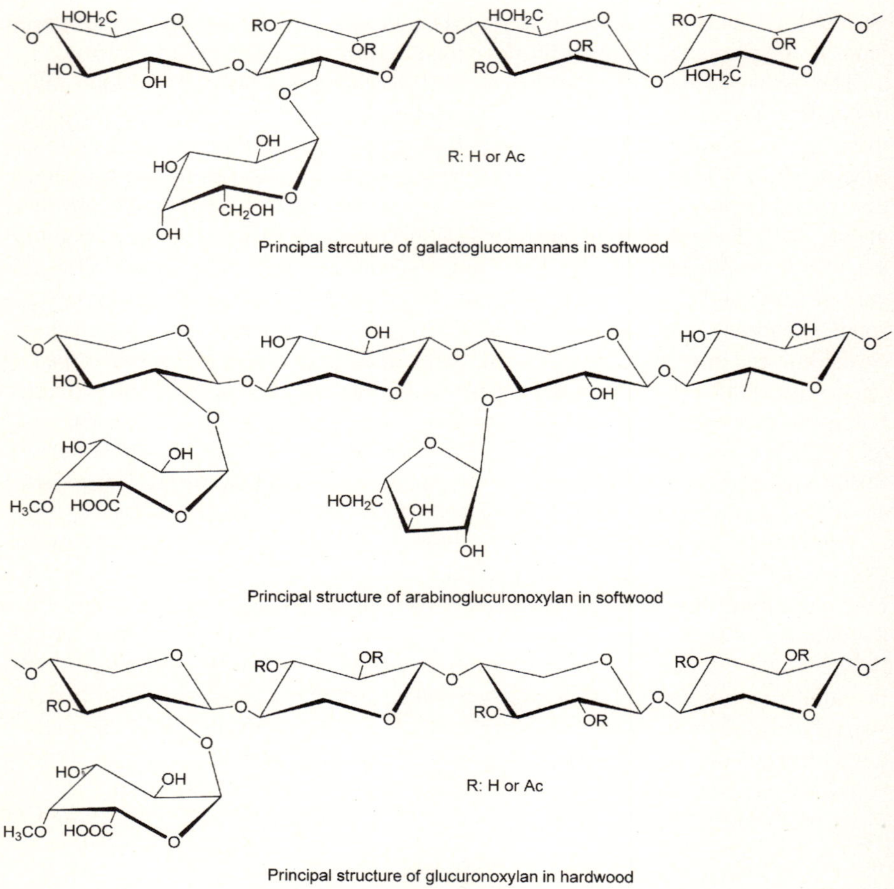

Fig. 8.3 Principal polysaccharides in woody hemicellulose. (Reproduced from Ref. [28] by permission of Wiley)

a high degree of polydispersity, polydiversity, and polymolecularity (a broad range of size, shape, and mass characteristics). However, the DP does not usually exceed 300 units whereas the minimum limit can be around 50 monomers, which are much lower than cellulose.

In addition, most sugar components in the hemicellulose can take part in the formation of lignin–carbohydrate complexes (LCC) by covalent linkages between lignin and carbohydrates [31, 32]. The most frequently suggested LCC-linkages in native wood are benzyl ester, benzyl ether, and glycosidic linkages [33]. The benzyl ester linkage is alkali-labile and may, therefore, be hydrolyzed during the alkaline pretreatment. The latter two linkages are alkali-stable and would survive from the hydrolysis during alkaline pretreatment.

Table 8.2 Major hemicellulose component in softwood and hardwood [29, 30]

Wood	Hemicellulose type	Amount (% on wood)	Composition			
			Units	Molar ratio	Linkage	DP
SW	Galactoglucomannans	10–15	β-D-Man*p*	4	1→4	100
			β-D-Glc*p*	1	1→4	
			β-D-Gal*p*	0.1	1→6	
			Acetyl	1		
	Arabinoglucuronoxylan	7–10	β-D-Xyl*p*	10	1→4	100
			4-O-Me-α-D-Glc*p*A	2	1→2	
			β-L-Araf	1.3	1→3	
HW	Glucuronoxylan	15–30	β-D-Xyl*p*	10	1→4	200
			4-O-Me-α-D-Glc*p*A	1	1→2	
			Acetyl	7		
	Glucomannan	2–5	β-D-Man*p*	1–2	1→4	200
			β-D-Glc*p*	1	1→4	

8.2.2.3 Lignin

Of the three major biopolymers that constitute wood, lignin is distinctly different from the other macromolecular polymers [34]. Lignin is an amorphous, cross-linked, and three-dimensional polyphenolic polymer that is synthesized by enzymatic dehydrogenative polymerization of 4-hydroxyphenyl propanoid units [35, 36]. The biosynthesis of lignin stems from the polymerization of three types of phenylpropane units as monolignols: coniferyl, sinapyl, and *p*-coumaryl alcohol [37, 38]. Figure 8.4 depicts these three structures. It has been identified that lignin from softwood is made up of more than 90 % of coniferyl alcohol with the remaining being mainly *p*-coumaryl alcohol units. Contrary to SW, lignin contained in hardwood is made up of varying ratios of coniferyl, sinapyl, and typically lesser amounts of *p*-coumaryl alcohol type of units.

The polymerization process is initiated by an enzyme-catalyzed oxidation of the monolignol phenolic hydroxyl groups to yield free radicals. A monolignol free radical can then couple with another monolignol to generate a dilignol. Subsequent nucleophilic attack by water, alcohols, or phenolic hydroxyl groups on the benzyl carbon of the quinone methide intermediate, restores the aromaticity of the benzene ring. The generated dilignols then undergo further polymerization to form protolignin.

Although the exact structure of protolignin is unknown, improvements in methods for identifying lignin-degradation products and advancements in spectroscopic methods have enabled scientists to elucidate the predominant structural features of lignin. Table 8.3 showed the typical abundance of common linkages and functional groups found in softwood lignin [39, 40].

The property of polydispersity, just as with hemicellulose, characterizes lignin as well. The DP for softwood lignin is approximately 60–100 and the molecular weight is in excess of 10,000 [41, 42].

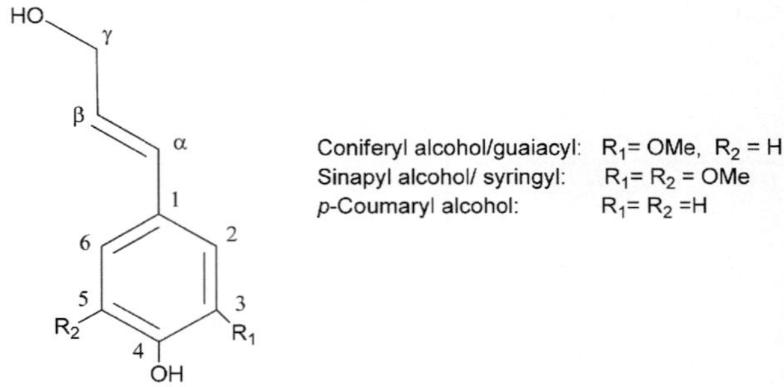

Fig. 8.4 Three building blocks of lignin. (Reproduced from Ref. [28] by permission of Wiley)

Table 8.3 Proportions of different types of linkages connecting the phenylpropane units in softwood lignin. (Reproduced from Ref. [28] by permission of Wiley)

Linkage type	Dimer structure	~Percentage
β-O-4	Phenylpropane β-aryl ether	50
β-5	Phenylcoumaran	9–12
5-5	Biphenyl	15–25
5-5/α-O-4	Dibenzodioxicin	10–15
4-O-5	Diaryl ether	4
β-1	1,2-Diaryl propane	7
β-β	β-β-linked structures	2

Lignin in wood behaves as an insoluble three-dimensional network. It plays an important role in the cell's endurance and development, as it affects the transport of water, nutrients, and metabolites in the plant cell. It acts as binder between cells creating a composite material that has a remarkable resistance to impact, compression, and bending [26].

Lignin is much less hydrophilic than either cellulose or hemicelluloses, and it has a general effect of inhibiting water adsorption and fiber swelling. Solvents that have been identified to significantly dissolve lignin include low molecular alcohols, dioxane, acetone, pyridine, dimethyl sulfoxide and select ionic liquids. Furthermore, it has been observed that at elevated temperatures, thermal softening of lignin takes place, which allows depolymerization reactions of acidic or alkaline nature to accelerate [43].

8.3 Chemical Pretreatment of Lignocellulosic Biomass

8.3.1 Dilute Acid Pretreatment (DAP)

Among the numerous pretreatment techniques, dilute acid pretreatment (DAP) has been shown as a leading pretreatment process that is currently under commercial development. DAP can significantly reduce lignocellulosic recalcitrance by disrupting the composite material linkage, such as the covalent bonds, hydrogen bonds, and van der Waals forces [44]. The most widely used and tested approaches in DAP are based on dilute sulfuric acid (H_2SO_4) since it is inexpensive and effective [45, 46].

However, nitric acid [47], hydrochloric acid (HCl) [48], and phosphoric acid [49] have also been examined. In addition, it was shown that sulfur dioxide (SO_2) was also an efficient acid catalyst in the DAP, especially for softwood [50–52]. However, there are certain drawbacks with such an approach. It is difficult to handle SO_2 (gas) at large scales, as safety issues may constitute a concern, and it is also a more expensive option as compared to similar alternatives such as using H_2SO_4.

8.3.1.1 Process Description

DAP is usually performed over a temperature range of 120–210 °C, with acid concentration typically less than 4 wt%, and residence time from several minutes to an hour [53]. In the DAP pretreatment, the combined severity (CS) is used for an easy comparison of pretreatment conditions and for facilitation of process control, which relates to the experimental effects of temperature, residence time, and acid concentration [54]. Lower CS is beneficial for the hemicellulose to hydrolyze to oligomers and monomers while higher CS could bring these monomers to furfurals, which are inhibitors for the subsequent enzymatic hydrolysis [55]. In order to maximize the efficiency of pretreatments, several studies have proposed a two-step procedure for DAP of SW [45, 56]. The conditions in the first step are less severe and serve to hydrolyze the hemicelluloses resulting in a high recovery of hemicellulose-derived fermentable sugars in the pretreatment effluent. By separating the solid and liquid phases after the first step, it is possible to minimize the degradation of hemicellulosic sugars to furfural and hydroxymethylfurfural (HMF). The solid material recovered from the first step is treated again under more severe conditions which promotes the enzymatic digestibility of cellulose fibers.

The DAP offers good performance in terms of recovering hemicellulose sugars but there are also some drawbacks. The dilute acid applied in the process could cause corrosion that mandates expensive materials of construction, such as hastelloy steel and ceramic valves. The neutralization of acid before the fermentation results in the formation of solid waste. In addition, the hemicellulose sugars might be further degraded to furfural and HMF, which are strong inhibitors to microbial fermentation [57]. Furthermore, most of the reported work used materials with significant size reduction, which consumes additional energy. Previous report indicated that grinding the materials to 1 mm accounted for 33 % of the power requirement of the entire process [58]. However, this is not practical in large-scale production. In addition, the detoxification step is required in DAP when running high solids pretreatment, which adds additional cost to the process.

8.3.1.2 Mode of Action

The main reaction that occurs during acid pretreatment is the hydrolysis of hemicellulose. Hemicellulose mainly xylan is hydrolyzed to fermentable sugars during DAP [59]. Solubilized hemicelluloses (oligomers) can be subjected to hydrolytic reactions producing monomers, furfural, HMF, and other (volatile) products in acidic environments [60, 61]. Recently, Sannigrahi et al. [62] have demonstrated that pseudo-lignin can be generated solely from carbohydrates without significant contribution from lignin during DAP especially under high severity pretreatment conditions. Further analysis indicates that pseudo-lignin is in spherical droplet form and has carbonyl, aromatic, methoxy, and aliphatic structures.

During DAP, it is generally accepted that the majority of the hemicellulose are removed initially, followed by the hydrolyzation of cellulose and subsequently some solubilization of Glu through the course of DAP [63–65]. Foston et al. [65] stated that cellulose degradation pathway can be viewed as acid catalyzed, thermally accelerated polysaccharide hydrolysis by chain scission within the fibril structure from either a crystalline or amorphous region of cellulose. This process consists of two major stages: The initial stage was regarded as rapid hydrolytic attack on the amorphous chain segments while the latter stage takes place on the crystal surfaces [66, 67]. Sannigrahi et al. [68] observed an increase in the relative proportion of cellulose I_β accompanied by a decrease in the relative proportion of both cellulose I_α and para-crystalline region from dilute acid pretreated Loblolly pine. This suggested that the types of lignocellulosic materials and pretreatment conditions influence cellulose crystalline allomorphs and para-crystalline contents during DAP.

DAP does not lead to significant delignification. Recent studies revealed an increase in the degree of condensation of lignin, during the DAP. The increase in degree of condensation is accompanied by a decrease in β-O-4 linkages which are fragmented and subsequently recondensed during the high-temperature acid-catalyzed reactions [68, 69]. In addition, studies also indicated that lignin balls (or lignin droplets) were formed during DAP. These lignin droplets originated from lignins and possible lignin carbohydrates complexes [70, 71].

8.3.1.3 Dilute Acid Pretreatment of Softwood

SW are generally considered as being much more refractory than (HW) or agricultural residues. This is due to the fact that SW have a more rigid structure and contain more lignin. However, various conditions for SW DAPs have been investigated (Table 8.4), which were performed using H_2SO_4 or SO_2.

The effect of the pretreatment is usually evaluated by the cellulose conversion yield during subsequent enzymatic hydrolysis process. Cellulose conversion yield is defined as the ratio of sugars liberated in the enzymatic hydrolysis to the theoretical value based on the sugars available in the raw material [77]. Some recent results on cellulose conversion yields from softwood are also shown in Table 8.4. It can be seen that the cellulose conversion yields of DAP treated SW are less than 65 %, which are generally lower than hardwood species [5]. This is, in part, due to the fact that SW have a more rigid structure and contains more lignin [12]. It should be noted that the addition of surfactant (i.e., Tween 80) in the post-DAP-treated substrate could enhance the cellulose conversion by 30 % [76]. During the hydrolysis, the surfactant

Table 8.4 DAP investigations using various softwoods as raw material

Wood species	Acid catalyst	Temperature (°C)	Time	Cellulose conversion yield (%)
Lodgepole pine [72]	4 wt% SO_2	200 °C	5 min	~60
White pine [73]	1.23 wt% H_2SO_4	220 °C	5 min	~65
Lodgepole pine [74]	4 wt% SO_2	200 °C	5 min	~65
Loblloy pine [68]	0.5–1.0 wt% H_2SO_4	180–200 °C	2–10 min	-
Radiata pine [5]	0.5–12 wt% SO_2	215 °C	3 min	57–60
Lodgepole pine [75]	4 wt% SO_2	200 °C	5 min	~63
Loblloly pine [76]	1 wt% H_2SO_4	180 °C	30 min	~52 with Tween
Loblloly pine [76]	5 wt% SO_2	180 °C	30 min	~63 with Tween

was added simultaneously with the enzyme. The surfactant concentration ranged 1–3 g/L [78]. This was attributed to the fact that the surfactant could change the nature of the substrate by increasing the available cellulose surface or by removing inhibitory lignin [79]. The surfactant could also increase the stability of the enzymes and reduce enzyme denaturation during the hydrolysis [80, 81]. Moreover, the surfactant could facilitate desorption of enzymes from substrate [82]. It should be noted, as indicated in Table 8.4, the cellulose conversion yields (52–63 %) is still low even with the addition of Tween 80. However, this research at least afforded a way to enhance the cellulose conversion yield through the addition of surfactant. Further study might be needed in the selection of effective surfactant.

8.3.2 Alkaline Pretreatment

Alkaline pretreatment is one of major chemical pretreatment technologies receiving numerous studies. It employs various bases, including sodium hydroxide (NaOH) [83], calcium hydroxide (lime) [84], potassium hydroxide (KOH) [85], aqueous ammonia [86], ammonia hydroxide [87], and NaOH in combination with hydrogen peroxide or others [88–90]. Among these alkaline pretreatments, lime has received much more attentions since it is inexpensive (about 6 % cost of NaOH), has improved handling, and can be recovered easily by using carbonated wash water [91].

8.3.2.1 Process Description

In comparison with other pretreatment technologies, alkali pretreatment usually uses lower temperatures and pressures and even ambient conditions. Pretreatment time, however, is recorded in terms of hours or days which are much longer than other pretreatment processes. In the alkaline pretreatment, the residual alkali could be reused through the chemical recycle/recovery process, which may make the system more complex due to the need for chemical recovery [92, 93]. The particle size of the biomass is typically 10 mm or less [57]. A significant disadvantage of alkaline pretreatment is the conversion of alkali into irrecoverable salts and/or the incorporation of salts into the biomass during the pretreatment reactions so that the treatment of a large amount of salts becomes a challenging issue for alkaline pretreatment [92]. The effectiveness of alkaline pretreatment varies, depending on the substrate and treatment conditions. In general, alkaline pretreatment is more effective on hardwood, herbaceous crops, and agricultural residues with low lignin content than on softwood with high lignin content [94]. In addition, in comparison with KOH and lime, pretreatment with NaOH was found to be more efficient for the subsequent enzymatic hydrolysis [92].

8.3.2.2 Mode of Action

Alkaline pretreatment is basically a delignification process, in which a significant amount of hemicellulose is solubilized as well. The major effect is the removal of lignin from the biomass, thus improving the reactivity of the remaining polysaccharides. In addition, the alkaline pretreatment could swell cell wall and improve cell wall accessibility for the subsequent enzymatic hydrolysis. The action mechanism is believed to be saponification of intermolecular ester bonds crosslink hemicellulose and lignin [92]. The presence of these LCC linkages is believed to prevent selective solubilization and removal of the wood components such as hemicelluloses and lignin in biorefining processes [85, 95]. Therefore saponification leading to the cleavage of these linkages and the expose of cellulose microfibrils can increase enzymatic digestibility of cellulose. Acetyl groups and various uronic acid substitutes are also removed by alkali, thereby increasing the accessibility of hemicellulose and cellulose to enzymes [96]. He et al. [97] recently characterized hemicelluloses from untreated and dilute NaOH-treated rice straws by FTIR spectroscopy. The results revealed that the dilute NaOH pretreatment did not change hemicellulose structure significantly, but it altered certain functional groups and linkages. For instance, the decrease in the hydroxyl stretching and C–OH banding peaks representing hemicellulose hydroxyl groups, as well as the reduction in the carbonyl stretching region attributed to hemicellulose acetyl and uronic ester groups were observed by different researchers [64, 97]. In addition, a decrease in the contents of β-glycosidic linkages between hemicellulose sugar units was reported in the literature [97]. Furthermore, the degraded hemicellulose could also form furfural and HMF in the hydrolysates, but the amount is much lower than that with DAP [98]. In addition, alkaline pretreatment decreases the DP of cellulose and causes swelling of cellulose, leading to an increase in its internal surface area [99]. This makes cellulose more accessible for enzymes in the subsequent hydrolysis stage. In terms of cellulose crystallinity change during the alkaline pretreatment, research indicated that the amorphous regions suffered greater peeling reactions than the crystalline regions, and the occurrence of the peeling actions of the amorphous regions leads to an increase of cellulose crystallinity [100]. During the alkaline pretreatment, lignin suffered delignification, which is rather similar to chemical pulping technologies [39, 57].

8.3.2.3 Alkaline Pretreatment of Softwood

Similar to DAP, alkaline treatment has been less effective on softwood than for hardwood, herbaceous plants or agricultural residues at the same process conditions because of the generally higher lignin content of wood. Zhu et al. [101] reported that a cold NaOH pretreatment could achieve about 70 % enzymatic hydrolysis Glu yield from spruce when pretreatment was conducted at -15 °C in a 7 % (w/v) NaOH solution with 12 % (w/v) urea for 24 h. However, Mirahmadi et al. [102] obtained only 35.7 % cellulose conversion yield when treated spruce with 7.0 % (w/w) NaOH for 2 h at 5 °C. In addition, research revealed that the addition of air/oxygen to the reaction mixture could enhance the cellulose conversion yield and improve the delignification of the biomass, especially highly lignified materials [85].

8.3.3 Wet Oxidation Pretreatment

Wet oxidation is an oxidative pretreatment method that employs oxygen or air as catalyst. It allows reactor operation at relatively low temperatures and short reactor times [103]. It has been proven to be an efficient method for solubilization of hemicelluloses and lignin and to increase digestibility of cellulose, specially.

8.3.3.1 Process Description

Typically, the procedure for wet oxidation consists of drying and milling lignocellulosic biomass to obtain particles that are 2 mm in length, to which water is added at a ratio of 1 L to 6 g biomass. A compound, usually Na_2CO_3, is introduced to the mixture to reduce the formation of by-products. Na_2CO_3 addition has been shown to decrease formation of inhibitory compounds by maintaining pH in the neutral to alkaline range. Air is pumped into the vessel until a pressure of 10–12 bar is reached. This method of pretreatment is performed at 170–200 °C for a range of 10–20 min [104, 105]. The addition of air/oxygen at temperatures above 170 °C makes the process exothermic reducing the total energy demand. In general, low formation of inhibitors and efficient removal of lignin are achieved with wet oxidation pretreatment. On the other hand, cost of oxygen and catalyst are considered one of the main disadvantages for wet oxidation development technologies [2].

8.3.3.2 Mode of Action

Wet oxidation can be used to fractionate lignocellulosic material by solubilizing hemicellulose and removing lignin [106, 107]. During wet oxidation, lignin is oxidized to carbon dioxide, water, and carboxylic acids [40, 43]. The amount of lignin removed after pretreatment ranges from 50 to 70 % depending on type of biomass pretreated and the conditions used [108]. The by-product formed in the oxidation, including succinic acid, glycolic acid, formic acid, acetic acid, phenolic compounds, and furfural, were much lower than the DAP [94]. In addition, the crystalline structure of cellulose is opened during the wet oxidation pretreatment, facilitating the enzymatic hydrolysis on the downstream process [94].

8.3.3.3 Wet Oxidation Pretreatment of Softwood

Although wet oxidation pretreatment is considered a promising technology for converting biomass into biofuels, it was rarely applied on softwood species. Palonen et al. [103] reported a 79 % cellulose conversion yield obtained from wet oxidation pretreatment of spruce. This pretreatment was performed at 200 °C for 10 min. This cellulose conversion yield was much higher than DAP and alkaline pretreatment of similar softwood species.

8.3.4 Sulfite Pretreatment to Overcome Recalcitrance of Lignocellulose (SPORL)

Recently Zhu et al. developed SPORL pretreatment for robust and efficient conversion of biomass through enzymatic saccharification [109]. During the SPORL pretreatment, the wood chips were pretreated in an aqueous sulfite solution followed by mechanical size reduction using disk refining. The terms sulfite and bisulfite are used interchangeably in the SPORL because the active reagents in the pretreatment liquor can be sulfite (SO_3^{2-}), bisulfite HSO_3^-, or a combination of two of the three reagents, sulfite (SO_3^{2-}), bisulfite HSO_3^-, and sulfur dioxide (SO_2, or H_2SO_3), depending on the pH of the pretreatment liquor at a pretreatment temperature [110]. The pretreatment liquor can be prepared and recovered using existing industrial practices as described elsewhere [111]. The pH of the solution can be easily controlled by the amount of SO_2 absorbed. SO_2 can be substituted by other acids, such as H_2SO_4, HCl, oxalic acid, and acetic acid (such as the acetic acid released from acetyl groups during pretreatment of hardwood or agricultural residues).

8.3.4.1 Process Description

The development of the SPORL process is based on the fundamental understandings of sulfite pulping [109]. Usually the SPORL pretreats the woodchips in an aqueous sulfite solution at 160–180 °C and pH 2–4 for about 30 min. The woodchips are then fiberized (size-reduced) using a disk mill to generate fibrous substrate for subsequent saccharification and fermentation. With low pretreatment cost, excellent substrate digestibility, along with sulfite pulping and chemical recovery, and disk refining technologies that have long been practiced in the pulp and paper industry, and existing industry infrastructure and commercial markets for high-value co-products from pretreatment-dissolved hemicellulose sugars and lignin, SPORL has low environmental and technological barriers and risks [112].

8.3.4.2 Mode of Action

Since the SPORL process is based on the sulfite pulping, this pretreatment chemistry is also similar to sulfite pulping. The major chemistry related to hemicellulose, cellulose, and lignin can be summarized as follows:

- A considerable amount of hemicellulose degradation and removal takes place during the pretreatment, as evidenced by the predominant Xyl content in pretreated effluent [113].
- The degrees of polymerization of xylan [114–117] and cellulose [118] are reduced.
- Sulfonation of lignin increases the hydrophilicity of lignin, which may promote the aqueous enzyme process.
- The degrees of dissolution of hemicellulose, degradation of cellulose, and sulfonation and condensation of lignin are increased as reaction time and temperature increases, and pH decreases [111, 119].

It should be noted that the production of fermentation inhibitors HMF and furfural in the SPORL is significantly lower than those in dilute acid, which is favorable to the fermentation of pretreatment-dissolved sugars from cellulose and hemicellulose. Excellent performance of the SPORL with different wood species indicates that this process may be tree species independent [109].

8.3.4.3 SPORL Applied on Softwood

Unlike DAP, dilute alkaline pretreatment, and wet oxidation pretreatment, SPORL was proved to be efficient for softwood species. Zhu et al. [109] investigated the combination of a sulfite treatment with mechanical size reduction by disk refining to enhance enzymatic hydrolysis of SW. This study was the first to establish this novel pretreatment process. Pretreatment conditions of spruce chips (20 %, w/v) that produced optimal cellulose conversion during enzymatic hydrolysis (>90 %) were treatment with 8–10 wt% bisulfite and 1.8–3.7 wt% H_2SO_4 for 30 min at 180 °C. Nearly all hemicellulose was removed, which exposed the underlying cellulose fraction to enzymatic attack. Additionally, furfural and HMF were produced in minimal concentrations, about 1 and 5 mg/g untreated wood, respectively. In addition, similar results were also observed with Lodgepole pine and red pine [109, 120].

8.3.5 Organosolv Pretreatment

Organosolv pretreatment is a promising pretreatment strategy, since it has demonstrated its potential for lignocellulosic materials [121]. Numerous organic or aqueous solvent mixtures can be utilized, including methanol, ethanol, acetone, ethylene glycol, and tetrahydrofurfuryl alcohol, in order to solubilize lignin and provide treated cellulose suitable for enzymatic hydrolysis [122]. Comparing to other chemical pretreatments, the main advantage of organosolv process is the recovery of relatively pure lignin as a by-product [122], which can be used as a substitute for polymeric materials, such as phenolic powder resins, polyurethane foams, and epoxy resins [123]. In some studies, these mixtures are combined with acid catalysts (HCl, H_2SO_4, oxalic, or salicylic) to break hemicellulose bonds. A high yield of Xyl can usually be obtained with the addition of acid. However, this acid addition can be avoided for a satisfactory delignification by increasing process temperature (above 185 °C) [124]. Usually in the organosolv pretreatment, high lignin removal (>70 %) and minimum cellulose loss (less than 2 %) could be achieved [121].

8.3.5.1 Process Description

Although several organic solvents can be applied in the organosolv pretreatments, the low-molecular weight alcohols with lower boiling points such as ethanol and methanol are favored solvent mainly due their low prices. The preferred conditions of organosolv process depend on the nature of the feedstock being processed, but will generally be in the following ranges: a cooking temperature of 180–195 °C, a cooking time of 30–90 min, an ethanol concentration of 35–70 % (w/v), and a liquor-to-solid ratio ranging from 4:1 to 10:1. The pH of the liquor might range from 2.0 to 3.8.

Compared with other pretreatments, organosolv pretreatment has some advantages as follows: (1) Organic solvents are always easy to recover by distillation and recycled for pretreatment; (2) the chemical recovery in organosolv pulping processes can isolate lignin as a solid material and carbohydrates as a syrup, both of which show promise as chemical feedstocks [125–127]. It seems that organosolv pretreatment is feasible for biorefinery of lignocellulosic biomass, which considers the utilization of all the biomass components. However, there are inherent drawbacks to the organosolv pretreatment. Organic solvents are always expensive, so it should be recovered as much as possible, but this causes increase of energy consumption. In addition, organosolv pretreatment must be performed under extremely tight and efficient control due to the volatility of organic solvents. No digester leaks can be tolerated because of inherent fire and explosion hazard [127]. This could also increase the capital cost. Moreover, removal of solvents from the system is necessary using appropriate extraction and separation techniques, for example, evaporation and condensation, and they should be recycled to reduce operational costs. Solvents need to be separated because they might be inhibitory to enzymatic hydrolysis and fermentative microorganisms [3]. The pretreated solids always need to be washed with organic solvent previous to water washing in order to avoid the reprecipitation of dissolved lignin, which leads to cumbersome washing arrangements.

Table 8.5 Organosolv pretreatment using various softwoods

Wood species	Solvent and catalyst	Temperature (°C)	Time (min)	Cellulose conversion yield (%)
Lodgepole pine [132]	1.1 wt% H_2SO_4, 65 % ethanol (v/v)	170	60	93–97
Radiata pine [133]	0.9 wt% H_2SO_4, 50 % acetone (v/v)	195	5	~99
Lodgepole pine [134]	1.1 wt% H_2SO_4, 65 % butanol (v/v)	170	60	~95
Loblolly pine [128]	1.0 wt% H_2SO_4, 65 % ethanol (v/v)	170	60	~70
Pitch pine [135]	1.0 wt% H_2SO_4, 50 % ethanol (v/v)	150–180	20	~95
Douglas fir [136]	1.0 wt% H_2SO_4, 50 % ethanol (v/v)	181–202	15–40	~80

8.3.5.2 Mode of Action

During the organosolv pretreatment, the largest component, cellulose, is partially hydrolyzed into smaller fragments that still remain insoluble in the liquor. Recently, Sannigrahi et al. [128] revealed that the degree of cellulose crystallinity increases and the relative proportion of para-crystalline and amorphous cellulose decreases after the organosolv pretreatment of Loblolly pine. The second largest component, hemicellulose, is hydrolyzed mostly into soluble components, such as oligosaccharides, monosaccharides, and acetic acid. Acetic acid lowers the liquor pH, stimulating acid-catalyzed hydrolysis of the other components. Some of the pentose sugars are subsequently dehydrated under the operating conditions to form furfural [129]. The third major polymer component, lignin, is hydrolyzed under the conditions employed in the process primarily into lower molecular weight fragments that dissolve in the aqueous ethanol liquor. In addition, studies [130] on the depolymerization of the lignin in macromolecule occurs primarily through cleavage of β-O-4 linkages which significantly influences delignification of SW. Moreover, lignin condensation was reported much lower when compared with DAP [131], owing in part to the countering effect of organic solvents that retain the lignin components in solution and slow recombination of macromolecules.

8.3.5.3 Organosolv Pretreatment of Softwood

Generally, the organosolv pretreatment was efficient on the bioconversion of softwood. After the pretreatment, the cellulose conversion yield during the subsequent enzymatic hydrolysis could be as high as 99 %, which is much higher than other chemical pretreatements, namely DAP, alkaline, and wet oxidation pretreatments (Table 8.5).

8.3.6 Ionic Liquids (ILs) Pretreatment

Ionic liquids (ILs) has recently received extensive research attention on the cellulose dissolution [137–142]. Some ILs show promise as efficient and "green", novel cellulose solvents. They can dissolve large amounts of cellulose at considerable mild conditions, and feasibility of recovering nearly 100 % of the used ILs to their initial purity makes them attractive [143]. After the ILs pretreatment, the precipitated cellulose is washed thoroughly with water to remove the ILs. No negative effect of the residual ILs was reported on the subsequent cellulose hydrolysis and fermentation [44]. As cellulose solvents, several ILs possess several advantages over regular volatile organic solvents of biodegradability, low toxicity, broad selection of anion and cation combinations, low hydrophobicity, low viscosity, enhanced electrochemical stability, thermal stability, high reaction rates, low volatility with potentially minimal environmental impact, and non-flammable property.

The dissolution mechanism of cellulose in ILs involves the oxygen and hydrogen atoms of cellulose hydroxyl groups in the formation of electron donor–electron acceptor (DA) complexes which interact with the ILs [144]. Upon interaction of the cellulose-OH and ILs, the hydrogen bonds are broken, resulting in opening of the hydrogen bonds between molecular chains of the cellulose [144]. These interactions result in the dissolution of cellulose. Solubilized cellulose can be recovered by rapid precipitation with some anti-solvents such as water, ethanol, methanol, or acetone. The recovered cellulose was found to have the same DP and polydispersity as the initial cellulose, but significantly different macro- and micro-structure, especially the decreased degree of crystallinity [145]. The previously used ILs include 1-*n*-butyl-3-methylimidazolium chloride (BMIMCl) [146], 1-allyl-3-methylimidazolium chloride (AMIMCl) [147], 3-methyl-*N*-bytylpyridinium chloride (MBPCl), and benzyldimethyl (tetradecyl) ammonium chloride (BDTACl) [143]. It should be noted that the presence of water significantly hampers the dissolution efficiency of ILs. Thus, the water content in the wood chips should be decreased prior to the pretreatment [148]. In addition, an IL can be recovered after regeneration of cellulose with water or water/acetone mixture. The solvent added to the IL should be evaporated prior to its reuse in the next extraction cycle [148].

Application of ILs has opened new ways for the efficient utilization of lignocellulosic materials in such areas as biomass pretreatment and fractionation. However, there are still many challenges in putting these potential applications into practical use, for example, the high cost of ILs, regeneration requirement, lack of detailed toxicological data and knowledge about basic physico-chemical characteristics and action mode on hemicellulose and/or lignin contents of lignocellulosic materials, and inhibitor generation issues. Further research is required to address such challenges.

8.3.7 Ozonolysis

Ozone treatment is one way of reducing the lignin content of lignocellulosic wastes. This results in an increase of the *in vitro* digestibility of the treated material, and unlike other chemical treatments, it does not produce toxic residues. Ozone can be used to degrade lignin and hemicellulose in many lignocellulosic materials such as wheat straw [149], bagasse, green hay, peanut, pine [150], cotton straw [151], and poplar sawdust [152]. Research indicated [153] ozone is highly reactive toward compounds incorporating conjugated double bonds and functional groups with high electron densities. Therefore, the moiety, most likely to be oxidized in ozonization of lignocellulosic materials, is lignin due to its high content of C=C bonds. Thus, during the ozonolysis, the degradation is mainly limited to lignin. Ozone attacks lignin releasing soluble compounds of less molecular weight, mainly organic acids such as formic and acetic acid [153]. The main advantages linked to this process are the lack of any degradation products that might interfere with subsequent hydrolysis or fermentation and the reactions occurring at ambient temperature and normal pressure. Furthermore, the fact that ozone can be easily decomposed by using a catalytic bed or increasing the temperature means that processes can be designed to minimize environmental pollution. A drawback of ozonolysis is that a large amount of ozone is required, which can make the process expensive and less applicable [154]. However, recently Hu et al. [155] demonstrated that a lower charge of ozone could be used to enhance the enzymatic digestibility of cellulose, if the ozone-treated biomass was not washed and the in-situ generated acids were employed in a subsequent DAP.

8.4 Summary

The effects of different chemical pretreatment technologies on the structure of lignocellulose are summarized in this section. In addition, the environment impacts of these pretreatments are also briefly discussed. Some directions and perspectives are also proposed for the future chemical pretreatment technologies.

8.4.1 Pretreatment Effect on the Structure of Lignocellulose

Most of the chemical pretreatment technologies that have been described herein are effective on one or more factors that contribute to lignocellulosic recalcitrance, as shown in Table 8.6. Table 8.7 summarizes the main advantages and disadvantages of these pretreatment technologies. Each method discussed shows the ability to take the complex carbohydrate and depolymerize the substrate to a lower fraction for enzymatic saccharification in the subsequent step. There are a number of feasible routes, each of which has their own merits and disadvantages, and consequences on the enzymatic hydrolysis.

Table 8.6 Effect of different chemical pretreatment technologies on the structure of lignocellulose [2, 11, 44]

	Increases accessible surface area	Cellulose Decrystal-lization	Hemicellulose solubilization	Lignin removal	Generation of inhibitor compounds	Lignin structure alteration
DAP	H	–	H	L	H	H
Alkali	H	–	L	M	H	H
Wet oxidation	H	L	L	M	H	H
SPORL	H	L	H	H	L	H
Organosolv	H	L	H	H	H	H
ILs	H	H	H	H	L	L
Ozonolysis	H	L	L	H	L	H

H high-effect; *M* moderate-effect; low-effect; – no effect

8.4.2 Environmental Impact of Chemical Pretreatment Technologies

Some studies were conducted on the analysis of environmental impact of chemical pretreatment technologies. For instance, the life-cycle assessment (LCA) was used to evaluate the impact of chemical pretreatment technologies on the environment. LCA is a conceptual framework and methodology for the assessment of environmental impacts of product systems on a cradle-to-grave basis [158]. Analysis of a system under LCA encompasses the extraction of raw materials and energy resources from the environment, the conversion of these resources into the desired products, the utilization of the product by the consumer, and finally the disposal, reuse, or recycle of the product after its service life [159]. The LCA approach is an effective way to introduce environmental considerations in process and product design or selection. Based on LCA studies, the chemical pretreatment for bio-ethanol production technologies can be compared. Energy production and utilization cycles based on cellulosic biomass have near-zero greenhouse gas emissions on a life-cycle basis [160]. Biomass utilization into ethanol production offers environmental benefits in terms of nonrenewable energy consumption and global warming impact [161].

8.4.3 Future Directions and Perspectives

Most of the leading chemical pretreatment technologies that have been described herein are effective on one or more factors that contribute to lignocellulosics recalcitrance. Despite much research that has been dedicated to understanding the chemistry and the plant cell wall structure changes during various pretreatment technologies, the insufficient knowledge of cell wall structure, ultra structure, and pretreatment effects still limits the economics and effectiveness of pretreatment. For instance, the biological and chemical properties of plants are very complex in terms of composition, structure, and ultra-structure [162]. Although researchers have put significant

Table 8.7 Summary of various chemical pretreatments of lignocellulosic biomass [2, 129, 156, 157]

Pretreatment process	Advantages	Disadvantages
DAP	Hydrolyzes hemicellulose to xylose and other sugars; alters lignin structure	High cost; equipment corrosion; formation of toxic substances
Alkali	Removes hemicelluloses and lignin; increases accessible surface area	Long residence times required; irrecoverable salts formed and incorporated into biomass
Wet oxidation	Increase accessible surface area; removes lignin and hemicellulose to an extent	Expensive
SPORL	Slight degradation of cellulose, nearly complete solubilization of hemicellulose; partial delignification and lignin sulfonation	Possible need great capital investment
Organosolv	Hydrolyzes lignin and hemicelluloses	Solvents need to be drained from the reactor, evaporated, condensed, and recycled; high cost
ILs	Lignin and hemicellulose hydrolysis; ability to dissolve high loadings of different biomass types; mild processing conditions (low temperatures)	High solvent costs; need for solvent recovery and recycle
Ozonolysis	Reduces lignin content; does not produce toxic residues; increase accessible surface area; cost effective; does not cause formation of inhibitory compounds	Does not modify hemicelluloses; large amount of ozone required; expensive

effort into optimizing the pretreatment effectiveness, the fundamental science behind these optimizations is still not fully understood. Furthermore, there has been a lack of mechanistic understanding of the ultrastructural and physicochemical changes occurring within the cell wall at the molecular level and the cellular/tissue scale during various pretreatment technologies. It is thus essential to understand the effects of pretreatment on plant cell walls at a more fundamental level, in order to develop a cost-effective pretreatment technology with maximum fermentable sugar recovery, minimum inhibitor production and energy input, low demand of post-pretreatment processes, and low capital costs for reactors, water, and chemicals. In addition, advances in the analytical chemistry would provide useful tools to investigate the cell wall deconstruction and understand the recalcitrance during the pretreatment process [163, 164].

Acknowledgments The authors are grateful for the financial support from the US Department of Energy (DOE biorefinery project: DE-EE0003144) for these studies.

References

1. Karmakar A, Karmakar S, Mukherjee S (2010) Properties of various plants and animals feedstocks for biodiesel production. Bioresour Technol 101:7201–7210
2. Alvira P, Tomas-Pejo E, Ballesteros M, Negro MJ (2010) Pretreatment technologies for an efficient bioethanol production process based on enzymatic hydrolysis: a review. Bioresour Technol 101:4851–4861
3. David K, Ragauskas AJ (2010) Switchgrass as an energy crop for biofuel production: A review of its ligno-cellulosic chemical properties. Energy Environ Sci 3:1182–1190
4. Hamelinck CN, van Hooijdonk G, Faaij APC (2005) Ethanol from lignocellulosic biomass: techno-economic performance in short-, middle- and long-term. Biomass Bioenergy 28:384–410
5. Galbe M, Zacchi G (2007) Pretreatment of lignocellulosic materials for efficient bioethanol production. Adv Biochem Eng/Biotechnol 108:41–65
6. Fatih Demirbas M (2009) Biorefineries for biofuel upgrading: a critical review. Appl Energy 86:S151–S161
7. Ragauskas AJ, Williams CK, Davison BH, Britovsek G, Cairney J, Eckert CA, Frederick WJ Jr, Hallett JP, Leak DJ, Liotta CL, Mielenz JR, Murphy R, Templer R, Tschaplinski T (2006) The path forward for biofuels and biomaterials. Science 311:484–489
8. Dalgaard T, Jorgensen U, Olesen JE, Jensen ES, Kristensen ES (2006) Looking at biofuels and bioenergy. Science 312(5781):1743–1744
9. Leonowicz A, Matuszewska A, Luterek J, Ziegenhagen D, Wojtas-Wasilewska M, Cho NS, Hofrichter M, Rogalski J (1999) Biodegradation of lignin by white rot fungi. Fungal Genet Biol 27(2–3):175–185
10. U.S. Department of Energy's Office of Biological and Environmental Research. https://public.ornl.gov/site/gallery/detail.cfm?id=248&topic=53&citation=&general= &restsection=public. Accessed 26 March 2012
11. Brodeur G, Yau E, Badal K, Collier J, Ramachandran KB, Ramakrishnan S (2011) Chemical and physicochemical pretreatment of lignocellulosic biomass: a review. Enzyme Res 2011: 1–17
12. Galbe M, Zacchi G (2002) A review of the production of ethanol from softwood. Appl Microbiol Biotechnol 59:618–628
13. Brosse N, Dufour A, Meng X, Sun Q, Ragauskas A (2012) Miscanthus: a fast - growing crop for biofuels and chemicals production. Biofuels Bioprod Bioref 6:580–598
14. Ingram T, Wormeyer K, Lima JC, Bockemuhl V, Antranikian G, Brunner G, Smirnova I (2011) Comparison of different pretreatment methods for lignocellulosic materials. Part I: conversion of rye straw to valuable products. Bioresour Technol 102:5221–5228
15. Hallac BB, Ragauskas AJ (2011) Analyzing cellulose degree of polymerization and its relevancy to cellulosic ethanol. Biofuels Bioprod Biorefin 5(2):215–225
16. Faulon J, Carlson GA, Hatcher PG (1994) A three-dimensional model for lignocellulose from gymno-spermous wood. Org Geochem 21:1169–1179
17. Klemm D, Heublein B, Fink HP, Bohn A (2005) Cellulose: fascinating biopolymer and sustainable raw material. Angew Chem 44:3358–3393
18. Thygesen A, Oddershede J, Lilholt H, Thomsen AB, Stahl K (2005) On the determination of crystallinity and cellulose content in plant fibers. Cellulose 12(6):563–576
19. Newman RH (2004) Homogeneity in cellulose crystallinity between samples of Pinus radiata wood. Holzforschung 58(1):91–96
20. Atalla RH, VanderHart DL (1984) Native cellulose: a composite of two distinct crystalline forms. Science 223(4633):283–285
21. Horii F, Hirai A, Kitamaru R (1987) Transformation of native cellulose crystals from cellulose I(β) to I(α) through solid-state chemical reactions. Macromolecules 20(6):1440–1442
22. Sugiyama J, Persson J, Chanzy H (1991) Combined infrared and electron diffraction study of the polymorphism of native celluloses. Macromolecules 24(6):2461–2469

23. Larsson PT, Hult EL, Wickholm K, Pettersson E, Iversen T (1999) CP/MAS 13C-NMR spectroscopy applied to structure and interaction studies on cellulose I. Solid State Nucl Magn Reson 15(1):31–40
24. Stephens CH, Whitmore PM, Morris HR, Bier ME (2008) Hydrolysis of the amorphous cellulose in cotton-based paper. Biomacromolecules 9:1093–1099
25. Wagberg L, Annergren GO (1997) Physicochemical characterization of papermaking fibers. In: Baker CF (ed) The fundamentals of papermaking materials. Transaction of the 11th fundamental research symposium. Pira International, Cambridge, pp 1–82
26. Harmsen PFH, Huijgen WJJ, Bermúdez López LM, Bakker RRC (2010) Literature review of physical and chemical pretreatment processes for lignocellulosic biomass. Energy Research Center of the Netherlands, the Amsterdam
27. Krassig H, Schurz J (2002) Ullmann's encyclopedia of industrial chemistry, 6th edn. Wiley Weinheim
28. Pu Y, Zhang D, Singh PM, Ragauskas AJ (2008) The new forestry biofuels sector. Biofuels Bioprod Biorefin 2:58–73
29. Willför S, Sundberg A, Pranovich A, Holmbom B (2005) Polysaccharides in some industrially important hardwood species. Wood Sci Technol 39:601–617
30. Willför S, Sundberg A, Hemming J, Holmbom B (2005) Polysaccharides in some industrially important softwood species. Wood Sci Technol 39:245–257
31. Barakat A, Winter H, Rondeau-Mouro C, Saake B, Chabbert B, Cathala B (2007) Studies of xylan interactions and cross-linking to synthetic lignins formed by bulk and end-wise polymerization: a model study of lignin carbohydrate complex formation. Planta 226:267–281
32. Bunzel M, Ralph J, Lu F, Hatfield RD, Steinhart H (2004) Lignins and ferulate-coniferyl alcohol cross-coupling products in cereal grains. J Agric Food Chem 52:6496–6502
33. Lawoko M, Henriksson G, Gellerstedt G (2006) Characterization of lignincarbohydrate complexes (LCCs) of spruce wood (Picea abies L) isolated with two methods. Holzforschung 60:156–161
34. Brunow G, Lundquist K, Gellerstedt G (1999) Lignin: analytical methods in wood chemistry. In: Sjostrom E (ed) Pulping and papermaking. Springer, Berlin, pp 77–124
35. Davin LB, Lewis NG (2005) Lignin primary structures and dirigent sites. Curr Opin Biotechnol 16:407–415
36. Halpin C (2004) Redesigning lignin for industry and agriculture. Biotechnol Genet Eng Rev 21:229–245
37. Boerjan W, Ralph J, Baucher M (2003) Lignin biosynthesis. Annu Rev Plant Biol 54:519–546
38. Pu Y, Jiang N, Ragauskas AJ (2007) Ionic liquid as a green solvent for lignin. J Wood Chem Technol 27:23–33
39. Chakar FS, Ragauskas AJ (2004) Review of current and future softwood kraft lignin process chemistry. Ind Crops Prod 20:131–141
40. Huang F, Singh PM, Ragauskas AJ (2011) Characterization of milled wood lignin (MWL) in Loblolly pine stem wood, residue, and bark. J Agr Food Chem 59:12910–12916
41. Meister JJ (2002) Modification of lignin. J Macromol Sci Polym Rev C42(2):235–289
42. Guerra A, Xavier A, Hai A, Filpponen I, Lucia L, Argyropoulos D (2006) Different wood species offer different yields, lignin structures, and molecular weights when isolated with the same method. J Agr Food Chem 54:9696–9705
43. Pandey MP, Kim CS (2011) Lignin depolymerization and conversion: a review of thermochemical methods. Chem Eng Technol 34:29–41
44. Li C, Knierim B, Manisseri C, Arora R, Scheller HV, Auer M, Vogel KP, Simmons BA, Singh S (2010) Comparison of dilute acid and ionic liquid pretreatment of switchgrass: biomass recalcitrance, delignification and enzymatic saccharification. Bioresour Technol 101:4900–4906
45. Nguyen QA, Tucker MP, Keller FA, Eddy FP (2000) Two stage dilute-acid pretreatment of softwoods. Appl Biochem Biotechnol 84–86:561–576
46. Kim JS, Lee YY, Park SC (2000) Pretreatment of wastepaper and pulp mill sludge by aqueous ammonia and hydrogen peroxide. Appl Biochem Biotechnol 84/86:129–139

47. Brink DL (1994) Method of treating biomass material. US Patent 5,366,558
48. Goldstein IS, Easter JM (1992) An improved process for converting cellulose to ethanol. TAPPI J 75(8):135–140
49. Israilides CJ, Grant GA, Han YW (1978) Sugar level, fermentability, and acceptability of straw treated with different acids. Appl Environ Microbiol 36(1):43–46
50. Monavari S, Galbe M, Zacchi G (2009) Impact of impregnation time and chip size on sugar yield in pretreatment of softwood for ethanol production. Bioresour Technol 100:6312–6316
51. Söderström J, Pilcher L, Galbe M, Zacchi G (2002) Two-step steam pretreatment of softwood with SO_2 impregnation for ethanol production. Appl Biochem Biotechnol 98–100:5–21
52. Tengborg C, Stenberg K, Galbe M, Zacchi G, Larsson S, Palmqvist E, Hahn-Hägerdal B (1998) Comparison of SO_2 and H_2SO_4 impregnation of softwood prior to steam pretreatment on ethanol production. Appl Biochem Biotechnol 70–72:3–15
53. Taherzadeh MJ, Karimi K (2007) Enzyme-based hydrolysis processes for ethanol from lignocellulosic materials: a review. Bioresources 2(4):707–738
54. Lloyd T, Wyman CE (2003) Application of a depolymerization model for predicting thermochemical hydrolysis of hemicellulose. Appl Biochem Biotechnol 105:53–67
55. Kabel MA, Bos G, Zeevalking J, Voragen AG, Schols HA (2007) Effect of pretreatment severity on xylan solubility and enzymatic breakdown of the remaining cellulose from wheat straw. Bioresour Technol 98:2034–2042
56. Soderstrom J, Galbe M, Zacchi G (2005) Separate versus simultaneous saccharification and fermentation of two-step steam pretreated softwood for ethanol production. J Wood Chem Technol 25:187–202
57. Mosier N, Wyman C, Dale B, Elander R, Lee YY, Holtzapple M, Ladisch M (2005) Features of promising technologies for pretreatment of lignocellulosic biomass. Bioresour Technol 96:673–686
58. Wooley R, Ruth M, Glassner D, Sheehan J (1999) Process design and costing of bioethanol technology: a tool for determining the status and direction of research and development. Biotechnol Prog 15:794–803
59. Hendriks AT, Zeeman G (2009) Pretreatments to enhance the digestibility of lignocellulosic biomass. Bioresour Technol 100:10–18
60. Fengel D, Wegener G (1984) Wood: chemistry, ultrastructure, reactions. De Gruyter, Berlin
61. Ramos LP (2003) The chemistry involved in the steam treatment of lignocellulosic materials. Quim Nova 26(6):863–871
62. Sannigrahi P, Kim DH, Jung SW, Ragauskas AJ (2011) Pseudo-lignin and pretreatment chemistry. Energy Environ Sci 4:1306–1310
63. Lloyd TA, Wyman CE (2005) Combined sugar yields for dilute sulfuric acid pretreatment of corn stover followed by enzymatic hydrolysis of the remaining solids. Bioresour Technol 96:1967–1977
64. Kumar R, Mago G, Balan V, Wyman CE (2009) Physical and chemical characterizations of corn stover and poplar solids resulting from leading pretreatment technologies. Bioresour Technol 100:3948–3962
65. Foston M, Ragauskas AJ (2010) Changes in lignocellulosic supramolecular and ultrastructure during dilute acid pretreatment of Populus and switchgrass. Biomass Bioenergy 34:1885–1895
66. Cao S, Pu Y, Studer M, Wyman CL, Ragauskas AJ (2012) Chemical transformations of Populus trichocarpa during dilute acid pretreatment. RSC Advances 2:10925–10936
67. Emsley AM, Heywood RJ (1997) On the kinetics of degradation of cellulose. Cellulose 4:1–5
68. Sannigrahi P, Ragauskas AJ, Miller SJ (2008) Effects of two-stage dilute acid pretreatment on the structure and composition of lignin and cellulose in loblolly pine. Bioenergy Res 1:205–214
69. Kobayashi T, Kohn B, Holmes L, Faulkner R, Davis M, Maciel GE (2011) Molecular-level consequences of biomass pretreatment by dilute sulfuric acid at various temperatures. Energy Fuels 25:1790–1797

70. Donohoe BS, Decker SR, Tucker MP, Himmel ME, Vinzant TB (2008) Visualizing lignin coalescence and migration through maize cell walls following thermochemical pretreatment. Biotechnol Bioeng 101:913–925
71. Selig MJ, Viamajala S, Decker SR, Tucker MP, Himmel ME, Vinzant TB (2007) Deposition of lignin droplets produced during dilute acid pretreatment of maize stems retards enzymatic hydrolysis of cellulose. Biotechnol Prog 23:1333–1339
72. Kumar L, Chandra R, Chung PA, Saddler J (2010) Can the same steam pretreatment conditions be used for most softwoods to achieve good, enzymatic hydrolysis and sugar yields? Bioresour Technol 101:7827–7833
73. Grethlein HE, Allen DC, Converse AO (1984) A comparative study of the enzymatic hydrolysis of acid-pretreated white pine and mixed hardwood. Biotechnol Bioeng 26:1498–1505
74. Ewanick SM, Bura R, Saddler JN (2007) Acid-catalyzed steam pretreatment of lodgepole pine and subsequent enzymatic hydrolysis and fermentation to ethanol. Biotechnol Bioeng 98(4):737–746
75. Tu M, Chandra RP, Saddler JN (2007) Recycling cellulases during the hydrolysis of steam exploded and ethanol pretreated lodgepole pine. Biotechnol Prog 23:1130–1137
76. Huang F, Ragauskas AJ (2012) Dilute H_2SO_4 and SO_2 pretreatments of loblolly pine wood residue for bioethanol production. Ind Biotechnol 8(1):22–30
77. Ouyang J, Dong Z, Song X, Lee X, Chen M, Yong Q (2010) Improved enzymatic hydrolysis of microcrystalline cellulose (Avicel PH101) by polyethylene glycol addition. Bioresour Technol 101:6685–6691
78. Alkasrawi M, Eriksson T, Börjesson J, Wingren A, Galbe M, Tjerneld F, Zacchi G (2003) The effect of Tween-20 on simultaneous saccharification and fermentation of softwood to ethanol. Enzyme Microb Technol 33:71–78
79. Helle SS, Duff SJB, Cooper DG (1993) Effect of surfactants on cellulose hydrolysis. Biotechnol Bioeng 42:611–617
80. Kaar WE, Holtzapple M (1998) Benefits from tween during enzymatic hydrolysis of corn stover. Biotechnol Bioeng 59:419–427
81. Kim MH, Lee SB, Ryu DDY(1982) Surface deactivation of cellulase and its prevention. Enzyme Microb Technol 4:99–103
82. Park JW, Takahata Y, Kajiuchi T, Akehata T(1992) Effects of nonionic surfactant on enzymatic-hydrolysis of used newspaper. Biotechnol Bioeng 39:117–120
83. Carrillo F, Lis MJ, Colom X (2005) Effect of alkali pretreatment on cellulase hydrolyiss of wheat straw: kinetic study. Process Biochem 40:3360–3364
84. Chang VS, Nagwani M, Kim CH (2001) Oxidative lime pretreatment of high-lignin biomass—poplar wood and newspaper. Appl Biochem Biotechnol 94:1–28
85. Chang VS, Holtzapple MT (2000) Fundamental factors affecting biomass enzymatic reactivity. Appl Biochem Biotechnol 84/86:5–37
86. Kim TH, Kim JS, Sunwoo C (2003) Pretreatment of corn stover by aqueous ammonia. Bioresour Technol 90:39–47
87. Prior BA, Day DF (2008) Hydrolysis of ammonia-pretreated sugar cane bagasse with cellulase, beta-glucosidase, and hemicellulase preparations. Appl Biochem Biotechnol 146:151–164
88. Mishima D, Tateda M, Ike M (2006) Comparative study on chemicals pretreatments to accelerate enzymatic hydrolysis of aquatic macrophyte biomass used in water purification processes. Bioresour Technol 97:2166–2172
89. Saha BC, Cotta MA (2006) Ethanol production from alkaline peroxide pretreated enzymatically accharified wheat straw. Biotechnol Prog 22:449–453
90. Saha BC, Cotta MA (2007) Enzymatic saccharification and fermentation of alkaline peroxide retreated rice hulls to ethanol. Enzyme Microb Technol 41:528–532
91. Peters MS, Timmerhaus KD (1991) Plant design and economics for chemical engineers, 4th edn. McGraw-Hill, New York
92. Zheng Y, Pan Z, Zhang R (2009) Overview of biomass pretreatment for cellulosic ethanol production. Int J Agr Biol Eng 2:51–68

93. Pavlostathis SG, Gossett JM (1985) Alkaline treatment of wheat straw for increasing anaerobic biodegradability. Biotechnol Bioeng 27:334–344
94. Kumar R, Wyman CE (2009) Effects of cellulase and xylanase enzymes on the deconstruction of solids from pretreatment of poplar by leading technologies. Biotechnol Prog 25:302–314
95. Balakshin M, Capanema E, Gracz H, Chang HM, Jameel H (2011) Quantification of lignin-carbohydrate linkages with high-resolution NMR spectroscopy. Planta 233:1097–1110
96. Wan CX, Zhou YG, Li YB (2011) Liquid hot water and alkaline pretreatment of soybean straw for improving cellulose digestibility. Bioresour Technol 102:6254–6259
97. He YF, Pang YZ, Liu YP, Li XJ, Wang KS (2008) Physicochemical characterization of rice straw pretreated with sodium hydroxide in the solid state for enhancing biogas production. Energy Fuels 22:2775–2781
98. Taherzadeh MJ, Karimi K (2008) Pretreatment of Lignocellulosic wastes to improve ethanol and biogas production: a review. Int J Mol Sci 9:1621–1651
99. Fan LT, Gharpuray MM, Lee YH (1987) Cellulose hydrolysis biotechnology monographs. Springer, Berlin
100. Wu L, Arakane M, Ike M, Wada M, Takai T, Gau M, Tokuyasu K (2011) Low temperature alkali pretreatment for improving enzymatic digestibility of sweet sorghum bagasse for ethanol production. Bioresour Technol 102:4793–4799
101. Zhu JY, Pan XJ (2010) Woody biomass pretreatment for cellulosic ethanol production: technology and energy consumption evaluation. Bioresour Technol 101:4992–5002
102. Mirahmadi K, Kabir MM, Jeihanipour A, Karimi K, Taherzadeh M (2010) Alkaline pretreatment of spruce and birch to improve bioethanol and biogas production. Bioresources 5(2):928–938
103. Palonen H, Thomsen AB, Tenkanen M, Schmidt AS, Viikari L (2004) Evaluation of wet oxidation pretreatment for enzymatic hydrolysis of softwood. Appl Biochem Biotechnol 117:1–17
104. Martín C, Marcet M, Thomsen AB (2008) Comparison between wet oxidation and steam explosion as pretreatment methods for enzymatic hydrolysis of sugarcane bagasse. Bioresources 3(3):670–683
105. Pedersen M, Meyer AS (2009) Influence of substrate particle size and wet oxidation on physical surface structures and enzymatic hydrolysis of wheat straw. Biotechnol Prog 25(2):399–408
106. Banerjee S, Sen R, Pandey RA, Chakrabarti T, Satpute D, Giri BS, Mudliar S (2009) Evaluation of wet air oxidation as a pretreatment strategy for bioethanol production from rice husk and process optimization. Biomass Bioenergy 33:1680–1686
107. Martín C, Klinke HB, Thomsen AB (2007) Wet oxidation as a pretreatment method for enhancing the enzymatic convertibility of sugarcane bagasse. Enzyme Microb Technol 40:426–432
108. Martín C, Marcet M, Thomsen AB (2008) Comparison between wet oxidation and steam explosion as pretreatment methods for enzymatic hydrolysis of sugarcane bagasse. Bioresources 3(3):670–683
109. Zhu JY, Pan XJ, Wang GS, Gleisner R (2009) Sulfite pretreatment (SPORL) for robust enzymatic saccharification of spruce and red pine. Bioresour Technol 100:2411–2418
110. Ingruber O (1985) Sulfite science part I: sulfite pulping cooking liquor and the four bases. In: Ingruber O, Kocurek M, Wong A (eds) Sulfite science and technology. The joint textbook committee of the paper industry, 3rd edn. TAPPI/CPPA, Atlanta, pp 3–23
111. Bryce JRG (1980) Sulfite pulping. In: Casey JP (ed) Pulp and paper: chemistry and chemical technology, 3rd edn. Wiley, New York, pp 291–376
112. Wang GS, Pan XJ, Zhu JY, Gleisner R, Rockwood D (2009) Sulfite pretreatment to overcome recalcitrance of lignocellulose (SPORL) for robust enzymatic saccharification of hardwoods. Biotechnol Prog 25(4):1086–1093
113. Janson J, Sjostrom E (1964) Behaviour of xylan during sulphite cooking of birchwood. Sven Papperstidn 67:764–771

114. Meier H (1962) On the behavior of wood hemicelluloses under different pulping conditions. Part I. Birch hemicellulose. Sven Papperstidn 65:299–305
115. Meier H (1962) On the behavior of wood hemicelluloses under different pulping conditions. Part II. Spruce hemicellulose. Sven Papperstidn 65:589–594
116. Pfister K, Sjostrom E (1977) The formation of monosaccharides and aldonic and uronic acids during sulphite cooking. Pap Puu 59:711–720
117. Sundman J (1950) Sockerutlosningen vid sulfitcellulosakoket. Pap Puu 32:267–274
118. Heuser E (1950) Trends in fundamental research in the cellulose and wood pulp field. TAPPI 33:118–124
119. Hall L, Stockman L (1958) Sulfitkokning vid olika aciditet. Sven Papperstidn 61:871–880
120. Zhu JY, Zhu W, Obryan P, Dien BS, Tian S, Gleisner R, Pan XJ (2010) Ethanol production from SPORL-pretreated lodgepole pine: preliminary evaluation of mass balance and process energy efficiency. Appl Microb Biotechnol 86:1355–1365
121. Papatheofanous MG, Billa E, Koullas DP, Monties B, Koukios EG (1995) Two stage acid-catalyzed fractionation of lignocellulosic biomass in aqueous ethanol systems at low temperatures. Bioresour Technol 54:305–310
122. Zhao X, Cheng K, Liu D (2009) Organosolv pretreatment of lignocellulosic biomass for enzymatic hydrolysis. Appl Microb Biotechnol 82:815–827
123. Zhang YH (2008) Reviving the carbohydrate economy via multi-product lignocellulose biorefineries. J Ind Microb Biotechnol 35:367–375
124. Duff SJB, Murray WD (1996) Bioconversion of forest products industry waste cellulosics to fuel ethanol: a review. Bioresour Technol 55:1–33
125. Lora JH, Aziz S (1985) Organosolv pulping: a versatile approach to wood refining. TAPPI J 68:94–97
126. Johansson A, Aaltonen O, Ylinen P (1987) Organosolv pulping method and pulp properties. Biomass 13:45–65
127. Aziz S, Sarkanen K (1989) Organosolv pulping—a review. TAPPI J 72:169–175
128. Sannigrahi P, Miller SJ, Ragauskas AJ (2010) Effects of organosolv pretreatment and enzymatic hydrolysis on cellulose structure and crystallinity in Loblolly pine. Carbohydr Res 345:965–970
129. Kumar P, Barrett DM, Delwiche MJ, Stroeve P (2009) Methods for pretreatment of lignocellulosic biomass for efficient hydrolysis and biofuel production. Ind Eng Chem Res 48:3713–3729
130. Meshgini M, Sarkanen KV (1989) Synthesis and kinetics of acid-catalyzed hydrolysis of some α-aryl ether lignin model compounds. Holzforschung 43:239–243
131. Evtiguin DV, Neto CP, Silvestre AJD (1997) Condensation reactions of lignin during oxygen delignification under acidic conditions. J Wood Chem Technol 17:41–55
132. Pan X, Xie D, Yu RW, Lam D, Saddler JN (2007) Pretreatment of lodgepole pine killed by mountain pine beetle using the ethanol organosolv process: fractionation and process optimization. Ind Eng Chem Res 46:2609–2617
133. Araque E, Parra C, Freer J, Contreras D, Rodríguez J, Mendonça R, Baeza J (2008) Evaluation of organosolv pretreatment for the conversion of *Pinus radiata D.* Don to ethanol. Enzyme Microb Technol 43:214–219
134. Del Rio LF, Chandra RP, Saddler JN (2010) The effect of varying organosolv pretreatment chemicals on the physicochemical properties and cellulolytic hydrolysis of mountain pine beetle-killed lodgepole pine. Appl Biochem Biotechnol 161:1–21
135. Park N, Kim HY, Koo BW, Yeo H, Choi IG (2010) Organosolv pretreatment with various catalysts for enhancing enzymatic hydrolysis of pitch pine (*Pinus rigida*). Bioresour Technol 101:7057–7064
136. Mabee WE, Gregg DJ, Arato C, Berlin A, Bura R, Gilkes N, Mirochnik O, Pan X, Pye EK, Saddler JN (2006) Updates on softwood-to-ethanol process development. Appl Biochem Biotechnol 129/132:55–70
137. Fukaya Y, Hayashi K, Wada M (2008) Cellulose dissolution with polar ionic liquids under mild conditions: required factors for anions. Green Chem 10:44–46

138. Kosan B, Michels C, Meister F (2008) Dissolution and forming of cellulose with ionic liquids. Cellulose 15:59–66
139. Mikkola JP, Kirilin A, Tuuf JC (2007) Ultrasound enhancement of cellulose processing in ionic liquids: from dissolution towards functionalization. Green Chem 9:1229–1237
140. Rogers RD (2002) Ionic liquids: not so green? Chem Eng News 80:4–5
141. Rogers RD, Holbrey JD, Spear SK (2004) Ionic liquids as green solvents: Engineering bioactive cellulose materials. Abstr Pap Amer Chem Soc 227:U310
142. Swatloski RP, Spear SK, Holbrey JD (2003) Ionic liquids as green solvents for the dissolution and regeneration of cellulose. Abstr Pap Amer Chem Soc 225:U288
143. Heinze T, Schwikal K, Barthel S (2005) Ionic liquids as reaction medium in cellulose functionalization. Macromol Biosci 5:520–525
144. Feng L, Chen ZJ (2008) Research progress on dissolution and functional modification of cellulose in ionic liquid. J Mol Liq 142:1–5
145. Zhu SD (2008) Perspective used of ionic liquids for the efficient utilization of lignocellulosic materials. J Chem Technol Biotechnol 83:777–779
146. Dadi AP, Schall CA, Varanasi S (2006) Enhancement of cellulose saccharification kinetics using an ionic liquid pretreatment step. Biotechnol Bioeng 95:904–910
147. Dadi AP, Schall CA, Varanasi S (2007) Mitigation of cellulose recalcitrance to enzymatic hydrolyis by ionic liquid pretreatment. Appl Biochem Biotechnol 137:407–421
148. Sun N, Rahman M, Qin Y, Maxim ML, Rodriguez H, Rogers RD (2009) Complete dissolution and partial delignification of wood in the ionic liquid 1-ethyl-3-methylimidazolium acetate. Green Chem 11:646–655
149. Ben-Ghedalia D, Miron J (1981) The effect of combined chemical and enzyme treatment on the saccharification and in vitro digestion rate of wheat straw. Biotechnol Bioeng 23:823–831
150. Neely WC (1984) Factors affecting the pretreatment of biomass with gaseous ozone. Biotechnol Bioeng 26:59–65
151. Ben-Ghedalia D, Shefet G (1983) Chemical treatments for increasing the digestibility of cotton straw. J Agr Sci 100:393–400
152. Vidal PF, Molinier J (1988) Ozonolysis of lignin—improvement of in vitro digestibility of poplar sawdust. Biomass 16:1–17
153. Garcia-Cubero MA, Gonzalez-Benito G, Indacoechea I, Coca M, Bolado S (2009) Effect of ozonolysis pretreatment on enzymatic digestibility of wheat and rye straw. Bioresour Technol 100:1608–1613
154. Contreras S (2002) Degradation and biodegradability enhancement of nitrobenzene and 2,4-dichlorophenol by means of advanced oxidation processes based on ozone. PhD Thesis, University of Barcelona, Spain
155. Sannigrahi P, Hu F, Pu Y, Ragauskas A (2012) Novel oxidative pretreatment of Loblolly pine, sweetgum, and miscanthus by ozone. J Wood Chem Technol, 32:361–375
156. Saratale GD, Oh SE (2012) Lignocellulosics to ethanol: the future of the chemical and energy industry. Afr J Biotechnol 11(5):1002–1013
157. Modenbach AA, Nokes SE (2012) The use of high-solids loadings in biomass pretreatment-a review. Biotechnol Bioeng 109(6):1430–1442
158. Tan RR, Culaba AB, Purvis MRI (2002) Application of possibility theory in the life-cycle inventory assessment of biofuels. Int J Energy Res 26:737–745
159. Azapagic A, Clift R (1999) The application of life cycle assessment to process optimization. Comput Chem Eng 23:1509–1526
160. Lynd LR, Cushman JH, Nichols RJ, Wyman CE (1991) Fuel ethanol from cellulosic biomass. Science 251(4999):1318–1323
161. Tonon SMT, Brown F, Luchi A, Mirandola A, Stoppato, Ulgiati S (2006) An integrated assessment of energy conversion processes by means of thermodynamic, economic and environmental parameters. Energy 31:149–163
162. Cosgrove DJ (2005) Growth of the plant cell wall. Nat Rev Mol Cell Biol 6:850–861

163. Foston M, Hubbell CA, Samuel R, Jung S, Fan H, Ding S-Y, Zeng Y, Jawdy S, Davis M, Sykes R, Gjersing E, Tuskan GA, Kalluri U, Ragauskas AJ (2011) Chemical, ultrastructural and supramolecular analysis of tension wood in *Populus tremula* x alba as a model substrate for reduced recalcitrance. Energy Environ Sci 4:4962–4971
164. DeMartini JD, Pattathil S, Avci U, Szekalski K, Mazumder K, Hahn MG, Wyman CE (2011) Application of monoclonal antibodies to investigate plant cell wall deconstruction for biofuels production. Energy Environ Sci 4:4332–4339

Part IV
Physicochemical Pretreatment

Chapter 9
Response Surface Optimization of Hot-Water Pretreatment for Enzymatic Hydrolysis of Hybrid Poplar: First Step of Bioconversion of Woody-Biomass to Value-Added Bioproducts

Jing Dai and Armando G. McDonald

Abstract In this study, the target product was the generation of sugars from woody biomass that can be the substrate for conversion into value-added chemicals, such as polyhydroxyalkanoates, lactic acid, succinic acid, etc. In order to release sugars from wood economically, wood needs to be pretreated to enhance the enzymatic hydrolysis of cellulose and hemicelluloses. The primary goal of this study was to determine the optimal condition to obtain fermentable monosaccharides from hydrolysates of hybrid poplar by a hot-water pretreatment (150–210 °C, 0–30 min). The pretreatment conditions were optimized using a response surface methodology (RSM) on a 2^3 full central composites design was performed by varying on temperature, reaction time, and solid loading. After pretreatment, the solid residue was subsequently treated with a cellulase preparation, and released sugars were quantified by HPLC. The total sugar yield was applied as response variable to the RSM. The optimal pretreatment condition for producing sugars was 200 °C, 18 min, and 20 % solid loading.

Keywords Hybrid poplar · Hot-water pretreatment · Enzymatic hydrolysis · Sugars · Response surface methodology

9.1 Introduction

Lignocellulosic biomass includes a wide range of carbon-rich resources, which can be utilized as feedstock for production of many industrial products ranging from lumber, paper, chemicals, biofuels, and value-added biodegradable polymers [1]. The technique for conversion of lignocellulosic biomass to fuel ethanol has been well developed. Cellulosic ethanol, however, has economic barriers to overcome to be economically competitive [2, 3]. Therefore, upgrading the conversion of cellulosic biomass to higher value products such as polyhydroxyalkanoates (PHA)

A. G. McDonald (✉) · J. Dai
Renewable Materials Program, Department of Forest, Rangeland and Fire Sciences, University of Idaho, Moscow, Idaho 83844-1132, USA
e-mail: armandm@uidaho.edu

Z. Fang (ed.), *Pretreatment Techniques for Biofuels and Biorefineries*, Green Energy and Technology,
DOI 10.1007/978-3-642-32735-3_9,

would gain better commercial value compared to cellulosic ethanol. The cost of the carbon substrate reportedly contributes more than 40 % of the production cost of PHA [4–6]. The use of inexpensive renewable agricultural materials such as woody biomass as feed stocks could be a tremendous advantage to the economics of PHAs production.

Hybrid poplar, as a short rotation fast growing wood species with low lignin content, has been highlighted as a good biomass resource for fuel and chemical production [7, 8]. Xylose is the main constituent of hardwood hemicellulose (acetyl-4-*o*-methylglucuronoxylan). Recently, studies showed that xylose can be obtained via a pretreatment process using dilute sulfuric acid (H_2SO_4) [9, 10]. Cellulose in wood is present as a semi-crystalline polymer and is comprised of glucose building blocks linked by β-o-4 linkages which can be cleaved by acids or enzymes. The major proportion of cellulose exists in the crystalline form. However, cellulose is more susceptible to degradation in its amorphous form [11]. Thus, breaking down cellulose crystalline structure to make it more accessible to cellulase enzymes usually requires pretreatment with heat, long reaction time, and addition of catalysts [8]. Lignin as a bonding component in wood is an inhibitor for hydrolysis or further fermentation process [11]. A pretreatment process can not only depolymerize lignin structure, but also remove some lignin in wood [12, 13]. Enzymatic hydrolysis is the most common method for converting woody biomass to sugars. Compared with acid hydrolysis, enzymatic hydrolysis yields no fermentation inhibitors such as furfural and it does not need neutralization and detoxification [2]. The only disadvantage of enzymatic hydrolysis is longer reaction times required for releasing the sugars. However, enzymatic hydrolysis is a better choice if further fermentation or bioconversions are required to produce value-added chemicals.

Hot-water pretreatment with controlled pH has been shown to improve enzymatic digestibility of lignocellulosic biomass [2, 14]. Acetic acid and other organic acids are released from the hemicelluloses, which help autocatalyze hemicellulose hydrolysis and disrupt cellulose and lignin structure. Unlike the anaerobic fermentation for cellulosic ethanol production, organic acids, such as acetic acid, are not considered as inhibitors during PHA biosynthesis but used as carbon source for PHA production [6]. The pH of the pretreatment liquor needs to be between 4 and 7 to minimize decomposition of sugars [2]. For the purpose of scale-up or industrial production, determining the optimal pretreatment conditions by using statistical approach is important. The experimental design works for variety of species, chemical reagents, temperature, and reactor features.

The aim of this study was to find optimal conditions to obtain total sugars (mainly glucose and xylose) by enzymatic hydrolysis via a hot-water pretreatment. A response surface methodology (RSM) was chosen to determine the optimal pretreatment conditions for sugar concentrations in enzymatic hydrolysates. Reaction time, temperature, and solid loading were the three variables tested in this design.

9.2 Materials and Methods

9.2.1 Raw Materials

Hybrid poplar (Potlatch Corp., ID, USA) was milled to <40 mesh using a Wiley mill (Thomas Scientific, NJ, USA), vacuum dried, and stored in sealed plastic bags (moisture content of 4.6 %). Chemical composition analysis was determined using procedures described in [15].

9.2.2 Hot-Water Pretreatment

Pretreatment was conducted in a 76 mL pressure reactor, (Model 4740, Parr Instrument Co., IL, USA) connected to a temperature controlled block heater built in-house. Wood meal (5.00 g) was introduced to the reactor to which water was added giving a solid loading range from 20.0 to 46.8 % and sealed. The reaction temperature ranged from 170 to 210 °C. A 2^3 full factorial design for temperature, time, and solid loading was conducted (Table 9.1). An additional temperature probe was used for controlling the outside temperature of the reactor vessel. The reaction vessel took 5 and 10 min, respectively, to reach 170 and 200 °C. After pretreatment, the vessel was placed in an ice-water bath to quench the reaction. The pretreated samples then were washed with hot-water (200 mL, 90 °C) [2, 3] to extract out the sugars and acids generated and centrifuged (10,000 rpm) to separate the solid and liquid fractions. The liquid fraction was named pre-liquid (PL) and pH was measured. The solid residue collected was used for enzymatic hydrolysis trials.

9.2.3 Enzymatic Hydrolysis

Enzymatic saccharification was done on the pretreated solid wood following the LAP method 009 [16]. The hydrolysis was conducted in 250 mL Erlenmeyer flasks in an oil bath for 3 days in citrate–Na_2PO_4 buffer (pH 4.8, 100 mL), at 50 °C, and with magnetic stirring (200 rpm). The pH was adjusted using 4 M NaOH. The enzyme loading for experimental design samples was 1 mL cellulase (612 u/(g mL), Fisher Scientific, IL, USA). Samples were taken every 24 h to determine sugar content by high performance liquid chromatography (HPLC).

Two commercial enzyme solutions (Cellic CTec2 (1.238 g/mL) and HTec2 (1.209 g/mL), Novozymes North America Inc., NC, USA) were also evaluated as received. Enzyme loading was based on solution weight % (100 × *g* enzyme solution/g wood)). The enzyme loadings used here were 1.5 % (CTec2 0.06 mL, HTec2 0.06 mL), 3 % (CTec2 0.12 mL, HTec2 0.12 mL), 6 % (CTec2 0.24 mL, HTec2 0.25 mL), and 30 % (CTec2 1.21 mL, HTec2 1.24 mL).

Table 9.1 2^3 Factorial experimental design varying on reaction temperature, time, and solid loading

Experiment	Variables			Coded levels		
No	Temperature (°C) *X*1	Time (min) *X*2	Solid loading (%) *X*3	Temperature (°C) *X*1	Time (min) *X*2	Solid loading (%) *X*3
1	170	10	20	−1	−1	−1
2	200	10	20	1	−1	−1
3	170	30	20	−1	1	−1
4	200	30	20	1	1	−1
5	170	10	40	−1	−1	1
6	200	10	40	1	−1	1
7	170	30	40	−1	1	1
8	200	30	40	1	1	1
9	160	20	30	−1.68	0	0
10	210	20	30	1.68	0	0
11	185	3.2	30	0	−1.68	0
12	185	36.8	30	0	1.68	0
13	185	20	13.2	0	0	−1.68
14	185	20	46.8	0	0	1.68
15	185	20	30	0	0	0
16	185	20	30	0	0	0
17	185	20	30	0	0	0
18	185	20	30	0	0	0
19	185	20	30	0	0	0
20	185	20	30	0	0	0

9.2.4 Experimental Design

A RSM was used to determine the optimal pretreatment condition for producing maximum total reducing sugars. The method has been described in other studies [8, 17, 18]. The design was based on a 2^3 full factorial central composite design (CCD) and was conducted using Design Expert 8.0 software (Stat-Ease, Inc. MN, USA). The experiment conditions with corresponding codes are listed in Table 9.1. The three variables were temperature, reaction time, and solid loading with six repeated experiments in the central point (185 °C, 20 min, 30 %). Since the total sugar was a dependent variable, all the three variables were coded to real independent variables. The independent variables were calculated as (condition of the run-condition at central point)/ step change of the variable. Therefore, the coded values were $X1$(temp-185)/15, $X2$(time-20)/10, and $X3$(solid-30)/10.

9.2.5 Analytical Methods

Sugars were quantified by HPLC using two Rezex RPM columns in series (7.8 mm × 30 cm, Phenomenex, Torrance, CA, USA) and a Waters HPLC (Waters,

Milford, MA, USA) equipped with differential refractive index detector (ERC-5710, ERMA), on elution with water (0.5 mL/min) at 85 °C. Aliquot portions of hydrolysates (6 mL) were centrifuged and the supernatant (5 mL) was transferred to a test tube containing inositol as an internal standard (1 mL, 0.5 mg/mL), mixed, deionized (column-containing Amberlite IR-120 H^+ (0.5 mL) and Amberlite IR-402 OH- (0.5 mL) resins), and filtered (0.45 μm).

Acetic acid was quantified by HPLC using a Rezex ROA organic acid column (7.8 mm × 30 cm, Phenomenex, Torrance, CA, USA) and a Waters HPLC (Waters, Milford, MA, USA) equipped with differential refractive index detector (ERC-5710, ERMA), on elution with 0.005 N aqueous H_2SO_4 (0.5 mL/min) at 65 °C. An aliquot of hydrolysate (1 mL) was taken and filtered (0.45 μm) into an HPLC vial.

The total reducing sugar yield (%) for each sample was calculated as Eq. 9.1. Since the maximum sugar yield was detected after 3 days hydrolysis, the third day total sugar yield was used in the response surface optimization analysis.

$$\text{Total sugar yield } (\%) = \frac{\text{sum of sugars concentrations (mg/mL)} \times \text{buffer volume100 (mL)} \times 100\,\%}{\text{wood dry weight 5,000 (mg)}} \quad (9.1)$$

9.3 Results and Discussion

9.3.1 Response Surface Model for Total Sugars Yield

Chemical composition analysis of the poplar wood was shown to consist of 49 % glucan, 21 % xylan, 1.5 % galactan, 1.0 % arabinan, 2.5 % mannan (total 75 % carbohydrate), 22 % Klason lignin, 2 % extractives, and 0.8 % ash.

In the hydrolysates, five reducing sugars (glucose, xylose, galactose, arabinose, and mannose) were measured. Neither furfural nor hydroxymethylfurfural were detected by HPLC in the hydrolysates and therefore not deemed in sufficient quantity to inhibit fermentation. In this preliminary trial, using a readily available cellulase enzyme, the sugars and acetic acid yields in both PL and enzymatic hydrolysates are listed in Table 9.2. In the PL, xylose was the main sugar while very little amount of other sugars could be detected. However, glucose was the major sugar released followed by xylose after enzymatic hydrolysis. Experiment 2 (200 °C, 10 min, 20 % solid loading) gave the highest total sugars yield (34 %). While experiment 11 (185 °C, 3.2 min, 30 % solid loading) gave the lowest sugars yield (14.8 %) which also had least total sugars in the PL. This result indicated that experiment 11 was not a severe pretreatment condition due to short reaction time (3.2 min).

The acetyl group is readily released from 4-*o*-methylglucuronoxylan as acetic acid during pretreatment [19] and therefore was quantified. Acetic acid concentrations in both PL and enzymatic hydrolysates were <1.2 mg/mL (Table 9.2) and below the level (5 mg/mL) at which it could act as an inhibitor for fermentation [8, 9, 20,

Table 9.2 Analysis of components in pre-liquid (PL) and enzymatic hydrolysates

Components in pre-liquid			Components in enzymatic hydrolysates				
Experiment No	Total sugars in PL (mg/mL)	Acetic acid (mg/mL)	Glucose (mg/mL)	Xylose (mg/mL)	Other sugars[a] (mg/mL)	Acetic acid (mg/mL)	Total sugars yield[b] (%)
1	1.49	0.33	4.27	1.71	0.31	0.33	15.56
2	4.98	0.24	8.18	3.27	0.64	0.06	34.14
3	2.99	0.32	5.15	2.06	0.46	0.34	21.32
4	4.35	0.97	7.21	2.89	0.61	0.26	30.12
5	1.63	0.26	4.26	1.70	0.38	0.28	15.94
6	4.45	0.78	7.18	2.87	0.60	0.21	30.20
7	2.60	0.38	5.21	2.09	0.45	0.27	20.70
8	3.16	1.15	6.83	2.73	0.56	0.29	26.56
9	1.32	0.20	4.60	1.84	0.32	0.20	16.16
10	3.23	1.18	6.37	2.55	0.52	0.26	25.34
11	1.03	0.28	4.32	1.72	0.35	0.32	14.84
12	3.58	0.70	6.27	2.50	0.53	0.26	25.76
13	4.48	0.37	7.63	3.05	0.68	0.31	31.68
14	3.26	0.56	5.71	2.28	0.43	0.28	23.36
15	3.74	0.74	6.74	2.70	0.54	0.32	27.44
16	3.04	0.42	6.83	2.73	0.56	0.32	26.32
17	3.44	0.73	6.75	2.70	0.58	0.28	26.94
18	3.14	0.59	6.75	2.70	0.52	0.23	26.22
19	3.05	0.45	6.60	2.64	0.55	0.29	25.68
20	3.19	0.65	6.42	2.57	0.51	0.28	25.38

[a]Other sugars means the sum of galactose, arabinose, and mannose concentrations
[b]Total sugars yield means the sum of total sugars in pre-liquid and total sugars in enzymatic hydrolysates divided by wood dry weight

21]. Compared with an acid pretreatment, a hot-water pretreatment generates much less acetic acid during the process [2]. The pH in all experiments which was about 4 together with the low acetic acid concentrations observed will result in limited sugar degradation. The highest acetic acid levels were observed at a pretreatment of 200 °C for 30 min. This suggests that pretreatment temperature and reaction time were important factors. A hot-water wash process can help reduce acetic acid and other inhibitors levels generated from pretreatment [2, 3].

Considering further fermentation or reaction for producing PHA, total sugars yield from enzymatic hydrolysis would be a major target. Therefore, RSM used total sugars yield as response variable. Before determining the optimization pretreatment condition, an RSM model was conducted using 20 experiments (Tables 9.1 and 9.2) (total sugars yield, %). Figure 9.1 is a three-dimensional (3D) plot that modeled the pretreatment conditions for total sugars yield in a curved surface and predicted the optimal sugars yield at fixed variable (solid loading, 30 %). This is a direct view of the data generated from this experimental design. Since 30 % solid loading was the central point in the design, it was selected to be the fixed variable and the other two more significant variables (based on analysis of variance (ANOVA) results) were displayed in the response surface. From Fig. 9.1, the highest surface occurred at 200 °C and a reaction time around 20 min.

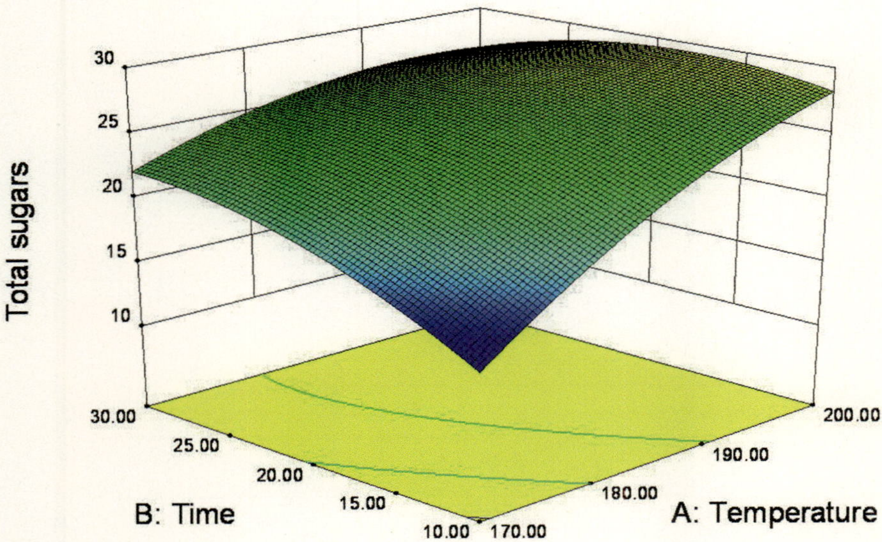

Fig. 9.1 Three-dimensional plot of RSM for total sugars yield based on 2^3 CCD (solid loading was set to 30 % and the maximum predicted response was 31.4 % of total sugar yield at 200 °C and 18.1 min)

The modeling results are shown in Fig. 9.2 and Eq. 9.2.

$$Y = a0 + a1 \times A + a2^2 \times B^2 + a12 \times A \times \mathrm{B} \quad (9.2)$$

Equation 9.2 was established based on ANOVA results in Table 9.3. Temperature (*A*) was shown to be the most significant variable while time (*B*) in the second order was significant and the interaction term of temperature × time had influence on total sugars yield (95 % significant level). Since solid loading was the least effective variable based on ANOVA results, the 3D plot was given on the other two variables (temperature and time). The *R*-square of the model was 0.86, which was acceptable to give a decent prediction on total sugars yield with appropriate pretreatment condition. Repeated experiments on the optimal pretreatment condition were done to support the model. Model coefficients were generated by fitting Y to the least squares of variables *A*, *B*, and *C* (solid loading).

9.3.2 Response Surface Optimization for Total Sugars Yield

The optimization of total sugars yield was conducted based on the model generated in Eq. 9.2. The 2D contour plot (Fig. 9.2) gave the optimization result based on the quadratic model (response surface). The maximum predicted response was 32.6 % of total sugar yield (pretreated at 200 °C for 18.2 min and 20 % solid loading followed by cellulase treatment). Pretreatment examinations were conducted at the optimal condition (200 °C, 18 min, 20 %) to confirm the predicted model. Heating the reactor

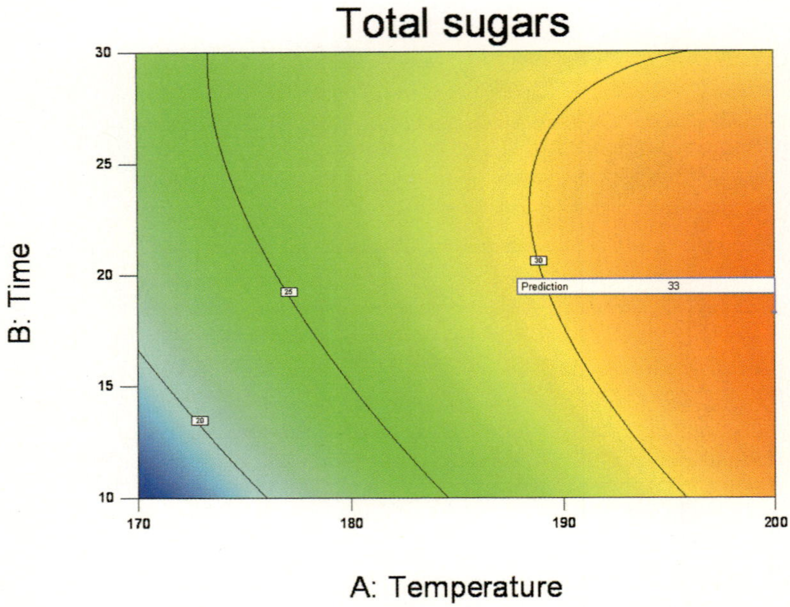

Fig. 9.2 The contour plot of RSM for optimization of total sugars yield (the maximum predicted response was 32.6 % of total sugar yield at 200 °C for 18.2 min and 20 % solid loading)

Table 9.3 Analysis of variance for total sugars yield (quadratic model)

Source	Sum of squares	DF	Mean square	*F* value	*P*-value (Prob > *F*)	Coefficient	
Model	498.658	9	55.406	6.8080	0.0030	*a*0	26.276
A	290.059	1	290.060	35.6405	0.0001	*a*1	4.609
B	32.988	1	32.988	4.0533	0.0718	*a*2	1.554
C	34.584	1	34.584	4.2494	0.0662	*a*3	−1.591
A^2	37.826	1	37.826	4.6477	0.0565	$a1^2$	−1.620
B^2	45.619	1	45.619	5.6054	0.0394	$a2^2$	−1.779
C^2	8.621	1	8.621	1.0593	0.3276	$a3^2$	0.774
AB	41.314	1	41.314	5.0764	0.0479	*a*12	−2.273
AC	6.589	1	6.589	0.8095	0.3894	a13	−0.908
BC	0.048	1	0.048	0.0059	0.9403	a23	−0.078
Residual	81.385	10	8.139				
Corrected Total	580.043	19					

R-square = 0.86, adjusted *R*-square = 0.74

to 200 °C took about 10 min and the total sugar yield was 34 % on average, which was a little higher than model predicted value. So this model was confirmed to be reliable for determining pretreatment condition. Chemical composition analysis indicated that a complete hydrolysis would theoretically yield 75 % total sugars. However, results from these 20 experiments followed by cellulase treatment did not achieve the theoretical maximum yield. This is likely due to the low enzyme activity

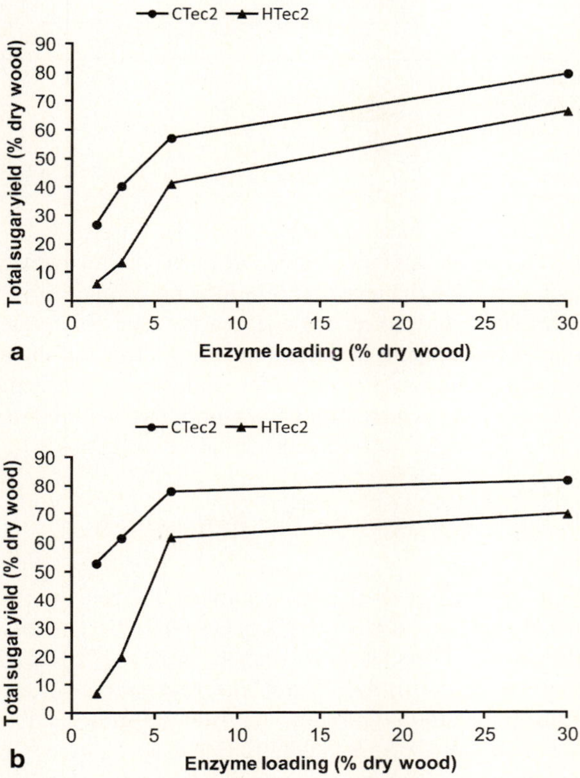

Fig. 9.3 Total sugars yield (% dry wood) from enzymatic hydrolysis at optimal hot-water pretreatment condition (200 °C, 18 min, 20 %) with different enzyme loadings (CTec2 and HTec2) at (a) 2 days and (b) 3 days

of the commercial cellulase used in this preliminary trial. Therefore, two industrial enzymes were chosen to conduct further enzymatic hydrolysis trials on the optimal pretreatment condition.

9.3.3 Enzyme-Loading Examination Under Optimal Pretreatment Condition

Due to the poor performance of the original cellulase preparation, two industrial enzymes specifically designed for cellulosic ethanol production were evaluated, namely CTec2 (cellulase and xylanase) and HTec2 (xylanase). Results using these enzymes on the pretreated material approached the maximum theoretical yield (Fig. 9.3). After 2 days of enzymatic hydrolysis, the mixed enzyme (CTec2) can extract almost 60 % total sugars in wood at 6 % enzyme loading. When using higher enzyme loadings of 30 %, the mixed enzyme can reach maximum sugar yield in 2 days. The xylanase (HTec2) preparation could also yield 60 % total sugars within 2 days, which indicated this enzyme had cellulase activity.

Table 9.4 Bioproducts from sugars

Bioproduct	Market price (US $/kg)	Reference
Ethanol	1.14	22
Succinic acid	5.9–8.8	23
Lactic acid	1.5–1.9	23
PLA	1.9–6.6	24
PHA	4.4–6.1	24

The total sugars yield after 3 days enzymatic hydrolysis reached 78 % sugar yield using the mixed enzyme at 6 % loading, while the xylanase achieved 58 % sugars. Therefore, the mixed enzyme can be used for further saccharification on hybrid poplar at a loading between 3 and 6 %. To note, the enzyme loading appears high since it is in a stabilizing buffer solution but it is actually a dilute protein solution (actual protein concentration is proprietary information).

9.3.4 Opportunities for Bioproducts

Work by Kazi et al. [22] had estimated the cost of ethanol production from ligncellulosic biomass at US $1.14/kg (US $0.90/L), and this was dependent on feedstock and enzyme costs (assumed enzyme price as US $0.23/kg of ethanol produced). Due to the poor returns for ethanol, alternate uses of sugars for bio-products offer higher value propositions and some are listed in Table 9.4. For example, succinic acid market price is at US $5.9–8.8/kg [23] and can be used in a variety of food products and as a building block in polymers. Another valued bioproduct, PHA, has a market price at US $4.4–6.1/kg [24]. PHA can be produced from low value sugars and organic acids [1, 4] derived from woody-biomass feedstocks, rather than current practices of using refined sugar as a carbon source, which offers significant financial advances in reducing PHA manufacturing costs by 50 % [1]. Future work will investigate the use of these hydrolysates for the manufacture of PHA using mixed microbial consortia [6].

9.4 Conclusions

A simple hot-water pretreatment on hybrid poplar was achieved and optimized by using a response surface methodology on a 2^3 central composite design. The optimized pretreatment condition (temperature 200 °C, time 18 min, solid loading 20 %) was used in further experiments. A subsequent cellulase/xylanase hydrolysis step resulted in yielding high-level of sugars. Thus, pretreatment was shown to be an important step for cost effective enzymatic hydrolysis of wood with low levels of inhibitory side-products. Future work will focus on enzyme loading, conversion to targeted bioproducts (PHA), energy balance, and cost.

Acknowledgments The project was supported by USDA—National Institute of Food and Agriculture—Wood Utilization Research program grant 2010-34158-20938. The enzymes CTec2 and HTec2 were kindly provided by Dr. Kurt Creamer at Novozymes North America Inc., Franklinton, NC, USA.

References

1. Keenan TM, Nakas JP, Tanenbaum SW (2006) Polyhydroxyalkanoate copolymers from forest biomass. J Ind Microbiol Biotechnol 33:616–626
2. Kim Y, Mosier NS, Ladisch MR (2009) Enzymatic digestion of liquid hot water pretreated hybrid poplar. Biotechnol Prog 25(2):340–348
3. Nagle N, Elander R, Newman M, Rohrback B, Ruiz R, Torget R (2002) Efficacy of a hot wash process for pretreated yellow poplar to enhance bioethanol production. Biotechnol Prog 18:734–738
4. Reis MAM, Serafim LS, Lemos PC, Ramos AM, Aguiar FR, Van Loosdrecht MCM (2003) Production of polyhydroxyalkanotes by mixed microbial cultures. Bioprocess Biosyst Eng 25:377–385
5. Choi J, Lee SY (1997) Process analysis of economic evaluation of poly(3-hydroxybutyrate) production by fermentation. Bioprocess Eng 17:335–342
6. Coats ER, Loge FJ, Wolcott MP, Englund K, McDonald AG (2007) Synthesis of polyhydroxyalkanoates in municipal wastewater treatment. Water Environ Res 79(12):2396–2403
7. Hamelinck C, Hooijdonk G, Faaij A (2005) Ethanol from lignocellulosic biomass: techno-economic performance in short-, middle- and long-term. Biomass Bioenergy 28(4):384–410
8. Kim HY, Lee JW, Jeffries TW, Choi IG (2011) Response surface optimization of oxalic acid pretreatment of yellow poplar (*Liriodendron tulipifera*) for production of glucose and xylose monosaccarides. Bioresour Technol 102:1440–1446
9. Hu R, lin L, Liu T, Liu S (2010) Dilute sulfuric acid hydrolysis of sugar maple wood extract at atmospheric pressure. Bioresour Technol 101:3586–3594
10. Yamaguchi T, Aoyama M (2010) Hydrolysis of Tilia japonica wood for production of a fermentable substrate. Cellul Chem Technol 44(7–8):293–298
11. Kumar P, Barrett DM, Delwiche MJ, Stroeve P (2009) Methods for pretreatment of lignocellulosic biomass for efficient hydrolysis and biofuel production. Ind Eng Chem Res 48:3713–3729
12. Mok WS-L, Antal MJ (1992) Uncatalyzed solvolysis of whole biomass hemicelluloses by hot compressed liquid water. Ind Eng Chem Res 31:1157–1161
13. Weil J, Sarikaya A, Rau SL, Goetz J, Ladisch C, Brewer M, Hendrickson R, Ladisch MR (1997) Pretreatment of yellow poplar sawdust by pressure cooking in water. Appl Biochem Biotechnol 68:21–40
14. Mosier NS, Hendrickson R, Ho NWY, Sedlak M, Ladisch MR (2005) Optimization of pH controlled liquid hot water pretreatment of corn stover. Bioresour Technol 96:1986–1993
15. Jain T, Van Gerpen J, McDonald AG (2010) Production of fuel ethanol from woody biomass. J Biofuels 1:109–114
16. Brown L, Torget R (1996) Enzymatic saccharification of lignocellulosic biomass, chemical analysis and testing task laboratory analytical procedures (LAP 009). NREL Biomass Program
17. Perez J, Ballesteros I, Ballesteros M, Saez F, Negro M, Manzanares P (2008) Optimizing liquid hot water pretreatment conditions to enhance sugar recovery from wheat straw for fuel-ethanol production. Fuel 87(17–18):3640–3647
18. Lee J, Rodrigues R, Jeffries T (2009) Simultaneous saccharification and ethanol fermentation of oxalic acid pretreated corncob assessed with response surface methodology. Bioresour Technol 100(24):6307–6311

19. Fengel D, Wegener G (1989) Wood: chemistry, ultrastructure, reactions. Walter de Gruyter, New York
20. Delgenes J, Moletta R, Navarro J (1996) Effects of lignocelluloses degradation products on ethanol fermentations of glucose and xylose by *Saccharomyces serevisiae*, *Zymomonas mobilis*, *Pichia stipitis*, and *Candida shehatae*. Enzyme Microb Technol 19(3):220–225
21. Takahashi C, Takahashi D, Carvalhal M, Alterthum F (1999) Effected of acetate on the growth and fermentation performance of *Escherichia coli* K011. Appl Biochem Biotechnol 81(3):193–203
22. Kazi FK, Fortman J, Anex R, Kothandaraman G, Hsu D, Aden A, Dutta A (2010) Techno-economic analysis of biochemical scenarios for production of cellulosic ethanol. National Renewable Energy Laboratory, Technical Report NREL/TP-6A2–46588
23. Pastor M, Pellegrino JL, Carole TM (2004) Industrial bioproducts: today and tomorrow. www.energetics.com/resourcecenter/products/studies/Pages/Bioproducts-opportunities.aspx. Accessed 02 May 2012
24. de Guzman D (2010) http://www.icis.com/Articles/2010/10/25/9402443/bioplastic-development-increases-with-new-applications.html. Accessed 07 May 2012

Part V
Gasification, Liquefaction and Biogas

Chapter 10
Biomass Pretreatments for Biorefinery Applications: Gasification

Mania Abdollahi-Neisiani, Jean-Philippe Laviolette, Rouzbeh Jafari and Jamal Chaouki

Abstract Biorefinery is the object of significant research and development efforts due to the scarcity of economically viable crude oil, renewable energy source, and its environmental benefits. This has prompted chemical corpo-rations to look for alternative sources of carbon and hydrogen to produce chemicals, biologics, and other products such as biomass and waste matter. Two main reaction pathways are currently explored for biorefinery: thermochemical and biochemical. The thermochemical pathway proposes significantly higher reaction rates compared to current biological processes that use non-genetically modified organisms. One of the thermochemical pathways for biomass conversion is gasification which is a decomposition of solid fuels at high temperatures and oxygen-lean atmosphere. The successful development of biomass gasification processes requires addressing several critical technical difficulties including biomass diversity, feedstock treatment, gasification mechanism and reactions, gasifier types, and their performances. This chapter reviews key features of biomass gasification as a pretreatment for biorefining which can be used as a practical guide for gasification process. This chapter consists of six sections that include types of biomass for gasification, their properties, and pretreatment steps; gasification mechanism and reactions; syngas cleaning and conditioning; different gasifier, their characteristics, and modeling.

Keywords Biomass gasification · Pretreatment · Gasifier · Gas cleaning · Tar removal · Catalytic gasification

J. Chaouki (✉) · M. Abdollahi-Neisiani · J.-P. Laviolette · R. Jafari
Department of Chemical Engineering, Ecole Polytechnique de Montréal,
C.P. 6079, succ. Centre Ville, Montréal, H3C 3A7, Canada
e-mail: jamal.chaouki@polymtl.ca

M. Abdollahi-Neisiani
e-mail: mania.abdollahineisiani@polymtl.ca

J.-P. Laviolette
e-mail: jean-philippe.laviolette@polymtl.ca

R. Jafari
e-mail: rouzbeh.jafari@polymtl.ca

Z. Fang (ed.), *Pretreatment Techniques for Biofuels and Biorefineries*,
Green Energy and Technology,
DOI 10.1007/978-3-642-32735-3_10,

10.1 Introduction

The beginning of industrial civilization has triggered the frantic use of non-renewable fossil-fuel resources (coal first, followed by oil and gas), which has grown worldwide ever since. Today, the price of fossil fuels is increasing due to depleting "conventional" resources, rising demands from developing countries and the establishment of a low-carbon economy. In this context, significant investments and research are focusing on the development of new processes to extract energy and goods from renewable resources, such as biomass.

Biomass is a carbonaceous matter, known as a renewable energy source from living or recently living organisms. Examples include forest residue, agricultural wastes, and even, municipal solid waste. To convert biomass, two main reaction pathways are currently considered: biochemical and thermochemical. Gasification is a thermo-chemical pathway, which transfers the combustion value of the solid fuel to the gas phase whose composition maximizes its chemical energy rather than sensible heat. Syngas, a mixture of CO and H_2, is one of the products of the gasification process, which could be used as a fuel or building block for many hydrocarbons. The products derived from syngas can be divided into three categories: (1) chemicals, such as ammonia and methanol; (2) transportation fuels, such as synthetic natural gas and synthetic diesel; (3) and energy feedstock, such as methane. Currently, syngas is produced mainly from fossil fuels; however, there is a growing interest in generating "green chemicals" and "green fuels" from the gasification process.

To successfully design an industrial gasification process, a thorough knowledge of biomass pretreatment, gasification reaction kinetics, and reactor technologies is essential. This chapter discusses the subject of biomass gasification through a detailed review of the scientific and industrial literature.

10.2 Types of Biomass for Gasification

The biomass properties have significant effects on the product compositions (through conservation of mass) and the gasification conditions (corrosion, slagging, etc.). Biomass includes any form of non-fossilized and biodegradable material derived from living species, such as plants and animals. The term "biomass" designates a very wide spectrum of substances, which includes products, byproducts, residues, and wastes from industries (forestry, agricultural, food, etc.) and municipalities [1, 2]. Each type of biomass is characterized by its specific physical and chemical properties: moisture, heating value, bulk density, chemical composition as well as ash and volatile contents. The biomass properties determine its performance as a fuel in gasification or any other process. Furthermore, the availability of the types of biomass as well as their properties is a function of a geographical location.

One of the most available biomass is wood, but it is a valuable material due to its current applications as a construction material. Wood residues (sawdust, bark, and misshapen pieces), however, have very little market value and are therefore prime candidates to be used as gasification feedstock. Other types of industrial residues (agricultural, forestry, etc.) could also be used as feedstock: husks from rice, coffee or coconut, bagasse from sugar cane, and verge grass. Energy cropping, such as poplar, sugar cane, and sweet sorghum, which consists of growing biomass specifically for fuels, is also another interesting possibility in renewable energy and the agricultural sector.

10.2.1 Biomass Properties

One important aspect of biomass is its potential for carbon offsetting. The energy conversion of biomass does not contribute to CO_2 emission or global warming. CO_2 released during biomass conversion is equivalent to the quantity it absorbed during its growth. In other words, energy production from biomass is almost carbon neutral. The veracity of this statement is, however, debatable as the calculation of the carbon balance associated with biomass conversion is complex: Many parameters must be considered, such as growth, harvest, transportation, and pretreatments. Yet the use of biomass fuels can result in the displacement of carbon dioxide emissions that are ordinarily released when using fossil fuels. *This displacement will depend entirely on the efficiency with which the biomass energy can be produced and used.*

Most biomass materials contain very low amounts of sulfur, chlorine, in comparison to most fossil fuels. Biomass conversion or its use as energy has the potential to lower pollutant emissions.

Biomass chemical composition is obtained through ultimate (also called elemental) and proximate analysis. Ultimate analysis yields the composition of biomass in terms of carbon, hydrogen, nitrogen, oxygen, and sulfur content (mass %). On the other hand, proximate analysis reports the composition in terms of volatile matter, ash, moisture, and fixed carbon. There are separate ASTM standards for measurements of individual components of biomass, such as volatile matter, ash, moisture, and fixed carbon.

- *Volatile matter* refers to the condensable and non-condensable vapor released during the early stages of biomass heating. It depends strongly on the heating rate and final temperature. There are different ASTM standards for the measurement of volatile matter; each standard is specific to different types of biomass.
- *Ash* is the inorganic residue composed mainly of silica, iron, calcium, magnesium, sodium, and sometimes potassium. It is the remaining material once biomass has been completely consumed in the reactor.
- *Fixed carbon* is an important parameter in biomass gasification, because its conversion is a limiting reaction, and it is used to size the gasifier.
- The *heating value* of the biomass is the energy chemically bound in the biomass with respect to a reference state. The best common ones are the lower heating value (LHV) where the reference state is water in its gas state and the higher heating value (HHV) where the reference state is water in its liquid state.

The biomass properties detailed above have significant effects on gasification conditions and product compositions: The basis on which they are measured and reported (wet, dry, or dry-ash-free basis) is very important. For example, low ash content improves thermal balance and reduces operating problems due to slagging and sintering. Water vapor is also an essential key component in gasification reactions; at high levels, it may negatively affect the process thermal balance. Biomass-containing sulfur, chlorine (may be present in low amounts), and alkali metals can also lead to the formation of corrosive components. Most biomass also contains nitrogen that can form gas-phase ammonia, which can oxidize and form NO_x.

10.2.2 Feedstock Pretreatments for Gasification

Pretreatment of the biomass feedstock is the first step in the gasification process. It is essential in some cases due to the process characteristics, but it generally helps to improve the quality of the fuel by homogenizing its size, moisture content, and density. For the operation of continuous processes, biomass storage is required, yet it may lead to issues related to decomposition and self-heating. Pretreatment can help solve or minimize the impact of these issues. It includes several processes, such as size reduction (milling, grinding, and pulverization), screening, drying, torrefaction, pelletization, and pyrolysis.

10.2.2.1 Drying and Size Reduction

The biomass feedstock moisture and ash content have an effect on the gasification product composition, the process energy balance, and the reactor operating condition. Different gasification processes require different moisture levels; high moisture content in biomass will, however, reduce the gasifier temperature and have negative effects on gasification efficiency and the quality of the products. To reduce moisture, driers are used and the most common type is the rotary drier. Nevertheless, the use of steam drying techniques is also increasing due to its easy integration into existing systems. Other techniques, such as biomass mechanical dewatering and leaching with water, have been demonstrated to reduce ash content efficiently [3]. Since water vapor plays an essential role in the chemical reactions of the gasification process, there is a trade-off between moisture removal (beneficial to the process energy balance) and syngas composition.

Biomass particle-size reduction techniques include milling, grinding, and pulverization, which make drying, transfer, and biomass storage easier. Their use depends mainly on the requirements of the gasification system as well as the biomass characteristics. For example, to pulverize particles to an average size below 0.2 mm, a vibration mill is recommended [4]. There are also some new methods being developed, like freezing pulverization or explosion disintegration, which are suitable for materials that could not be pulverized by conventional methods; however, their power consumption is relatively high.

10.2.2.2 Torrefaction

Torrefaction is a biomass thermal treatment process at low temperature (200–300 °C) in the absence of oxygen and mostly at near atmospheric pressure. It is a mild pyrolysis process that destroys the fibrous structure of biomass, increasing its calorific value as well as its hydrophobic nature to improve biomass stability during storage. Since torrefied biomass is more friable compared to its original state, energy consumption for biomass particle-size reduction is lower. It has also been shown that torrefied biomass particles can be fluidized more smoothly compared to untreated biomass [5]. There are two main torrefaction methods: (1) the wet process [6] and (2) the dry process. In wet torrefaction, biomass is treated with hot compressed water resulting in three groups of products: solid fuel, aqueous compounds, and gases. Dry torrefaction is an intensive drying at higher temperatures [7]. One of the drawbacks of the wet process is the necessity to separate the excess water from the torrefied biomass and liquid by-products with relatively high organic and mineral contents. Dry torrefaction is typically performed at temperatures in the range of 230–300 °C in the absence of oxygen, at near atmospheric pressure and a relatively low particle heating rate (lower than 50 °C/min). The mass and energy yield from the original biomass to the torrefied biomass is strongly dependent on torrefaction temperature, reaction time, and biomass type.

10.2.2.3 Pyrolysis

Biomass offers some advantages as a feedstock due to its high volatile content and high char reactivity. Compared to fossil fuels, however, biomass is a solid with a low heating value, containing much less carbon and more oxygen. In general, it is important to remove oxygen when producing fuel from biomass because high oxygen content results in low energy density. Biomass contains about 40–60 wt % oxygen compared to less than 1 wt % for fossil fuel. Pyrolysis can be used as a pretreatment process to reduce the oxygen content in biomass to lower levels compared to torrefaction depending on the operating condition.

Biomass pyrolysis yields three products: semi-char solid, liquid bio-oil (condensable gases), and a non-condensable gaseous fraction. The non-condensable gaseous fraction is composed mainly of hydrogen, carbon dioxide, carbon monoxide, and methane. Following the biomass feedstock pyrolysis, the solid and liquid products of pyrolysis can be mixed and fed as slurry to the gasifier, which is advantageous for entrained bed reactors [8]. However, bio-oils mainly contain double carbon bound chemicals. These materials tend to polymerize and become increasingly viscous, which is detrimental to its stability during storage.

It is also important to note that the pyrolysis conditions have an effect on the reactivity of the char produced during gasification. Therefore, the proper selection of pyrolysis conditions is the key to ensure the gasification process will benefit [9].

As discussed in this section, there are several methods for biomass pretreatment, which can be combined for the benefit of the gasification process. Each method has its specific advantages as well as disadvantages, and the best selection depends on the feeder characteristics and gasifier type.

10.3 Biomass Gasification

Gasification is the conversion of solid fuels into gas fuels in an *oxygen-lean atmosphere.* Pyrolysis, gasification, and combustion are classified as three separate thermo-chemical conversion processes. However, both pyrolysis and combustion take place during gasification. In gasification, at least four different stages of reactions are involved: (1) pyrolysis, (2) combustion of solid char and other gases, (3) gasification of char, and (4) tar cracking.

Different types of gasifiers must deal with these reactions. The contribution of these reactions upon the final gaseous product as well as the specific conversion of each reaction depends on the operating conditions (temperature, pressure, etc.), the biomass characteristics (chemical composition, particle size, etc.), and gasifier type. The fundamental gasification reactions (reduction reactions) are endothermic. The necessary energy is supplied by the exothermic combustion reactions. Compared to the original solid fuel, the produced gaseous fuel is easy to clean, to transport, and if cleaned properly it can be used in fuel cells as well as burned in gas turbines, furnaces, boilers, and reciprocating engines [10].

As mentioned earlier, syngas is one of the products of gasification, which is an important source for valuable chemicals and energy, such as hydrogen, diesel (through Fischer–Tropsh synthesis), electricity (through combustion), fertilizer (through ammonia production), and methanol.

In the gasification process, heavier hydrocarbons, called tar, are produced along with syngas. Tar consists of high-molecular weight components, usually rich poly aromatic hydrocarbons (PAH). The presence of tar in syngas can cause serious problems in syngas applications. Condensation, fouling, and the polymerization of tar are general problems that can arise. The amount of undesirable products in the gas depends mainly on the design of the gasifier, feedstock characteristics as well as the gasifier operating conditions. Once the raw syngas leaves the gasifier, it goes through treatment processes, such as *gas cleaning* of dust or particulates (by cyclone, fabric or electrostatic filters, or solvent scrubber), *conditioning* (with the use of shift reaction to adjust the molar ratio of CO and H_2), and *separation* (to refine the syngas stream from tar and other catalyst poisons).

10.3.1 Reaction Kinetics

A good understanding of the gasification reaction kinetics and gasifier hydrodynamics is essential for the design, operation, and optimization of gasification processes. This section will discuss the main chemical reactions that govern gasification and the different kinetic models in the scientific literature that are available.

During a gasification process, the biomass particles undergo the following reaction steps:

1. Drying: The biomass particles are heated and dried on entering the reactor (endothermic step).
2. Pyrolysis (thermal cracking): As they reach high temperatures, the biomass particles undergo pyrolysis and decompose to gas and solid char (endothermic step).
3. Gasification:
   a. Combustion: The char and gases (condensable and non-condensable) react with oxygen to produce H_2O, CO, CO_2 (exothermic step).
   b. Gasification: Where the produced gas and solid char from previous steps react with the gasifying agent and each other (endothermic step).
   c. Tar cracking: The condensable gas decomposes (thermal cracking) to smaller molecular weight components (endothermic).

The level of oxygen in the gas can be set in order for the system to be autothermal. These steps are generally modeled in series, but it is widely accepted that there are no sharp boundaries between them. Table 10.1 lists the important reactions taking place during gasification.

10.3.1.1 Pyrolysis Kinetic

Pyrolysis is the cracking of hydrocarbon molecules into smaller gas molecules without any major reaction with air or any other gasifying medium. The kinetic information of pyrolysis is crucial for the design and scale-up of any gasification process. Extensive investigations have been done on the kinetics of biomass devolatilization in an inert atmosphere. Table 10.2 shows the most widely used kinetic schemes of biomass pyrolysis.

Earlier kinetic models consisted of simple, single first-order reaction schemes to describe the total volatile yield. Later, more complicated two- and three-step reaction networks containing parallel and series reactions were introduced by different authors [22, 24]. These are empirical models whose parameters are calculated by fitting experimental data generally derived from thermo-gravimetric measurements. Since most of the biomass is composed of cellulose, hemi-cellulose, and lignin, the most accurate models are reported to be a three independent parallel reactions model [25–29].

Most modeling efforts with the three independent parallel reactions model have been conducted using experimental data from a single heating rate [25, 30–33]. The effect of heating rate on the pyrolysis yield is, however, significant because the kinetic parameters derived from a single heating rate cannot be confidently extrapolated to other heating rates. Radmanesh and Chaouki proposed an improved model for biomass pyrolysis, which is applicable to different heating rates [27]. The kinetic parameters were calculated from experimental data obtained at relatively low heating rates (maximum heating rate was $50°$ C/min). Therefore the extrapolation of these kinetic models to actual gasification process conditions yields significant

Table 10.1 Gasification reactions

Reaction		Reference
Carbon reactions		
Boudouard	$C + CO_2 \rightarrow 2CO + 172\ kJ/mol$	[11]
Water–gas	$C + H_2O \rightarrow CO + H_2 + 131\ kJ/mol$	[12]
Hydro-gasification	$C + 2H_2 \leftrightarrow CH_4 - 74.8\ kJ/mol$	
	$C + \frac{1}{2}O_2 \rightarrow CO - 111\ kJ/mol$	[13]
Oxidation reactions		
	$2\left(\frac{\eta+1}{\eta+2}\right)C + O_2 \rightarrow \frac{2\eta}{\eta+2}CO + \frac{2}{\eta+2}CO_2$	[13]
	$CO + \frac{1}{2}O_2 \rightarrow CO_2 - 284\ kJ/mol$	[14]
	$CH_4 + 2O_2 \rightarrow CO_2 + 2H_2O - 803\ kJ/mol$	[14]
	$H_2 + \frac{1}{2}O_2 \rightarrow H_2O - 242\ kJ/mol$	[15]
Shift reaction	$CO + H_2O \leftrightarrow CO_2 + H_2 - 41.2\ kJ/mol$	[16]
Methanation reaction		
	$2CO + 2H_2 \rightarrow CH_4 + CO_2 - 247\ kJ/mol$	
	$CO + 3H_2 \rightarrow CH_4 + H_2O - 206\ kJ/mol$	
	$CO_2 + 4H_2 \rightarrow CH_4 + 2H_2O - 165\ kJ/mol$	
Steam reforming reactions		
	$CH_4 + H_2O \leftrightarrow CO + 3H_2 + 206\ kJ/mol$	
	$CH_4 + \frac{1}{2}O_2 \leftrightarrow CO + 2H_2 - 36\ kJ/mol$	
Tar reactions		
Tar cracking	$\text{tar (gas)} \rightarrow \vartheta_{H_2}H_2\ (g) + \vartheta_{CH_4}CH_4\ (g) + \vartheta_{CO}CO\ (gas) + \vartheta_{CO_2}CO_2\ (gas) + \vartheta_{tar}tar_{inert}$	[17]
Tar combustion	$CH_{1.522}O_{0.0228} + 0.867O_2 \rightarrow CO + 0.761H_2O$	[18]

uncertainties since, in reality, biomass temperature increases very rapidly from ambient temperature to about 800–1,000 ° C (in less than 1 s) as it is fed into the gasifier [27]. Consequently, the actual heating rates applied to biomass particles in pyrolysis systems are significantly higher than 50 ° C/min or even 100 °C/min.

There are only a few studies on pyrolysis at high heating rates (1,000 °C/s) and the resulting gas production [34–37]. Although it is very important to have knowledge about pyrolysis kinetic gasification design and optimization, it is very difficult to obtain reliable data for kinetic constants that can be used for a wide range of biomass at different heating rates. Most models are derived from cellulose pyrolysis experiments, and the available models in the literature are only applicable to specific conditions.

10.3.1.2 Gasification Reaction Kinetics

In the gasification step that follows pyrolysis, several parallel reactions occur:

- *Char gasification* involves the reaction between char and steam, carbon dioxide, hydrogen, and oxygen (R1, R2, R3, and R4 shown in Table 10.1). These reactions

Table 10.2 Summary of the proposed kinetics for biomass pyrolysis

	Biomass type	Kinetic scheme	Ref.
1	Cellulose	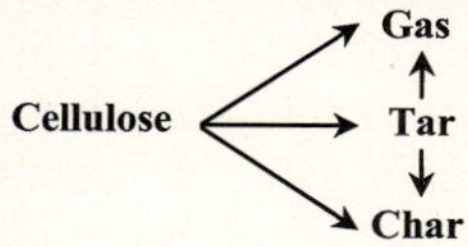	[19]
2	Cellulose	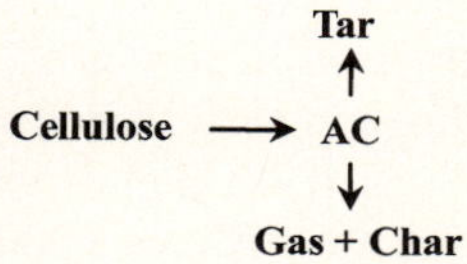	[20]
3	Wood	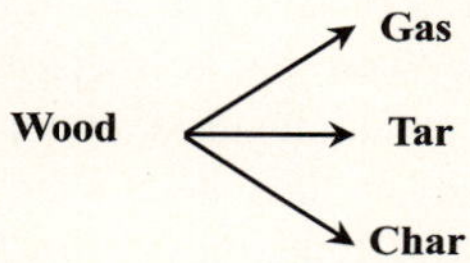	[21]
4	Wood	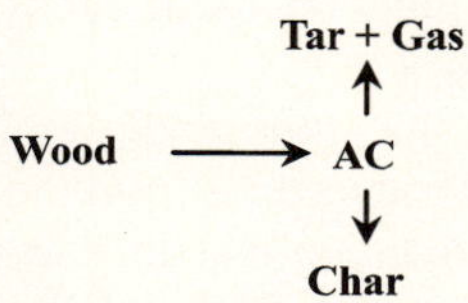	[22]
5	Wood	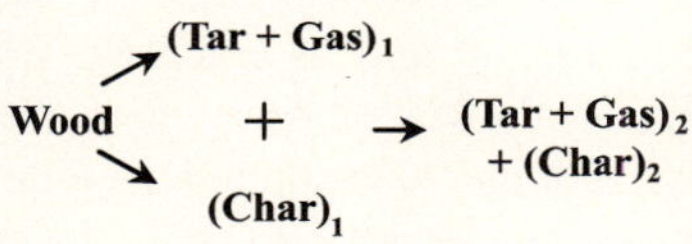	[23]
6	Cellulose	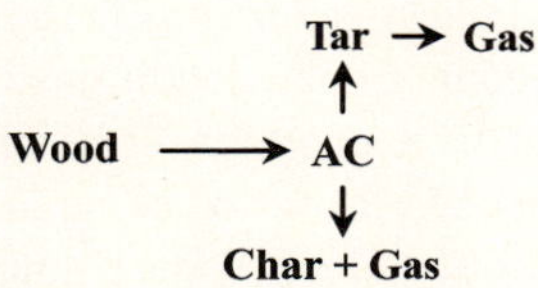	[24]

are endothermic, except for those involving O_2 and H_2 which are exothermic. The rate of these reactions depends on the reactivity of char and the gasifying medium: Oxygen is the most reactive species followed by steam and carbon dioxide. Char oxidation reactions are so fast that most of the oxygen is used in this specific reaction step. The relative reaction rates of the gasification reactions are estimated by Walker et al. [38]:

$$R_{C+O_2} \gg R_{C+H_2O} \gg R_{C+CO_2} \gg R_{C+H_2}$$

- Char is usually assumed to be pure carbon for simplification. In reality, it is composed of small amounts of hydrocarbon. Biomass char is generally more porous and reactive compared to coal char, so its reaction should be considered different [39].

- Water–gas (R2) reaction involves hydrogen, which affects char and steam reaction negatively as shown by Barrio et al. [40]. The continuous removal of hydrogen from the reactor is necessary in order to accelerate water–gas reactions.
- The gasification of char with carbon dioxide (known as Boudouard reaction—R1) is a relatively slow reaction. The rate of this reaction is negligible below 1,000k [41].
- Water–gas shift reaction (R8) is an important kinetic step in the gas phase. It controls the production and the ratio of hydrogen and carbon monoxide, which is critical for downstream processes. It is a slightly exothermic equilibrium reaction with negligible sensitivity to pressure. Above 1,000 °C, it reaches equilibrium fast but a heterogeneous catalyst is required to reach equilibrium at lower temperatures. Probstein and Hicks showed that at lower temperatures the reaction has a higher equilibrium constant, which means a higher hydrogen yield with low reaction rates [42]. Different catalysts have been tested and employed for water–gas shift reaction, like copper promoted catalysts for the temperature range of 300–510 °C, and a copper–zinc–aluminum oxide catalyst for the 180–270 °C range in commercial applications [43].
- As mentioned before, one of the products of gasification is a condensable heavy hydrocarbon, known as tar. Produced tar from the pyrolysis reactions undergoes further cracking and polymerization reactions to produce lighter or heavier hydrocarbons. Several studies have been done on the secondary pyrolysis reactions, which involve the fate of tar and its cracking. Boronson et al. and Liden et al. have reported separately the kinetic parameters of tar cracking derived from wood [17, 44]. Rath and Staudinger also studied the tar cracking kinetics of birch wood particles in a thermo-gravimetric analyser and a coupling of thermo gravimetric analysis (TGA) with a tubular reactor [45]. They showed that the extent of tar cracking is not only dependent on the conditions in the reactor (temperature and residence time) but also on the temperature at which the tar was formed [46]. Most of the kinetic models proposed for tar cracking are based on a single step, first-order reaction. Among different kinetics, the results of Boronson et al. show comparable rates and are the most used. The kinetics of tar cracking has also been studied in another approach. Due to the complexity of the tar, several researchers have studied its cracking and decomposition reactions using a model-biomass–tar compound, such as phenol, toluene, naphthalene, 1-methylnaphthalene, and so on. In most of the proposed kinetics, a first-order reaction for tar cracking was used.

10.4 Catalytic Gasification

The use of a catalyst for gasification is not essential, but it can increase gasification efficiency by reducing tar content or other unpleasant products, such as methane. The application of a catalyst promotes tar cracking at lower temperatures or promotes a steam-reforming reaction, which is a reaction between methane and steam in the temperature range of 700 °C–1,100 °C to produce syngas. The catalyst can be used directly in the gasifier or the secondary reactor downstream of the gasifier [47, 48]. There are different criteria for developing or choosing a proper catalyst, such as being inexpensive, effective, and resistant to attrition, carbon fouling, and sintering. Catalysts used in tar cracking can be classified into three main groups: alkali metals, non-metallic oxides, and supported metallic oxides.

10.4.1 Alkali Metals

Alkali metals, such as sodium and potassium, can promote a tar cracking reaction. Catalysts, such as potassium carbonate and sodium carbonate, are effective in methane production and are available in most biomass ashes. Potassium, however, is well-known for agglomeration in fluidized beds. Studies on the effects of inorganic salts ($MgCl_2$, NaCl, $FeSO_4$, and $ZnCl_2$) on the pyrolysis of cellulose have shown that $MgCl_2$ does not change the overall pyrolysis but has some slight effects on gas production. For the other tested catalysts (NaCl, $FeSO_4$, and $ZnCl_2$), an increase in char production was reported [49].

10.4.2 Non-metallic Oxides

Calcined dolomite is the most popular and most investigated material as a tar cracking catalyst. Dolomite is a calcium magnesium ore with a general chemical formula $CaMg(CO_3)_2$. This catalyst is relatively inexpensive, abundant, and disposable. Aznar, Corella and their research groups, in several publications, have investigated the effect of the in-bed use of dolomite, which can decrease the tar level from 6.5 %wt to 1.3 %wt [50]. The addition of 3–10 % calcined dolomite to biomass feed decreases the tar level by 40 % and improves gas quality significantly [51, 52]. Although dolomite has proved to be effective in terms of tar reduction, it has some critical limitations. Dolomite in its naturally occurring form is not very active in tar cracking, and it needs to be calcined. Calcination of dolomite involves its decomposition and the elimination of CO_2 to form the MgO-CaO complex. Calcination reduces the surface area of the dolomite catalyst, and makes it more friable resulting in severe catalyst attrition and fine-particle production. While using dolomite in a fluidized bed, dust entrainment due to the eroding of soft dolomite particles necessitates the continuous feeding of dolomite into the reactor by mixing it with biomass fuel [51]. This requires a major gas-cleaning operation.

Another popular catalyst in this group is olivine, which could be an alternative for dolomite. Olivine is a mineral, which contains magnesium, iron oxide, and silica. In terms of attrition, olivine has certain advantages over dolomite. However, Corella et al reported that dolomite was 1.4 times more active than olivine in biomass gasification with air, but dolomite generated $\sim$ 4–6 times more particulates in the gasification gas than olivine [53]. Nickel has also been used with an olivine support, which led to improved catalytic activity compared to unsupported olivine catalysts [54, 55].

10.4.3 Supported Metallic Oxides

Popular catalysts in this group are Ni-based catalysts. A wide variety of Ni-based steam reforming catalysts are commercially available, and they are widely applied in the petrochemical industry. The literature contains numerous studies reporting the use of commercial Ni-based catalysts for tar cracking, which has been shown to be very effective in increasing synthesis gas yield. However, the use of the Ni catalyst in the gasifier is limited due to its fast deactivation caused by coke, chlorine, and alkali metals that may be present in the gasifier (from biomass ash). So far, the use of a nickel catalyst as an additive to the gasifier has had little success.

All these catalysts, when used in-situ, are not promising due to the combination of coking and friability. Using them in secondary beds is more effective. The duration of most-reported catalyst tests has been quite short, especially considering the long activity requirements for expensive catalysts, such as Ni, to be economical. Although downstream gas cleaning methods are reported to be very effective in tar reduction; catalyst deactivation due to impurities in the gas outlet of the gasifier makes catalytic tar cracking economically unfeasible.

10.5 Product Gas Cleanup and Conditioning

The raw products of biomass gasification contain contaminants, including particles, organic impurities (tar) alkali metals, chlorine, nitrogen, and sulfur compounds (such as H_2S, CS_2, COS, AsH_3, PH_3, HCl, NH_3, and HCN). These contaminations can block the downstream units, such as gas coolers and engines, and also interfere with the catalyst used in the production of synthetic fuels. Therefore, they have to be completely removed or significantly reduced before utilizing the gas depending on the application of interest. Also, the purpose of the conditioning system is to adjust the components to the appropriate ratio. Depending on the type of feedstock, its composition, and the type of gasification product application, there are different types of gas cleaning and conditioning that can be categorized as physical cleaning, such as cyclone, filters, and wet scrubbers' application, or chemical cleaning, such as catalytic cracking, thermal reforming, shift hydrolysis, and hydrogenation.

10.5.1 Particulate Removal Technologies

Particulate impurities in the product gas typically originate from ash, dust, carry-over bed materials, and unconverted char. The particulates can cause corrosion and plugging in the downstream process equipment. The most commonly used techniques for particulate removal are cyclones for the initial cleaning of larger particles, barrier filters (low and high temperature operations), electrostatic filters (ESP), wet scrubbers, and alkali salts. The application of the proper removal technique depends on the concentration of impurities, particle size distribution, and the particulate tolerance of the downstream application.

10.5.1.1 Barrier Filter

Barrier filters work by presenting a physical barrier in the path of tar and particulates while allowing the clean gas to pass through. Their surface can be coated with appropriate catalytic agents to facilitate tar cracking [55]. There are two types of filters: candle and fabric.

Candle filters are porous, ceramic, or metallic and can work at temperatures as high as 900 °C. The porosity of the material is chosen so that the finest particles do not pass through. Particles failing to pass through the filter barrier deposit on the wall and form a layer of solids called a "filter cake". Candle filters have been shown to fail within a short period of operation, which makes them economically undesirable [56, 57]. The main reasons for filter failure in the case of gasification are as follows: (1) thermal stress or shock due to char combustion on the filter surface; and (2) dust and ash deposition between the candles, which increases the pressure drop over the filter. To improve the lifetime of traditional candle filters, impurities, such as alkalis, chlorine, and sulfur, can be removed by sorbents prior to filter application, which can reduce the risk of weakening due to high-temperature corrosion [58]. Fixed beds with sorbents, such as bentonites or bauxites, can be used to remove alkalis, and CaO sorbents can be used to remove sulfur and chlorine, as CaS and $CaCl_2$ upstream from the particulate candle filter.

Fabric filters are made of woven fabric and, unlike candle filters, can operate only in lower temperatures (<350 °C). The condensation of tar on the fabric is a major problem if the gas is cooled excessively.

10.5.1.2 Wet Scrubber

In this method, water is sprayed on the gas, which makes the particles and tars collide, creating large droplets and separate from the gas stream by using cyclones. The tar liquid can be re-injected into the gasifier, and water may be regenerated by stripping the tar away. This method produces a large amount of waste water with a high organic content, removes a large fraction of the carbon and hydrogen stored in tars, and also reduces the gas temperature to near ambient temperatures, which results in a loss of thermal efficiency.

There are some commercial methods of wet scrubbing available, such as OLGA and TARWTC technologies, which use oil as a scrubbing liquid [59, 43]. The oil with separated tars can be recirculated to the gasifier so the energy in tar can be recovered. In this method, however, cooling the gas prior to cleaning is still required.

10.5.1.3 Alkali Remover

Compared to fossil fuels, biomass has a high concentration of alkali salts, and their removal from the product gas is a very important step in biomass gasification. Alkali salts will condense below 600 °C, which causes serious corrosion problems. If the temperature of the gas decreases below 600 °C, the alkali salts condense and can be separated in a cyclone or filters. However, in some applications, gas cannot be cooled so alumosilicates, such as bauxite, kaolinite, bentonite, and naturally occurring zeolite, can be used for alkali removal at temperatures up to 700 °C [59].

10.5.1.4 Electrostatic Precipitators

Electrostatic precipitators (ESP) are used to remove fine solids and liquid droplets from gas stream; however, they are not very efficient in terms of tar removal at high temperature. In order to have an efficient tar removal, gas stream should be quenched before ESP. In wet ESP, gas is ionised upon passing between high voltage electrodes and a grounded electrode. The produced ions attach themselves to dust particles or tar and water droplets. The charged particles and droplets are attracted to the grounded electrode, flowing to the bottom of the ESP where they are collected.

10.5.2 Tar Removal

Tars in the product gas can be tolerated in some systems where the gas is used as a fuel in applications, such as burners. However, in most of applications, tars in the raw product gases, even at low concentrations, can create major handling and disposal problems. Two basic approaches have been used to remove tars from product gas streams: (1) physical removal technologies similar to those used for particulate removal, such as wet scrubbers, ESP; and 2) catalytic and thermal tar-reduction methods where tars are converted to permanent gases. The catalytic approaches can potentially destroy tars in either vaporized or condensed state. The second approach is discussed below.

10.5.2.1 Thermal Cracking

By increasing the gasifier temperature all organic compounds will crack to smaller hydrocarbons (~1,200 °C). Oxygen or air can be added to the gasifier to allow partial combustion of the tar to raise its temperature. Using electrical arc plasma for tar cracking is another option. It is a simple technique but it produces gas with lower energy content.

10.5.2.2 Catalytic Cracking

This technique can be applied in the gasifier or in a secondary reactor. The gasifier is commercially used in many plants for the removal of undesired elements from product gas. This method is explained in more detail in Sect. 4. For most syngas applications, the optimization of operating conditions, catalytic gasifier material or additives combined with secondary hot gas cleaning (catalytic cracking) are the most preferred methods.

10.5.3 Inorganic Impurities

Inorganic compounds, such as Cl and S-containing material, HCN, COS, and ammonia, can be removed by means of physical and chemical washing techniques. Removal of ammonia from biomass is necessary in many situations since they are converted to NO_x when gas is burned. Ammonia can be removed by catalytic destruction using catalysts similar to those used for tar cracking and also by wet scrubbing where low temperature product gases are acceptable. Catalysts, such as dolomite, nickel-based steam reforming catalysts, and iron-based catalysts, have been used for ammonia removal with $> 99\ \%$ efficiency. Catalytic removal is economically an attractive option since it has the potential to remove tar and ammonia from product gas while keeping its heat [60].

In systems where product gas is first cooled, ammonia can be removed by wet scrubbing. The ammonia recovered from the scrubber is reinjected into the gasifier to reduce ammonia production through equilibrium.

H_2S can be removed by different sorptions, such as metal oxides, Cu- and Ca-based sorbents. For example, the Selexol process uses dimethyl ethers of polyethylene glycol and the Rectisol process employs methanol as a solvent to remove H_2S and COS and simultaneously remove CO_2 from syngas [61].

10.5.4 CO_2 Removal

Carbon dioxide can be removed from syngas by chemical and physical absorption with a washing liquid or by adsorption with solid absorption. The choice for chemical or physical absorption (or a combination of both) depends on the partial pressure in the gas. For chemical absorption in commercial processes substituted amines are used, while solvents, like methanol, polyethylene glycol, and dimethyl ether, are used for physical absorption. The CO_2 concentration can be removed to approximately 0.1 vol% by these processes. When the syngas contains significant concentrations of other gases besides H_2 and CO_2, adsorption on solid materials, such as silica gel, active carbon, zeolites, and molecular sieves, is preferred.

As discussed earlier, the produced syngas often contains high amounts of problematic impurities, such as sulfur, chlorine, and alkalis. The gasification process has several cleaning units and its total efficiency depends on the heat management of the various steps. One of the most important challenges for an efficient gas cleaning process is developing a hot gas cleaning technology, which works at or close to the gasifier temperature. These techniques include the development of novel particulate removal techniques, an improved catalyst for tar cracking to produce a tar- and particulate-free product gas and a higher degree of process integration. For example,

Table 10.3 Characteristics of different categories of the gasification process

Application criteria	Updraft	Downdraft	Cross-draft	Fluidized bed	Entrained flow
Fuel moisture (%)	60 max	25 max	10–20	20–30	35
Ash-dry basis (%)	25 max	6 max	05–1.0		Slurry feed: 20 max dry ash: 40
Ash melting temp (C)	>1000	>1250	–		
Feedstock size (mm)	5–100	20–100	5–20	6–10	< 100 μm
Application range (MW h)	2–3	1–2	–	< 25	< 50
Gas exit temp (C)	200–400	700	1,250	900–1,050	1,250–1,000
Tar (g/N m^3)	30–150	0.015–3.0	0.01–0.1	1–3	–
Gas LHV (MJ/N m^3)	5–6	4.5–5.0	4.0–4.5	–	–
Hot-gas efficiency (%)	90–95	85–90	75–90	–	–
Turndown ration	5–10	3–4	2–3	–	–

References: [43, 62]

one possible way to increase the efficiency of hot gas cleaning and also reduce the cost is to decrease the number of gas cleaning stages by combining different physical and chemical processes in the same equipment, such as catalytic tar cracking in a particular barrier filter. This was first proposed for combustion applications [62], but further applied to the gasification process by many research groups, such as the deposition of the Ni/MgO catalyst onto the pore walls in α-alumina in a candle [63] and a catalytically active fixed bed in a cylindrical catalytic filter element [64]. Still, there is continuous research and development being done to improve particulate filtration, various sorbents and associated equipment to achieve a high efficiency of gas cleaning, especially at high temperatures.

10.6 Types of Gasifiers

The chemical composition of the gasification product strongly depends on the type of gasifier. Different reactors have been used to perform the gasification process. The different types of reactors can be categorized based on the specifics of the solid transportation in the reactor or the means by which the gasifying agent is introduced to them. The main characteristics, advantages, and limitations of the most widely used gasifiers are summarized in Table 10.3.

10.6.1 Fixed Bed Gasifier

This type of reactor consists of a cylindrical reactor with a fixed bed of solid fuel. The gasifying agent is injected upward or downward through the reactor. These simple reactors can operate at high carbon conversion, for a long solid residence time, at

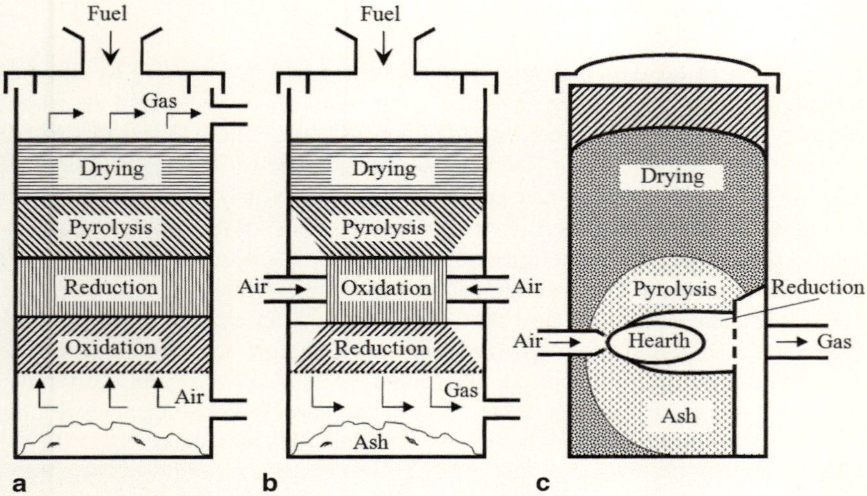

Fig. 10.1 Fixed bed gasifiers: (**a**) updraft, (**b**) downdraft, (**c**) cross-draft

low gas velocity and they are suitable for small scale applications, that is, <10 MW [65]. Fixed bed gasifiers are classified into three groups according to the way the biomass and gasifying agent enter the reactor.

10.6.1.1 Updraft Gasifier

Updraft gasifiers are the oldest and simplest type of gasifier. In these reactors, the gasifying gas travels upward while the solid fuels move downward as shown in Fig. 10.1a. The major advantage of this type of gasifier is its simple structure and design, low capital cost, and high char burn-out, which leads to low gas exit temperatures and high equipment efficiency, as well as the possibility to process feedstock of various shapes [66].

On the other hand, poor heat and mass transfer can increase the risk of "channelling" in the equipment, which may lead to an oxygen breakthrough and rapid gas-phase combustion reactions (and, possibly, an explosion). Fuels that are prone to agglomeration during gasification are not suitable for these types of reactors due to poor heat and mass transfer. However, high-ash, high-moisture or low-volatile feedstocks are suitable fuels for updraft gasifiers. Also, there are problems associated with high tar production that require gas cleaning operations. This is of minor importance, though, if the gas is used for direct heat applications, where the tar simply can be burnt. Nonetheless, this is not recommended for engine applications.

10.6.1.2 Downdraft Gasifier

The problem of tar entrainment in the gas stream had been solved by designing downdraft gasifiers where the gasifying agent enters from the top and the gaseous products leave the reactor through a bed of hot ash at the bottom. A schematic of this reactor is shown in Fig. 10.1b. The main advantage of downdraft gasifiers lies in the *possibility* of producing a tar-free gas from high volatile fuels which will be suitable for engine applications. This is due to the fact that most of the tar is cracked when passing through the hot ash before exiting the reactor. A major drawback of this type of gasifier is its inability to operate with a wide range of fuels. In particular, fluffy, low density materials can cause flow problems, like an excessive pressure drop, requiring the solid fuel to be pelletized or briquetted before use. Downdraft gasifiers also suffer from slagging when operating with fuels characterized by high ash content. Compared to updraft gasifiers, downdraft systems show lower efficiencies resulting from a lack of internal heat exchange as well as the LHV of the gas.

10.6.1.3 Cross-draft Gasifier

This type of gasifier is suitable for low ash fuels. Unlike downdraft and updraft types, it releases the product from its sides. Air at high velocity enters the gasifier through a nozzle at a certain height above the grate creating a very high temperature zone. The product gas exits from the opposite sides of the gasifier. Start-up time is much faster (5–10 min) compared to other moving bed gasifiers, which improves the response to load changes. Due to relatively high temperature zones on a cross-draft gasifier, the product gas is low in tar, high in carbon monoxide, and low in hydrogen and methane.

10.6.2 Fluidized Bed Gasifier

Fluidized bed reactors have been used extensively for coal gasification. Fluidized bed gasifiers are characterized by excellent heat and mass transfer, which facilitates the control of the bed temperature. Furthermore, this allows the use of a wide variety of fuels, such as fluffy and fine grained materials without the need of pre-treatment. However, problems with solid feeding and fly-ash sintering in the gas channels can occur with some biomass fuels.

There are two major categories of fluidized beds for biomass gasification: bubbling (BFB) and circulation fluidized beds (CFB). These two types of gasifiers are characterized by significantly different gas and solids hydrodynamics. CFBs are operated with a higher gas velocity (3–5 m/s) compared to BFB (0.4–1 m/s). In CFBs, the solid particles are entrained along a tall tubular section (called a riser), which allows long gas and solid residence times. CFB gasifiers at atmospheric pressure have proven very reliable with a variety of feedstock and are relatively easy to scale up from a few MWh up to 100 MW h [67].

The family of fluidized bed reactors include other relatively new concepts for gasifiers: chemical looping, dual gasifier, and an internally circulating fluidized bed. The main motivation driving the development of these reactors is to reduce the syngas dilution in nitrogen or carbon dioxide. In chemical looping gasification, a solid oxygen carrier is circulated continuously between two fluidized bed reactors: (1) a biomass fluidized bed gasifier and (2) a fluidized bed to oxidize the oxygen carrier. In the gasifier, the oxygen-carrier releases the oxygen for the gasification reactions before being entrained back to the second fluidized bed where it is re-oxidized [68]. The concept of circulating solids between two fluidized beds has also been extensively studied for CO_2 capture [69]. There is currently an operating industrial plant in Gussing (Austria), which consists of one circulating fluidized bed combustor and a bubbling fluidized bed gasifier. This system has, however, its own drawbacks. For example, the amount of char combustion in the gasifier may not be sufficient to provide the required heat for endothermic reactions. Co-firing may therefore be necessary [70].

10.6.3 Entrained-Flow Gasifier

Entrained flow gasifiers consist of a co-current plug flow reactor and have been extensively studied and used for coal gasification. The gas and particle residence times in the reactor are very short (a few seconds), which requires very small size feedstock (100 μm) and very high temperature (>1,000 °C) to maximize conversion. There are two types of entrained flow gasifiers: slagging and non-slagging. In the slagging gasifier, the ash melts in the gasifier, flows down the walls of the reactor and finally leaves the reactor as a liquid slag. This type of entrained flow gasifier is preferred for biomass. In a non-slagging gasifier, the walls are free of slag, which is suitable for feedstock with low ash content [71]. Some of the advantages of entrained flow gasifiers are low tar and methane content, high carbon conversion, and ash existing as slag. Furthermore, the reactor can be operated, with a wide variety of feedstock at high pressure and temperature. Due to some problems associated with biomass molten ash and the use of very fine biomass particles, however, the use of an entrained flow biomass gasifier has been rather limited. One example of an entrained flow gasifier application for biomass is the Choren process with a maximum capacity of 45 MW h and using wood as feedstock [72].

10.6.4 Plasma Gasifier

Plasma gasifiers use a plasma gun to create an intense electric arc between two electrodes. The temperature of the arc can reach extremely high (>13,000 °C). Biomass, on the other hand, is fed at lower temperatures (2,700–4,500 °C), but still sufficiently high to crack heavy hydrocarbons. Due to its high operating temperature, the plasma

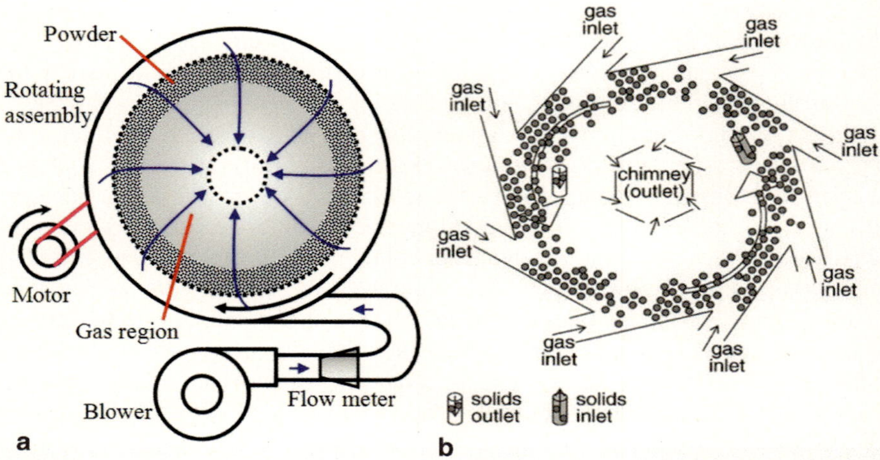

Fig. 10.2 Sketch of a (a) conventional rotating fluidized bed and (b) novel rotating fluidized bed. (Reprinted from ref. [74], Copyright 2008, with permission from Elsevier)

gasifier can crack harmful products, such as furan and dioxin, making it suitable for MSW and other types of waste products.

10.6.5 Rotating Fluidized Bed Gasifier

As previously discussed, heat and mass transfer play an important role in the gasification process. As such, fluidized beds are advantageously characterized by high heat and mass transfer rates that will result in temperature homogeneity and the rapid mixing of particulate materials. Heat and mass transfer are maximized by increasing the gas fluidization velocity, but the gas velocity cannot be indefinitely increased. With increasing superficial gas velocity the solid hold up in the bed region decreases causing a short solid residence time and a low conversion or reactor volume increase. To overcome these limitations of conventional fluidized bed gasifiers, the concept of a rotating fluidized bed (RFB) has been introduced. RFB reactors were first patented by Horgan and Morrison in 1979 for a coal combustion application in a centrifugal fluidized bed [73]. RFB consists of a cylindrical gas distributor chamber rotating around its axis (shown in Fig. 10.2). The rotating motion of the cylinder is transferred to the particles via fraction and gas is injected inward through the gas distributor. The particles are fluidized uniformly under the action of two opposite forces: the radially inward drag force exerted by the injected gas and the radially outward centrifugal force. The minimum fluidization velocity increases with increasing the reactor rotation speeds (rising centrifugal force magnitude). The magnitude of the forces, which can be much higher than gravity, depends on the operating conditions: solid rotating velocity and gas injection velocity.

The rotating motion of RFBs may cause difficulties in design and operation, like severe vibrations of the reactor during operation. To overcome these difficulties, a new concept of a rotating fluidized bed has been proposed where the geometry of the reactor is fixed and the fluidizing gas is injected tangentially via multiple gas inlet slots at the fluidization chamber wall. As a result, the tangential drag force will induce the solid particles into a rotating motion as well as produce a radially outward centrifugal force [74]. The RFB reactor can operate at much higher gas velocities and solid hold-up compared to the fluidized bed. Due to high attrition the novel RFB reactor is suitable for processes where solid is a reactant, like biomass gasification [75].

As discussed above there are several options when choosing and designing a gasifier. The choice of one type of gasifier over another is, however, determined by the type of fuel, its size, the moisture content, the physical limitations of the reactor, its production capacity and the final use of the product gas since one type is not necessarily suitable for the full range of capacities. The different gasifier characteristics are summarized in Table 10.3.

10.7 Gasifier Reaction Modeling

Gasifier performance could be expressed in terms of gasification efficiency and product quality, such as heating value and the amount of desired product gas. Mathematical modeling and simulation of the gasification process provides qualitative information on the gasifier performance, such as the effect of feedstock, design parameters, and operating conditions of the process even though it is not very accurate. A mathematical model or simulation is, however, useless if it cannot reproduce experimental data with acceptable deviation [14]. Gasifier models could be classified into the following groups: thermodynamic equilibrium, kinetic, artificial neural network and computational fluid dynamics (CFD).

10.7.1 Equilibrium Models

The equilibrium model predicts the maximum yield when the reactants are in contact for an infinite time without taking into account the reactor type and size [76]. In reality, the products leave the reactor before having the opportunity to reach equilibrium so this type of model only provides the ideal yield. For practical applications, therefore, the use of the kinetic model is more realistic. At higher temperatures ($>$1,500 K), however, the use of the equilibrium model is more effective. There are two types of equilibrium modeling approaches: (1) stoichiometric or the use of the equilibrium constant; (2) non-stoichiometric or the minimization of Gibbs free energy method. In the stoichiometric model, all the chemical reactions and species involved are considered. For a known reaction mechanism this method predicts the maximum yield of all the products and the possible limiting behavior of the reactor. In the

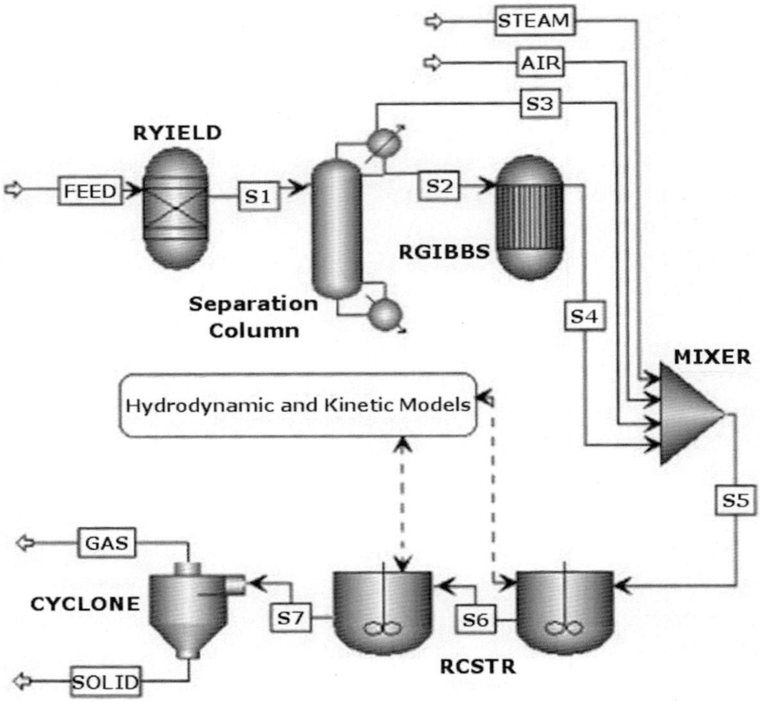

Fig. 10.3 Comprehensive simulation diagram for the fluidized bed gasification process. (Reprinted from Ref. [83], Copyright 2008, with permission from Elsevier)

non-stoichiometric method, however, no knowledge of the reaction mechanism is required. This method is based on the minimization of total Gibbs energy with only the input of feed composition and is suitable for fuels, like biomass, although the chemical reaction mechanism is not very clear.

Also process simulators such as Aspen Plus, Simulink [77] and, Cycle Tempo [78] have been used to evaluate mass and energy flows and to perform economic and environmental evaluations. Gasification process simulation were done by considering thermodynamic equilibrium by minimizing the Gibbs free energy [77, 79]. Mathieu et al. modeled wood gasification in a fluidized bed using Aspen Plus [80]. The model was based on the minimization of the Gibbs free energy and the process was uncoupled in pyrolysis, combustion, Boudouard reaction and gasification. Other simulations were proposed based on using two equilibrium reactors in series with a selective by-pass over the first one [80] or by breaking up biomass into its constituting elements (C, H, N, O, S, Cl, ash, and moisture) dealing with them separately [81–84]. Mitta et al. modeled a fluidized-bed tire gasification plant with air and steam using Aspen Plus [85]. Their gasification model was divided into three different stages: drying, devolatilization-pyrolysis and gasification-combustion. Nikoo and Mahinpey developed a model capable of predicting the performance of an atmospheric fluidized-bed gasifier [83]. They used both built-in Aspen Plus reactor models and external FORTRAN subroutines for hydrodynamics and kinetics to simulate the gasification process. Other authors have worked with Aspen Plus to model the gasification process for coal and biomass. Yan and Rudolph developed a model for a compartmented fluidised-bed coal gasifier process [86], Sudiro et al. modeled the gasification process to obtain synthetic natural gas from petcoke [87]. Abdelouahed proposed a compressive model for dual fluidized bed gasifier modeling with ASPEN Plus [88]. A comprehensive review on Biomass gasification simulation also provided by Puig-Arnavat et al. [90] (Fig. 10.3)

10.7.2 Kinetic Models

Gas composition exiting a gasifier often varies from the composition predicted by equilibrium models [76]. This is caused by the fact that products exit from the reactor before reaching an equilibrium state and thus demonstrates the need for kinetic models to simulate gasifier behavior. Gasification reactions are divided into three categories: drying, devolatilization, and gasification. The time taken for drying and devolatilization is much faster than gasification of char. Some models assume the first two steps to be instantaneous and that the rate of char gasification controls the overall process [89, 90].

Kinetic models provide information on the progress of the reaction by taking into account the reactor type, size, and its hydrodynamics. In the kinetic model, the reaction kinetic is solved simultaneously with bed hydrodynamics, mass and energy balance to achieve the gas, and tar and char yield at specific operating conditions. Unlike other models, the kinetic model is sensitive to the gas-solid mixing and the flow pattern in the gasifier. Based on the process, this type of model can be divided into three groups: (1) fluidized bed; (2) fixed bed; and (3) entrained flow.

10.7.2.1 Fluidized Bed Kinetic Model

The kinetic model of the fluidized bed gasifier consists of reactor hydrodynamics, which define the transport phenomena of the gasification medium through the system and solid mixing behavior. There are several versions of fluidized bed hydrodynamic models [91–93]:

1. The two-phase model in a bubbling fluidized bed, which considers bubble and emulsion phases.
2. The three-phase model in a bubbling fluidized bed consists of bubbling, cloud, and emulsion phases.
3. The core-annulus model for circulating fluidized beds where the core is the upward flow of gas and solid in the center and the annulus is the downward flow of gas and solid close to the wall.

4. The compartment models in which the fluidized bed is divided into slices or horizontal sections.

These types of fluidized bed models avoid the complexity of gas-solid dynamics but still keep the fluid-dynamic effects by considering different regions and phases throughout the reactor, which are described by semi-empirical correlations [91, 92, 93, 94 and 95].

Gas flow through the bed can be modeled as follows:

1. Bubble phase as plug flow and emulsion phase as ideally mixed gas.
2. Both phases as ideally mixed gases.
3. Both phases as plug flow with mass transfer between two phases.
4. Upward gas in the core as plug flow and solid backflow in the annulus [95].

There are three ways to describe conversion models for single char particles:

1. Shrinking core model;
2. Shrinking particle model; and
3. Uniform conversion model [96, 97].

Shrinking core and shrinking particle models are both surface reaction models where the fast reaction takes place as soon as the reactant gas reaches the external surface of the particle. In the shrinking particle model, however, the ash peels off instantaneously and the particle shrinks during reaction. In the shrinking core model, the particle size stays constant since the ash remains attached to the particle and becomes an additional heat and mass resistance. In the uniform conversion model the reaction takes place all over the char particle uniformly.

Different types of fluidized bed modeling have been applied for coal and biomass gasifiers [90, 98–102]. The following section presents essential equations for a one-dimensional steady-state model of a bubbling fluidized bed gasifier. The fluidized bed is divided into two regions, a dense zone and a free board. The gas flow in the dense part consists of bubble and emulsion phases, which deals with drying, pyrolysis, and gasification of biomass. The freeboard is free of solid particles and only gas phase reactions continue from the bed. The mass balance for the emulsion and bubble phases can be expressed as (10.1) and (10.2):

Emulsion phase:

$$-(1-\delta_b)\varepsilon_{mf}\frac{d}{dz}(C_{ie}u_e) - K_{be}(C_{ib}-C_{ie}) + (1-\delta_b)(1-\varepsilon_{mf})\Sigma_{g-s}R_i + (1-\delta_b)\varepsilon_{mf}\Sigma_{g-g}R_i = 0 \tag{10.1}$$

Bubble phase:

$$-\delta_b\frac{d}{dz}(C_{ib}u_b) + K_{be}(C_{ib}-C_{ie}) + \gamma_b\delta_b\Sigma_{g-s}R_i + (1-\gamma_b)\delta_b\Sigma_{g-g}R_i = 0 \tag{10.2}$$

The boundary conditions for the above equations are defined in the feeding zone, which is the gas composition, predicted using the pyrolysis kinetics.

For solid hydrodynamics, a concurrent back mixing model is considered [95]. Based on the CCBM, (10.3) could be written for the char particles.

Char balance in the ascending phase:

$$F_{as,0}\frac{dX_{as}}{dz} + K_w(C_{ds,c} - C_{as,c})Af_{as} + Af_{as}\Sigma_{g-s}R_i = 0 \qquad (10.3)$$

Char balance in the descending phase:

$$-F_{ds,0}\frac{dX_{ds}}{dz} + K_w(C_{as,c} - C_{ds,c})Af_{as} + A(1 - f_{as})\Sigma_{g\text{-}s}R_i = 0 \qquad (10.4)$$

The bubble and emulsion equations, which give the gas concentration profile in the bed, and the char balance equation, which provides the char conversion profile in the bed, will be solved simultaneously for all gas species to obtain produced gas compositions and yields in the bed.

Based on this assumption there won't be any solid in the freeboard and it will be considered as a plug flow reactor. The mass balance for each gas species in this region can be written as (10.5):

$$\frac{d(u_g C_{i,g})}{dz} = \sum\nolimits_{g-g} R_i \qquad (10.5)$$

10.7.2.2 Updraft Kinetic Model

The major assumptions regarding the moving bed (updraft gasifier) model are as follows:

- There is no temperature or concentration distribution radially.
- The solid hydrodynamic is considered as a plug flow flowing downward.
- The gas flows upward as a plug flow.
- The mass transfer between two phases takes place by diffusion [14].

The mass balance equation of jth-gas species can be written as (10.6):

$$u_g\frac{d\rho_{g,j}}{dz} = D_{g,j}\frac{d^2\rho_{g,j}}{dz^2} + R_{m,j} \qquad (10.6)$$

Also, the energy balance of the gasifier in the z direction is expressed as follows:

$$\rho_g C_g u_g\frac{dT}{dz} = \lambda g\frac{d^2T}{dz^2} + Q_{\text{gasification}} + Q_{\text{conv}} + Q_{\text{rad}} + Q_{\text{mass}} \qquad (10.7)$$

These equations will be solved simultaneously with the appropriate kinetics discussed in section 3.

10.7.2.3 Entrained Flow Gasifier

The reactor is considered as a one-dimensional plug flow reactor under steady-state conditions [94]. The gas phase is considered as perfectly mixed radially and the solid particles are distributed uniformly in the radial direction. The mass balance for solid and gas components can be described as follows [103]:

$$\frac{dW}{dL} = -N_v A \sum r_k(T_S, L) \tag{10.8}$$

$$\frac{dF_{g,i}}{dL} = \pm N_v A \sum \nu_{1k} r_k(T_S, L) \tag{10.9}$$

10.7.3 CFD Models

CFD modeling solves a set of equations for the conservation of mass, momentum, and energy simultaneously to give the gasifier temperature, the product concentration, and the hydrodynamic parameters at different locations. Due to the complexity of the gasification process, however, there are not many CFD models available for this process and most of them must use fitting parameters and major assumptions for areas where accurate information is not available. Most of the CFD models are for coal gasification and combustion in entrained flow reactors since gas–solid flow is less complex compared to fluidized bed reactors [104] and [105]. A typical CFD model for gasification consists of a set of sub-models for different reactions and phenomena, such as drying biomass particles, devolatilization (pyrolysis), secondary pyrolysis, and char oxidation [106]. There are also other sophisticated subroutines for the destruction of solid fuels during gasification and combustion, which could be coupled with transport phenomena of the gasifier [107]. Due, however, to considerable computational times for CFD models, particularly when chemical reactions are involved, this type of modeling is not very common for fluidized bed gasifiers.

10.7.4 Neural Network Models

Neural network models basically rely on a large number of experimental data [108, 109]. This model connects the input and output of a process unit with less knowledge of the system phenomena compared to the equilibrium and kinetic approaches. Guo et al. developed a neural network model for biomass gasification in a fluidized bed and emphasized its success and applicability for this process.

Symbols and Nomenclature

A = Cross-sectional area of the gasifier (m^2)
C = Molar concentration (mol/m^3)
C_p = Specific heat

D = Diffusion coefficient (m^2/s)
F = Molar flow rate (mol/s)
$Q_{gasification}$, Q_{conv}, Q_{rad}, and Q_{mass} = Energy transfer due to gasification, convection, radiation (kW/m^3 of bed)
R = Reaction rate (kg/m^3 s)
X = Char conversion
u = Gas velocity (m/s)
W = Solid flow rate (kg/s)

Subscripts
as = Ascending phase
b = Bubble
be = Bubble-emulsion
c = Char
ds = Descending phase
f = Freeboard
g = Gas phase
i = Gaseous components in the product gas
k = Reaction number
w = Wake

References

1. Panwar NL (2011) Biomass for domestic and agro industrial application. Nova Science Publishers, New York
2. Vassilev V, Baxter D (2012) An overview of the organic and inorganic phase composition of bio-mass. Fuel 94:1
3. Skoulou V, Kantarelis E, Arvelakis S, Yang W, Zabaniotou A (2009) Effect of biomass leaching on H2 production, ash and tar behavior during high tem-perature steam gasification (HTSG) process. Int J Hy-drogen Energy 34(14):5666–5673
4. Kobayashi N, Guilin P, Kobayashi J, Hatano S, Itaya Y, Mori S (2008) A new pulverized biomass utilization technology. Powder Technol 180:272–283
5. Bergman PCA, Boersma AR, Kiel JHA, Prins MJ, Ptasinski KJ, Janssen FJJG (2004) Torrefaction for entrained flow gasification of biomass. Second world biomass conference. ETA-Florence and WIP-Munich, Rome, pp 679–682
6. Pels JR, Bergman PCA (2006) proof of prin-ciples—Phase 1, Report ECN, Petten, (NL) ECN-E-06-021
7. Prins MJ, Ptasinski KJ, Janssen FJJG (2006) More efficient biomass gasification via torrefaction. Energy 31:3458–3470
8. Xiao R, Chen X, Wang F, Yu G (2010) Py-rolysis treatment of biomass for entrained-flow gasifi-cation. Appl Energy 78:1
9. Okumura Y, Hanaoka T, Sakanishi K (2009) Effect of pyrolysis conditions on gasification reactivity of woody biomass-derived char. Proc Combust Inst 32(2):2013–2020
10. Rezaiyan J, Nicholas P (2005) Gasifica-tion technologies: a primer for engineers and scientists. CRC Press, Boca Raton
11. Van Den Aarsen F, Beenackers A, Swaaij WV (1985) Wood pyrolysis and carbon dioxide char gasification kinetics in a fluidized bed. Fundamentals of biomass thermochemical conversion. Elsevier, London

12. Hemati M, Laguerie C (1988) Détermination de la cinétique de vapogazéification de charbon de bois en thermobalance. Entropie 142:29–40
13. Janse AMC (1998) Combustion kinetics of char obtained by flash pyrolysis of pine wood. Ind Eng Chem Res 37(10):3909–3918
14. Souza-Santos MLd (2004) Solid fuels combustion and gasification: modeling, simulation, and equipment operation. Marcel Dekker, New York
15. Jensen A, Johnsson JE, Andries J, Laughlin K (1995) Formation and reduction of NOx in pressurized fluidized bed combustion of coal. Fuel 74(11):1555–1569
16. Parent JD, Katz S (1948) Equilibrium compositions and enthalpy changes for the reaction of carbon, oxygen, and steam. Inst Gas Tech Res Bull 2
17. Boronson ML, Howard JB, Longwell LP, Peters WA (1987) Product yields and kinetics from the vapor phase cracking of wood pyrolysis tars. AIChE J 35(1):120–128
18. Hamel S, Krumm W (2001) Mathematical modelling and simulation of bubbling fluidised bed gasifiers. Powder Technol 120(1–2):105–112
19. Shafizadeh F, Chin P (1977) Thermal de-terioration of wood, ACS Symposium series. American Chemical Society, Washington, DC
20. Bradbury AGW, Sakai Y, Shafizadeh F (1979) A kinetic model for pyrolysis of cellulose. J Appl Polym Sci 23(11):3271–3280
21. Thurner F, Mann U (1981) Kinetic investi-gation of wood pyrolysis. Ind Eng Chem Process Des Dev 20(3):482–488
22. Koufopanos CA, Lucchesi A, Maschio G (1989) Kinetic modelling of the pyrolysis of biomass and biomass components. Can J Chem Eng 67:75–84
23. Koufopanos CA et al (1991) Modeling of the pyrolysis of biomass particles. Studies on kinetics, thermal and heat transfer effects. Can J Chem Eng 69:907–915
24. Di Blasi C (1997) Influences of physical properties on biomass devolatilization characteristics. Fuel 76(10):957–964
25. Gronli M, Varhegyi G, Blasi CD (2002) Thermogravimetric analysis and devolatilization kinet-ics of wood. Ind Eng Chem Res 41(17):4201–4208
26. Mészáros E, Jakab E, Várhegyi G, Szep-esváry P (2004) Comparative study of the thermal be-havior of wood and bark of young shoots obtained from an energy plantation J Anal Appl Pyrolysis 72:317
27. Radmanesh R, Chaouki J (2006) A uni-fied lumped approach in kinetic modeling of biomass pyrolysis. Fuel 85(9):1211–1220
28. Bassilakis R, Carangelo RM, Wójtowicz MA (2001) A comparative kinetic study on the pyroly-sis of three different wood species. Fuel 80:1765
29. Teng H, Wei YC (1998) Thermogravimet-ric studies on the kinetics of rice hull pyrolysis and the influence of water treatment. Ind Eng Chem Res 37:3806–3811
30. Meszaros E, Várhegyi G, Jakab E (2004) Thermogravimetric and reaction kinetic analysis of biomass samples from an energy plantation. Energy Fuel 18:497–507
31. Varhegyi G, Antal MJ, Szekely T (1989) Kinetics of the thermal decomposition of cellulose, hemicellulose, and sugarcane bagasse. Energy Fuel 3:329–335
32. Orfao JM, Antunes FJA, Figueiredo JL (1999) Pyrolysis kinetics of lignocellulosic materials three independent reactions model. Fuel 78:349–358
33. Sorum L, Gronli MG, Hustad JE (2001) Pyrolysis characteristics and kinetics of municipal solid wastes. Fuel 80:1217–1227
34. Hajaligol MR, Howard JB, Longwell JP, Peters WA (1982) Product compositions and kinetics for rapid pyrolysis of cellulose. Ind Eng Chem Process Des Dev 21:457–465
35. Nunn TR, Howard JB, Longwell JP, Pe-ters WA (1985) Product compositions and kinetics in the rapid pyrolysis of sweet gum hardwood. Ind Eng Chem Proc Des Dev 24:836–844
36. Simmons G, Gentry M (1986) Kinetic for-mation of CO, CO2, H2, and light hydrocarbon gases from cellulose pyrolysis. J Anal Appl Pyrolysis 10:129–138
37. Banyasz JL, Lyons-Hart J, Shafer KH (2001) Gas evolution and mechanism of cellulose py-rolysis. Fuel 80:1757–1763
38. Walker PL, Rusinko F, Austin LG (1959) Gas reactions of carbon. Adv Catal 11:133–221

39. Encinar JM, González JF, Rodriguez JJ, Ramiro MJ (2001) Catalysed and uncatalysed steam gasification of eucalyptus char: influence of variables and kinetic study. Fuel 80:2025–2036
40. Barrio M, Gøbel B, Rimes H, Henriksen U, Hustad JE, Sørensen LH (2000) Steam gasification of wood char and the effect of hydrogen inhibition on the chemical kinetics, Prog Thermochem Biomass Convers 1:32–46
41. Di Blasi C (2009) Combustion and gasifi-cation rates of lignocellulosic chars. Prog Energy Combust Sci 35(2):121–140
42. Probstein RF, Hicks RE (2006) Synthetic fuels. Dover Publications, New York, pp 98–99.
43. Boerrigter H, Rauch R (2005) Syngas production and utilization. In: Knoef HAM (ed) Biomass gasification handbook, chap. 10. Biomass Technology Group, Enschede
44. Liden AG, Berruti F, Scott DS (1988) A kinetic model for the production of liquids from the flash pyrolysis of biomass. Chem Eng Comm 65(1):207–221
45. Rath J, Staudinger G (2001) Vapour phase cracking of tar from pyrolysis of birch wood. Therm Sci 5(2):83–94
46. Rath J, Staudinger G (2001) Cracking re-actions of tar from pyrolysis of spruce wood. Fuel 80(10):1379–1389
47. Sutton D, Kelleher B, Ross JRH (2001) Review of literature on catalysts for biomass gasification. Fuel Process Technol 73:155–173
48. Dayton D. (2002) A review of the literature on catalytic biomass tar destruction. National Re-newable Energy Laboratory, New York
49. Varhegyi G, Antal MJ (1988) Simultane-ous thermogravimetric-mass spectrometric studies of the thermal decomposition of biopolymers. 1. Avicel cellulose in the presence and absence of catalysts. Energy Fuel 2:267–272
50. Corella J, Herguido J, Gonzalez-Saiz J, Alday FJ, Rodrıguez-Trujillo JL (1988) Fluidized bed steam gasification of biomass with dolomite and with a commercial FCC catalyst. Research in thermo-chemical biomass conversion. Elsevier, London, pp 754–765
51. Olivares A, Aznar MP, Caballero MA (1997) Biomass gasification: produced gas upgrading by in-bed use of dolomite. Ind Eng Chem Res 36:5220–5226
52. Narvaez I, Orio A, Aznar MP (1996) Bio-mass gasification with air in an atmospheric bubbling fluidized bed. Effect of six operational variables on the quality of the produced raw gas. Ind Eng Chem Res 35(7):2110–2120
53. Corella J, Toledo JM, Padilla R (2004) Olivine or dolomite as in-bed additive in biomass gasi-fication with air in a fluidized bed: which is better? En-ergy Fuel 18(3):713–720
54. Courson C, Makaga E, Petit C, Kienne-mann A (2000) Development of Ni catalysts for gas production from biomass gasification reactivity in steam- and dry-reforming. Catal Today 63:427
55. Swierczynski D, Courson C (2006) Char-acterization of Ni–Fe/MgO/Olivine catalyst for fluidized bed steam gasification of biomass. Chem Mater 18:4025
56. Rapagna S, Gallucci K, Di MM (2010) Gas cleaning, gas conditioning and tar abatement by means of a catalytic filter candle in a biomass fluid-ized-bed gasifier. Bioresour Technol 101:7123–7130
57. Gardner B, Davidson M, Guan X, Hendrix H, Shirley B (2001) Commercial readiness of hot-gas filtration for pressurised combustion. In: Dahlin R, Landham C, Spain J (eds) Proceedings, 16th Interna-tional Conference on FBC, Reno. Southern Research Institute, Wilsonville
58. Sharma SD, Dolan M, Ilyushechkin AY, McLennan KG, Nguyen T, Chase D (2010) Recent developments in dry hot syngas cleaning processes. Fuel 89:817–826
59. Knoef HAM (2000) Inventory of biomass gasifier manufacturers and installations. Final report to the European Commission, Contract DIS/1734/98–NL, Biomass Technology Group B.V., University of Twente, Enschede, The Netherlands
60. Punjak WA (1988) High temperature in-teractions of alkali vapours with solids during coal combustion and gasification, Ph.D. thesis, The Uni-versity of Arizona, USA
61. Iversen HL, Gøbel B (2004) Update on gas cleaning technologies for biomass gasification gas for different applications. Biomass Gasification Group, Denmark

62. Higman C, Van Der Burgt M (2008) Gasi-fication, 2nd edn. Gulf Professional Publishing/Elsevier, Oxford
63. Saracco G, Specchia V (1994) Catalytic inorganic membrane reactors: present experience and future opportunities, Catal Rev Sci Eng 36(2):305
64. Ma L, Verelst H, Baron GV (2005) Inte-grated high temperature gas cleaning: tar removal in biomass gasification with a catalytic filter. Catal Today 105:729–734
65. Nacken M, Ma L, Engelen K, Heidenreich S, Baron GV (2007) Development of a tar reforming catalyst for integration in a ceramic filter element and use in hot gas cleaning. Ind Eng Chem Res 46:1945–1951
66. Carlos L (2005) High temperature air/steam gasification of biomass in an updraft fixed batch type gasifier, Ph.D. thesis, Royal Institute of Technology, Energy Furnace and Technology, Stock-holm, Sweden
67. Maniatis K (2001) Progress in biomass gasification: an overview. In: Bridgwater AS (ed) Pro-gress in thermochemical biomass conversion, vol I. Blackwell, London
68. Aghabararnejad M, Chaouki J, Patience GS (2011) The development of a novel Cu-Mn oxygen carrier for the chemical looping gasification of biomass, International conference on circulating fluid-ized beds and fluidization
69. Chen S, Wang D, Xue Z, Sun X, Xiang W (2011) Calcium looping gasification for high-concentration hydrogen production with CO2 capture in a novel compact fluidized bed: simulation and op-eration requirements. Int J Hydrogen Energy 36(8):4887–4899
70. Corella J, Toledo JM, Molina G (2007) A review on dual fluidized-bed biomass gasifiers. Ind Eng Chem Res 46(21):6831–6839
71. Van Der Drift A, Boerrigter H, Coda B (2004) Entrained flow gasification of biomass: ash behaviour, feeding issues and system analyses, Re-port C-04-039. Energy Research Centre of The Neth-erlands (ECN), Petten, The Netherlands, p 58
72. Dautzenberg K, Hanf J (2008) Biofuel chain development in Germany: organisation, opportunities, and challenges, Energy Policy 36:485–489
73. Horgan JJ, Morrison FH (1979) Centrifu-gal combustor with fluidized bed and construction thereof.
74. De Wilde J, De-Broqueville A (2008) Ex-perimental investigation of a rotating fluidized bed in a static geometry, Powder Technol 183:426–435
75. Abdollahi NM, Chaouki J (2010) Biomass gasification in rotating fluidized bed. Conference on fluidization, Halifax, Nova Scotia, Canada
76. Li XT, Grace JR, Watkinson AP, Lim CJ, Ergudenler AE (2001) Equilibrium modeling of gasifi-cation free energy minimization approach and its ap-plication to a circulating fluidized bed coal gasifier. Fuel 80:195–207
77. Baratieri M, Baggio P, Bosio B, Grigiante M, Longo GA (2009) The use of biomass syngas in IC engines and CCGT plants: a comparative analysis. Appl Therm Eng 29(16):3309–3318
78. Toonssen R, Woudstra N, Verkooijen AHM (2008) Exergy analysis of hydrogen production plants based on biomass gasification. Int J Hydrogen Energy 33(15):4074–4082
79. Paviet F, Chazarenc F, Tazerout M (2009) Thermo chemical equilibrium modelling of a biomass gasifying process using ASPEN PLUS. Int J Chem Reactor Eng 7:A40
80. Mathieu P, Dubuisson R (2002) Performance analysis of a biomass gasifier. Energy Convers Manage 43(9–12):1291–1299
81. Toonssen R, Woudstra N, Verkooijen AHM (2008) Exergy analysis of hydrogen production plants based on biomass gasification. Int J Hydrogen Energy 33(15):4074–4082
82. Hannula I, Kurkela E (2010). A semi-empirical model for pressurised air-blown fluidised-bed gasification of biomass. Bioresour Technol 101(12):4608–4615
83. Nikoo MB, Mahinpey N (2008) Simulation of biomass gasification in fluidized bed reactor using ASPEN PLUS. Biomass Bioenergy 32(12):1245–1254
84. Wang Y, Dong W, Dong L, Yue J, Gao S, Suda T, Xu G (2010) Production of middle caloric fuel gas from coal by dual-bed gasification technology. Energy Fuel 24(5):2985–2990
85. Mitta N, Ferrer-Nadal S, Lazovic A, Perales J, Velo E, Puigjaner L (2006) Modelling and simulation of a tyre gasification plant for synthesis gas production. 16th European symposium

on computed aided process engineering and 9th international symposium on process systems engineering
86. Yan H, Rudolph V (2000) Modeling a compartmented fluidized bed process using ASPEN PLUS. Chem Eng Commun 183:1–38
87. Sudiro M, Zanella C, Bressan L, Fontana M, Bertucco A (2009) Synthetic natural gas (SNG) from petcoke: model development and simulation, 9th International conference on chemical and process engineering, Rome, Italy
88. Abdelouahed L, Authier O, Mauviel G, Corriou JP, Verdier G, Dufour A (2012) Detailed modelling of biomass gasification in dual fluidized bed reactors under Aspen Plus. Energy Fuel
89. Puig-Arnavat M, Bruno JC, Coronas A (2010) Review and analysis of biomass gasification models. Renewable Sustainable Energy Rev 14(9):2841–2851
90. Radmanesh R, Chaouki J, Guy C (2006) Biomass gasification in a bubbling fluidized bed reac-tor: experiments and modeling. AIChE J 52(12):4258−4272
91. Kunii D, Levenspiel O (1991) Fluidization engineering, 2nd edn. Elsevier, Amsterdam
92. Grace JR (1986) Fluid beds as chemical reactors. In: Geldart D (ed) Gas fluidization technology, Wiley, Chichester, pp 285–339
93. Davidson F (1985) Davidson JF, Clift R, Harrison D (eds) Fluidization, 2nd edn. Academic Press, London
94. Wen CY, (1979) Chaung TZ Entrainment coal gasification modeling. Ind Eng Chem Proc Des Dev 18(4):684–695
95. Radmanesh R, Mabrouk R, Chaouki J, Guy C (2005) The effect of temperature on solids mix-ing in a bubbling fluidized bed reactor. Int J Chem Reactor Eng 3
96. Sorensen LH, Saastamoinen J, Hustad JE (1996) Evaluation of char reactivity data by different shrink-ing-core models. Fuel 75(11):1294
97. Di Blasi C (1996) Heat, momentum and mass transport through a shrinking biomass particle exposed to thermal radiation. Chem Eng Sci 51(7):1121–1132
98. Ma RP, Felder RM, Ferrell JK (1988) Modelling a pilot-scale fluidised bed coal gasification reactor. Fuel Process Technol 19:265–290
99. Ross DP, Yan HM, Zhong Z, Zhang DK (2005) A non-isothermal model of bubbling fluidised-bed coal gasifier. Fuel 84:1469–1481
100. Yan HM, Heidenreich C, Zhang DK (1999) Modelling of bubbling fluidised bed coal gasifi-ers. Fuel 78:1027–1047
101. Jiang H, Morey RV (1992) A numerical model of a fluidized bed biomass gasifier. Biomass Bioenergy 3:431–447
102. Fiaschi D, Michelini M (2001) A two-phase one-dimensional biomass gasification kinetics model. Biomass Bioenergy 21:1211–1232
103. Vamvuka D, Woodburn ET, Senior PR (1995) Modelling of an entrained flow coal gasifier—1. Development of the model and general predictions. Fuel 74(10):1452–1460
104. Fletcher DF, Haynes BS, Christo FC, Joseph SD (1997) Combustion modelling for an en-trained flow biomass gasifier. International conference on CFD in mineral and metal processing and power generation
105. Marklund M, Tegman R, Gebart R (2007) CFD modelling of black liquor gasification: identification of important model parameters. Fuel 86:1918–1926
106. Di Blasi C (2008) Modeling chemical and physical processes of wood and biomass pyroly-sis. Prog Energy Combust Sci 34(1):47–90
107. Syred N, Kurniawan K, Griffith T, Gral-ton T, Ray R (2007) Development of fragmentation models for solid fuel combustion and gasification as subroutines for inclusion in CFD codes. Fuel 86(14):2221–2231
108. Kalogirou SA (2001) Artificial neural networks in renewable energy systems applications: a review. Renewable Sustainable Energy Rev 5:373–401
109. Guo B, Li D, Cheng C, Lü Z-a, Shen Y (2001) Simulation of biomass gasification with a hy-brid neural network model. Bioresour Technol 76:77–83

Chapter 11
Biomass Pre-Treatments for Biorefinery Applications: Pyrolysis

Jean-Remi Lanteigne, Jean-Philippe Laviolette and Jamal Chaouki

Abstract Biorefineries are small integrated plants aiming at the recovery of specific biomass wastes via their conversion to high-value biofuels and chemicals. Pyrolysis is among the promising technologies to achieve this goal. Three major factors influence the development of a pyrolysis process: the type of biomass, process operating conditions and choice of reactor technology. In this chapter, pyrolysis as a solution to sustain biorefineries is reviewed. The chapter first discusses the various biomass feedstocks and their important characteristics. Secondly, the pyrolysis concepts and kinetics are reviewed in light of their importance in process design and modelling. The chapter also discusses the influence of several process conditions and reactor technologies on pyrolysis reaction and pyrolysis products behaviour. Finally, strategies for product optimization and avoiding purity issues are analyzed. The emphasis of this chapter is put on technologies that have been developed at commercial scale.

Keywords Biomass · Pyrolysis · Biorefinery · Pre-treatments · Bio-oil · Bio-char · Kinetics · Hydrodynamics

11.1 Introduction

In the present day, various technologies are presented as feasible to sustain biorefineries [1, 2]. Two main pathways are often highlighted: the thermochemical [3, 4] and the biochemical pathways [5]. Thermochemical pathways involve the decomposition of matter at high temperature in the absence (pyrolysis) or presence (gasification) of oxygen. On the other hand, single- and multi-step alcoholic fermentations are the main focus of biochemical process development and involve the digestion of matter by microorganisms.

The development of both thermochemical and biochemical processes faces many challenges. Cellulose fermentation processes are characterized by slow reaction rates and low overall yield for non-genetically modified microorganisms [5]. On the other hand, reaching high yield and selectivity remains an issue for both gasification and pyrolysis [4]. However, the thermochemical pathway offers a significant advantage over

J. Chaouki (✉) · J.-R. Lanteigne · J.-P. Laviolette
Chemical Engineering Department, École Polytechnique de Montréal, C.P. 6079, succ. Centre-Ville, Montréal, QC, Canada H3C 3A7,
e-mail: jamal.chaouki@polymtl.ca

Z. Fang (ed.), *Pretreatment Techniques for Biofuels and Biorefineries*, Green Energy and Technology,
DOI 10.1007/978-3-642-32735-3_11,

the biochemical processes: reaction rates are high and offer the potential for high product throughput, which is essential to develop a commercially viable industry. Nevertheless, there is an increasing interest in using both pathways in biorefineries such that their respective advantages are exploited.

Gasification is a multi-step process in the context of biorefineries: it yields a synthesis gas rich in hydrogen and carbon monoxide that requires further synthesis to produce 'biorefinables' [6]. The second recombination process is performed at mild temperatures with patented catalysts [3, 6], and achieving high conversion as well as high selectivity remains a challenge to this day.

On the other hand, pyrolysis potentially offers interesting techno-economic advantages over gasification since it is a single-step process operating at lower temperature that yields three products: non-condensable gas, condensable gas (oil) and char [6]. Pyrolysis processes may therefore require significantly less process equipment compared to gasification. Produced from biomass pyrolysis, bio-char have direct applications as activated carbon [7]. Furthermore, bio-oil can be further refined to produce specialty chemicals and/or biofuels in dedicated plants (biorefineries) that this chapter will discuss in more detail.

Together with products' market value, the operating scale also determines the feasibility of biomass pyrolysis and gasification pathways for biomass pre-treatments for biorefineries. It has been repeatedly demonstrated that gasification is sustainable at very large scale. However, considering that biomass availability is geographically limited, pyrolysis may be better suited for smaller distributed biorefineries. This chapter will discuss biomass pre-treatments for pyrolysis processes as well as pyrolysis as a pre-treatment for further biorefining. The pyrolysis process products and operability depend on several factors including (1) the type of biomass (chemical and physical characteristics), (2) the pyrolysis process operating conditions and (3) the type of reactor (gas/solid hydrodynamics and heat/mass transfer).

11.2 Types of Biomass

Most of the biomass feedstocks can be classified into three families as defined by the US Department of Energy [8]: forestry, agriculture and municipal. Table 11.1 summarizes key chemical and physical characteristics of the main feedstocks that are considered for biorefineries.

Through the conservation of mass, the biomass chemical composition determines the chemical elements present in the three pyrolysis products: non-condensable gas, condensable gas and char. The presence of specific chemical elements in each product fraction is determined by the pyrolysis conditions.

Environmental and purity standards restrict the presence of oxygen, nitrogen, sulfur and inorganics in the pyrolysis products. During pyrolysis, the tendency of producing an aqueous phase generally increases with increasing biomass–oxygen fraction (dry weight) since water is produced [16]. The presence of oxygen may also lead to the production of acids, which are detrimental to the oil stability. On

Table 11.1 Summary of available biomass feedstock for biorefineries

	Biomass family	Forestry/pulp and paper			Agriculture			Municipal	
	Feedstock	Bark and wood residue	Black liquor	Perennial crops	Corn and grains	Oilseeds and plants	Manures	Municipal Solid Waste (MSW) (RDF)	Municipal biosolids
Elemental composition (dry wt%)	C	50–55	30–35	40–55	N/A	60–65	35–40	40–45	40–50
	H	5–7.5	4–6	5–7.5	N/A	7.5–10	5–7.5	5–7.5	5–7.5
	O	40–45	35–40	30–45	N/A	20–25	30–35	35–40	25–35
	N	0.5–1	Trace	3–5	N/A	3–5	2–5	<1	<1
	S	Trace	1–2	<1	N/A	Trace	<1	<1	<1
	Inorganics/trace	3.5–4 (bark)	20–25 (sodium)	3–6	N/A	5–7.5	20–25	10–15	25–35
Molecular composition (dry wt%)	Cellulose	40–45	N/A	30–35	4–5	10–20	N/A	60–70	N/A
	Hemicellulose	25–35	N/A	20–40	4–5	10–20	N/A	10–15	N/A
	Lignin	25–35	N/A	5–20	–	–	N/A	2–5	N/A
	Starch	–	N/A	–	75–77	10–20	N/A	–	N/A
	Extractibles	3–15	N/A	10–20	12–15	40–50	N/A	2–5	N/A
	Inorganics/trace	3.5–4 (bark)	N/A	3–6	1.5–2.5	5–10	N/A	10–15	N/A
	Moisture (wt%)	20–80	40–50	10–70	10–20	5–15	20–70	15–20	5–98
	References	[9–11]	[12]	[10]	[10]	[13, 14]	[10]	[10, 15]	[14]

Note 1: Wood energy crops as per bark & wood residue
Note 2: Agricultural crops as per perennial crops

the other hand, sulfur and nitrogen are not present in biomass in large amounts as shown in Table 11.1, but they will, nonetheless, be present in the products. In this case, the pyrolysis products may need post-treatment since sulfurous compounds are corrosive, while nitrogen affects reactivity as well as pollutant emissions (e.g. fuel-bound NO_X). Moreover, these species are also problematic when performing bio-oil upgrade. Finally, inorganics in the biomass and pyrolysis products may represent a risk of slagging and sintering.

Furthermore, the biomass physical properties strongly influence the gas/solid hydrodynamics as well as the heat/mass transfer in the pyrolysis reactor such that it affects the pyrolysis products (respective yield of the three pyrolysis products and their composition). The important physical properties include the shape of the biomass feedstock, its particle size and moisture fraction. These properties will determine the required biomass physical transformations or pre-treatments.

11.2.1 Biomass Species

11.2.1.1 Bark and Wood Residues from the Pulp and Paper Sector

In 2007, forest mills in the United States of America (USA) produced about 86.7 million dry tons of primary mill residues [17], which were composed mainly of bark, sawdust, wood chips and shavings. Of this amount, over 35 million dry tons of wood residues were used as combustibles and could have been used as feedstock for biorefineries. Wood pyrolysis has been shown to generate high-value products, such as bio-char (promising activated carbon) and bio-oil. Ensyn and DynaMotive are two companies running commercial-scale pilot plants, which convert wood residue via fast pyrolysis. There are many incentives to develop in situ biorefineries (close to pulp and paper plants) in order to avoid significant issues related to transportation and storage.

11.2.1.2 Black Liquor

Black-liquor pyrolysis has been the subject of several studies, but the main efforts have been invested towards gasification. This has been motivated by the fact that black-liquor pyrolysis generates too much solid char [18], which would need to be burned to release the inorganics. The advantage of gasification is that it includes the char combustion process. Thus far, pyrolysis has been mostly considered in the scientific literature as a precursor step to gasification. Consequently, it will not be considered as a potential feedstock for pyrolysis aiming at biorefineries.

11.2.1.3 Wood Energy Crops

The idea of cultivating trees strictly for energy and biorefining purposes has been proposed. Certain fast growing tree species such as cottonwood, aspen and eucalyptus can grow at rates of around 1 m/yr or even more. The short-rotation woody crop (SRWC) technique can be used to reach yields of about 10 dry metric tons of woody crops per hectare per year can be achieved. However, the economic viability of SRWC is very fragile due to the high costs of preparation and fertilization of the sites [19].

Depending on the maturity of the woody crops, chemical composition will remain close to that of wood and bark (see Sect. 11.2.1.1). The main difference will arise due to the leaves and trimmings, which will accumulate dust and metabolic inorganics up to a few mass per cent during the trees' growth.

11.2.1.4 Perennial Herbaceous Crops

Perennial crops are vegetal—not edible for humans—which include among others: switchgrass, weeping lovegrass and Napier grass. Herbaceous crops are usually dedicated to alcoholic fermentation because of their high availability of complex sugar content, but pyrolysis of these vegetals has been shown to produce high oil yields [20]. The oil produced contained water-soluble and water-insoluble fractions. Moreover, significant amount of alkanes and phenolic compounds can be found in these oils [21] suggesting a high potential of perennial crops for specialty chemicals production from pyrolysis.

11.2.1.5 Corn and Grains

Alcoholic fermentation has been the main focus for these feedstocks with bioethanol as its main product. With the current problems surrounding worldwide food supply, it is not ethically and politically justifiable to use food as a fuel while certain countries suffer famine. However, food conservation and storage may sometimes be very difficult and some considerable amounts of corn and grains may become unfit for human consumption. Nevertheless, considering the high starch content and appreciable fermentation yields with this feedstock, it has been rarely studied in fields other than bioconversion.

11.2.1.6 Oilseeds and Plants

Contrary to corn and grains, oilseeds and their plants show very poor starch content. Many species, such as colza, are dedicated to the production of biodiesel. Although biodiesel production has been demonstrated as technically feasible at a large scale, it is not economically sustainable without government grants or incentives. Several studies on oilseeds and plants pyrolysis can be found in the scientific literature, which indicates a strong interest for this conversion technology. As an example, castor bean slow pyrolysis yields easily over 65 % oil with as low as 20 % solid residue [22]. Due to the high oil yield, there is interest in mixing these oils with diesel to produce blends for transportation fuels. However, for the same reasons that were brought in Sect. 11.2.1.5, these feedstocks should not be diverted from their primary function, namely food supply.

11.2.1.7 Agricultural Crops and Residues

The fruits and vegetables harvest and transformation processes yield many wastes: trimmings, hulls and shells. In 1995, the US Department of agriculture estimated that over 250 million dry tons of agricultural crops and residue were generated over a year in the country [23]. The chemical composition of agricultural crops and residues is very similar to that of perennial herbaceous plants (see Sect. 11.2.1.4). The interest in these feedstocks is reflected in the abundant literature found on agricultural crops and residues pyrolysis [24–26].

11.2.1.8 Animal Manures

Animal manures are used as fertilizers: their high urea, phosphorus and organic contents enrich soils dedicated to agriculture. Cattles are the main manure producers with production of over 200 million dry tons a year in the US (commercial broilers are showing comparable numbers) [23]. Because manure has a heterogeneous composition, thermal decomposition has gained interest to recover that feedstock. Cattle manure is more difficult to collect than poultry manure [10]. Therefore, poultry manure is considered as a good candidate for industrial pyrolysis, and the scientific literature has been mostly focused on this type of manure.

11.2.1.9 Municipal Solid Waste (MSW)

Municipal solid waste (MSW) management has become a major issue worldwide. Issues related to landfilling include: occupation of large areas, generation of greenhouse gases by digestion of the waste, generation of hazardous and refuse materials, etc. New recovery strategies to generate energy, such as incineration, have also given rise to many problems. In Europe, particularly in Germany, rotary kiln incinerators have been extensively studied and issues related to the high temperature have been reported: leaching of metals, emission of carcinogen compounds, emission of particulate matter, etc. Pyrolysis has been identified as a promising avenue for MSW management: its lower operating temperature and absence of oxygen decrease the pollutant emissions as well as the cost of post-treating flue gases.

11.2.1.10 Municipal Biosolids

Waste water treatment is a critical process for our society: waste water contains dissolved organics and inorganics as well as suspended solids and microorganisms that must be eliminated before the water can be released into the environment or purified further to be drinkable. The recovered waste forms sewage sludge, which is difficult to recycle. When dried, contaminants such as heavy metals limit its potential applications. Currently, it is common practice to incinerate sewage sludge with the similar disadvantages to MSW incineration (see Sect. 11.2.1.9). However, incineration could be replaced by more efficient technologies, such as pyrolysis.

11.2.2 Feedstock Pre-Treatment for Pyrolysis

Physical pre-treatments are the key to control feedstock properties, which significantly influence gas/solid hydrodynamics as well as heat/mass transfer in the pyrolysis reactor. Recommended feedstock pre-treatments depend on the initial biomass characteristics, the pyrolysis conditions as well as the reactor type. Pre-treatments also allow the homogenization of the feedstock characteristics with time.

Intrinsic feedstock properties such as specific heat, thermal conductivity and density (dry and true) cannot be easily modified and constitute limitations for thermal processes. On the other hand, feedstock moisture and particle size are the main physical parameters that can be adjusted to optimize the pyrolysis process performance.

Particle fluidizability has been correlated to its average size and density. Geldart classified particles into four groups (Geldart classes A, B, C and D) based on their fluidization behaviour at ambient conditions [27]. Figure 11.1 illustrates the Geldart classification of powders and indicates the properties of common feedstocks for biorefineries.

In the Geldart classification, class C (cohesive particles) and D (large particles) can be detrimental to gas/solid mixing as well as heat/mass transfer. For example, these particles cannot be easily and uniformly fluidized in a reactor: class C particles lead to channelling [28]. Furthermore, large particles (class D) are more subject to internal temperature gradients and species diffusion effects, which affect the final pyrolysis product distribution and composition. Species intra-particle diffusion increase the species exposure time to the pyrolysis conditions (additional time for reactions) while temperature gradients lead to uncertainties related to the characterization of the pyrolysis conditions.

Based on typical wood densities (various species) and Fig. 11.1 sawdust particles (50–500 μm) are classified as Geldart A. On the other hand, coarser bark or wood residue particles are classified as Geldart B or Geldart D.

11.2.2.1 Particle Size Reduction

For particle size reduction, it is preferable to process biomass with low moisture content (after a drying pre-treatment) since its brittleness is increased and higher

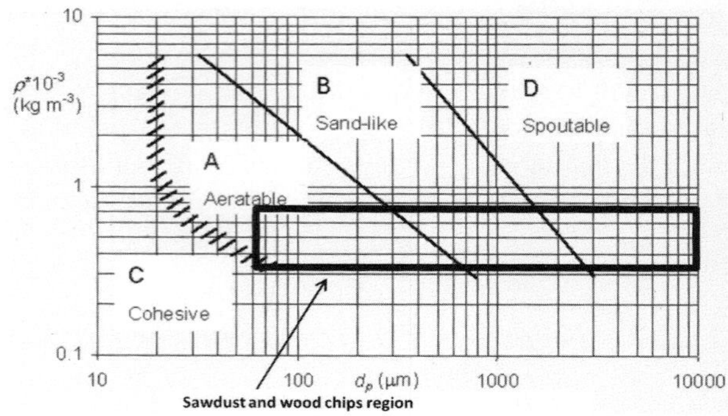

Fig. 11.1 Geldart classification of particles. (Reprinted and modified from Ref. [27], Copyright 1973, with permission from Elsevier)

shear forces are promoted. Taking that into consideration, the most common size reduction techniques are dry shredding and hammermilling. Dry shredding relies on rotating cutters: A geared roll is mounted with sharp designed metal cutters, which are regularly disposed on its surface. Larger wood pieces can this way be converted into wood chips. As smaller pieces will simply bypass the cutters, there is no need to separate the biomass feed before this step. Dry shredding can easily reduce biomass size down to wood chips-like particulate [10].

If a powder-like feedstock (<500 μm) is required for pyrolysis, further size reduction can be achieved with hammermills [10]. The principle is to grind a material until it reaches a minimal particle size. It is designed to limit particle size by the use of perforated plate outlet whose holes size determines the final average particle diameter. In a small drum, solid metal hammers are mounted on a central shaft. The metal hammers are rotated and the biomass material comes under the action of centrifugal force: the biomass is crushed between the hammers and the drum wall. The drum wall has grooves oriented perpendicular to those of the hammers extremities to maximize shear forces. By gravity, the fine particles percolate at the bottom of the drum where the perforated plate controls their exit in the outlet duct.

11.2.2.2 Particle Size Increase

Some feedstocks are characterized by low densities (such as bark and wood residues) and increasing their particle size may be necessary for some reactor technologies. In fluidized bed reactors, for example, particles with low terminal velocities may be rapidly entrained outside of the reaction zone resulting only in partial conversion. If the pyrolysis reactor technology requires larger particles, particle agglomeration techniques can be used. Pelletization, also referred to as densification, is a well-established process, and it is currently used to transform many MSW into denser particulate RDF (Refuse Derived Fuel) feedstocks to be directly used as fuels [10]. These same processes can also be used for biomass pre-treatment for biorefineries.

The most widely used equipment to produce pellets is the extruder [29]. It consists of one or two (partially overlapped) cylindrical ducts in which screws force deformable solids to flow with very high shear forces. At the end of the extruder, a die controls the pellet average size and a binding agent can be introduced with the solids in order to consolidate the agglomerates. Depending on the objective of the extrusion process, the design can consider multiple outlets to remove water.

Since pulp and paper industry wastes contain significant quantities of water, drying can become very expensive. In this instance, some compacting technologies can mechanically remove water from biomass while forming pellets or briquettes. Some extruders and other hydraulic or pneumatic presses perform compaction as well as removal of liquid water. By reaching high pressures using extrusion, Edwards [30] was able to compact a mix of bark and wood residue and lower its moisture content from 56.5 wt% down to 34.8 wt%.

11.2.2.3 Drying

Drying constitutes another relevant pre-treatment for bark and wood residues for most pyrolysis reactor technologies. Biomass moisture fraction is generally controlled by the use of rotary drum dryers in the industry [31]. It uses the same principles as conventional clothes dryers: a conditioned air stream with low humidity enters the drum and is charged with moisture evacuated from biomass. Industrial rotary drum dryers can reach volumes as high as 200 m^3. Using such drying equipment allows one not only to reach very low moisture fractions, but also attain a given moisture fraction set point that is desired for pyrolysis.

11.2.2.4 Sorting

Large pieces of inorganic material (e.g. metals and glass) can be present in significant quantities in MSW. Therefore, sorting of MSW is normally required since inert material will only consume useful volume, while some metal can act as catalyst to produce more pollutants [32].

11.2.2.5 Pre-Treatment for Sewage Sludge

In sewage sludge, organic matter is diluted in water and micro-scale inorganic elements can be present. Once the sludge has been chemically and biologically stabilized at the wastewater treatment plant (pH neutralization and microorganisms' denaturation), dewatering is the first step to recover municipal biosolids. Many techniques can be employed for dewatering and centrifugation is commonly used since it can easily yield suspensions of 20–25 wt% biosolids from 0.5 to 3 wt% diluted sewage sludge with an overall solids recovery of over 90 % [10]. Rotary drum filtration can also be used to remove water at lower levels. Depending on the operation, a controlled flow of air can be injected into the rotary drum filter, which can be used as a dryer to obtain moisture fractions below 10 wt% [10]. The wall of a rotary drum filter is meshed so the gas flow will be radial, as clogging of the filter is avoided by constant scrubbing mechanisms. As pure water can be obtained through the wastewater treatment process, the sludge sequesters impurities, which are then very difficult to remove.

11.3 Pyrolysis Reaction Kinetics

During pyrolysis, the feedstock is decomposed under relatively high temperature (300–1,000 °C) and anaerobic conditions into three products: non-condensable gas, condensable gas (oil and water) and char.

The non-condensable gas (at ambient temperature) is the lightest pyrolysis product and is composed of small hydrocarbons (C_1-C_4), carbon dioxide, carbon monoxide, hydrogen and other trace components. For biomass, the high fraction of molecular oxygen promotes the production of carbon monoxide and carbon dioxide. The fraction of trace components depends on the initial composition of the feedstock: sulfur, nitrogen, phosphorus and inorganics.

The second product is a condensable gas (at ambient temperature) whose composition can significantly vary depending on the process and biomass properties. The condensable gas fraction is characterized by two immiscible phases: an aqueous and an oily phase. Since molecular oxygen is present in biomass, aldehydes, ketones, alcohols and acids are formed during pyrolysis and condense with water to form the aqueous phase. The oily phase regroups all hydrocarbons that are immiscible with water. The remaining esters, phenols, thiols, nitriles, amines, and amides can be present in both aqueous and oily phases.

The last product is a solid phase that contains a carbon-rich powder (char or bio-char) and inorganics. For biomass pyrolysis, bio-char can have up to 20 wt% in molecular oxygen. Because of the initial fibrous structure of biomass (particularly in ligneous materials), the bio-char is characterized by a high pore volume and specific surface (after treatment and activation). It thus possesses interesting properties to be used as activated carbon [7].

During the pyrolysis of a specific feedstock, the respective yield and chemical composition of the above three products are governed by the pyrolysis reaction kinetics. Furthermore, the yields and chemical composition can be varied by adjusting the pyrolysis conditions in order to promote certain reactions.

11.3.1 Kinetics Characterization

Pyrolysis reaction kinetics are characterized by hundreds or thousands of parallel reactions and in series (solid and gas phase). Pyrolysis is governed by several chemical mechanisms: resonance, bond breaking, rearranging, dehydrogenation, cyclization etc. Due to its complexity, the entire chemical pyrolysis reaction network has not been characterized and pyrolysis kinetics models currently available in the scientific literature are highly simplified.

Since biomass is a solid phase macromolecular system, it is impossible to characterize its reaction kinetics with conventional gas-phase Arrhenius equations containing partial pressures or concentrations. Generally, each reaction step that is considered in the conceptualization of pyrolysis reactions has its own kinetics and parameters to describe its rate: a specific order of reaction and enthalpy of reaction. Pyrolysis kinetics can be expressed in the following general modified Arrhenius form [33]:

$$\frac{dm}{dt} = Ae^{-\frac{E}{RT}} f(m)^n \tag{11.1}$$

In Eq. (11.1), the rate of reaction is the function of a pre-exponential factor (A), an Arrhenius term containing an activation energy (E) and a linear function representing the weight of the decomposing sample ($f(m)$) to the power of the order of reaction (n). The weight function ($f(m)$) can be written in an absolute (n = 1) or normalized form. In the latter case, the non-dimensional term can be formulated in two ways: (1) normalized with respect to the weight of emitted volatiles or (2) the weight of decomposable material. In the first case (emitted volatiles), the rate of reaction will be referred to as the rate of devolatilization with the weight function $f(m)$ being equal to (1–m). In the second case (decomposable material), the weight function $f(m)$ will be equal to m. When expressed in an absolute form (not normalized), the weight function is also equal to m, but with appropriate weight units. The order of reaction (n) will depend on the reaction model and the biomass material. Generally, authors assume the reactions to be of first or second order. However, since most pyrolysis reaction models are global models, the apparent order of reaction is generally characterized by a value between 0.5 and 3. Few studies have experimentally evaluated pyrolysis kinetics by considering the order of reaction as an unknown [25, 34]. Similarly to the reaction order (n), the activation energy (E) and pre-exponential factor (A) depend on the biomass material as well as the characteristics of the reaction model.

Apart from the assumption related to the kinetic expression, several other factors may bias the measurement of kinetic parameters. In fact, heat and mass transfer are important phenomena to consider during experiments. Biomass is characterized by a very poor thermal conductivity combined with a high specific heat. Therefore, biomass particles may have a significant internal temperature gradient when heated at high rates [35]. When conducting laboratory scale pyrolysis kinetics experiments, static systems (thermogravimetric analysers (TGA)) are often employed where the biomass particles remain immobile. The use of TGA minimizes attrition such that the particles remain intact throughout the experiments and the internal mass transfer is limited. Unfortunately, this may not be representative of industrial pyrolysis systems and the derived reaction kinetics will not be accurate when applied at the industrial scale.

11.3.2 Reaction Models

Many simplified pyrolysis global reaction models were proposed in the scientific literature. Babu [36] proposed to regroup these conceptualizations of pyrolysis kinetics into three categories: (1) single-step models, (2) independent components models and (3) parallel and series reactions models. Figure 11.2 illustrates these models.

11.3.2.1 Single Decomposition Step Models (1-Step Models)

The simplest pyrolysis models consider a single decomposition step (Fig. 11.2a). Biomass decomposition directly yields a stream of bio-char, bio-oil and non-condensable gas. These models have the advantage of simplicity and possess a limited number of parameters. These models can be accurate for a limited range of pyrolysis conditions where the temperature is constant (isothermal system) or relatively low (<450 °C, thus conventional pyrolysis) and the product composition does not vary significantly.

In reality, biomass decomposition is more complex. In the case of non-isothermal systems at high heating rates, it may appear as if different components are reacting at different temperatures with their specific reaction kinetics. The notion of 'pseudo-component' emerges from that behaviour. Even if a system is operated isothermally at high temperature, pyrolysis will happen during the heating step where different products composition will be obtained. Single decomposition step models are thus only suitable in specific pyrolysis situations and are generally inadequate to reproduce industrial fast pyrolysis behaviour.

11.3.2.2 Independent Components Models

Independent components models incorporate the notion of 'pseudo-component'. Figure 11.2d gives a typical example of pyrolysis with multiple decomposition levels. The combination of 'pure' components decomposing in the same system at different temperatures would be equivalent to the reaction pattern in Fig. 11.2b. As the temperature of the biomass particles increases, the stability of the solid macromolecular matrices changes accordingly. This variation is mostly due to the release of low molecular weight compounds as previously explained. The reactivity of the particulate material will thus vary during pyrolysis and this behaviour is similar to having different components decomposing with specific reaction kinetics.

The pyrolysis of wood, for example, can be described by the combination of hemicellulose, cellulose and lignin pyrolysis kinetics, which all show single or two steps decomposition [37]. In this case, the combination of the decomposition steps for each of the three individual components can accurately model the overall pyrolysis behaviour of wood. However, ideal cases like wood are rare and the concept of 'pseudo-component' may not be useful for other biomass such as manure and MSW, for example, which are highly heterogeneous. One major disadvantage of this second

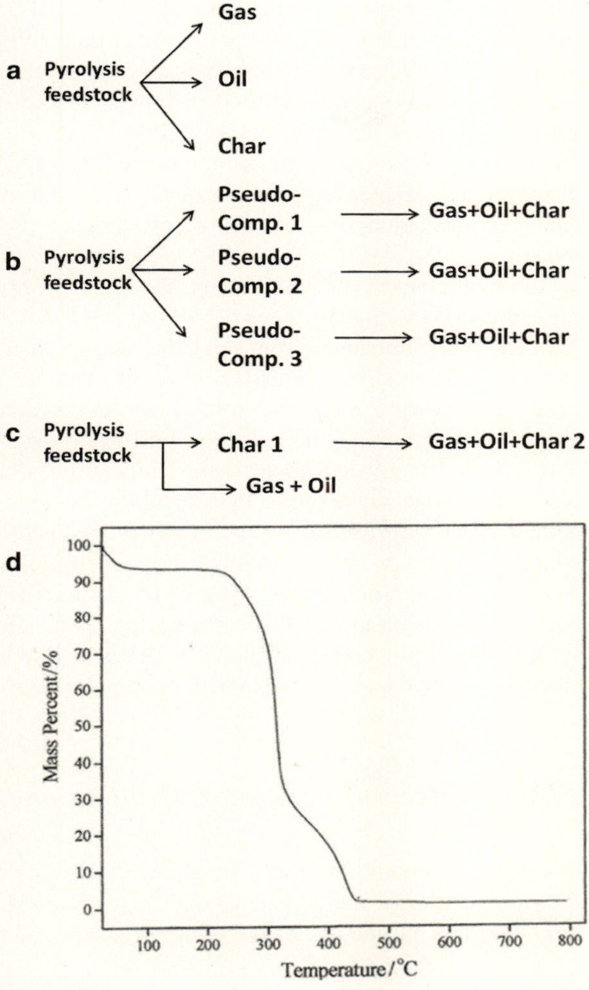

Fig. 11.2 Pyrolysis conceptualizations with a typical thermogravimetric pyrolysis curve

type of models is its complexity and the large number of reaction parameters involved. With materials having a variable composition from one provider to another, the value of these parameters will change as well as the weighting of each 'pseudo-component' species.

11.3.2.3 Parallel and Series Reactions

In these models, a feedstock first reacts to yield volatiles and intermediate solid products (Fig. 11.2c). The following steps then depend on the feedstock and the model. In most models, the volatiles undergo thermal cracking while the intermediate solid products further decompose into other final and/or intermediate products and so on. For highly heterogeneous materials, where the notion of 'pseudo-component' loses its physical meaning, this conceptualization appears more realistic. However, the main disadvantage of this type of model is the high number of parameters and the difficulty to segregate the different reactions experimentally to clearly estimate their kinetics parameters. Considering all the possibilities for the different reactions under various conditions, this family of models will lead to a myriad of different models.

Table 11.2 lists the kinetics parameters of several models available in the scientific literature along with their range of validity. Note that the parameters reported in Table 11.2 were all obtained from static thermogravimetric experiments. Furthermore, some of these models assumed an order of reaction, while pyrolysis reactions are in reality characterized by multiple elemental reactions with their specific reaction order. Moreover, it has been demonstrated that the product yield during biomass pyrolysis is strongly dependent on the temperature, while all the models listed in Table 11.2 assume that the product yields follow the non-isothermal or isothermal TGA temperature profile used to evaluate the kinetics. Therefore, none of these models can confidently reproduce the variability of products yields with respect to temperature. Also, there is significant uncertainty as to whether these kinetic expressions will be accurate when extrapolated to industrial pyrolysis conditions. To model industrial scale pyrolysis processes, it is critical to develop reliable and robust pyrolysis models based on experimental data obtained at representative conditions.

11.3.3 Effects of Pyrolysis Conditions on the Kinetics

The pyrolysis conditions affect the global kinetics by promoting specific elemental reactions. The main operational parameters for pyrolysis are temperature (and heating rate), pressure, co-feeding of different feedstocks and presence of catalysts.

11.3.3.1 Temperature and Heating Rate

Pyrolysis is governed by many parallel and series reactions characterized by their specific kinetics and the relative importance of each of these reactions will depend on the temperature of the system [41, 42]. Also, a slow heating process implies that the biomass remains at every temperature for a longer time period. As pyrolysis kinetics and heat transfer compete, pyrolysis occurs during the heating of the particles and might even be completed (at thermodynamic equilibrium) before reaching the temperature set-point. At low heating rate, more decomposition happens at low temperature such that more bio-char and less volatiles (condensable (bio-oil) and non-condensable gases) are produced.

Table 11.2 Pyrolysis kinetics parameters for selected materials from literature

Biomass	Model	A (min^{-1})	E (kJ/mol)	n	Temperature range of validity (°K)	Reference
Poplar	One step decomposition	2.14×10^{12}	153.9	1	<673	[22]
Wheat straw	Three pseudo-components linear combination	2.57×10^{12}	69	2.3	<873	[24]
		3.97×10^{7}	78	0.65	<873	
		3.17×10^{6}	80	2.7	<873	
Rice husk	Two pseudo-components linear combination	1.02×10^{2}	33.1	1.5	<623	[26]
		3.3×10^{1}	28.3	2	623–823	
Rice husk	Two pseudo-components linear combination	7.25×10^{3}	30	0.91	<640	[25]
		5.14×10^{2}	16.3	0.3	640–813	
Cellulose	Two pseudo-components linear combination	4.69×10^{5}	82.7	1	<623	[38]
		1.33×10^{23}	282	2	623–673	
Cellulose	One step decomposition	1.6×10^{10}	244	1	<623	[35]
Cellulose	Two pseudo-components linear combination	7×10^{7}	126	1	–	[39]
		4×10^{17}	234	1	–	
Lignin	Two pseudo-components linear combination	5.39×10^{4}	67	1	–	[39]
		2.1×10^{5}	70.7	2	–	
Klason lignin	One step decomposition	1.21×10^{12}	156.5	1.53	<1000	[40]

One of the main mechanisms controlling the interaction between the temperature and the heating rate is the stabilization and reorganization of the macromolecular solids. Thermal decomposition brings lighter molecules to unbind from the solids (biomass or waste in the present case) to form a volatile phase. In parallel, this creates physicochemical instabilities that lead to a molecular rearrangement. The kinetics associated to these intra-molecular modifications then inhibits the volatile formation kinetics. If the heating rate is slow, stabilization occurs and higher char yield is obtained. On the other hand, heating faster will impede stabilization and volatile production will be promoted. Temperature has a different effect on the pyrolysis products. Char production decreases with increasing temperature and the yield of gas increases (both condensable and non-condensable). The extent of the gas thermal cracking determines the yield of non-condensable gas and average molecular weight of the volatile fraction. Thermal cracking kinetics becomes important with increasing temperature and gas residence time.

11.3.3.2 Pressure

Pressure influences the equilibrium reactions and therefore affects the volatile products composition: the condensable (bio-oil) and non-condensable gases. It has also been shown that pressure can promote other gas–solid reactions involving moisture, hydrogen, carbon dioxide and possibly other gaseous species. The mechanisms involved remain unclear, but chemical reactions such as the Boudouard reaction are suspected:

$$C_{(s)} + CO_{2_{(g)}} \rightleftarrows 2CO_{(g)} \tag{11.2}$$

Moisture, which is inevitably present during biomass pyrolysis, has also been shown to influence the volatile yield and composition [43]. These reactions are characterized by fast kinetics only under specific conditions: they generally occur in pressurized gasification and under high temperatures, while pyrolysis is typically performed at milder temperatures.

The solids will not be significantly influenced by pressure if an inert gas is employed [44, 45]. Experimentation is still the best method to characterize the effect of pressure on the pyrolysis products since it is specific to the biomass feedstock used.

11.3.3.3 Co-Feeding

One of the most evident synergic pyrolysis behaviours has been demonstrated by Brebu et al. [16]. By co-feeding pine cones with waste polyolefin, they showed that it was possible to significantly decrease the overall char yield and increase the amount of volatile produced. For binary blends of pine cones and polyethylene (PE), polypropylene (PP) or polystyrene (PS), the char yield decreased (by over 6 % units for PE) compared to the calculated value by linear combination, hence revealing synergic behaviour. When a blend of pine cones and the three polymers (PE/PP/PS) was pyrolyzed in the ratio 3:4:2:1, the synergy was even more significant. The reported char yield was lower than the average calculated value by 10 % units and the liquid product yield increased by over 11 % units. However, while these types of synergies are desired, they have been rarely observed. One typical example is the co-pyrolysis of biomass with coal. Weiland et al. [46] explored the possible synergetic interactions between coal and biomass in pyrolysis. Unfortunately, the interaction was almost nonexistent. Linear combination explained most of the variations with the various blends of coal/biomass.

11.3.3.4 Catalysts

Pyrolysis of biomass with catalysts has been widely studied and they are sometimes used to tailor the yields of the pyrolysis products. However, catalysis chemistry is extremely complex and only a very few research groups in the world can explain catalysis mechanism for specific reactions and in highly controlled conditions. Thus, understanding (in a fundamental mechanistic way) the effects of adding catalysts on the evolution of pyrolysis products distribution and their composition is not currently feasible. The effects of catalysis on pyrolysis reactions are therefore determined empirically and undesired behaviours were often observed: significant drops in liquid yield have been the most common [47].

11.4 Types of Pyrolysis

As previously discussed, the pyrolysis conditions affect the global kinetics by promoting specific elemental reactions. Pyrolysis processes have therefore been classified with respect to the prevailing conditions during the reaction used to maximize the yield of one or more of these products. Conventional or slow pyrolysis, fast or ultrafast pyrolysis and vacuum pyrolysis are the three main categories of pyrolysis process operation.

11.4.1 Conventional or Slow Pyrolysis

Heat transfer to the biomass particles is generally the main limitation in industrial pyrolysis. Biological and organic polymeric materials have poor thermal conductivity, but high specific heats. Therefore, a limitation is reached in pyrolysis processes when heating a feedstock to high temperatures due to the high temperature dependence of the reaction kinetics (expressed as an Arrhenius law). The limitation arises at the specific temperature where the decomposition rate becomes greater than the heating rate. Acknowledging that pyrolysis is an overall endothermic reaction, increasing temperature from that point is very difficult.

Conventional pyrolysis is also referred to as slow pyrolysis because of the low heating rates (6–60°C/min [36]). The peak pyrolysis rate will be reached at a relatively low temperature and the limited heat transfer will result in moderate pyrolysis temperatures (300–700°C). These reaction conditions promote bio-char production and minimize volatiles (non-condensable and condensable gases). As the temperature of the pyrolysis process is increased, the weight fraction of volatile increases: this effect is governed by the resonance mechanism. The release of lighter molecules from a macromolecular matrix generates instabilities that are dispersed within this matrix in order to stabilize its structure. As temperature increases, the instability gains in magnitude. In slow pyrolysis, biomass is kept at constant moderate temperatures, such that the macromolecule has time to reach a new stable form with a new composition that will handle higher internal energy without decomposing (thermodynamic equilibrium), hence limiting the release of volatile. This is where conventional (slow) pyrolysis differentiates from fast pyrolysis. Volatile and char yields therefore depend on this resonance kinetics.

11.4.2 Fast and Ultrafast Pyrolyses

Fast and ultrafast pyrolyses are performed under high heating rates (600–12,000°C/min [36]), which are many orders of magnitude higher than those of conventional pyrolysis. Thus, the peak rate of decomposition is reached at higher temperatures compared to slow pyrolysis. Under these conditions, the macromolecular reorganization kinetics are slower than the volatile release kinetics. Consequently, the bio-char yield is significantly lower compared to slow pyrolysis, while the volatile yield is higher. Since the bio-oil (condensable gases) products are of higher interest for biorefineries, the emerging industrial biomass pyrolysis processes ideally target fast and ultrafast pyrolysis processes. Meanwhile, bio-char obtained at a higher temperature show a greater specific surface, which is another motivation for operating at very high heating rate and temperature. In addition, higher heating and reaction rates allow higher biomass process rates or the use of smaller, more compact systems: both of these aspects increase process profitability.

11.4.3 Vacuum Pyrolysis

The third category of pyrolysis process is referred to as vacuum pyrolysis. This pyrolysis process is performed under vacuum conditions independent of the heating rate (slow and fast). Under vacuum, the heavier products in the gas phase are entrained out of the reacting environment without having time to crack into smaller molecules. For that reason, vacuum pyrolysis oils contain high molecular weight components and are consequently tarry and more viscous than common pyrolysis oils. Because of their specific molecular composition, vacuum pyrolysis oils are of great interest for specialty chemicals production in biorefineries. It is nevertheless a great challenge to produce industrial scale vacuum environments. The company Pyrovac (Laval, Quebec, Canada) attempted to operate a commercial and large-scale industrial vacuum pyrolysis plant in the 1990s and failed due to high operating costs.

11.5 Reactor Technologies

The choice of reactor technology is critical to follow the desired kinetic pathway in pyrolysis. The emphasis of this section will be on selected reactor technologies that have been successfully demonstrated at large scale with a pilot-scale or larger unit. Several reactor concepts have been demonstrated in the scientific literature at small-scales, but they will not be considered.

The mass and energy balances are the main fundamental and semi-empirical tools to design reactors when coupled to reaction kinetics. These balances involve heat and mass transfer equations, which are dependent on the system gas/solid hydrodynamics. Several handbooks are dedicated to reactor design with various hydrodynamics models to represent these equipments. However, the objective of this section is not to review these models in details. Considering that heat transfer is the limiting step in pyrolysis reactors, the emphasis is put on heat transfer parameters evaluation.

11.5.1 Bubbling Fluidized Bed (BFB)

Fluidized beds are widely used in the chemical industry for catalytic cracking and other processes. In fluidized bed reactors, a gas stream (inert gas for pyrolysis) is forced through a bed of powdered material from a distributor plate that supports the bed. At low gas velocities, the bed of particles is non-moving and this is referred to as a fixed bed. As the gas velocity is increased, the drag forces applied on the particles increase until minimum fluidization velocity is reached: the bed is 'supported' by the gas and behaves like a fluid. If the gas velocity is increased further, bubbles are formed at the distributor plate and rise through the bed of solids similar (but not identical) to air bubbles in water. Bubbles promote the circulation of solids to ensure a uniform temperature throughout the fluidized bed. A bubbling fluidized bed (BFB) reactor is designed in a way to avoid the entrainment of particles outside the reactor (also called elutriation). The bed zone is narrower to promote the circulation of particles and the formation of bubbles. The gas exits the bed to enter a freeboard zone and a higher diameter disengagement region where the gas velocity is significantly reduced. In the disengagement region, particles that would be entrained in the bed and freeboard zones fall back into the bed by gravity. There is an appreciable amount of established scientific literature on fluidized beds [48, 49]. BFB reactors are used for fast pyrolysis by the company Dynamotive, which is operating a pilot-scale unit to convert biomass into bio-oil. The unit has been reported to process at up to 100 tons of biomass per day [50]. Figure 11.3a gives a global view of a possible biomass pyrolysis unit with a BFB.

11.5.1.1 Operability

Biomass particles are generally difficult to fluidize, such that a denser and more homogeneous inert particle media (generally sand) is employed as a fluidization media to improve transport phenomena. Bench and pilot-scale continuous operation of a pyrolysis BFB has been demonstrated by Dynamotive. Char removal from the reactor can be an issue: if char is very fragile, its particle size will decrease within the bed by attrition. When char particle size reaches a critical particle value, it is entrained out of the fluidized bed reactor and it must be separated from the gas and recovered via a cyclone. Therefore, the disengagement region must be carefully designed to allow char particles to exit the reactor once they are sufficiently small. The main advantages of BFBs for pyrolysis applications include a uniform reaction temperature (minimizes the formation of cold/hot spots in the bed) and capability to operate the reactor continuously (continuous biomass feeding). On the other hand,

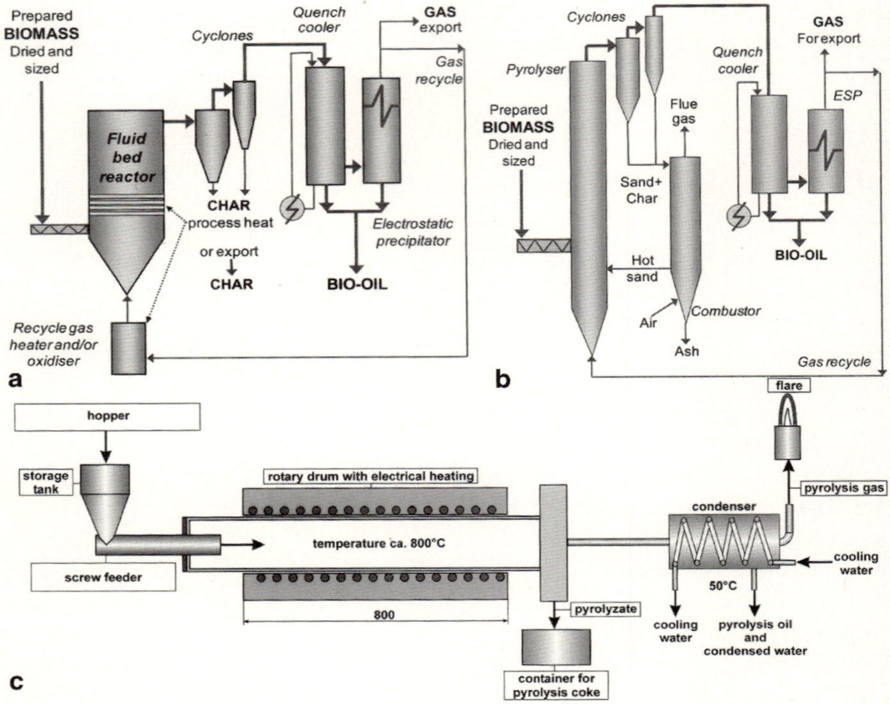

Fig. 11.3 Pyrolysis units global schemes: **a** Bubbling fluidized bed [41], **b** circulating fluidized bed [41] and **c** rotary drum reactor [51]. (Reprinted from Ref. [41], Copyright 2012, with permission from Elsevier; Reprinted from Ref. [51], with permission from Prof. Dr-Ing. Roman Weber)

the main disadvantage of BFBs is that the volatile will be mixed with the inert fluidizing gas. Therefore, the bio-oil can be recovered, but the non-condensable gas is diluted such that it can hardly be used as a primary energy source. Thus, energy must be obtained from the solid char, which is significantly detrimental to the process profitability.

11.5.1.2 Hydrodynamics

The fluidization gas acts as a heating media and promotes mass and heat transfer by inducing a movement on the solids particles as well as removing the volatile from the bed. The BFB can be divided into three phases: (1) the bubble phase (dilute phase—low solids fraction), (2) the emulsion phase (dense phase—high solids fraction) and (3) the cloud phase. The cloud phase is at the interface between the bubble and emulsion phases such that the local solids fraction is between dilute and dense [48]. Several gas-phase (1-phase, multiple-phase, multiple regions etc.) and solid-phase (counter-current back mixing etc.) hydrodynamic models are available in the scientific literature [49]. These models can be coupled with pyrolysis kinetics (reviewed in Sect. 11.3) to estimate the yield of products. These models have been reviewed in detail in several publications [49].

When designing a BFB pyrolysis system, one could desire to minimize the fluidization gas flow to facilitate post-pyrolysis separation of the products. However, the superficial gas velocity also affects the reaction rates since it influences the heat and mass transfer. When temperature is sufficiently high, the pyrolysis reaction characteristic time becomes shorter than the heating characteristic time, such that heat transfer is the limiting step. In this case, the particles reaction rate (and residence time) is determined by the convection heat transfer to the biomass particles in the fluidized bed; and the convection coefficient can be calculated from the following correlation [52]:

$$Nu_{bed} = \frac{h_{bp} d_p}{k_g} = 0.033\ {Re_p}^{1.33} \quad \text{for } 0.1 < Re_p < 100 \tag{11.3}$$

In Eq. (11.3), the overall fluidized bed Nusselt number (Nu_{bed}) is a function of the particle Reynolds number (Re_p):

$$Re_p = \frac{\rho_g \left(U - U_p\right) d_p}{\mu_g} \tag{11.4}$$

The convection coefficient from Eq. (11.3) is averaged over the bed of particles and it is shown to increase with increasing slip velocity (U-U_p). As demonstrated by Avidan and Yerushalmi [53], the slip velocity (U-U_p) for BFBs is equal to the superficial gas velocity. This is the case because the average particle velocity is zero: solids circulate within the bed (negligible or limited entrainment) and particles flow co-current or counter-current with the gas. Therefore, the fluidization gas velocity should be sufficiently high to maximize reaction rates and the yield in volatiles: there is therefore a trade-off associated with the selection of the fluidization velocity.

Note that Eq. (11.3) has been shown to yield a more accurate estimation of the convection coefficient than the typical correlations involving the Prandtl number [54]. Furthermore, Eqs. (11.3) and (11.4) should be used by considering the inert (sand) fluidization media, in which case the use of the inert material properties is generally sufficiently accurate (the biomass particles are highly diluted in the inert media). Basic heat transfer estimations with Eq. (11.3) and (11.4)[1] suggest that operating a fluidized bed in the bubbling regime widely promotes fast pyrolysis rather than conventional pyrolysis.

To model biomass pyrolysis in a BFB, less importance is generally given to the bubble characterization since the fluidization gas is inert. The modelling is therefore focused on the dense emulsion phase, which contains the solid biomass particles. If the fluidized bed temperature is uniform and the inert (sand) particles do not leave

---

[1] Using 500 μm square wood particles showing average properties with nitrogen as a fluidization medium operating at atmospheric pressure and 773°K. Heating rate estimated at 250°K temperature difference between the gas and particle with the lumped-capacitance method.

the bed, the inert particles temperature can be assumed equal to the gas temperature. In this case, the heat balance strictly involves the biomass particles.

11.5.1.3 Feedstock Pre-Treatment

Dewatering and drying should be considered, since heating the biomass may only vaporize the water while delaying the biomass pyrolysis reactions and increasing the operation costs. Also, studies suggest that steam explosion of biomass could influence the quality of pyrolysis products [55].

Particle size reduction is also important to minimize heat transfer limitations and internal temperature gradients in biomass particles. The Biot number criterion of 0.1 suggests that the maximum wood chips size that should be fed to a BFB to avoid heat transfer limitations (and product yields issues) is ~2 cm (by estimation with typical wood properties). This calculation assumes that the wood chips have a thickness five times lower than their diameter (parallelepiped shape). Particle size reduction is therefore recommended for biomass particles that are larger than that value. Heterogeneous feedstocks such as MSW would not be recommended with this technology since pyrolysis operation temperatures do not promote slagging, in opposition to gasification. Undesired particulate would then need to be removed from the bed, which is a difficult operation in BFBs, considering the wide particle size and densities distribution of the inorganics.

11.5.2 Circulating Fluidized Bed (CFB)

A Circulating Fluidized Bed (CFB) works at higher superficial gas velocities than a BFB to increase the slip velocity and heat transfer to the particles. Due to the high gas velocities, the particles are entrained outside the bed region (called the riser) and cyclones are used downstream to separate the particles from the gas and return them to the bed. During biomass pyrolysis, char and inert (generally sand) particles are present. Depending on the design of the reactor, the char/sand particles mix can be transported to a second reactor where the char is burned. In this case, the hot sand is returned to the riser and the combustion flue gases can be directly fed to the riser as an inert fluidization and heating medium (see Fig. 11.3b).

Ensyn is a well-known company that operates a commercial scale CFB biomass pyrolysis system. Their largest pilot plant can process up to 100 bone-dry tons of biomass per day. Sand is co-fed with biomass in the riser, while the residual char and sand are recovered in a separate BFB combustor where the char is burned. The hot sand is then fed into the riser.

11.5.2.1 Operability

The main advantage of CFBs over BFBs is that char can be easily separated from the sand and the gas. Similarly to BFBs, the non-condensable gas cannot be considered as a primary heat source for the process: it is diluted into the inert fluidization media. To supply the process with heat, bio-char is burned instead. As this solid pyrolysis product showed an interesting potential [7, 56] for an eventual commercialization, choosing a CFB greatly affects the overall profitability of the pyrolysis process.

11.5.2.2 Hydrodynamics

The apparent density in CFBs is lower compared to BFBs such that heat transfer calculations are generally performed by assuming single particles convection heating. Since particles are entrained outside the fluidized bed, the relative gas velocity with respect to the particles is their terminal velocity (U_t). Therefore, the particle Reynolds number is calculated using:

$$Re_p = \frac{\rho_g U_t d_p}{\mu_g} \tag{11.5}$$

Two main individual systems must be designed to conduct pyrolysis in a CFB: (1) the riser where pyrolysis occurs and (2) the combustor where char is burned and the flue gases heat the sand media. In the Ensyn technology, the combustor consists of a BFB. The important design calculations for this component are the mass and energy balances: It must be considered that the biomass and char particles do not circulate at the same rate as sand particles. To design the riser section, very fast heat transfer from the sand to the gas can be assumed such that an average gas temperature can be estimated. This gas temperature is then used to calculate the biomass transient heating over its residence time. The convection coefficient can be estimated with the correlation of Ranz [57]:

$$Nu_p = \frac{h_p d_p}{k_g} = 2 + 0.6 {Re_p}^{0.5} {Pr_g}^{0.33} \tag{11.6}$$

Referring to the calculations made for BFBs, fast pyrolysis is the type of pyrolysis promoted in CFBs; lower gas velocities would not even allow reaching that regime. Similarly to BFB pyrolysis, the gas flow rate in CFBs must be carefully chosen as well as the biomass and sand circulation rates. Moreover, sand and biomass particle size must be chosen to control the sand/char flow rate ratio. If the gas flow rate in the combustor is too high, the sand temperature may be too low, which would lower the temperature in the riser where pyrolysis occurs. Likewise, a high gas flow rate in the riser would also lower the temperature at which pyrolysis occurs. CFB design considerations have been reviewed in several publications [49].

11.5.2.3 Feedstock Pre-Treatment

It was mentioned previously that the biomass particle size has to be homogenized and controlled in order to obtain a specific sand-to-char flow rate ratio. In their process, Ensyn uses wood sawdust that has been dried to a low moisture fraction as a pre-treatment. Fine particles $d_p \leq 44\ \mu m$, Geldart C are generally required for CFB technologies to promote a homogeneous fluidization. It is critical that this feedstock does not contain particulate metals and inorganics; otherwise, it might contaminate the sand. Highly heterogeneous materials such as MSW, RDF and sewage sludge are thus not recommended with this technology.

11.5.3 Entrained-Flow Reactor

In entrained-flow reactors, free-falling particles are entrained downwards by a gas flow. For gasification, entrained-flow reactors are a promising technology as demonstrated in 2002–2003 by Shell and BTG with woody biomass as feedstock [58]. Preliminary tests for biomass pyrolysis applications have shown very high non-condensable yields [59]. It is believed that the geometry of this type of reactor technology promotes important bio-oil thermal cracking. Since the purpose of pyrolysis for biorefineries is to produce higher molecular-weight chemicals, this technology will not be covered.

11.5.4 Rotary Drum Reactor

Rotary kilns have been used for decades to generate energy from wastes. The operation of this type of reactor has been demonstrated in a continuous mode at industrial scale and many problems have been reported. The very high temperatures during incineration promote NO_x and SO_x production as well as dioxins and furans, which are carcinogens. Moreover, leaching and slagging can affect rotary kiln incinerators operability. Notwithstanding these reports, the interest for using rotary drum reactors for pyrolysis applications is currently growing [15]. Operation of the reactor at lower temperature and without oxygen will likely decrease pollutant emissions as well as minimize risks of leaching and slagging. In most cement plants around the world, rotary kilns can have impressive dimensions (over 100 m long). For pyrolysis applications, some pilot-scale units were built and operated to process waste.

Typically, a rotary drum pyrolysis reactor processes raw materials such as MSW, RDF and waste tires. It consists of a horizontal cylindrical reactor rotating at a certain speed in order to mix the bed of materials and promote transport phenomena. For this application, as oxygen must be purged, heat supply is generally indirect, that is, gas burners are mounted underneath the rotary drum and the flue gases are circulated around the drum in a blanket-like chimney. Figure 11.3c shows a schematic of a rotary drum pyrolysis system with an electric heater.

11.5.4.1 Operability

Rotary kilns are usually operated in full continuous mode, which can complicate its application to pyrolysis. The reactor is purged from oxygen during the process start-up, but no inert gas is fed to the reactor during operation. The major advantage of this mode of operation is that all of the pyrolysis products can be recovered, including the combustible non-condensable gas that is characterized by significantly less dilution compared to the other reactors considered in this chapter. Optimally, the non-condensable gases are used as the primary energy source for the reaction to heat the rotary drum.

Char is also recoverable, but conventional (slow) pyrolysis has the disadvantage of producing less bio-oil. The lower heating rates observed in rotary drum reactors will also complicate its continuous operation: most technologies in this field operate in batch mode. Thus, profitability will be achieved by combining multiple batch pyrolysis units to increase production and benefit volume discounts for the equipment.

11.5.4.2 Hydrodynamics

Rotational speed and mixing are probably the most important parameters to consider when designing a rotary kiln pyrolysis system since it governs the heat and mass transfer. Mellmann [60] produced a complete review on rotary drum hydrodynamics, highlighting the flow patterns and regimes as a function of the particulate material friction coefficient, the drum filling level and the Froude number. The Froude number is defined as:

$$Fr = \frac{\omega^2 R}{g} \tag{11.7}$$

In Eq. (11.7), ω is the angular rotational speed and R is the drum radius. With increasing Froude number, the motion patterns vary from (1) a slipping motion (sliding, then surging as motion subtypes), (2) cascading motion (slumping, then rolling, then cascading) and (3) cataracting motion (cataracting, then centrifugating). According to Mellmann [60], reaching the cascading motion regimes is preferable to achieve good mixing and higher homogeneity. If the reactor is properly operated, it can be assumed that the temperature gradient in the bed is negligible. In this case, heat transfer to the particles results from drum wall convection and radiation to the particles. Generally, it is more convenient to estimate the heat transfer coefficient experimentally, since it is specific to the studied system. In fact, the heat transfer coefficient can typically vary from 25 W/m^{2}K up to over 200 W/m^{2}K if radiation becomes important. Experiments are needed for that estimation also because baffles are often added on the drum inner wall in order to amplify mixing. As stated previously in this section, rotary drum pyrolyzers can operate at various regimes, depending on their design, to promote very slow pyrolysis up to the fastest conventional pyrolysis heating rates.

11.5.4.3 Feedstock Pre-Treatment

Feedstock does not need much pre-treatment for rotary drums. Moisture control could be required, but the feedstock particle size can be kept relatively high (on the order of centimetres). Also, the purification of MSW into RDF is not necessary for operation. Metals and inorganics removal is related more likely to useful volume: at lower temperature and in the absence of oxygen, undesired catalytic reactions will show significantly lower kinetics than if gasification was performed. For that reason, rotary drum reactors are preferred for MSW and municipal biosolids pyrolysis. Woody biomass is not likely to fit to this technology, because the liquids yields are particularly low [60] and that fluidized bed technologies have already been proven successful at industrial scale.

11.6 Pyrolysis Products Optimization

The sustainability of biorefineries is very sensitive economically and their main objective is to generate the pyrolysis products that maximize the plant profitability.

11.6.1 Pyrolysis Products for Biorefineries

Bio-oil is considered as the most commercially valuable pyrolysis product if it contains specialty chemicals in high percentages. Therefore, maximizing bio-oil production can be an objective during process development. Table 11.3 summarizes the main specialty chemicals present in bio-oil obtained from the pyrolysis of selected feedstocks along with reactor type and operating conditions.

Fatty acids, phenolic compounds and aldehydes/ketones are among the high-value chemicals used for crops. On the other hand, woody biomass generates high amounts of levoglucosan, which is a promising biopolymer building block. Also, catechol and its derivatives have a high market value. In addition, hard wood has different pyrolysis behaviour than soft wood through their bio-oil composition (Table 11.3: pine wood vs oak wood).

Chemical species containing double bonds and oxygen can be detrimental to the stability of the bio-oil. These chemicals polymerize with time and increase the bio-oil viscosity as well as modify other properties. Acetic acid is one common pyrolysis product which can cause this phenomenon. Acetic acid is derived from cellulosic biomass and process optimization will generally focus on reducing its production. Moisture fraction has been shown to influence acetic acid production [64]. Another method consists of using metal oxides catalysts [47].

11.6.2 Bio-Oil Applications

The interest for bio-oil resides in its complexity, but this complexity implies an important obstacle: selectivity. Bio-oil composition can be easily modified, but voluntarily promoting the production of one specific chemical species over another is

Table 11.3 Selected feedstocks with their bio-oil main compounds

	Feedstock	Pine bark	Pine wood	Pine wood	Oak wood	Switchgrass	Switchgrass	Sewage sludge
	Reference	[61]	[61]	[63]	[62]	[62]	[64]	[65]
	Reactor	Auger	Auger	Auger	Auger	Stirred reactor	Fluidized bed	Fixed bed
	Temperature (K)	723	723	723	723	873	823	723
	Pressure (atm)	1	1	1	1	6.8	1	1
	Moisture (%)	10	10	dried	10	8.4	10	7
	Bio-oil yield (%)	43.5	52	50	53	37	37	13
Specialty chemical	*Estimated price*	Yield based on global bio-oil weight, otherwise specified						
Catechol	80 USD/kg	4.46 %	3.79 %	<1 %	2.25 %	–	–	–
Levoglucosan	>100 USD/kg	21.50 %	14.20 %	6.37 %	21.60 %	–	–	–
2-Propanone	1,50 USD/L	–	–	–	–	8.78 %	2.95 %	–
2-Butanone	1,60 USD/L	–	–	–	–	11.49 %	–	–
Acetic acid	0,55 USD/L	–	–	–	–	18.91 %	6.85 %	–
Acetic acid, methyl ester	1,20 USD/L	–	–	–	–	5.47 %	–	–
Hydroxyacetaldehyde	1,50 USD/L	–	–	–	–	–	9.14 %	–
9-octadecenoic acid	1,50 USD/L	–	–	–	–	–	–	4.01 %
Tetradecanoic acid	1,00 USD/L	–	–	–	–	–	–	4.31 %
Hexadecanoic acid	1,50 USD/L	–	–	–	–	–	–	15.70 %

very challenging. There are essentially three markets in which the pyrolysis bio-oil can be categorized: high-value chemicals, biofuels and chemical precursors. If the pyrolysis process is designed to produce a low molecular weight bio-oil, it could be used as a chemical precursor for further recombination to yield heavier and higher value products. On the other hand, high molecular weight bio-oils are often unstable chemically due to double chemical bonds [66].

Compounds found in the bio-oil include: acids, aldehydes, anhydrosugars, hydrocarbons, saccharides, alcohols, ketones, furans, phenols etc [3]. The oxygen weight fraction in bio-oil is thus high at approximately 45 % [66]. As a result, bio-oil is likely not suited to be directly used as a fuel. However, co-feeding of polyolefins can help controlling this parameter efficiently [16]. Besides the possibility of producing biofuel through co-pyrolysis of different species, it is also interesting to observe important variations in specific chemical species in this process. This is only one of the numerous modifications that can be implemented on the pyrolysis process, in order to upgrade the products. Nevertheless, it must be kept in mind that pyrolysis has the great advantage of being a single-step process, so it must conserve this characteristic through its modification.

11.6.3 Bio-Char Applications

Char is sometimes used as fuel for pyrolysis reactors. However, the market value of bio-char must be considered before taking decisions. Activated carbon sells on the market at around 1 USD per kg, which is comparable to the value of some chemicals found in bio-oil. Bio-oil and bio-char both need post-processing transformation in order to yield their valuable products.

Activated carbon production implies carbonization and chemical activation. Typically, carbonization consists of a very slow pyrolysis process, which yields very high amounts of char. The principal characteristic of activated carbon is a very high specific surface, generally over 500 m^2/g. To be competitive with the actual commercial carbons, this specific surface must be attained and it has recently been demonstrated as feasible [7]. Since 2006, the number of publications on bio-char activity has increased considerably [56].

11.6.4 Dealing with Foreign Elements in Pyrolysis Products

Biomass and MSW contain oxygen, nitrogen, sulfur, halogens, metals and other elements whose concentrations in the pyrolysis products must be controlled as per existing standards.

Sulfur and nitrogen have been shown to mostly cluster in the bio-oil phase [21]. Sulfur has been successfully avoided in pyrolysis oil from coal in the past [67] by the use of lime (CaO). Similarly, the same experiments showed that the oxygen weight fraction in oil could be as well significantly decreased. The main disadvantage of this technology is that the char will remain charged with calcium and that the gas phase will mostly receive this excess of sulfur [68]. However, the gas phase can be post-treated efficiently (scrubbing) and considering the high-level of inorganics and contaminants in the solid phase, it will not likely be used as bio-char or as activated carbon.

Adding CaO in the pyrolysis environment can also significantly inhibit the formation of liquid chlorinated organics in the bio-oil [69]. The co-feeding of CaO in MSW pyrolysis could eventually become widespread as this additive is relatively cheap and available.

Post-processing of bio-oil to remove oxygen, nitrogen and sulfur is also possible through hydrodeoxygenation using metal catalysts (and an H_2 stream) [70]. For biofuels production, this post-treatment process is desirable, as the oxygen weight fraction can be reduced below 1 %. Furthermore, the effect of this post-treatment on the bio-oil chemical composition has not been investigated such that there is a possibility that it also generates higher value chemicals.

Most metals and metal oxides cluster in the solid phase due to the low temperature associated with pyrolysis, which remains below metal sublimation or melting temperatures. Metals can also be absorbed by activated carbon produced from bio-char [56].

Nomenclature

Abbreviations

A	= Pre-exponential factor [s^{-1}] for a 1st order reaction
d	= Diameter
E	= Activation energy [$Jmol^{-1}$]
Fr	= Froude number
g	= Standard gravity [$m\ s^{-2}$]
h	= Heat transfer constant [$Wm^{-2}K^{-1}$]
k	= Thermal conductivity [$Wm^{-1}K^{-1}$]
m	= Dimensional & non-dimensional weight
n	= Order of reaction, non-dimensional
Nu	= Nusselt number
Pr	= Prandtl number
R	= Universal gas constant OR Radius [$Jmol^{-1}K^{-1}$] OR m
Re	= Reynolds number
t	= Time
T	= Temperature
U	= Velocity [$m\ s^{-1}$]

Symbols

μ	= Viscosity [Pa s]
ρ	= Density [kgm^{-3}]
ω	= Angular velocity [s^{-1}]

Subscripts

bed = Fluidized bed
bp = Bed of particles (in rotary drums)
g = Gas phase
p = Particle
t = Terminal (velocity)

References

1. Demirbas A (2009) Biorefineries: for biomass upgrading facilities. Springer-Verlag, London
2. Kamm B, Gruber P, Kamm M (2010) Biorefineries—industrial processes and products. Wiley
3. Brown R, Stevens C (2011) Thermochemical processing of biomass: conversion into fuels chemicals and power. Wiley
4. Crocker M (2010) Thermochemical conversion of biomass to liquid fuels and chemicals. R Soc Chem
5. Waldron K (2010) Bioalcohol production: biochemical conversion of lignocellulosic biomass. Taylor & Francis
6. Basu P (2010) Biomass gasification and pyrolysis: practical design and theory. Elsevier
7. Azargohar R, Dalai A (2006) Biochar as a precursor to activated carbon. Appl Biochem Biotechnol 131(1–3):762–773
8. US Department of Energy (2012) Feedstock types. Consulted on February 1st, 20112, sur. http://www1.eere.energy.gov/biomass/feedstocks_types.html
9. Harkin J, Rowe J (1971) Bark and its possible uses. U.S. Department of Agriculture, Forest Service, Madison
10. Klass D (1998) Biomass for renewable energy, fuels, and chemicals. Academic Press
11. Tillman D (1981) Wood combustion: principles, processes and economics. Academic Press
12. Gavrilescu D (2008) Energy from biomass in pulp and paper mills. Environ Eng Manage J 7(5):537–546
13. Onay O, Beis S, Kockar O (2001) Fast pyrolysis of rape seed in a well-swept fixed-bed reactor. J Anal Appl Pyrolysis 58–59, pp 995–1007
14. Rodrigues I, Coelho J, Carvalho M (2012) Isolation and valorisation of vegetable proteins from oilseeds plants: methods, limitations and potential. J Food Eng 109:337–346
15. Li A, Li X, Li S, Ren Y, Shang N, Chi Y (1999) Experimental studies on municipal solid waste pyrolysis in a laboratory scale rotary-kiln. Energy 24:209–218
16. Brebu P, Ucar S, Vasile C, Yanik J (2010) Co-pyrolysis of pine cone with synthetic polymers. Fuel 89(8):1911–1918
17. USDA Forest Service (2007) Timber product output database. Consulted on January 2012, on forest inventory analysis. http://fia.fs.fed.us
18. Bhattacharya P, Parthiban V, Kunzru D (1986) Pyrolysis of black liquor solids. Ind Eng Chem Process Des Dev 25(2):420–426
19. Langholtz M, Carter D, Rockwood D, Alavalapati J (2007) The economic feasibility of reclaiming phosphate mined land with short rotation woody crops in Florida. J For Econ 12:237–249
20. Boateng A, Daugaard D, Goldberg N, Hicks K (2007) Bench-scale fluidized-bed pyrolysis of switchgrass for bio-oil production. Ind Eng Chem Res 46:1891–1897
21. Imam T, Capareda S (2012) Characterization of bio-oil, syn-gas and bio-char from switchgrass pyrolysis at various temperatures. J Anal Appl Pyrol 93:170–177
22. Singh R, Shadangi K (2011) Liquid fuel from castor seeds by pyrolysis. Fuel 90:2538–2544

23. National Agricultural Statistics Service (1995) Agricultural statistics. U.S. Department of Agriculture, Washington D.C.
24. Mani T, Murugan P, Abedi J, Mahinpey N (2010) Pyrolysis of wheat straw in a thermogravimetric analyzer: effect of particle size and heating rate on devolatilization and estimation of global kinetics. Chem Eng Res Des 88:952–958
25. Mansaray K, Ghaly A (1999) Kinetics of the thermal degradation rice husks in nitrogen atmosphere. Energy Sources 21:773–784
26. Sharma A, Rao R (1999) Kinetics of pyrolysis of rice husk. Bioresour Technol 67:53–59
27. Geldart D (1973) Types of gas fluidization. Powder Technol 7(5):285–292
28. Bi H, Grace J, Zhu J (1995) Regime transitions affecting gas-solids suspensions and fluidized beds. Chem Eng Res Des 73(2):154–161
29. Kanury A (1994) Combustion characteristics of biomass fuels. Combust Sci Technol 97(4–6):469–491
30. Edwards W (1991) In: Klass D (ed) Energy from biomass and wastes XV. Chicago, Institute of gas technology, p 503
31. Mujumdar A (2007) Handbook of industrial drying. CRC Press.
32. Joung HT, Seo YC, Kim KH, Seo YC (2006) Effects of oxygen, catalyst and PVC on the formation of PCDDs, PCDFs and dioxin-like PCBs in pyrolysis products of automobile residues. Chemosphere 65:1481–1489
33. Wendlandt W (1982) Thermal analysis. Anal Chem 54(5):97R–105R
34. Chang YM (1996) On pyrolysis of waste tire: degradation rate and product yields. Resour Conserv Recy 17:125–139
35. Varhegyi G, Bobaly B, Jakab E, Chen H (2011) Thermogravimetric study of biomass pyrolysis kinetics: a distributed activation energy model with prediction tests. Energy Fuels 25(1):24–32
36. Babu B (2008) Biomass pyrolysis: a state-of-the-art review. Biofuels, Bioprod Biorefin 2(5):393–414
37. Gasparovic L, Korenova Z, Jelemsky L (2010) Kinetic study of wood chips decomposition by TGA. Chem Pap 64(2):174–181
38. Rao R, Sharma A (1998) Pyrolysis rates of biomass materials. Energy 23(11):973–978
39. Roberts AF (1970) A review of kinetics data for the pyrolysis of wood and related substances. Combust Flame 14(2):261–272
40. Jiang G, Nowakowski DJ, Brigwater AV (2010) A systematic study of the kinetics of lignin pyrolysis. Thermochim Acta 498:61–66
41. Bridgwater A (2012) Review of fast pyrolysis of biomass and product upgrading. Biomass Bioenergy 38:68–94
42. Williams P, Besler S, Taylor D (1990) The pyrolysis of scrap automotive tyres: the influence of temperature and heating rate on product composition. Fuel 69(12):1474–1482
43. Demirbas A (2004) Effect of initial moisture content on the yields of oily products from pyrolysis of biomass. J Anal Appl Pyrol 71(2):803–815
44. Chen G, Sjostrom K, Bjornbom E (1992) Pyrolysis/gasification of wood in a pressurized fluidized bed reactor. Ind Eng Chem Res 31(12):2764–2768
45. Pindoria R, Megaritis A, Messenbock R, Dugwell D, Kandiyoti R (1998) Comparison of the pyrolysis and gasification of biomass: effect of reacting gas atmosphere and pressure on Eucalyptus wood. Fuel 77(11):1247–1251
46. Weiland N, Means N, Morreale B (2012) Product distributions from isothermal co-pyrolysis of coal and biomass. Fuel 94:563–570
47. Mullen C, Boateng A, Mihalcik D, Goldberg N (2011) Catalytic fast pyrolysis of white oak wood in a bubbling fluidized bed. Energy Fuels 25:5444–5451
48. Kunii D, Levenspiel O (1991) Fluidization engineering, 2nd edn. Butterworth-Heinemann
49. Yang WC (2003) Handbook of fluidization and fluid-particle systems (Chemical Industries). CRC Press
50. Dynamotive Energy (2012) Corporate history. consulted on March 1st, 2012, on Dynamotive, the evolution of energy. http://www.dynamotive.com/about/corporate-history/

51. Clausthal University of Technology, Institute of Energy Process Engineering and Fuel Technology, Prof. Dr.-Ing. R Weber. http://www.ievb.tu-clausthal.de
52. Kothari A (1967) Master Thesis. Illinois Institute of Technology, Chicago
53. Avidan A, Yerushalmi J (1985) Solids mixing in an expanded top fluid bed. AICHe J 31(5):835–841
54. Watanabe I, Yong C, Hasatani M, Yushen X, Naruse I (1991) Gas to particle heat transfer in fast fluidized beds. Circulating Fluidized Bed Technology III. Pergamon Press, pp 283–287)
55. Wang H, Srinivasan R, Yu F, Steele P, Li Q, Mitchell B (2011) Effect of acid, alkali and steam explosion pretreatments on characteristics of bio-oil produced from pinewood. Energy Fuels 25:3758–3764
56. Lima I, Boateng A, Klasson K (2010) Physicochemical and adsorptive properties of fast pyrolysis bio-chars and their steam activated counterparts. J Chem Technol Biotechnol 85:1515–1521
57. Ranz W (1952, May) Friction and transfer coefficients for single particles and packed beds. Chem Eng Prog 48(5):247–253
58. Van der Drift A, Boerrigter H, Coda B, Cieplik M, Hemmes K (2004) Entrained flow gasification of biomass. ECN
59. Sun S, Tian H, Zhao Y, Sun R, Zhou H (2010) Experimental and numerical study of biomass flash pyrolysis in an entrained-flow reactor. Bioresour Technol 101:3678–3784transportation fuels from biomass: chemistry
60. Mellmann J (2001) The transverse motion of solids in rotating cylinders—forms of motion and transition behaviour. Powder Technol 118:251–270
61. Fantozzi F, Colantoni S, Bartocci P, Desideri U (2007) Rotary kiln slow pyrolysis for syngas and char production from biomass and waste—Part II: introducing product yields in the energy balance. J Eng Gas Turbines Power 129:908–913
62. Ingram L, Mohan D, Bricka M, Steele P, Strobel D, Crocker D, Mitchell B, Mohammad J, Cantrell K, Pittman Jr CU (2008) Pyrolysis of wood and bark in an auger reactor: physical properties and chemical analysis of the produced bio-oils. Energy Fuels 22:614–625
63. Thangalazhy-Gopakumar S, Adhikari S, Ravindran H, Gupta RB, Fasina O, Tu M, Fernando SD (2010) Physicochemical properties of bio-oil produced at various temperatures from pine wood using an auger reactor. Bioresour Technol 101:8389–8395
64. He R, Ye X, English B, Satrio J (2009) Influence of pyrolysis condition on switchgrass bio-oil yield and physicochemical properties. Bioresour Technol 100:5305–5311
65. Sanchez ME, Menendez JA, Dominguez A, Pis JJ, Martinez O, Calvo LF, Bernad PL (2009) Effect of pyrolysis temperature on the composition of the oils obtained from sewage sludge. Biomass Bioenergy 22:922–940
66. Mohan D, Pittman Jr C, Steele P (2006) Pyrolysis of wood/biomass for bio-oil: a critical review. Energy Fuels 20:848–889
67. Khan M (1987) Production of high quality liquid fuels from coal by mild pyrolysis of coal/lime mixtures. Energy Fuels 5(2):185–231
68. Lin L, Khang S, Keener T (1997) Coal desulfurization by mild pyrolysis in a dual-auger coal feeder. Fuel Process Technol 53:15–29
69. Cho MH, Jung SH, Kim JS (2010) Pyrolysis of mixed plastic wastes for the recovery of benzene, toluene, and xylene (BTX) aromatics in a fluidized bed and chlorine removal by applying various additives. Energy Fuels 24:1389–1395
70. Huber G, Iborra S, Corma A (2006) Synthesis of transportation fuels from biomass: chemistry, catalysts and engineering. Chem Rev 106:4044–4098

Chapter 12
Improvements of Biomass Gasification Process by Plasma Technologies

Philip G. Rutberg, Vadim A. Kuznetsov, Victor E. Popov, Alexander N. Bratsev, Sergey D. Popov and Alexander V. Surov

Abstract The chapter is dedicated to a promising method of biomass treatment-plasma gasification. Increased temperatures and energy supply allows significantly in-crease the range of wastes and other carbonaceous materials which could be efficiently processed. Fea-tures of plasma usage in updraft and downdraft bio-mass gasification are described. Several promising re-newable energy sources (wood, energy crops, wastes of livestock, and poultry industry) are examined for the usage in downdraft plasma gasification. The correlation of key parameters of biomass plasma gasification was studied in thermodynamic equilibrium approach along with syngas usage for liquid fuel production. Institute for Electrophysics and Electric Power RAS experimental installation is described. Its primary component is a downdraft plasma gasifier for processing of biomass and wastes. Its technical characteristics and functionali-ty are described. A brief survey of existing pilot and in-dustrial projects is given. Methods of energy supply into plasma chemical reactor are described. The review of powerful plasma torches for industrial application is represented. Experimental procedures and test results on biomass gasification by air-plasma are presented as well as the comparison with the calculated data.

Keywords Plasma · Gasification · Syngas · Plasma torch · Gasifier · Energy balance · Efficiency · Biomass · Renewable energy · Alternative energy

12.1 Introduction

The importance of the decrease of anthropogenic impact on the environment rises dramatically nowadays. In particular, it is a problem of carbon dioxide emissions [1]. CO_2 is the basic component of combustion products of widely used kinds of fuel; it possesses high radiative forcing and is one of the main (and the most dangerous) greenhouse gases [2]. In 2010, the global emission of carbon dioxide increases on

P. G. Rutberg (✉) · V. A. Kuznetsov · V. E. Popov · A. N. Bratsev · S. D. Popov · A. V. Surov
Institute for Electrophysics and Electric Power RAS (IEE RAS),
Dvortsovaya nab., 18, St.-Petersburg, 191186, Russia
e-mail: rc@iperas.nw.ru

Z. Fang (ed.), *Pretreatment Techniques for Biofuels and Biorefineries*,
Green Energy and Technology,
DOI 10.1007/978-3-642-32735-3_12,

~5.9 % and for the first time exceeds 9 Pg per year [3]. The increase in emission is mainly caused by economic growth of developing economics in which even during a global economic crisis the CO_2 emission increased [3]. Economic growth and increase in the living standards of the population invariably lead to growth of energy consumption per capita. According to IEA [4], in 2009 about 80.9 % of mankind energy demands were provided by fossil fuel combustion, about 5.8 %—nuclear energy,about 2.3 % hydroenergetics, and about 10.2 %—energy of biofuels and waste, which in total is 509 EJ. The world's reserve of fossil fuels is about 35,094 EJ (~23.6 %—oil, ~20.4 %—natural gas, ~56.0 %—coal) [5]. By estimations [6] world oil production will peak before 2020, coal before 2030, and natural gas around 2040. Unfortunately, the comprehension of not only the importance, but also the fact of finiteness of fossil fuels occurs extremely slowly among people defining directions of development both on regional and global level. The citation ideally illustrates the situation: "perpetual growth is often held as a pious belief and fundamental assumption for economists" [7]. The only possibility for humanity to prevent impending energy crisis is usage of renewable energy sources: biomass, solar energy, wind, tidal and wave power, hydroenergetics, and thermal power. Electricity generation from solar energy is very expensive that is why construction of large power plants is unlikely. The wind farms have low efficiency and their applicability is limited. Nowadays, the natural resources for development of hydroenergetics are almost exhausted. In spite of biosphere limits on bioresources generation governed by necessity to sustain the ecological balance and growth of mankind food demand the biomass is one of the most promising types of renewable energy sources. According to the forecasts [8] in 2050, the global potential of biomass energy will be about 1,135–1,300 EJ (without the use of seaweed as biomass), while the world consumption will reach ~826 EJ (on average under different scenarios). Energy use of biomass does not increase CO_2emission, as far as the whole carbon dioxide, formed after bioenergy use, is absorbed by green plants in the process of biomass formation. Other renewable sources hardly will be widely used; in particular, nuclear fusion energy can be practically used not until the end of the century. Nuclear power cost will rapidly increase due to safety issues. Thus, it is clear that if the cultivation requirements of bioresources are met, then the bio-energetic development is one of the most promising ways to form a sustainable and independent economics for both developed and developing countries.

12.2 Biomass Sources

All organics substances originating from living nowadays or recently lived organisms are consideredas a biomass. For example, coal and oil are not a biomass as they were formed from organisms living millions years ago. On the other hand, a municipal and industrial waste can be considered as a biomass because an essential part of their organic weight is such kinds of a biomass as wood, rubbers, food waste, and other materials of an organic nature. Clearly, a biomass is one of the most diverse (by quantity of representatives) class of fuels. Therefore, determination of the perspectives of their energy use demands examination of energy characteristics and chemical composition along with the rates of biomass formation and expenditures connected with its production. We will examine four common types of biomass: wood, energy crops, animal waste, and poultry waste.

12.2.1 Wood Waste

Wood, perhaps, is one of the most common types of biomass. Approximately 40.5 % of annually extracted forest resources are used for obtaining the roundwood, ~6.3 % in paper manufacturing, ~45.1 % as a fuel, and about 8.1 % at charcoal production (total wood consumption ~2.48 Gt) [9]. The average density of wood with moisture of 20 % (~624 kg/m^3 [10]) was used to estimate mass of wood by its volume. The wood (moisture ~20 %) consumption during pyrolysis aimed on charcoal production was estimated by proximate analysis results on total yield of ash and fixed carbon (~12.1 % [11]). Wood, compared to fossil fuels, basically does not contain sulfur and other ecologically unfriendly elements. However, its use in the wood industry is more economically sound than its energy application. Therefore, it is more expedient to process lumbering wastes, residues, and municipal waste mostly consisting of woody materials.

12.2.2 Energy Crops

Development of the biofuel production processes had begun in the nineteenth century; in the beginning of the twentieth century interest to the biofuels had died out because of the rapid growth of cheaper fossil fuels usage; developments in this field have been resumed again due to the oil crisis in 1973 [12]. Today, the development of biofuels production and using technologies is driven by: increase in prices of energy resources, fossil fuels depletion, and also CO_2 emission issue.

Energy crops are plants which have been cultivated as a source of energy. Basically they are represented as herbaceous or woody fast-growing plants, for example switchgrass [13] and willow [14]. Algae are one of the most promising biofuels. Fertile lands are not required for their cultivation, and they grow in virtually any kind of water [15].

Brazil was the major producer of biofuels until 2000; however, by 2008 its world output has increased from ~20 billion liters a year to almost 75 billion, basically due to the rapid development of the bio-energy technologies in the USA and Canada [16].

One of the most promising sources of biomass is *Panicum virgatum*, well-known as switchgrass. Let us examine it as a characteristic representative of energy crops.

12.2.3 Livestock and Poultry Waste

The cattle and poultry breeding are the vital branches of agriculture which overall production is closely associated with population and its living standards. Local demand on their production grows with the increase of the population density. Partly therefore a shift toward larger farms which do not produce their own forage is observed recently in this sector; it made the manure disposal issue very important. Its mishandling leads to bacteriological contamination of ground waters [17]. Meanwhile manure is a biomass and it can be used as a renewable energy source.

The manure composition can significantly vary depending on disposal technologies, even if it is produced by the same group of animals. Therefore, determination of average composition of waste demands separate investigation. Here, for example, cow and chicken manure will be studied.

12.2.4 Comparison of their Suitability for Plasma Gasification

Determination of the most promising types of biomass for the plasma gasification is a big challenge. For its solution it is necessary to carry out researches in absolutely different areas: gasification technologies (taking into account special features of plasma usage), wood industry, agriculture, and also other industries producing biomass as end- or by-product. Therefore, in this work the problem of prospectivity is considered not among the vast diversity of biomass types, but among several representatives chosen for the consideration.

Let us discuss plasma gasification features versus autothermal one. Currently gasification utilizes only thermal plasma. Correspondingly, the plasma influence on the process is defined by its contribution to energy balance (the higher share of plasma energy is the grater difference it makes for the process). In the plasma injection zone of a reactor the temperatures grows significantly in comparison with the similar autothermal gasifiers that results in increase of chemical reaction rates. Content and yield of valuable products of gasification (hydrogen and carbon monoxide) increase due to additional energy injected with plasma. Moreover, high efficiency of energy introduction makes it possible to gasify feedstock by pure steam and carbon dioxide.

The gasifier type and the duration of main stages of gasification determine the syngas composition. The gasifier type defines the direction of mass stream of the process. Plasma is used in two general types of gasifiers: counter-current fixed bed (updraft), co-current fixed bed (downdraft). In the former, the gas stream (oxidant and gasification products) is directed upward, whereas in the latter it is downward, but in both cases the feedstock stream is directed downward. Solid gasification products mainly consist of inorganic components and are discharged from the bottom of the reactor.

In the updraft process, the organic matter during gasification is first devolatilized and then oxidized. Pyrolysis products (volatiles) are formed and leave the reactor

Table 12.1 The most important characteristics of several types of biomass

Feedstock	Wood waste	Switchgrass	Cattle manure	Chicken manure
Proximate analysis (wt %) (as received basis)				
Moisture	20.00	6.29	36.60	20.20
Ash	0.80	8.51	25.20	21.23
Volatile	67.90	70.26	31.60	54.18
Fixed carbon	11.30	14.94	6.60	4.39
Ultimate analysis (wt %) (dry, ash-free basis)				
Carbon	50.25	49.16	50.39	53.96
Hydrogen	6.09	6.36	5.77	5.60
Nitrogen	0.20	0.63	3.94	7.92
Sulfur	0.10	0.13	1.31	0.96
Oxygen	43.35	43.73	38.58	31.56
LHV (MJ/kg)	13.90	15.82	6.49	12.09
Adiabatic combustion temperature in dry air (K)	2,123	2,304	1,815	2,210
Oxygen consumption required for complete gasification (g/kg)	9.2	129.5	0	56.8
Oxygen consumption required for complete combustion (g/kg)	1,101	1,174	546	923
Maximal yield of chemical energy at complete gasification (MJ/kg)	17.84	17.20	8.51	14.05
Annual generation (t/km^2)	90–150	1,000–1,500	700–1,300	300–800
Harvesting cost[a] (¢/kg)	3.1–3.9	4.4	-0.5—0.8	-1.3
Reference	[11, 18, 19]	[20–22]	[23–26]	[27–30]

[a] Wood waste and switchgrass are energy resources which production requires expenses. On the other hand, cattle and chicken manure are wastes which disposal requires expenses, and a farmer has to pay for their treatment.

volume at rather cold zones where tars, water, and some permanent gases are not converted to syngas. Sometimes in such cases the separate unit for tar plasma conversion is used. Plasma usage in the updraft gasifier allows liquid slag discharge.

In the downdraft process, the organic matter during gasification is also first devolatilized and then oxidized. However, in this case the pyrolysis products pass through the high-temperature oxidizing zone, where they are converted by plasma. It results in considerable reduction of tar concentration of syngas and allows using plasma energy to increase H_2 and CO yield and content.

There are parameters permitting preliminary estimation of the prospects and suitability of raw materials for plasma gasification (see Table 12.1).

Proximate analysis data show that organic matter of chicken manure consists of more than 90 % (moisture + volatile matter) from substances turning into gaseous phase during the devolatilization stage, for other types of biomass this value is more than 80 %. In general, more the content of volatile matter, easier and more effective is its use in the downdraft plasma gasification [31]. In the updraft process, it becomes a disadvantage because volatile matter is not exposed to high-energy plasma flow and the produced syngas is heavily contaminated by tars [32]. Conversion of these tars is an important line of plasma usage development. For example, Europlasma is developing a reactor module comprising an autothermal gasifier (similar to updraft one on organization of mass flows) and plasma system for tar conversion, thus the most complete conversion occurs at energy consumption of ~1.8 MW, while syngas yield possesses ~10.2 MW of chemical energy [33]. Non-equilibrium plasma is used for conversion of synthesis gas with very low tar content (0.7–1.9 g/Nm^3). In order to decrease tar concentration by ~20 % it is required that 27–39 % of electricity is produced by a gasifier [34]. The content of tars in the downdraft plasma process is already significantly reduced because they pass through the high-temperature oxidizing zone and almost does not affect the energy balance [35].

Feedstock's fixed carbon/ash mass ratio is important for downdraft process since feedstock and oxidizer flows are co-current; the gasification rate decreases dramatically downstream. Concentration of an oxidizing component in gas phase and carbon in char-ash residue decrease, and multiplication of these concentrations determines the mass exchange rate. Thus, fuels with high carbon content in a char-ash residue are preferred for downdraft plasma and autothermal gasification processes. Wood waste (~93 % carbon in the char-ash residue) and switchgrass (~64 %) are the best raw materials according to this characteristic.

In the updraft process feedstock and oxidizer flow in counter-current configuration. The plasma flow at the inlet contacts with the extremely carbon depleted char-ash residue. This method versus the downdraft gasification allows achievement of higher carbon conversion level of char-ash residue. Moreover, position of the high temperature zone at the bottom of the reactor simplifies liquid slag removal and slag vitrification [36].

Data on proximate and ultimate analysis allow determination of oxygen consumption required for complete gasification of carbon and for complete combustion of feedstock. These values should be considered simultaneously with the heating value. Jointly they determine adiabatic combustion temperature having higher impact on practical value of a feedstock for energy industry than heating value. The fuel can have a significant heating value, but the higher amount of oxygen is required for its combustion, the larger amount of energy is spent for heating of neutral nitrogen at operation on air. The decrease of adiabatic combustion temperature is mainly caused by three factors: excessive moisture, high content of oxygen in a feedstock, and in a less degree, high ash content. Only moisture can be easily altered, its removal leads to a decrease of oxygen and hydrogen content in the feedstock. Updraft plasma gasification process compared to downdraft one allows decreasing of energy consumption for fuels with high ash and moisture if their removal is impossible or leads to a decline in economic viability of the whole process. However, usage of such types of feedstocks for energy needs usually is not rational.

One of the key parameters for all plasma processes is a relationship of energy and oxygen consumption for stoichiometric carbon gasification conditions. However, it is impossible to determine energy consumption without calculation of the gasification process. At the analysis initial stages we could examine ratios of chemical energy yield to a heating value and oxygen consumptions for combustion and gasification. The higher these values are the harder optimum parameters of plasma gasification are achieved. The most challenging feedstock in the given context is cattle manure (ratios: ~1.31 and ∞, respectively), and the simplest switchgrass (~1.09 and ~9.07).

It should be noted that estimations of chemical energy yield limit are correct if the downdraft plasma gasification or any other method with complete tar conversion is used. It is impossible to determine clearly the most promising feedstock by this value, because it is necessary to spend energy and funds for feedstock acquisition and to take into account transportation expenses. Both these parameters depend on used methods and technologies of feedstock collection and transportation. Assuming that thermal energy of gasification products in all cases amounts to 3 MJ in energy balance on 1 kg (this rough approximation is admissible for stoichiometric gasification of many types of feedstocks with LHV less than ~15 MJ/kg), synthesis gas will be used in the combined cycle with efficiency of ~60 % [37] and electricity cost amounts to 5¢/kW h, we define that treatment of chicken manure will be the most profitable. According to the energy balance (per 1 kg), energy consumption will be ~1.38 kW h (~6.9 ¢), electricity yield ~2.3 kW h (~11.7 ¢), and income from treatment of manure ~1.3 ¢. In this case, the income from treatment of 1 kg of feedstock will be ~6.1 ¢.

12.3 Numerical Simulation of Plasma Gasification

According to Forest Products Laboratory (U.S. Department of Agriculture) data [38], the efficiency of wood-fired power plants is 18–24 %. And wood is used as a fuel by millennia and is one of easiest to use energy resources. The efficiency of electricity generation can be increased up to ~29 % using gasification and combined cycle technologies, and using plasma gasification up to ~35 % [39, 40]. That is why the development of energy industry technologies based on plasma gasification is one of the most promising ways of evolution of energy use. During plasma generation, the molecules disintegrate to electrons, ions, atoms, and radicals [41], which makes it highly reactive. Plasma application leads to an increase in rates of chemical reactions, which in particular enhances the conversion in endothermal gas-phase processes [42]. Plasma is a universal oxidizer using for gasification of virtually any kind of feedstock including wood waste [43, 44], coal [45], RDF [46], etc. Though plasma has not yet widely used, such advantages as reduction of trace contaminants [47], tar conversion [48], high rates of heat exchange [49], and also high throughput at low plasma flow rate [50] attract attention to its use in pyrolysis and gasification. The main deficiency of plasma technologies is low level of their industrialization [51].

12.3.1 Calculation Methods

It is useful to perform numerical simulation of the process to now the prospects of plasma gasification of chicken manure. Calculation of equilibrium composition allows evaluating the yield limits of valuable gasification products at the set parameters (oxidant/feedstock ratio, temperature, and pressure) [52]. Deficiencies of the method are the restrictions imposed by assumptions: ideal mixing and unlimited residence time [53]. Nevertheless, the given approach accords satisfactory with the experimental data [39, 54–56].

Calculations of the equilibrium composition were implemented by means of the software Chemical WorkBench (Kinetic Technologies Ltd., http://www.kintech.ru/).

Possibility to supply energy with plasma almost completely removes the kinetic restrictions imposed on the process by low temperatures in autothermal modes. Actually, the plasma mode boundary lines are determined by the oxygen consumption required for oxidation of carbon of the feedstock to CO (so-called carbon boundary) and autothermal limit when additional energy is not required for achievement of the set parameters. For chicken manure, the carbon boundary is ~245 g/kg (mode 1, Fig. 12.1) for air and ~63.9 g/kg (mode 3) for steam plasma. Autothermal mode (temperature 1,500 K, pressure 1 atm) air gasification is achieved at oxidizer consumption ~2.16 kg/kg. Accordingly, all regimes between these two air consumptions are allothermal under equal conditions (temperature and pressure).

Let us examine modes of stoichiometric gasification of chicken manure by air and steam, and also a mode with an air consumption ~1.20 kg/kg (mode 2) that is between autothermal and stoichiometric modes, and a mode with the steam consumption ~0.314 kg/kg (mode 4), matching to the previous one by the amount of fed oxygen. Figure 12.1 shows the results of calculations. In calculations, the composition and heating value of a chicken manure specified in Table 12.1 were used and also the following composition of air was used: N_2—78.09, O_2—20.95, Ar—0.93, CO_2—0.03 %mol.

Approximation of synthesis rate was used for calculation of space velocity of Fischer–Tropsch process on Co–Mn/TiO_2 catalyst [57] at pressure 10 bar and temperature 523 K. Before Fischer–Tropsch synthesis, the syngas was cleaned from sulfur compounds and nitrogen oxides, part of CO was converted to H_2 by water–gas shift reaction to provide stoichiometric relation of synthesis—H_2/CO = 2 according to its chemical equation (12.1).

$$CO + 2H_2 \rightarrow -(CH_2) - +H_2O \tag{12.1}$$

The synthesis continued until CO content in convertible gas decreased to 0.1 %mol. The process rate was calculated according to equation (12.2) [57].

$$r_{CO} = k_P \times b_{CO} \times P_{CO} \times P_{H_2}/(1 + b_{CO} \times P_{CO}) \tag{12.2}$$

where, r_{CO} is CO conversion rate (mmol$_{CO}$/min g$_{cat}$), k_P and b_{CO} are kinetic parameters equal to 0.367 and 1.454, respectively, P_{CO} and P_{H_2}– are partial pressures of CO and H_2.

12.3.2 Discussion

Let us discuss presented plasma gasification modes of chicken manure with various consumptions and oxidants—1–4 (according to the data of Fig. 12.1). As it is seen the

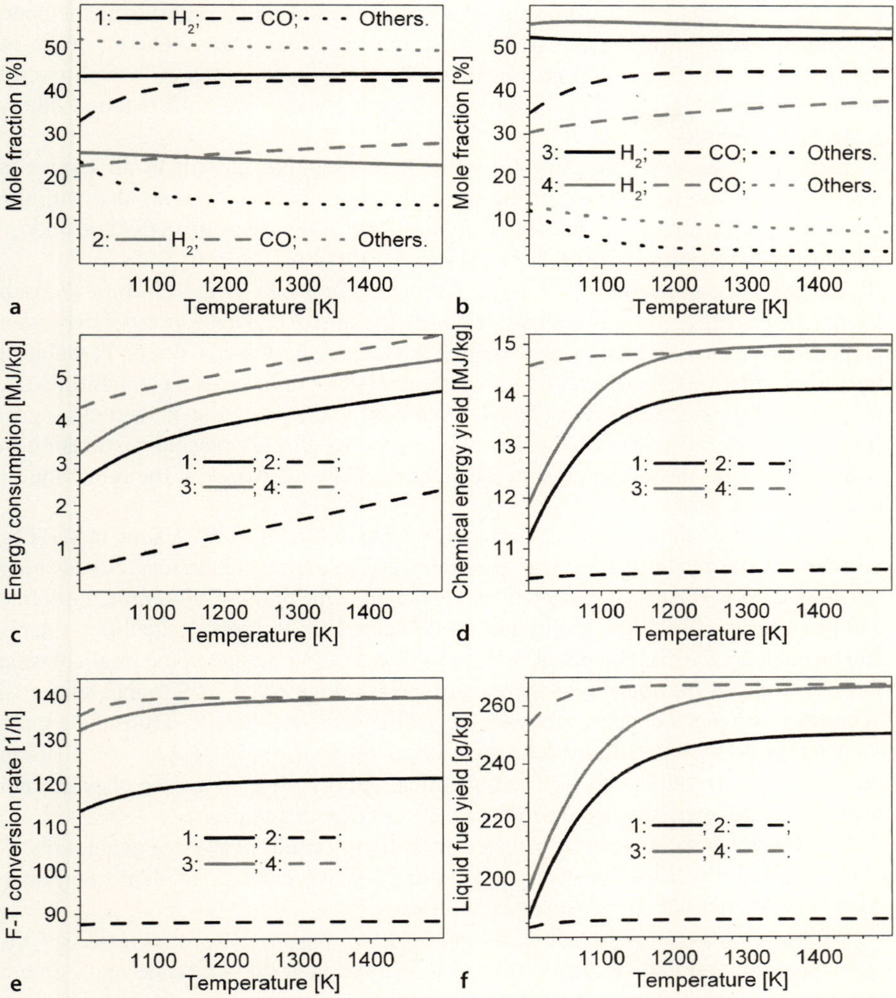

Fig. 12.1 Main parameters of integrated plasma gasification and Fischer–Tropsch system for chicken manure processing: **a** and **b** Composition of dry syngas; **c** Energy consumption per mass unit of feedstock (LHV basis); **d** Yield of syngas chemical energy per mass unit of feedstock (LHV basis); **e** Hourly space velocity of Fischer–Tropsch synthesis; **f** Yield of synthetic fuel per mass unit of feedstock. Data are represented for plasma gasification modes consuming: *1* ~245 g, *2* ~1.20 kg of air and *3* ~63.9 g, *4* ~314 g of steam per 1 kg of feedstock

CO content in syngas and the chemical energy yield increase while the temperature grows at the stoichiometric gasification mode. It is generally caused by the decrease of the graphite yield which becomes less than 0.1 g/kg only at temperatures more than ~1,400 K, and its conversion to carbon monoxide. On modes 3 and 4, change of composition is caused by the shift of equilibrium toward H_2 and CO_2 formation according to stoichiometry of water–gas shift reaction and insignificant methane content increase (up to ~1.2–1.6 %mol) at temperature decrease. The influence of the process temperature on a chemical energy yield is insignificant, as volumetric heating values of H_2 and CO are close, and during water–gas shift reaction one is replaced by another.

Increase of energy consumption with the temperature growth in all modes is mainly caused by the increase of the sensible heat of the system, and also with the graphite gasification on stoichiometric modes. The influence of this effect on energy balance upon reaching 1,200 K decreases to ~2 kJ/kg K.

The syngas composition is a major factor defining rate of the Fischer–Tropsch synthesis. During the water–gas shift reaction, the sum of concentrations of hydrogen and carbon monoxide in syngas decreases (steam during the conversion is replaced by carbon dioxide), therefore the more initial H_2/CO ratio close to stoichiometric, the higher is the rate of Fischer–Tropsch synthesiswhile using the converted syngas. It causes lower rates of synthesis for syngas produced by air plasma gasification in comparison with the steam plasma gasification. Rate decreases at the reduction of the plasma energy.

The content of contaminants (such as H_2S, COS, NH_3, HCN, HCl, soot, tars, BTX, volatile metals, and dust) in syngas should be limited to prevent the accelerated aging of catalyst [58]. The residual content and residence time of these contaminants define catalyst lifetime, therefore the higher space velocity is the more valuable products can be made by the catalyst pellets before its deactivation. Usage of the steam plasma instead of the air plasma allows increasing space velocity by 15–59 % that leads to a decrease of financial expenses for the catalyst pellets in the Fischer–Tropsch process.

Alterations in liquid fuel yield are caused by changes in syngas composition and resemble the dependence of the chemical energy yield; however, they are not proportional as methane is inert to the Fischer–Tropsch synthesis.

The energy consumption (associated with energy used for plasma generation) is about 5.0–6.2 kW h for the steam and 1.9–5.2 kW h for the air oxidant to produce 1 kg of synthetic fuels by plasma gasification.

Carbon dioxide is a byproduct of synthetic fuel production from biomass. CO_2 content in the residual gas can be very high (more than 90 %), while using steam plasma that makes it very attractive for innovative applications [59]. For example, CO_2 reforming of methane [60] or synthesis of CH_3OH by non-equilibrium electrocatalysis plasma reactor [61].

12.4 Implementation of Plasma Gasification

Nowadays, there are a lot of companies in the world which are engaged in the commercial advancement of plasma gasification technologies. The most successful among them are: AlterNRG [62], Integrated Environmental Technology LLC [63], Advanced Plasma Power [64], Plasco Energy Group [65], Pyrogenesys Canada Inc. [66], etc. These companies have created pilot plants of various sizes. Some of them are implementing the large commercial projects. Generally, these gasification processes are based on bath of molten slag (Integrated Environmental Technology LLC), usage of bath and with the subsequent plasma conversion of the crude syngas (Pyrogenesys), traditional gasification in the updraft process with plasma conversion of crude syngas (Plasco Energy Group, Advanced Plasma Power), or updraft process of plasma gasification (AlterNRG).

Metallurgical furnaces with the Joule heating of molten slag are used in the gasification on bath where waste is supplied. This area heated either by the electric arc ignited on the slag surface, or by the plasma stream from the plasma torch. Sometimes, the plasma torch is installed in the separate chamber at the outlet of the reactor. There the crude syngas mixes with a high-temperature oxidizer causing the conversion of tars. Advantage of this process is the possibility to create large single installations. The second advantage is the possibility to discharge the incombustible components of waste in the liquid phase which allows separating metallic and nonmetallic fractions. An essential disadvantage is the growth of the energy consumption to maintain slag in liquid phase.

In the updraft gasification process of the AlterNRG company, the metallurgical coke and limestone are supplied into the reactor together with waste. Plasma generators are installed in the bottom part of the gasifier and generally used for heating and maintaining the slag in a liquid phase (limestone is fed to reduce melting point). Oxygen is supplied into the gasification zone above the plasma injection zone. The syngas produced in the gasification zone passes through the upper colder layers of waste. There the syngas is polluted by tars, formed in pyrolysis zones. Evaporating water does not participate in the conversion process due to low temperatures. The outlet for syngas is at the top of the gasifier. It is an essential disadvantage of the up-draft process. Produced gas also requires either additional conversion or tar cleaning.

The common essential disadvantage of all these technologies is the usage either free-burning arcs or DC plasma torches. Low efficiency of energy transfer from an arc to gas $\sim$30 % is typical for free-burning arcs. The efficiency for DC plasma torches is about 60 % due to the high losses in power-supply system. In our opinion, the most effective systems generating plasma are AC plasma torches. Plasma torches and power-supply systems developed in Institute for Electrophysics and Electric Power RAS (IEE RAS) provide energy transmission efficiency of the electric energy from the electrical grid into the plasma energy about 90–94 %.

12.4.1 Methods of Energy Transfer

Now plasma technologies of treatment and gasification of organic substances develop in two directions differed by the way of energy transfer to the plasma-chemical reactor. One of them is based on the electric arcs burn directly in the reactor volume (transferred arc). Arcs close between the electrode injected into the reactor and electroconductive melt (molten slag) in the bottom part of the reactor. Electrodes can be made of graphite or metal. The advantage of this method is possibility to create a large single plasma-chemical reactors with power consumption of about 5–30 MW. Free-burning arcs are used in large-scale metallurgical installations which are almost ready to commercial operation.

The disadvantages are rapid wear of electrodes, considerable quantity of admixtures, and low-energy transfer coefficient of a free-burning arc into the processing substance. The efficiency of such installations, as a rule, does not exceed 30–35 %.

Stationary plasma torches are used for plasma generation in another method. The plasma-forming gas gains energy from the electric arcs burning in the discharge chamber of the plasma torch (non-transferred arc), and then arrives into the reactor volume.

The efficiency of heat exchange between the generated plasma and processed substance is significantly higher in comparison with the first method. The efficiency of plasma torches optimized for industrial applications exceeds 90 % and energy of a plasma jet almost completely transfers to the processed substance. That rather simplifies control over the chemical composition of syngas. Now, the application of this method is restrained by absence of sufficiently powerful plasma torches capable of generating plasma from oxidizing media (air, steam, etc.) in prolonged modes with high efficiency and lifetime of electrodes.

Powerful stationary electric arc plasma torches, meeting the requirements of plasma-chemical technologies, can be classified by current type: alternating current (AC) plasma torches [67–81] and direct current (DC) plasma torches [82–86]. They also can be divided by a type of plasma forming gas: neutral, reducing, or oxidizing. Devices differ in design and other features: type of discharge chambers, material and form of electrodes, a way of working gas supply, a principle of the arc stabilization, etc. Most plasma torch designs utilize a combination of methods of arc discharge stabilization comprising gas stream organization, arc contraction by the insulated inserts, and magnetic field. Electrode designs of plasma torches are rod, toroidal, ring, and tubular; in some cases the electric arc chamber is one of the electrodes. We will present only basic types due to enormous variety of designs.

12.4.1.1 Direct Current Plasma Torches

Until now, most gasification and pyrolysis systems use DC plasma torches. Figure 12.2a shows the typical schematic diagram of power-supply circuit. The ballast resistor is used to stabilize burning of DC arc that causes considerable real power losses.

The most widespread DC plasma torches are Westinghouse Plasma Corp plasma torches with cylindrical electrodes. Sketch of this type of plasma torch is presented in Fig. 12.2b.

Power ranges for plasma torches MARC-3A and MARC-11L are 130–300 kW and 300–800 kW [87, 88], respectively, which are most attractive for industry. These are the most advanced designs. Their efficiency of energy transfer from the arc to plasma (thermal efficiency) is 70–85 %. The total efficiency of the system including

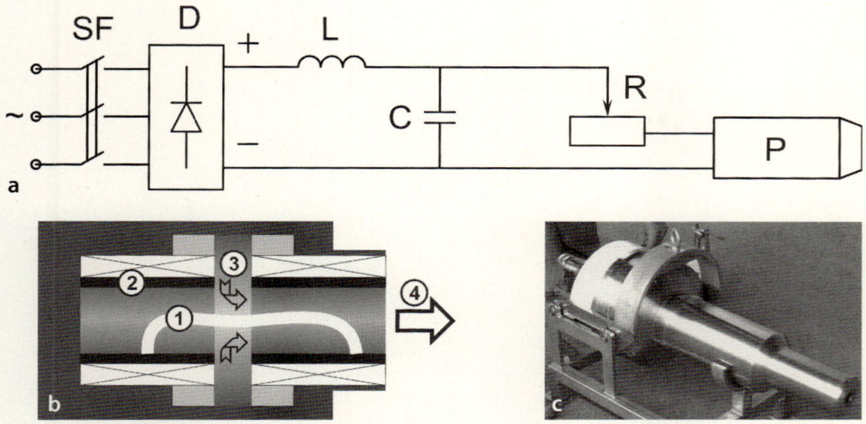

Fig. 12.2 DC plasma torches: **a** Typical power-supply schematic diagram of DC plasma torch (*P* plasma torch; *D* rectifier; *L–C* filter; *R* ballast resistor, *SF* automatic circuit breaker); **b** After [87] schematic representation of Westinghouse Plasma Corporation plasma torch (*1* plasma column; *2* electrode; *3* entering process gas; *4* heated process gas); **c** Photo of 300 kW Europlasma's plasma torch. (Reprinted from [89], with permission from the International Plasma Chemistry Society)

ohmic resistance losses in power-supply system is significantly lower. The lifetimes of electrodes for these models are 600 and 1,000 h, respectively.

EUROPLASMA designed a series of plasma torches of the same type which could be fed by CO_2, CO, CH_4, H_2, N_2, and their mixtures. Figure 12.2c shows a 300 kW plasma torch [89].

Series of "Linde" plasma torches are described in [83–85]. These devices are designed for short-time operation at high pressures and power up to 20 MW. These plasma torches use cylindrical copper electrodes. Some models have the working gas enthalpy up to 10 MJ/kg. Stabilized rectifiers power these devices. Voltage of power sources is up to 10 kV. Typically, the flow rates of cooling agents for ballast resistor and plasma torch are virtually equal. Pressure in the cooling system is about 4 MPa. Plasma torches MDC-200 and MDC-300 [86] use air as a working gas and operate in relatively short-time mode, they are intended for special experiments. Maximum working pressure of gas in the plasma torch chamber is 25 MPa, a range of operating currents 100–1,200 A, voltage 1.2–14 kV, thus power changes from 0.7 to 10.5 MW.

12.4.1.2 Alternating Current Plasma Torches

AC plasma torches are more promising for industrial applications and technologies requiring relatively high powers (e.g., waste treatment). Figure 12.3a represents the typical schematic diagram of a power-supply circuit. The power-supply system in this case is essentially cheaper and more reliable, rather than power-supply systems of DC plasma torches. Their maintenance is easier. The thermal efficiency of AC

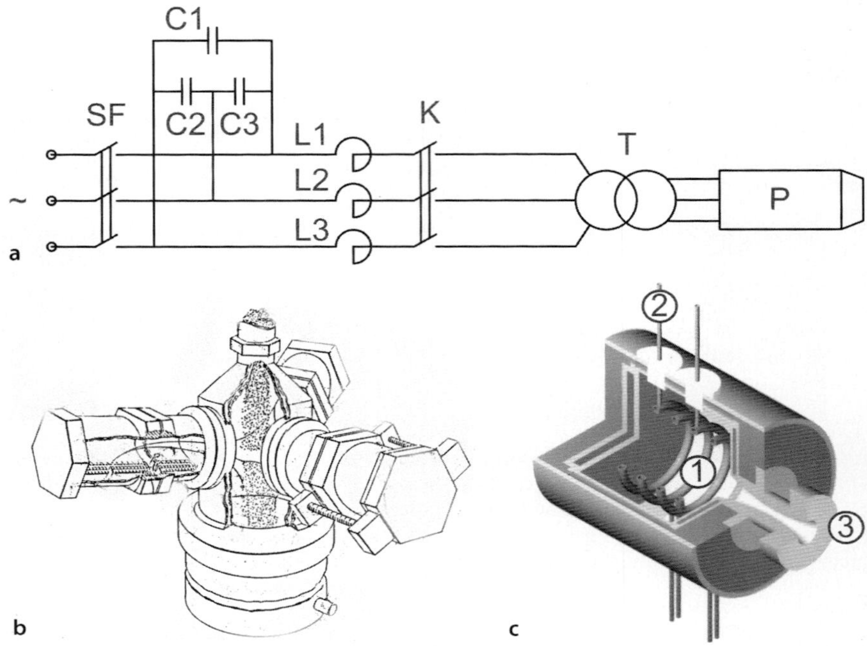

Fig. 12.3 AC plasma torches: **a** Schematic diagram of power-supply circuit of AC plasma torch (*P* plasma torch; *K* contactor; *T* standard step-up transformer; *L*1–*L*3 current limiting inductances; *C*1–*C*3 capacitor compensators; *SF* automatic circuit breaker); **b** Multi-phase electric arc Westinghouse heating system (schematic of the installation) [73]; **c** NOL electric arc heater (*1* ring electrodes; *2* current leads; *3* nozzle unit)

plasma torches is high. In AC power-supply systems, the losses do not exceed several percent as the reactive ballasts stabilizing an arc are used. Reactive power losses are minimized using standard capacitor compensators.

Existing AC plasma torches can be divided into three basic groups: the single-phase [72, 73], the three-phase single-chamber, and the three-phase multi-chamber plasma torches. In some cases, a DC plasma torch with linear circuit and cylindrical electrodes is taken as a basis for single-phase AC plasma torches, thus AC power source is used. Working gas is usually supplied tangentially into such system. Another option of a single-phase plasma torch is a construction with the central rod electrode and ring or toroidal electrode. Usually, the arc is stabilized by the magnetic field rotating the arc in the interelectrode gap. Such type plasma torches use axial supply of the working gas.

Multi-phase multi-chamber AC plasma torches comprise various combinations of several single-phase plasma torches using multi-phase AC electrical grid. There are constructions consisting of three separate single-phase plasma torches. It is possible to connect three single-phase plasma torches sharing one mixing chamber, while the connection configuration can be different. There are systems designed by this principle, for example, a multi-phase electric arc heating system (see Fig. 12.3b) described in [73].

Multi-phase single-chamber AC plasma torches are described in [70]. Their feature is installation of an electrode system of the plasma torch in a single-electric arc chamber. The electrode systems of multiphase single-chamber plasma torches can have the form of rings, toruses, or rods. In case of using toroidal or ring electrodes (Fig. 12.3c), generally the first and the last electrode are connected to the same phase. Electrodes are usually separated from each other by heat-resistant insulating pads. Stabilization and arc twirl are supported either by a magnetic field by means of the solenoid mounted on the plasma torch case, or by the creation of the tangential gas vortex setting the arc column on an axis of the electric arc chamber and moving the attachment points of the arc along the electrode surface.

Another widely used group of single-chamber multiphase plasma torches are plasma torches with rod electrodes. Several types of plasma torches, including the ones with rod electrodes have been developed, produced, and tested in IEE RAS. Research and development, and design experience are described in [67, 80, 90]. Figure 12.4a shows a single-chamber plasma torch with rod electrodes.

Ignition of several simultaneously burning AC arcs in a single chamber allowed creation of simple and reliable plasma torches transforming the electric current energy into plasma energy with high efficiency of 80–90 %.

Several different designs were developed. Tungsten or tungsten-containing rod electrodes were used for operation on inert gases, nitrogen, and hydrogen. Water-cooled copper tubular electrodes were used for operation on oxidizing media. Plasma torches with rod electrodes can be divided into three groups: with power up to 200 kW and 2 MW, and also working at short-time modes with power 4–50 MW [90]. Both types have the similar design comprising three main parts: case, arc chamber (nozzle), and electrode unit. Multi-phase mode of arc burning in the discharge chamber allows using low voltage of re-ignition due to the preliminary ionization of the discharge gap. Tungsten with additives of rear earth metals and compounds having low work function were used as an electrode material. The advantages of single-phase plasma torches with rod electrodes are: simple design, high efficiency providing by optimal relation of volume and surface area of the arc chamber, and also the possibility of electrode operation in the thermo-emission mode. In these systems, it is easier to stabilize burning of AC arcs.

A series of plasma torches with rail electrodes have been developed (Fig. 12.4b, 12.4c) [81]. A plasma torch with rail electrodes can provide stable operation with oxidizing (air) and neutral media (nitrogen, inert gases). The range of air flow rates varies from 15 to 70 g/s. Power input into the arcs varies from 100 to 700 kW. Thermal efficiency almost does not differ from the system (plasma torch and power supply) efficiency and is 70–95 %.

The basic principle of plasma torch operation is the rail-gun effect (arcs move along the electrodes in the field of their own current). The movement of arc attachment point along the electrode allows uniform distribution of thermal load, which gives the opportunity to use the water-cooling electrodes made of a fusible material with high thermal conductivity (copper tubes). The multi-phase single-chamber AC plasma

Fig. 12.4 Powerful single-chamber AC plasma torches: **a** Three-phase plasma torch of EDP type (*1* electrode tip; *2* insulator; *3* current lead; *4* gas supply loop); **b** Single-chamber three-phase plasma torch with rail electrodes (*1* electrode tip; *2* insulator; *3* current lead; *4* gas supply; *5* injector); **c** Photo of operating plasma torch with rail electrodes, power 500 kW

torch with rail electrodes uses an integrated single-phase high-voltage plasma torch of low power as an injector. It creates a plasma stream providing sufficient electron concentration in a zone of the minimal interelectrode gap for ignition of the basic arcs.

It allows stable ignition of arcs between the electrodes mounted with a gap up to 20 mm powering from the industrial grid with voltage about 380–500 V. Arcs fill the major part of the discharge chamber, moving in the longitudinal and transverse directions. The insulating layer is formed near the wall where cold gas moves, where concentration of charged particles dramatically decreases, and arcs extinguish. The above-described process repeats continually forming a low-temperature plasma jet with average mass temperature of about 1,500–6,500 K at the plasma torch nozzle.

Fig. 12.5 Prolonged lifetime high-voltage AC plasma torches: **a** Photo of high-voltage AC plasma torches with rod electrodes in cylindrical channels; **b** Schematic representation of single-phase high-voltage plasma torch with rod electrodes; **c** Photo of operating high-voltage plasma torch with power 600 kW (IEE RAS), plasma forming gas–air

High-voltage plasma torches with high thermal efficiency 80–95 % have been developed for operation with power up to 100 kW. These plasma torches have rod electrodes in cylindrical channels. Figure 12.5a represents their general view and design. High supply voltage 4–10 kV provides stable ignition and burning of the long arc.

Currently, AC plasma torches with cylindrical electrodes operating with high arc voltage drop (up to 5 kV) are the most promising ones. Figure 12.5b shows a plasma torch of this type.

The plasma torch except high efficiency has the following advantages: long lifetime of electrodes (more than 1,000 h) and possibility to change the plasma heat content over a wide range, providing at this a range of average mass temperature for air plasma from 1,500 to 7,500 K. Of special note is the ability to provide plasma temperature less than 2,000 K, which is claimed for some technological processes. Moreover, we have developed 6 MW AC plasma torches, but their electrode lifetime did not exceed 100 h [90].

Fig. 12.6 General view and schematic diagram of the experimental installation IEE RAS for plasma waste gasification: *1* reactor-gasifier; *2* main plasma torch; *3* auxiliary plasma torch (H_2O, CO_2); *4* auxiliary plasma torch for initial heating; *5* loading device; *6* device for slag discharge and cooling; *7* branch pipe for syngas removal; *8* afterburner; *9* ignition plasma torch; *10* cyclone; *11* gas-analysis system; *12* spray scrubber; *13* packed bed scrubber; *14* stack; *15* exhaust fan. Zones of: *I* accumulation; *II* evaporation; *III* pyrolysis; *IV* oxidation; *V* reduction; *VI* weak reaction rates; *VII* slag discharge

12.4.2 Large-Scale Experimental Installation

In the late 1990s of the 20th century, interest to gasification technologies of solid fuels has renewed against the background of fossil resources price rising. IEE RAS started researching in this field as it is one of perspective implementations of low-temperature plasma systems. The experimental installation for investigation of plasma gasification process (Fig. 12.6) has been created [91].

The reactor is the key element of this installation. This is a fixed bed downdraft apparatus gasifier. Solid fuel is loaded into the reactor from above (Fig. 12.6, zone I).

After that it is gradually moved downwards to a zone of slag discharge (VII) by gravity and due to fuel gasification in the lower layers of the reactor. The raw materials consistently pass a zone of evaporation (II), pyrolysis (III), oxidation (IV), and reduction (V); as a result, the organic mass and water are converted into the syngas. These processes are initiated and supported in the reactor by plasma streams generated by plasma torches. Gasifier has the loading device which allows portion loading of feedstock during the experiment (Fig. 12.6, pos. 5).

The reactor has three oxidant injection points along the shaft length. The first (top) point is meant for supplying the oxidizer of the moderate temperature (no more than ~400–500 °C), the second and the third points are used to supply the low-temperature plasma. Plasma can be supplied in each these two points from two plasma torches simultaneously. High-voltage AC air plasma torches with power up to 50 kW or a combination of air and steam (or CO_2) plasma torches are used on the installation.

Syngas is removed from the bottom part of the reactor. In addition, the possibility to supply plasma or other oxidant for the accelerated heating at the first stage of experiment is provided in this zone.

The created gasifier is designed for working under a pressure close to the atmospheric. The gas flow in the gasifier shaft is induced by pressure 0.3 ± 0.2 kPa below an atmospheric pressure created in the outlet branch (Fig. 12.6, pos. 7) by the exhaust fan positioned at the end of the processing chain (Fig. 12.6, pos. 15).

The bottom part of the reactor is equipped with a revolving grate (Fig. 12.6, pos. 6) for the slag removal. Below the revolving grate, the water valve is arranged which has two functions: the reaction chamber closure and explosive valve function.

The installation is designed to investigate the composition of syngas and process characteristics. Gas samples for the analysis are taken at the gasifier outlet (Fig. 12.6, pos. 7).

There are two sampling systems. The first one utilizes two sampling lines in automatic mode. Gas, pumped out from gas duct by the vacuum pump, passes the hot filter, the cyclone, the cooler, the fine gas cleaning filter, and then is analyzed by a time-of-flight mass spectrometer EMG-20–1 (Mettek, Russia).

The second system is designed for separation and measuring of water and tars of syngas. The first element in the system after sampling probe is the hot filter. Two methods of water content measurements based on condensation and absorption are realized. The volume of liquid condensed from the syngas stream during its cooling is measured along with the gas-flow rate and its temperature in the condensation method. In the absorption method, water and steam are completely absorbed from the syngas stream at constant flow rate and measured by the milligram scale. The dried gas sample is pumped to the quadrupole mass-spectrometer MKS Cirrus-300 (MKS Instruments, USA) for composition analysis. Condensed liquid is investigated on tar content.

Mass-spectrometers allow performing the continuous analysis of synthesis gas composition for both the macro- and micro-concentrations. The plasma gasification mode is corrected using these data along with the information about temperatures in the reactor and plasma flow rates. These data are recorded and subsequently analyzed to determine the mass and energy streams of processes and other parameters.

Control over the mass streams, temperatures, and pressures in the reactor and in other elements of the experimental installation allows the installation operation mode regulating. These parameters are continuously measured and recorded. The changing of oxidant flow rate and plasma torch power are the basic leverages of the process.

The produced syngas after samples are taken in branch pipe is subjected to combustion. The afterburner serves for this purpose. In the afterburner, the syngas is mixed with the required amount of air and burns out. The device for forced ignition is installed for flame stabilization and prevention of explosion risk. A single-phase high-voltage plasma torch of low power is used as such a device.

The exhaust gases from the afterburner pass through the gas treatment system to the stack and into the atmosphere. Wet method of cleaning is used. It comprises two consecutive devices: spray and packed bed scrubbers.

12.4.2.1 Experimental Procedure

The important stage of experiment is determination of moisture and composition of investigated fuel. The fuel moisture is determined by the method of long-time drying at air ambient of 105 °C. The total carbon and hydrogen content in a fuel is determined by ISO 625–96 technique.

Before the experiment, the gasifier shaft is completely filled with char coal. This kind of solid fuel is the most suitable for a stage of preliminary heating up as it possesses the low content of volatile matter.

Ignition of one of the gasifier's plasma torches is considered as the beginning of the experiment (timing is counted from this point). Supply of air plasma improves heating and firing of fuel. Heating up the reactor shaft takes at least 8 h and comprises several stages differing by oxidant flow rate. Gasifier lining has very high thermal inertia that is why practically all experiments are carried out at quasi-stationary mode.

The analysis of syngas composition begins at the final stage of preheating and the charcoal feeding is replaced by the investigated material feeding. The transient process from the charcoal gasification to the investigated fuel gasification starts when the investigated material reaches drying and pyrolysis zone. The transient process could be considered completely finished when the whole gasifier shaft is filled by this fuel and the products of its gasification. When the experimental program is completed the fuel feeding into the reactor stops and starts after-burning of fed materials and their remnants, and when they are burned out the installation cools.

Experiments on plasma gasification have been carried out for such fuels as charcoal, wood (of various types: woodchips, pressed sawdust, and chocks), coal, lignite, Refuse Derived Fuel (RDF), and others.

12.4.2.2 Experimental Results

Figure 12.7 and Table 12.2 represent the experimental and calculated data of investigation of the plasma gasification of biomass using wood waste as the example.

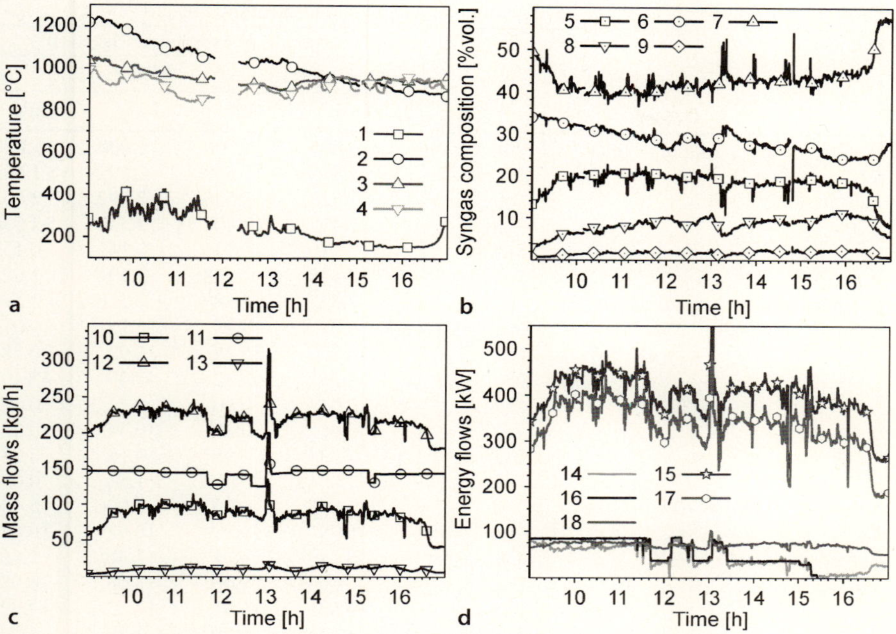

Fig. 12.7 Time dependence of main experimental parameters: **a** Temperature (*1* wall in pyrolysis zone; *2* wall in oxidizing zone; *3* wall in reduction zone; *4* syngas at the outlet); **b** Dry syngas composition (*5* H_2; *6* CO; *7* N_2; *8* CO_2; *9* the others); **c** Mass flows (*10* fuel; *11* air; *12* syngas; *13* steam); **d** Energy flows. (*14* energy losses; *15* fuel LHV; *16* plasma; *17* syngas LHV; *18* sensible heat)

Feedstock consumption was determined by differences of mass streams of major elements (carbon, hydrogen, oxygen, and nitrogen) assuming that ash content and the nitrogen content in wood are negligibly small. The fuel element composition was determined by mean values of flow rates of carbon, hydrogen, and oxygen of fuel for two intervals. The lower heating value (LHV) was estimated by this composition (by formula for an estimation of heating value of biomass [92]).

The mode of wood plasma gasification was observed on the interval 9:43–16:30. About 606.34 kg of wood was processed during this period according to the estimates, and actually about 609.98 kg of wood was loaded into the reactor in the course of the experiment. That validates the technique of experimental data processing.

The experimental results were compared with calculations for two regimes with constant air plasma flow rate and powers of plasma torches (see Table 12.2). The comparison showed that the data agreed well for the chemical energy yields and are satisfactory for syngas composition.

Calculations were carried out assuming adiabatic process and thermodynamic equilibrium of gasification products composition. On these modes, the composition of raw materials which matches to wood with moisture of 8–10 % also had been

Table 12.2 Comparison of averaged experimental data with the calculation results

Parameter			Time period of experiment (hh:mm)			
			9:30–11:36		13:36–15:12	
			Exp.	Calc.	Exp.	Calc.
Mass balance per 1 kg of feedstock (kg)	Inlet	Wood waste	1.000	1.000	1.000	1.000
		Air plasma	1.538	1.538	1.692	1.692
		Total	2.538	2.538	2.692	2.692
	Outlet	Syngas	2.428	2.402	2.549	2.536
		Steam	0.110	0.136	0.142	0.155
		Total	2.538	2.538	2.692	2.692
Syngas yield (Nm^3/kg of feedstock)			2.633	2.484	2.726	2.571
Syngas composition, (%vol)		H_2	20.01	19.45	18.25	19.01
		CO	30.87	35.26	26.92	30.78
		N_2	40.25	40.82	43.39	43.40
		O_2	0.01	0.00	0.22	0.00
		Ar	0.48	0.49	0.54	0.52
		CO_2	7.33	3.98	9.15	6.28
		CH_4	1.05	0.00	1.53	0.00
Syngas LHV (MJ/Nm^3)			5.558	6.001	5.036	5.440
Energy balance per 1 kg of feedstock (LHV basis)	Inlet	Wood waste	16.79	16.79	16.62	16.62
		Air plasma	3.21	3.21	1.46	1.46
		Total	20.00	20.00	18.08	18.08
	Outlet	Syngas	14.63	14.91	13.73	13.99
		Sensible heat	3.27	5.09	2.89	4.09
		Heat losses	2.10	–	1.46	–
		Total	20.00	20.00	18.08	18.08
Syngas LHV/Plasma energy ratio			4.565	4.651	9.381	9.555

determined. Well data agreement on the chemical energy yields was explained by the fact that the air plasma flow rate was more than stoichiometric in 2.8–3.0 times. It led to a considerable shortening of a reduction zone, therefore heat losses did not significantly affect syngas composition. More considerable difference in composition is caused by high methane stability and water–gas shift reaction at weak reaction rates and branch pipe sampling zones. As a whole, comparison results confirm usability of equilibrium approach for the estimation of plasma gasification key parameters.

12.5 Summary

The basic advantage of low-temperature plasma usage in biomass gasification is substantial growth of the hydrogen content and carbon monoxide content in syngas composition. At treatment of a chicken manure, the content of $H_2 + CO$ in syngas can be raised to ~97 % that allows an increase in the efficiency of its use in the Fisher–Tropsh process by 15–59 % and reach a specific yield of synthetic fuels ~240–260 g/kg, thus power inputs on the organization of the allothermal process will make only 4–5 MJ/kg. These parameters can be reached in downdraft plasma gasification. Plasma energy is used for liquid slag removal in updraft gasification, and it has rather less influence on syngas composition. The most effective way of plasma generation is by systems on a basis of AC plasma torches thanks to a long lifetime of electrodes (more than 1,000 h), an effective energy transfer of the discharge to the plasma forming gas (to 80–95 %), high power at work in long-time modes (to 2 MW), and low losses in a power-supply system (no more than ∼1–5 %). New AC plasma torches have arc voltage drops of about 1–10 kV and discharge currents of about 10–100 A. However, DC plasma torches are used more often. They typically have high currents (0.1–1 kA) and low voltage drops in the discharge (10–1,000 V) which are necessary for DC arc stabilization. Their main advantages are long-time operating experience and hence well-developed plasma torch models. Reliability of the executed estimations proves to be true by a good coordination with the experimental data which are obtained on the large-scale plasma gasifier. Total difference between mass balances was less than 1 %, and for energy balance less than 2 %. A series of long-time experiments on plasma gasification of wood proves that plasma gasification of biomass with use of AC plasma torches is ready for industrial implementation.

References

1. Juanjuan D, Wenping Z (2011) The analysis to the influence of carbon dioxide emissions in different countries. Energy Procedia 5:2426–2431. doi:10.1016/j.egypro.2011.03.417
2. Myhre G, Highwood EJ, Shine KP, Stordal F (1998) New estimates of radiative forcing due to well mixed greenhouse gases. Geophys Res Lett 25:2715–2718
3. Peters GP, Marland G, Le Quéré C, Boden T, Canadell JG, Raupach MR (2012) Rapid growth in CO2 emissions after the 2008–2009 global financial crisis. Nat Clim Change 2:2–4. doi:10.1038/nclimate1332
4. IEA (2011) Key World Energy Statistics 2011. http://www.iea.org/textbase/nppdf/free/2011/key_world_energy_stats.pdf. Accessed 14 Feb 2012
5. Whitney G, Behrens CE, Glover C (2011) U.S. fossil fuel resources: terminology, reporting, and summary. CRS report for Congress. http://assets.opencrs.com/rpts/R40872_20110325.pdf. Accessed 14 Feb 2012
6. Li M (2011) Peak energy and the limits to global economic growth. University of Utah. Annual-Report 2011. July 2011. http://www.econ.utah.edu/~mli/Annual%20Reports/Annual%20-Report%202011.pdf. Accessed 14 Feb 2012
7. Hook M, Sivertsson A, Aleklett K (2010) Validity of the fossil fuel production outlooks in the IPCC Emission scenarios. Nat Resour Res 19:63–81. doi:.1007/s11053-010-9113-1
8. Ladanai S, Vinterbäck J (2009) Global potential of sustainable biomass for energy. Swedish University of Agricultural Sciences, Uppsala
9. Hassan R, Scholes R, Ash N (2005) Ecosystems and human well-being: current state and trends. Island press, Washington DC
10. Dietenberger MA, Green DW, Kretschmann DE et al (1999) Wood handbook—wood as an engineering material. General Technical Report FPL–GTR–113. U.S. Department of Agriculture, Madison, WI
11. Green DW, Perry RH (2007) Perry's chemical engineers' handbook, 8th edn. McGraw-Hill, USA
12. History of Biofuels (2010) http://biofuel.org.uk/history-of-biofuels.html. Accessed 16 Feb 2012

13. Barney JN, DiTomaso JM (2010) Bioclimatic predictions of habitat suitability for the biofuel switchgrass in North America under current and future climate scenarios. Biomass Bioenergy 34:124–133. doi:10.1016/j.biombioe.2009.10.009
14. Keoleian GA, Volk TA (2005) Renewable energy from Willow Biomass Crops: life cycle energy, environmental and economic performance. Crit Rev Plant Sci 24:385–406. doi:10.1080/07352680500316334
15. Demirbas MF (2011) Biofuels from algae for sustainable development. Appl Energy 88:3473–3480. doi:10.1016/j.apenergy.2011.01.059
16. Bacovsky D, Mabee W, Worgetter M (2010) How close are second-generation biofuels? Biofuels, Bioprod Biorefin 4:249–252. doi:10.1002/bbb.222
17. Ribaudo M, Gollehon N, Aillery M (2003) Manure management for water quality: costs to animal feeding operations of applying manure nutrients to land. Agricultural Economic Report 824. U.S. Department of Agriculture. http://www.ers.usda.gov/publications/aer824/aer824.pdf. Accessed 14 Feb 2012
18. Kärkkäinen L, Matala J, Härkönen K, Kellomäki S, Nuutinen T (2008) Potential recovery of industrial wood and energy wood raw material in different cutting and climate scenarios for Finland. Biomass Bioenergy 32:934–943. doi:10.1016/j.biombioe. Accessed 01 Jan 2008
19. Yoshioka T (2011) Study on the feasibility of a harvesting, transporting, and chipping system for forest biomass resources in Japan. Agri-Biosci Monographs 1:1–60. doi:10.5047/agbm.2011.00101.0001
20. Reijnders L (2010) Transport biofuel yields from food and lignocellulosic C_4 crops. Biomass Bioenergy 34:152–155. doi:10.1016/j.biombioe.2009.10.004
21. Duffy MD, Nanhou VY (2001) Costs of producing switchgrass for biomass in Southern Iowa. Iowa State University Extension, Pm 1866. http://iowaswitchgrassdocs/pdf/Costs%20-of%20Switchgrass.pdf. Accessed 14 Feb 2012
22. Weiland NT, Means NC, Morreale BD (2012) Product distributions from isothermal co-pyrolysis of coal and biomass. Fuel 94:563–570. doi:10.1016/j.fuel.2011.10.046
23. Ghaly AE, Al hattab M (2012) An innovative farm scale biogas/composting facility for a sustainable medium size dairy farm. Am J AgricBiol Sci 7:1–14. doi:10.3844/ajabssp.2012.1.16
24. MacDonald JM, O'Donoghue EJ, McBride WD, Nehring RF, Sandretto CL, Mosheim R (2007) Profits, costs, and the changing structure of dairy farming. U.S. Department of Agriculture. http://www.ers.usda.gov/publications/err47/err47.pdf. Accessed 14 Feb 2012
25. Santoianni DA, Bingham MF, Woodard DM, Kinnell JC (2008) Power from animal waste—economic, technical, and regulatory landscape in the United States. Journal of Energy and Environment Conference 2:Paper #01
26. Hadrich JC, Wolf CA, Black JR, Harsh SB (2008) Incorporating environmentally compliant manure nutrient disposal costs into least-cost livestock ration formulation. J AgricAppl Econ 40:287–300
27. WA Government (2004) Environmental code of practice for poultry farms in western Australia. The Department of Environment, Australia
28. Roeper H, Khan S, Koerner I, Stegmann R (2005) Low-tech options for chicken manure treatment and application possibilities in agriculture. Proceedings Sardinia 2005, 10th International Waste Management and Landfill Symposium. Environmental Sanitary Engineering Centre, Italy
29. Giuntoli J, de Jong W, Arvelakis S, Spliethoff H, Verkooijen AHM (2009) Quantitative and kinetic TG-FTIR study of biomass residue pyrolysis: dry distiller's grains with solubles (DDGS) and chicken manure. J Anal Appl Pyrolysis 85:301–312. doi:10.1016/j.jaap.2008.12.007
30. Mkhabela TS (2004) Substitution of fertiliser with poultry manure: is this economically viable? Agrekon 43:347–356. doi:10.1080/03031853.2004.9523654
31. Popov VE, Bratsev AN, Kuznetsov VA, Shtengel SV, Ufimtsev AA (2011) Plasma gasification of waste as a method of energy saving. J Phys: Conf Ser 275:012015. doi:10.1088/1742–6596/275/1/012015
32. Zhang Q, Dor L, Fenigshtein D, Yang W, Blasiak W (2012) Gasification of municipal solid waste in the plasma gasification melting process. Appl Energy 90:106–112. doi:10.1016/j.apenergy.2011.01.041

33. Fourcault A, Marias F, Michon U (2010) Modelling of thermal removal of tars in a high temperature stage fed by a plasma torch. Biomass Bioenergy 34:1363–1374. doi:10.1016/j.biombioe.2010.04.018
34. Nair SA, Pemen AJM, Yana K et al (2003) Tar removal from biomass-derived fuel gas by pulsed corona discharges. Fuel Process Technol 84:161–173. doi:10.1016/S0378-3820(03)00053-5
35. Brattsev AN, Kuznetsov VA, Popov VE, Ufimtsev AA (2011) Arc gasification of biomass: example of wood residue. High Temp 49:244–248. doi:10.1134/S0018151X11010020
36. Moustakas K, Xydis G, Malamis S, Haralambous K-J, Loizidou M (2008) Analysis of results from the operation of a pilot plasma gasification/vitrification unit for optimizing its performance. J Hazard Mater 151:473–480. doi:10.1016/j.jhazmat.2007.06.006
37. Bassily AM (2008) Enhancing the efficiency and power of the triple-pressure reheat combined cycle by means of gas reheat, gas recuperation, and reduction of the irreversibility in the heat recovery steam generator. Appl Energy 85:1141–1162. doi:10.1016/j.apenergy.2008.02.017
38. U.S. Department of Agriculture (2004) Wood biomass for energy. http://www.fpl.fs.fed.us/-documnts/techline/wood-biomass-for-energy.pdf. Accessed 16 Feb 2012
39. Rutberg PG, Bratsev AN, Kuznetsov VA, Popov VE, Ufimtsev AA, Shtengel' SV (2011) On efficiency of plasma gasification of wood residues. Biomass Bioenergy 35:495–504. doi:10.1016/j.biombioe.2010.09.010
40. Kobayashi Y, Ando Y, Kabata T, Nishiura M, Tomida K, Matake N (2011) Extremely high-efficiency thermal power system-solid oxide fuel cell (SOFC) Triple combined-cycle system. Mitsubishi Heavy Ind Tech Rev 48:9–15
41. Rutberg PG, Kuznetsov VA, Bratsev AN, Popov VE, Shtengel' SV, Ufimtsev AA (2011) Use of carbon dioxide in the chemical synthesis technologies, plasma gasification and carbon production. IOP Conference Series: Mater Sci Eng19:012003. doi:10.1088/1757–899X/19/1/012003
42. Bratsev AN, Kuznetsov VA, Popov VE, Ufimtsev AA, Shtengel SV (2009) Estimation of perspectivity of steam-plasma methane conversion. High Temp Mater Processes: Int J 13:241–246. doi:10.1615/HighTempMatProc.v13.i2.120
43. Bratsev AN, Glezin IL, Kovshechnikov VB, Kumkova II, Kuznetsov VA, Popov VE, Shtengel SV, Ufimtsev AA (2007) Experimental research of air gasification of waste. The first results. Proceedings of 28th International Conference on Phenomena in Ionized Gases. Institute of Plasma Physics AS CR, Prague. pp 1848–1851
44. Kuznetsov VA, Bratsev AN, Kovshechnikov VB, Kumkova II, Popov VE, Shtengel SV, Ufimtsev AA (2007) Distinctive features of biomass gasification using ac plasma generators working on air. IEEE Pulsed Power Conference. Digest of technical papers 1976–2007. Omnipress, Madison, WI, USA, pp 1223–1226
45. Bratsev AN, Kuznetsov VA, Popov VE, Rutberg AP, Ufimtsev AA, Shtengel SV (2009) Experimental development of methods on plasma gasification of coal as the basis for creation of liquid fuel technology. High Temp Mater Processes: Int J 13:147–154. doi:10.1615/HighTempMatProc.v13.i2.30
46. Bratsev AN, Kumkova II, Kuznetsov VA, Popov VE, Shtengel' SV, Ufimtsev AA (2011) Air plasma gasification of RDF as a prospective method for reduction of carbon dioxide emission. IOP Conf Ser: Mater Sci Eng19:012004. doi:10.1088/1757–899X/19/1/012004
47. Mondal P, Dang GS, Garg MO (2011) Syngas production through gasification and cleanup for downstream applications—recent developments. Fuel Process Technol 92:1395–1410. doi:10.1016/j.fuproc.2011.03.021
48. Anis S, Zainal ZA (2011) Tar reduction in biomass producer gas via mechanical, catalytic and thermal methods: Rev Renew Sust Energ Rev 15:2355–2377. doi:10.1016/j.rser.2011.02.018
49. Bahng M-K, Mukarakate C, Robichaud DJ, Nimlos MR (2009) Current technologies for analysis of biomass thermochemical processing: Rev Anal Chim Acta 651:117–138. doi:10.1016/j.aca.2009.08.016
50. Gomez E, Rani DA, Cheeseman CR, Deegan D, Wise M, Boccaccini AR (2009) Thermal plasma technology for the treatment of wastes: A critical review. J Hazard Mater 161:614–626. doi:10.1016/j.jhazmat.2008.04.017

51. Yang L, Wang H, Wang H, Wang D, Wang Y (2011) Solid waste plasma disposal plant. J Electrostat 69:411–413. doi:10.1016/j.elstat.2011.05.007
52. Puig-Arnavat M, Bruno JC, Coronas A (2010) Review and analysis of biomass gasification models. Renew Sust Energ Rev 14:2841–2851. doi:10.1016/j.rser.2010.07.030
53. Morrin S, Lettieri P, Chapman C, Mazzei L (2012) Two stage fluid bed-plasma gasification process for solid waste valorisation: Technical review and preliminary thermodynamic modelling of sulphur emissions. Waste Management 32:676–684. doi:10.1016/j.wasman.2011.08.020
54. Loha C, Chatterjee PK, Chattopadhyay H (2011) Performance of fluidized bed steam gasification of biomass – Modeling and experiment. Energ Convers Manage 52:1583–1588. doi:10.1016/j.enconman.2010.11.003
55. Minutillo M, Perna A, Di Bona D (2009) Modelling and performance analysis of an integrated plasma gasification combined cycle (IPGCC) power plant. Energ Convers Manage 50:2837–2842. doi:10.1016/j.enconman.2009.07.002
56. Karamarkovic R, Karamarkovic V (2010) Energy and exergy analysis of biomass gasification at different temperatures. Energy 35:537–549. doi:10.1016/j.energy.2009.10.022
57. Atashi H, Siami F, Mirzaei AA, Sarkari M (2010) Kinetic study of Fischer–Tropsch process on titania-supported cobalt–manganese catalyst. J Ind Eng Chem 16:952–961. doi:10.1016/j.jiec.2010.04.005
58. van Steen E, Claeys M (2008) Fischer-Tropsch catalysts for the biomass-to-liquid process. Chem Eng Technol 2008:655–666. doi: 10.1002/ceat.200800067
59. Amouroux J, Siffert P (2011) Carbon dioxide: a raw material and a future chemical fuel for a sustainable energy industry. IOP Conference Series: Mater Sci Eng 19:012001. doi:10.1088/1757–899X/19/1/012001
60. Machrafi H, Cavadias S, Amouroux J (2011) CO2 reforming of methane: valorizing CO2 by means of dielectric barrier discharge. IOP Conference Series: Mater Sci Eng 19:012006. doi:10.1088/1757–899X/19/1/012006
61. Amouroux J, Cavadias S, Doubla A (2011) Carbon dioxide reduction by non-equilibrium electrocatalysis plasma reactor. IOP Conference Series: Mater Sci Eng 19:012005. doi:10.1088/1757–899X/19/1/012005
62. Alter NRG. http://www.alternrg.com/. Accessed 21 Feb 2012
63. InEnTec. http://www.inentec.com/. Accessed 21 Feb 2012
64. Advanced Plasma Power. http://www.advancedplasmapower.com/. Accessed 21 Feb 2012
65. Plasco Energy Group. http://www.plascoenergygroup.com/. Accessed 21 Feb 2012
66. PyroGenesis Canada Inc. http://www.pyrogenesis.com/. Accessed 21 Feb 2012
67. Rutberg PG (2003) Plasma pyrolysis of toxic waste. Plasma physics and controlled fusion 45:957–969. doi:10.1088/0741–3335/45/6/309
68. Rutberg PG, Safronov AA, Goryachev VL (1998) Strong-current arc discharge of alternating current. IEEE T Plasma Sci 26:1297–1306. doi:10.1109/27.725162
69. Electric arc heater develops very high temperatures (1960) Iron Steel Eng 37(11):149–150
70. Maniero DA, Kienast PF, Hirayama C (1966) Electric arc heaters for high-temperature chemical processing. Westinghouse engineer 26(3):66–72
71. Harry JE (1970) A power frequency plasma torch for industrial process heating. IEEE T Ind Gen Appl IGA-6:36–42. doi:10.1109/TIGA.1970.4181126
72. Iwata M, Shibuya M (1999) Effect on transferred ac arc plasma stability of increasing ambient temperature and superimposing pulse at current zero point. J Phys D: Appl Phys 32:2410–2415. doi:10.1088/0022–3727/32/18/312
73. Fey MG (1977) U.S. Patent 4013867
74. Wolf CB, Fey MG (1972) U.S. Patent 3705975
75. Reed JF, Peterson CW, Curry WH (1975) Electric heater development and performance data for a Mach 14 wind tunnel. J Spacecraft Rockets 12:308–313
76. Spongberg RM (1964) U.S. Patent 3140421
77. Reid JW (1960) U.S. Patent 2964678
78. Feldmeyer E, Schallus E (1960) U.S. Patent 2923811

79. Roots WK, Kadhim MA (1969) Measuring the electrothermal efficiency of a 50-Hz plasma torch. IEEE T Instrum Meas 18:150–156. doi:10.1109/TIM.1969.4313791
80. Rutberg PG (1970) Three-phase plasma torch. Some problems of investigation of gas-discharge plasma and creation of strong-current magnetic fields. Nauka, Leningrad (in Russian)
81. Rutberg PG, Safronov AV, Popov SD, Surov AV, Nakonechny GV (2005) Multiphase stutionary plasma generators working on oxidizing media. Plasma Phys Contr F 47:1681–1696. doi:10.1088/0741-3335/47/10/006
82. Jeništa J, Takana H, Nishiyama H et al (2011) Integrated parametric study of a hybrid-stabilized argon–water arc under subsonic, transonic and supersonic plasma flow regimes. J Phys D: Appl Phys 44:435204. doi:10.1088/0022-3727/44/43/435204
83. Eschenbach RC, Bryson DA, Sargent HB, Sarlitto RJ, Troue HH (1964) Characteristics of high voltage vortex-stabilized arc heaters. IEEE T Nucl Sci 11:41–46. doi:10.1109/TNS.1964.4323330
84. Paintes JH, Saeffer JF (1976) Performance and scaling characteristics of a Huels-type arc heater operating on hydrogen, helium or air. AIAA-14
85. Boatright WB, Sabol AP, Sebacher DI, Pinckney SZ, Guy RW (1976) Langley facility for tests at Mach 7 of subscale, hydrogen-burning, airframe-integratable, scramjet models. AIAA-11
86. Painter JH (1977) Hybrid arc air heater performance. AIAA-111
87. Westinghouse Plasma – Plasma torches. http://www.westinghouse-plasma.com/technology/-plasma-torches. Accessed 5 Mar 2012
88. NRG FOCUS – October 2011 issue. http://alternrg.com/sites/default/files/content/all/NRG%20Focus_Sept%202011_FN_LR_LK.pdf?phpMyAdmin=1,25SWTwdk48LH,ZtfJ1P24LAkc. Accessed 5 Mar 2012
89. Hacala A, Michon U (2009) Innovative industrial plasma torch for converting biomass into high purity syngas. International Plasma Chemistry Society 19. Conference proceedings. Bochum, Germany. 39. P2.14.04
90. Rutberg P (2009) Physics and technology of high-current discharges in dense gas media and flows. Nova Science Publishers, New York
91. Bratsev AN, Popov VE, Rutberg AF, Shtengel' SV (2006) A facility for plasma gasification of waste of various types. High Temp 44:823–828. doi:.1007/s10740-006-0099-7
92. Reed TB, Das A (1998) Handbook of biomass downdraft gasifier engine systems. Solar Research Institute, Golden CO

Chapter 13
Biogas Purifier for Japanese Rural Areas

Yoshiaki Kimura, Seiichi Yasui, Takahisa Hinata, Toshiyuki Imai and Hideyuki Takenaka

Abstract Currently, the biogas produced by biogas plants at dairy farms in Japan is a carbon-neutral energy. However, utilization of biogas has thus far been restricted solely to the farms where it is produced because there is no effective method of transporting unused biogas. Thus, there is a need to establish practical methods for biogas refinement and transport from operating systems. In this study, a biogas refining-compressing-filling facility using a gas membrane that would allow the use of surplus biogas produced by privately owned biogas plants was manufactured. Furthermore, field tests of biogas utilization systems (BGUS) made up of equipment that could use purified gas obtained from such a facility were performed. Finally, the possibility of a regional purified biogas system of Japan was validated in rural areas. The refining-compressing-filling facility was able to achieve a biogas Wobbe index of 49.2–53.8 and a combustion rate equivalent to 34–47 m/s. The total carbon load of the common portions of the BGUS was 102 t-CO^2 eq. Compared with the carbon load of the common portion of the biogas plant before introduction of the BGUS and of the gas utilizing equipment inside and outside the farm production system (209 t-CO^2 eq), a reduction of 107 t-CO^2 eq was achieved. The area's carbon dioxide emissions could be reduced through the standardization of biogas products through refinement; this would allow for the export of biogas outside of the system for use in common gas appliances. Currently, purified gas is locally produced and consumed as a source of carbon-neutral energy on dairy farms and adjacent residences. Packing the purified gas into tanks and supplying it to the town create the possibility of further reducing the carbon emissions of rural areas.

Keywords Biogas plant · Biogas purifier · Biogas utilization system (BGUS) · Gas membrane ·Wobbe index

---

Y. Kimura (✉) · T. Hinata · H. Takenaka
Hokkaido Central Agricultural Experiment Station, Hokkaido Naganuma, Japan
e-mail: kimura-yoshiaki@hro.or.jp

S. Yasui
Zukosha. Co. Ltd, Hokkaido Obihiro, Japan

T. Imai
Green Plan. Co. Ltd, Hokkaido Sappro, Japan

Z. Fang (ed.), *Pretreatment Techniques for Biofuels and Biorefineries*,
Green Energy and Technology,
DOI 10.1007/978-3-642-32735-3_13, © Springer-Verlag Berlin Heidelberg 2013

13.1 Introduction

Anaerobic fermentative treatment of livestock waste by biogas plants is more effective in lowering the environmental load than other methods. The biogas produced can be used as an energy source. Biogas from anaerobic digestion using livestock waste consists primarily of methane (typically 60 %) and carbon dioxide [1]. Other components can include oxygen and nitrogen, originating from air, sulfur compounds, particularly hydrogen sulfide, and water.

Approximately 80 biogas plants have been built in Japan as a means of effectively utilizing livestock waste. However, because many of the power generators installed in the plants to produce electricity—a representative method of utilizing biogas that has not been completely consumed by the facilities where it is produced—are foreign-made, repairing the generators when problems occur is difficult. Furthermore, the price of surplus electricity being sold is too low to recover the running cost. Use of biogas in Japan is premised on "consumption within the farm production system only used by farm management." Because of this, a method of transporting biogas outside the farm production system is needed to capture the full potential of this agricultural biogas.

It was assumed that effective utilization of biogas outside the farm production system could be accomplished by simple application to general gas equipment. However, the amount of biogas produced and the concentration of methane fluctuates daily, its caloric value is not stable, and there is residual hydrogen sulfide; thus, domestic gas equipment manufacturers have shown reluctance to directly use general gas equipment for biogas. Also, as part of the IGF21 Plan, the Ministry of Economy, Trade and Industry, Japan has been advancing the integration of the highly caloric natural gases, for use as town gas [2, 3]. Thus, there is need for standardization of high-caloric biogas and stabilization of its caloric value.

A method of resolving this problem would necessarily involve the construction of a system that purifies biogas, fills storage cylinders at high pressure with the gas, and distributes it within the region. Purified biogas is used as transportation fuel in a number of countries but in Europe, it has reached a major breakthrough in Sweden [4–6] and German [7]. Thus, it is necessary to introduce a system of equipment that carries out in a single effort the basic technological sequence of biogas refining for Japanese rural areas, standardized high calorification of the gas, compression (high pressurization of the gas), and flow to storage cylinders [8]. There is also a need to extensively troubleshoot the problems that could occur in the actual use of this biogas purifier and clarify measures on how to solve these problems for rural areas in Japan.

In this chapter, a biogas refining–compressing–filling (RCF) facility that uses surplus biogas produced by privately owned biogas plants was devised and evaluated in terms of greenhouse gas (GHG) reduction, and field tests of biogas utilization systems made up of equipment that using purified gas obtained from the facilities were performed. Thus, the possibility of a regional purified biogas system of Japan was validated in rural areas.

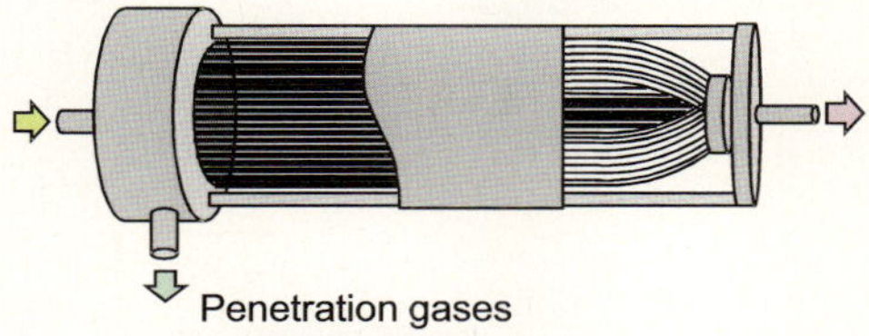

Fig. 13.1 Gas membrane biogas refinement using membranes. (Hollow fiber separation)

13.2 Biogas Purifier for Japanese Rural Areas

13.2.1 Biogas Purifier Using Gas Membrane

There are several techniques of biogas purification. Till date, membrane separation [6, 9, 10], Pressure Swing Adsorption (PSA) [5], absorption without chemical reaction, absorption with chemical reaction, and other methods have widely been used to purify or upgrade gas [11]. To separate out the methane in biogas, a membrane separation process combining multiple gas separation membranes (membrane modules) has been most extensively used at rural Japanese facilities because this method requires less maintenance work.

Figure 13.1 shows a gas membrane and a schematic of biogas refinement using a gas membrane used in this study (PRISM®, Air Products and Chemicals, Inc. [12]). The membrane module is a bundle of thin tubular membranes (hollow-fiber membrane), with membrane thicknesses of 1 μm or less. It is important for the temperature of the inflow gas to remain fixed every time as gas purification is carried out with the gas separation membrane. A thermal efficiency system for biogas refinement was used in the developed equipment. This system used "gas compression heat" and "cooler exhaust heat" to re-heat chilled biogas to ~60 °C, the ideal temperature for the gas separation membrane. In the development of this facility, the greatest separation efficiency was sought according to biogas composition.

Purification conditions that allow for the realization of high purity and purification efficiency were determined through simulations and laboratory testing. Consequently, operation conditions to refine raw biogas with a methane concentration of 55 % into purified gas at 100 m^3/day in 24 h include a biogas input temperature of 60 °C, pressure of 0.6 MPa, and flow rate of 10 m^3/h. The concentration of methane in biogas purified under these conditions was 91.4 %, and the methane recovery rate was 92.6 %.

13.2.2 Biogas Utilization System (BGUS)

Figure 13.2 shows a biogas utilization system (BGUS). As a measure to handle new energy supply and surplus biogas in agricultural regions, the system has an

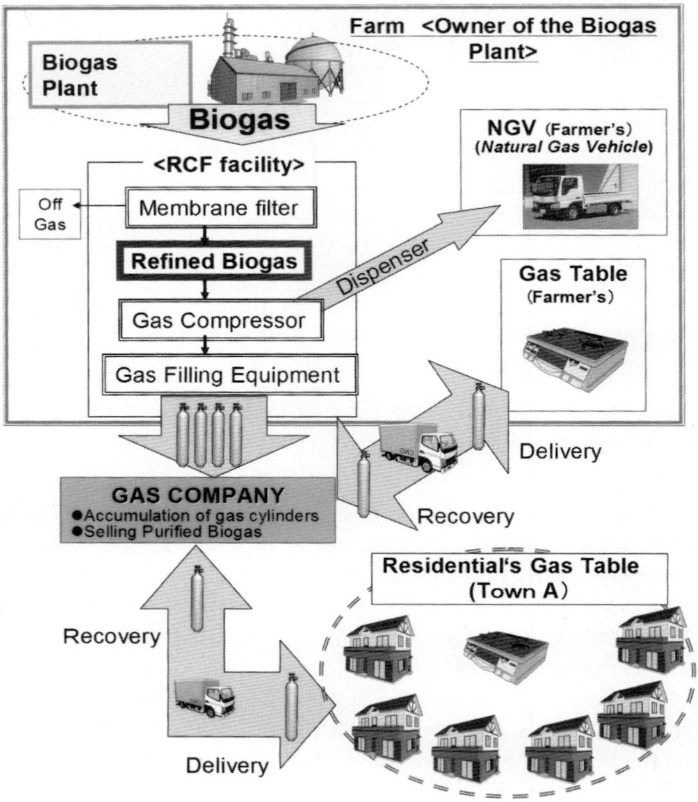

Fig. 13.2 Biogas utilization system (BGUS)

RCF component to produce refined gas. The BGUS also comprises equipment that consumes the refined gas. The system can be used to fill storage cylinders with surplus biogas and supply general households and businesses in the region with fuel.

13.2.3 High-pressure Gas Safety Act and City Gas Specifications in Japan

13.2.3.1 High-pressure Gas Safety Act

In Japan, legal matters pertaining to high-pressure gas were established in the "High-pressure Gas Safety Act [13]", which was enacted in 1951 to guarantee public safety. In order to prevent disasters, the Act regulates production, storage, sale, import, and transfer of high-pressure gas, and also promotes autogenous activities

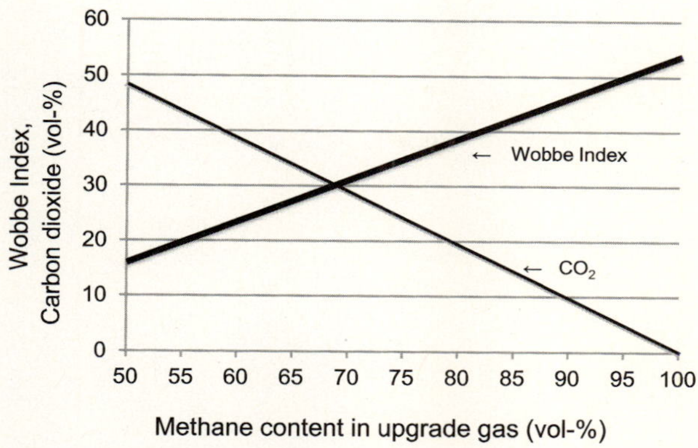

Fig. 13.3 Wobbe Index as a function of the carbon dioxide concentration in CH_4/CO_2 mixture

concerning high-pressure gas through private business people and the High Pressure Gas Safety Institute of Japan.

Manufacture of high-pressure gas is classified according to high-pressure gas processing capacity. Biogas (a flammable gas) manufacturers are classified as "Class 1 producers" if daily high-pressure gas production volumes are 100 Nm^3 or greater, and "Class 2 producers" if their daily output is below 100 Nm^3. Significant differences between regulations for "Class 1 producers" and "Class 2 producers" include the need for permission of gas production and for a resident safety controller. The refining-compressing-filling equipment developed in this research satisfied the "Transfer production facility" and "Class 2 producer" designations, which handle gas production volumes of 100 m^3/day or less. Required administrative procedures include the "manufacturing report", "maintenance of and compliance to technical standards", "implementation and documentation of periodic self-inspections," and "discontinuation report."

13.2.3.2 Town Gas Specifications

The upgrading process is basically a separation of CH_4 and CO_2 in the biogas, in order to obtain town gas quality with regard to the caloric value, Wobbe index (WI: the value was obtained by dividing the amount of heat released by the square root of the gas density), and relative density. In Japan, town gas is defined as gas fuel supplied by companies that conduct "General Gas Utility Business" under the Gas Business Act. Specifications regard the gas's heat capacity; the gas produced in the developed purification facility corresponds to town gas 12A specifications. Town gas 12A specifications are the common classification of town gas in Japan. The WI for town gas 12A is in the range of 49.2–53.8 and the combustion speed is in the range of 34–47 m/s.

Figure 13.3 illustrates how the WI increases and the relative density decreases as the methane content of the upgraded gas increases. A raw biogas composition of approximately 60 % methane and 40 % carbon dioxide is assumed.

For specific methane purity, the methane yield can be improved by recirculation of a part of the permeated CO_2-enriched gas. In the case of several modules connected

Fig. 13.4 RCF facility

in series, the best result is obtained with recirculation of only the permeated gas from the last module. Another way to maximize the methane yield and still obtain pipeline quality gas is to upgrade the biogas to a lower quality than required and then add propane in order to meet the specifications.

13.2.4 Refining-compression-filling (RCF) Facility

Figures 13.4 and 13.5 show the refining process of the RCF facility. The facility produces less than 100 Nm^3/day, making it a Class 2 producer according to the High Pressure Gas Safety Act [13]. It is also qualified by law to be a "mobile production facility". As this mobile production facility can be moved with a crane, the fixed asset tax does not apply. Biogas purification is carried out by membrane separation inside the facility. In order to create a standardized caloric quantity required for town gas 12A, a device that adjusts the caloric content by adding propane was installed. The gas at the completion of the refining process is temporarily stored in storage units below the RCF facility. High-pressure filling of transportable cylinders using high-pressure boosters makes it possible for the biogas to be utilized outside the farm production system as fuel for general households and compressed natural gas (CNG) vehicles.

13.3 On-site Field Testing at Japanese Rural Area

13.3.1 Biogas Production from Biogas Plant

The farm where the biogas purification facility was established is located in central Hokkaido (Hokkaido: northern part of Japan). A free-stall configuration is used for rearing and chaff is used for bedding. The biogas plant set up on the farm uses the

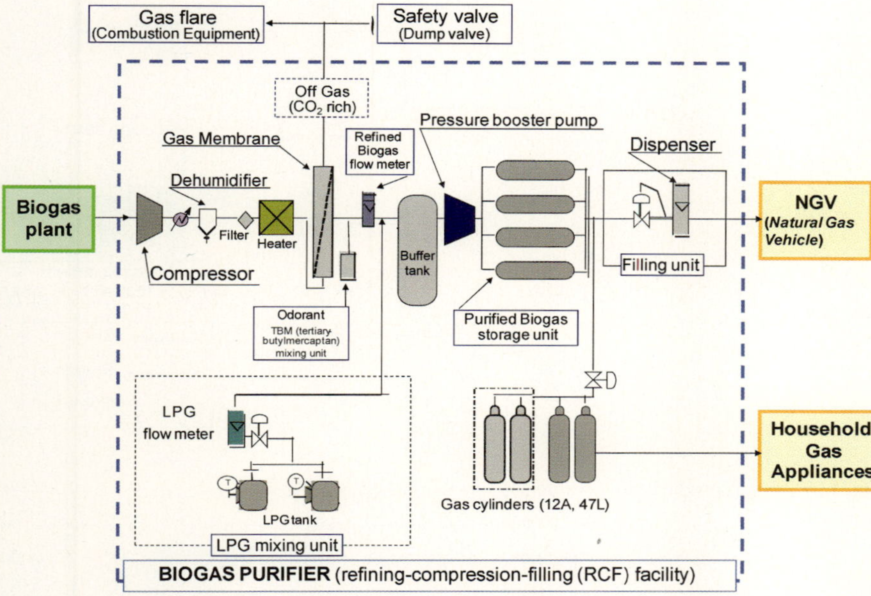

Fig. 13.5 Purification process by refining-compression equipment

manure/urine slurry of 250 dairy cows and parlor wastewater for raw material. The digestive slurry reservoir is 2840 m^3 (Fig. 13.6). The digestive slurry is dispersed as fertilizer over the ranch grasslands (80 ha) between April and May, July and August, and September and November.

The total volume of the anaerobic fermentation tank is 396 m^3 and the effective volume is 330 m^3. Operating conditions include a fermentation temperature at 55 °C and a hydraulic retention time (HRT) of 15 days. The daily material input is 22 m^3/day of dairy cow slurry (density assumption: 1 kg/L dairy cow slurry=1 kg/L water). Material is added to the fermenter twice after manure removal. This is done at 9:30 in the morning and 4:30 in the evening. In order to prevent hydrogen sulfide from affecting the biogas purification membrane film, the hydrogen sulfide concentration must be virtually 0. Desulfurization equipment that combines microbial desulfurization and dry desulfurization was used. After desulfurization, the biogas hydrogen sulfide concentration was virtually 0 ppm.

Figure 13.7 shows biogas production from an on-farm biogas plant. Table 13.1 shows the operation conditions in the anaerobic fermentation tank. In winter, the biogas produced was reduced to a low raw material supply volume, because slurry on the free-stall floor was frozen by the cold outside air.

The loading rate was 3.87 kg VS/m^3/day, the average amount of biogas produced was 450 ± 23 m^3/day, and the average methane concentration was 58 ± 1.8 %. The input materials and digestive slurry had a high ammonia nitrogen level compared to

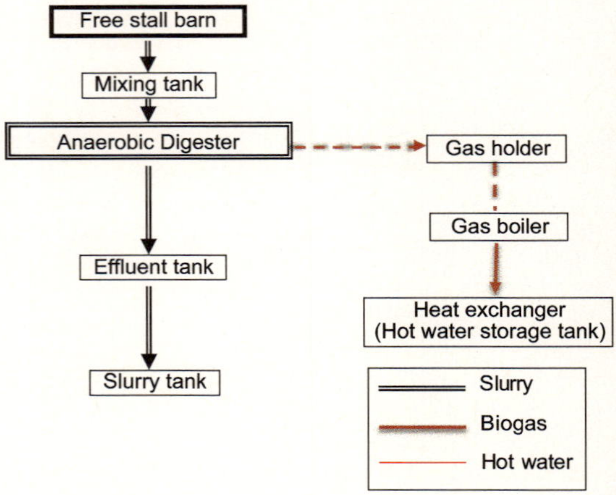

Fig. 13.6 Biogas plant flow at the farm

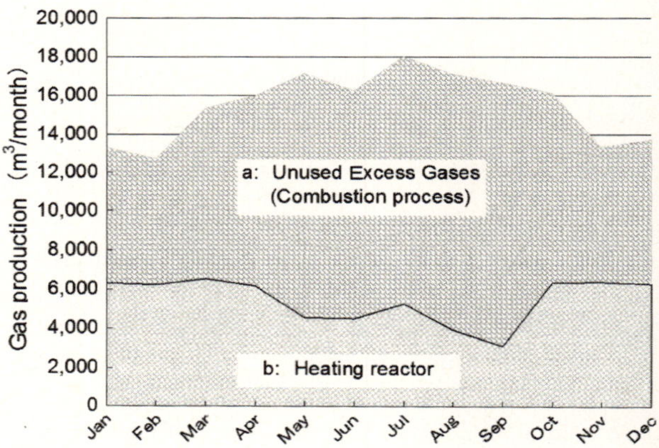

Fig. 13.7 Biogas production at the farm

Table 13.1 Daily average values for raw gas, off-gas, refined methane, and refined gas

	Raw material	Digested slurry
TS (%)	9.1 ± 1.2	6.5 ± 0.7
VS (%)	7.7 ± 0.7	5.2 ± 0.6
pH	6.7 ± 0.2	8.03 ± 0.2
T-N (mg/L)	2,607 ± 266	2,421 ± 129
NH_4-N (mg/L)	1,038 ± 94	1,144 ± 48
P_2O_5 (mg/L)	1,494 ± 230	1,578 ± 285
K_2O (mg/L)	3,088 ± 418	3,098 ± 157
VFA (mg/L)	4,169 ± 893	480 ± 352
AA (mg/L)	2,460 ± 585	392 ± 272
PA (mg/L)	920 ± 131	92 ± 65
P/A ratio	0.38 ± 0.04	0.21 ± 0.09

*Mean ± S.D.

** *TS* Total Solids; *VS* Volatile Solids; *T-N* Total nitrogen; NH_4^+-N Ammonia nitrogen; P_2O_5 Phosphorous oxide; K_2O Potassium oxide; *VFA* Volatile Fatty Acid; *AA* Acetic acid; *PA* Propionic acid *P/A* ratio Acetic acid/Propionic acid ratio

Table 13.2 Daily average values for raw gas, off-gas, refined methane, and refined gas

	Raw gas (biogas)	Off gas (waste gas)	Purified biogas	Heating value adjustment gas	Products gas
Influent gas (Nm^3)	216.0	120.0	96.0	–	97.0
Methane (Nm^3)	114.9	43.8	91.1	–	89.5
Carbon dioxide (Nm^3)	84.7	73.7	4.9	–	1.7
Nitrogen (Nm^3)	16.4	2.5	–	–	4.8
Propane (Nm^3)	–	–	–	1.0	1.0
TBM[a] (Nm^3)	–	–	Trace	–	Trace

[a]TBM: Tertiary-buty mercaptan

Table 13.3 Average composition of raw gas, off-gas, and refined gas

	Raw gas	Off gas	Purified biogas (product purity)
Methane (%)	53.2	32.1	94.5
Propane (%)	–	–	0.5
Dioxide (%)	39.2	64.8	Trace
Nitrogen (%)	7.6	3.1	5.0

*Product purity: O_2 free

the typical value, but the concentration of methane was the same as in other reported cases [14, 15].

Furthermore, lack of accumulation of propionic acid meant that the fermentation state was favorable. In Japan, the merits of methane fermentation include not only energy production, but also its role in odor mitigation countermeasures. Proprionic acid, one of the main malodorous organic acids produced in the dairy cow slurry due to the anaerobic fermentation process, was 95 % decomposed by methane fermentation processing. The odor intensity of the post-treatment digestive slurry was reduced to 1/50 of the input material. Thus, the refinement process provided highly effective odor control in neighboring villages when the digestive slurry was sprayed over grasslands.

13.3.2 On-site Field Testing at Japanese Rural Area

13.3.2.1 Operation of RCF Facility

Tables 13.2 and 13.3 show the results of refining testing. The average electrical energy consumed by the RCF facility during the testing period was 5.5 kWh. Using the RCF facility, about 44 % of the raw biogas (216.0 Nm^3/days) was refined to methane (96.0 Nm^3). To increase the methane concentration of the biogas after purification to about 94.5 % and to standardize the caloric content, about 0.5 % propane (1.0 Nm^3) was added. This resulted in refined gas (97.0 Nm^3) with a caloric content of 9,290 kcal/Nm^3 (38.9 MJ/m^3).

The gas at the final refining process had a WI of 49.3 and maximum combustion potential (MCP) of 34.3 m/s. These values are the minimum requirement within the specifications of town gas (WI: 49.2–53.8; MCP, 34–47 m/s). In addition, though the amount of raw gas produced per day was 216.0 Nm^3, the amount of refined gas was 97.0 Nm^3, which satisfied the condition set by the High-Pressure Gas Safety Act of less than 100 Nm^3 per day for Class 2 producers.

13.3.2.2 Application of the Refined Biogas to General Gas Equipment

Using the RCF facility, the following two tests were conducted from July 2007 to February 2008: (1) test of biogas refining using the RCF facility and (2) test of utilization of refined gas by gas equipment. Items measured were the concentration of methane, carbon dioxide and nitrogen in the raw biogas and refined gas, and the amount of refined gas produced after refining. For utilization tests of refined gas, we tested gas tables (IC3300CB2) and CNG trucks (ABF-SYE6T ([revised]) and measured the amount of refined gas consumed. Comparing the gross caloric value of liquefied petroleum gas (LPG) with refined biogas, we see that the value for LPG is 100.5 MJ/Nm^3compared to 38.6–39.1 MJ/Nm^3 for refined biogas, or roughly 40 % of LPG. Therefore, we enlarged the nozzle diameter of the gas equipment to maintain the same degree of combustion capability as LPG during usage. The results of the combustion tests show that the area near the combustion had concentrations less than 0 ppm CH_4 and 0–10 ppm CO, which were similar to that of LPG during combustion. The average amount of refined gas used by gas ranges was 0.41 Nm^3/day (150 Nm^3/yr). In addition, we tested the operation of CNG trucks under the conditions of 94 km as the usage distance of refined gas and an average speed of 56.4 km/h. The results show that the fuel consumption rate of the refined biogas was 10.6 km/Nm^3. Fuel consumption was almost the same as with commercial CNG.

13.4 Evaluation of Regional Utilization Model Revolving around Biogas Utilization System (BGUS)

13.4.1 Estimate of Energy Consumption by CNG Vehicles during Crop Production Activities at Model Farm

Based on the results obtained from onsite field testing, we selected Town A and established a regional utilization model that revolved around the BGUS. We evaluated the model's energy and environment aspects. We used the calculation of the process potential in terms of GHG reduction per year for the environmental assessment.

For our model, we selected a farm in Town A that has a plot with an area of 75 ha. The crop activity was cultivation of pasture grass (cut twice yearly). In this farm's system of cultivating mid-moisture grass silage, trucks are used for laying fertilizer, harvesting grass, and spreading soil improvement materials. The trucks' distance of movement were measured and collected using a geographic information system (GIS). The total distance of 4-t trucks involved in the production of pasture grass at the farm was 3,513 km/yr. The amount of diesel fuel consumed was 462.2 L/yr, and the amount of refined biogas consumed was 462.2 Nm^3/yr.

13.4.2 Players in Regional Utilization Model Revolving around BGUS

The players that are studied in the utilization model (subjects playing roles) are: (1) the farmer with the biogas plant installation, (2) the biogas vendor, and (3) consumers of the refined biogas. The farmer with biogas plant installation manages the biogas plant and is responsible for the refining of surplus gas and compression; that is, he is responsible up to the point of storing the refined gas in storage canisters. The biogas vendor is responsible for filling cylinders with refined biogas, delivering, installing, and recovering the cylinders, as well as modifying residential gas equipment. Consumers consume the gas.

The utilization model is centered on equipment that can be easily modified and introduced; this equipment was tested by onsite field testing.

The areas of use were established as "inside the farm production system" and "adjacent area and residence of the farm production system". The equipment "inside the farm production system" that consumed the refined biogas produced by the RCF facility included the farmer's residence as well as the farm operations. Otherwise, the gas vendor delivered cylinders filled with refined gas to the customers. The routes utilized the existing infrastructure.

13.4.3 Energy Evaluation of Regional Utilization Model Revolving around BGUS

Figure 13.8 shows the amount of refined biogas produced and consumed at the farm. Figure 13.9 shows the composition of gases produced at the farm and the percentage of consumption of refined biogas. The biogas plant produced an average of 185,000 Nm^3 biogas per year. Of the amount of biogas produced, about 35.3 % of the gas was consumed by gas boilers within the plant.

Raw biogas to be refined made up 42.5 % of the gas produced, and unused biogas (biogas that was not used in any stage of the process) made up 22.2 %. The amount of refined biogas completely treated by the RCF facility was about 35,000 Nm^3/yr (19.1 % of the total biogas produced). The amount of refined biogas used within the farm production system as energy substitute for LPG by kitchen gas appliances and CNG trucks was about 600 Nm^3/yr (CNG trucks, 462 Nm^3/yr; kitchen appliances, 150 Nm^3/yr).

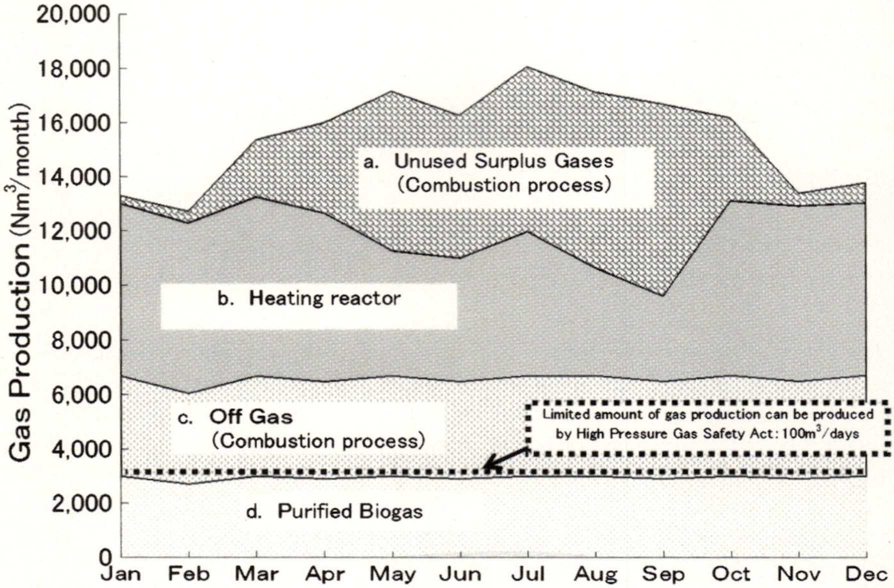

Fig. 13.8 Composition of refined gas produced and consumed at the farm

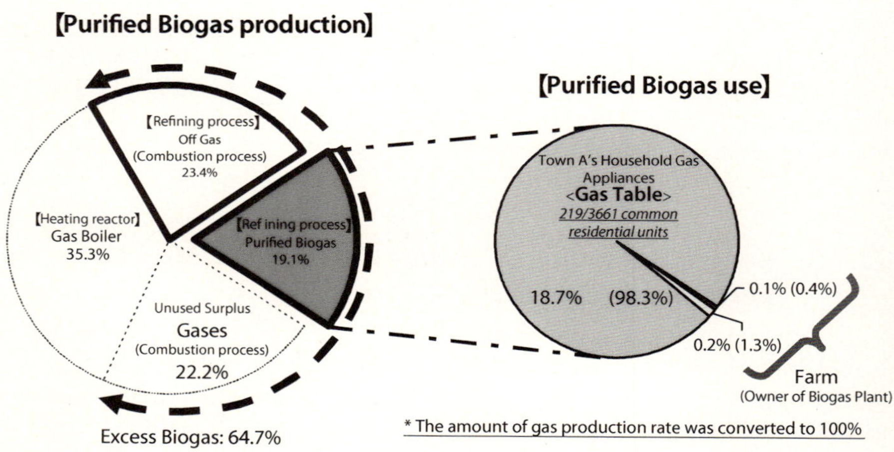

Fig. 13.9 Percentage and breakdown of refined gas produced and consumed in Town A

This amount made up 0.3 % of the refined gas produced in a year. The amount of refined gas transported outside the farm production system was 35,000 Nm^3/yr. This made it possible to supply 219 out of 3661 residences (as of November 2008) in Town A (roughly 6 % of all households).

13.4.4 *Environmental Evaluation of BGUS*

13.4.4.1 Specification of Process Potential Material Flow and Boundaries

Figure 13.10 shows the process potential material flow of the BGUS. There are three GHGs released during the course of BGUS: carbon dioxide (CO_2), methane (CH_4), and nitrogen monoxide (N_2O). Carbon dioxide shown by the symbol CO_2 in the diagram is emitted due to treatment of biomass (livestock waste in this case). It is not included in the calculation of the process potential because it is considered to be released to the atmosphere in the same amount as CO_2 that was absorbed and removed from the atmosphere in the short term by plants; thus, it is considered carbon neutral according to international agreements concerning emission of GHGs made by the Intergovernmental Panel on Climate Change (IPCC)[16]. Chemicals indicated by the following symbols CO_2 and CO_2 (carbon dioxide), CH_4 (methane), and N_2O (nitrogen monoxide) are gases produced during crop production and biogas plant operation. Please note that when the symbol CO_2 indicates the use of refined biogas as an alternative to fossil fuels by equipment, it is counted toward CO_2 reduction. The scope of the calculation of the process potential in terms of GHG reduction evaluation is shown in Fig. 13.10.

The main sources of GHGs produced in the processing of livestock waste by the biogas plant were divided roughly into three groups: (1) combustion of fossil fuels as power and heat sources for vehicles, (2) combustion of fossil fuels for the production of electricity when using commercial electricity, and (3) GHGs produced by the livestock waste itself by fermentation or sublimation. Further analysis shows that the process of carbon load generation can be divided into the following eight categories: (1) fuel combustion by equipment to transport livestock waste into the plant, (2) combustion of fossil fuels used by the biogas plant, (3) combustion of fossil fuels by equipment within the biogas plant, (4) deposit fermentation of solids after solid–liquid separation, (5) sublimation of digestive liquid during storage, (6) fuel combustion by equipment transporting digestive liquid and compost, (7) fuel combustion by equipment that spreads digestive liquid and compost, and (8) sublimation from digestive liquid and compost in fields after spreading. In addition, carbon loads were generated during (1) use of commercial electricity in the process of refining surplus biogas, (2) combustion of fossil fuels by farm trucks inside the farm production system, and (3) combustion of LPG fuel by kitchen equipment inside and outside the farm production system.

The calculation of the process potential evaluation of the BGUS modeled in Town A in terms of GHG reduction per year used the emission of carbon dioxide as its single indicator. It did not use other environmental indicators such as eutrophication of water in the region or acidification of the atmosphere in its evaluation. Below are the qualifications used in this calculation.

- Because energy involved in the delivery of biogas cylinders by the gas vendor was almost the same as that of the previous delivery system, it was not counted as a GHG that contributes to warming in the analysis.

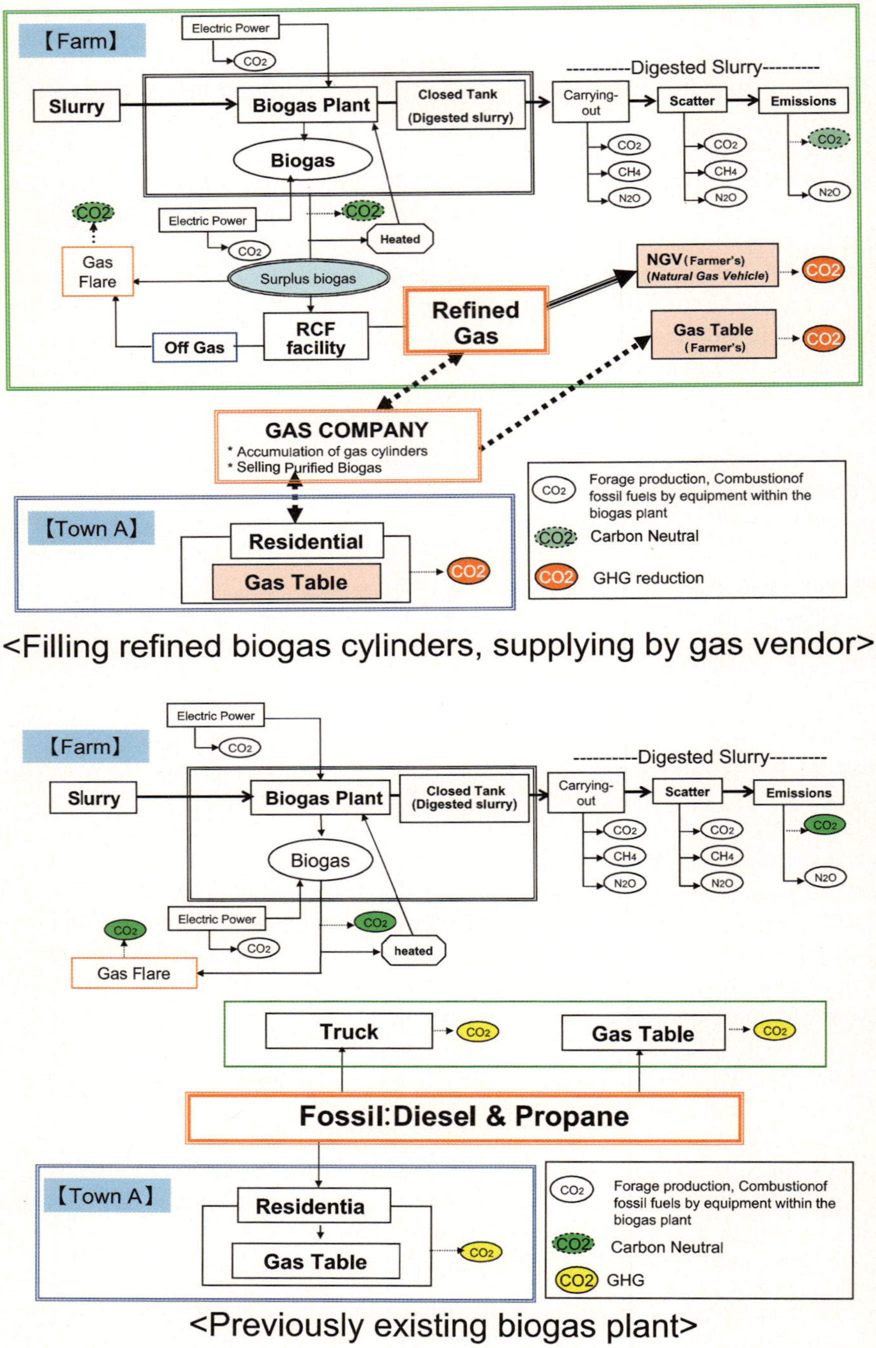

Fig. 13.10 Process potential flow of previously existing biogas plant and BGUS. (Filling refined biogas cylinders, supplying by gas vendor)

- Carbon dioxide produced when biogas was created by anaerobic fermentation of livestock waste and CO_2 produced by sediment fermentation were considered the same amount as CO_2 absorbed by plants during their growth, so they were considered carbon neutral and were not included in the count of GHGs.
- Besides equipment in the farm production system, there was equipment in the barn that used fossil fuels, such as boilers and tractors.

13.4.4.2 The Calculation of the Process Potential in Terms of GHG Reduction

Table 13.4 shows the results of the calculation of the process potential in terms of GHG reduction per year and the amount of GHGs reduced by the BGUS. The following GWP (Global Warming Potential) ratios were used: CO_2:1; CH_4:25; N_2O:298 [16]. The total carbon load of the shared portion of the previous biogas plant was 271 t-CO_2eq ((219,260 kg-CO_2eq (gas flare) + 6,042 kg-CO_2eq (Buying electricity for Biogas Plant) + 45,215 kg-CO_2eq (volatilization)).

Against this value, CO_2 reduction in the total carbon load of common portions from the biogas plant that introduced the BGUS was counted as being produced by ~6.3 % of Town A's households to which biogas was sent (which did not include the farm in Town A with the biogas plant installation) and by farm trucks and kitchen equipment that used refined biogas as substitute energy source in the biogas plant, besides the livestock waste treatment system that existed in the previous biogas plant.

The results of analysis show that the total carbon load of the common portions of the BGUS was 102 t-CO_2eq. Compared with the carbon load of the common portion of the biogas plant before introduction of the BGUS and of the gas-utilizing equipment inside and outside the farm production system (209 t-CO_2eq), a reduction of 107 t-CO_2eq was achieved.

13.5 Conclusion

In this chapter, a biogas refining–compressing–filling facility that uses surplus biogas produced by a privately owned biogas plant was devised and constructed, field tests of biogas utilization systems made up of equipment using purified gas obtained from the facilities were performed, and the possibility of a regional purified biogas system in rural areas of Japan was validated. Consequently, the refining–compressing–filling facility was able to reach the biogas Wobbe Index (WI) of 49.2–53.8 and combustion rate of 34–47 m/s (town gas 12A specification) in production volumes of high-pressure gas that qualified for class 2 producer status under the specifications of Japan's "High-pressure Gas Safety Act" (<100 Nm^3/day).

Additionally, the budget analysis results of a biogas utilization system modeled after Town A in Northern Japan showed a distribution of the purified gas such that 0.3 % of the purified gas produced by the biogas plant (approximately 35,000 Nm^3/yr) was used for running consumption and 98.3 % was distributed to the town's gas infrastructure, thereby satisfying the gas needs of 219 (6 %) of the 3,661 residences of Town A.

	CO_2(kg)	CH_4(kg)	N_2O(kg)
A. Previously existing Biogas plant (before the introduction of biogas utilization system (BGUS))			
(1) Farm (Owner of the Biogas Plant)			
#1 Biogas plant and digested slurry			
• Gas flare (Combusted: Surplus biogas: ***Carbon Neutral***)	*(219,260)**	-	-
• Buying electricity (Biogas Plant)	6,042	-	-
• Volatilization (from digested slurry in grass field, land application, slurry spreader)	45,215	0.27	21.62
<Subtotal>	51,257	0.27	21.62
GWP	51,257	7	6,443
GWP (CO_2 eq) (a)	**58 t- CO_2 eq**		
#2 Gas and diesel Equipment			
• Truck(Diesel) and Gas table(Propane)	70,973	-	-
GWP	70,973	0	0
GWP (CO_2 eq) (b)	**71 t- CO_2 eq**		
(2) Town A (216 common residential units)			
#1 Gas Equipment			
• Gas table	80,024	-	-
GWP	80,024	0	**0**
GWP (CO_2 eq) (c)	**80 t- CO_2 eq**		
Total GHG (A=a+b+c)	**209 t- CO_2 eq**		
B. Biogas Plant with biogas utilization system (BGUS)			
(1) Farm (Owner of the Biogas Plant)			
#1 Biogas plant and digested slurry			
• Gas flare(Combusted: Off Gas and Surplus biogas: ***Carbon Neutral***)	***(115,177)****	-	-
• Buying electricity (Biogas Plant)	6,042	-	-
• Volatilization (from digested slurry in grass field, land application, slurry spreader)	45,215	0.27	21.62
<Subtotal>	51,257	0.27	21.62
GWP	51,257	7	6,443
GWP (CO_2 eq) (a)	**58 t- CO_2 eq**		
#2 Biogas Purifiers (refining-compression-filling (RCF) facility)			
• Buying electricity (RCF facility)	43,642	-	-
GWP	43,642	0	0
GWP (CO2 eq) (b)	**44 t- CO_2 eq**		
#3 Refined Biogas Equipment (***Carbon Neutral***)			
• NGV(Truck) and Gas table (Instead of Purified Biogas)	***(1,407)****	-	-
GWP	0	0	0
GWP (CO_2 eq) (c)	**0 t- CO_2 eq**		
(2) Town A (216 common residential units)			
#1 Refined Biogas Equipment (Instead of Refined Biogas from LPG)			
• Gas table (***Carbon Neutra*** **l**)	***(80,024)****	-	-
GWP	0	0	0
GWP (CO_2 eq) (d)	**0 t-CO_2 eq**		
Total GHG (B=a+b+c+d)	**102 t- CO_2 eq**		
C. GHG reduction (c=A - B) (Effect on the introduction of BGUS)	**107 t- CO_2 eq**		

GWP: Global Warming Potential , CO_2: Carbon dioxide, CH_4: Methane, N_2O: Nitrous oxide

CO_2 eq: CO_2 equivalents, (@@,@@@)*:Carbon Neutral

GWP were used: CO_2:1; CH_4:25; N_2O:298 (IPCC, 2007)

Fig. 13.4 Biogas utilization system (BGUS)

The results of analysis show that the total carbon load of the common portions of the BGUS was 102 t-CO_2eq. Compared with the carbon load of the common portion of the biogas plant before introduction of the BGUS and of the gas-utilizing equipment inside and outside the farm production system (209 t-CO_2eq), a reduction of 107 t-CO_2eq was achieved.

The results show that the area's carbon dioxide emissions can be reduced through the standardization of Town Gas 12A and that refining biogas allows for the export of biogas outside of the system to be used by common gas appliances. Purified gas is locally produced and consumed as a source of carbon-neutral energy in dairy farming areas. Packing the purified gas into tanks and supplying it to the town makes possible the reduction of the area's carbon emissions.

References

1. Hil DT (1984) Methane productivity of the major animal waste types. Trans ASAE 26(3):841–848
2. Oda H (2009) Structural changes in city gas production in Japan: diversification of gas production as a result of gas family integration by IGF21. WGC Final00594
3. Omura M, Takeuchi Y, Matui K, Kikuchi S (2005) The preliminary consideration on transportation and storage of excess gas from the biogas plant for the individual dairy farm, J Soc Agric Struct, Japan 35:211–220
4. Jönsson O, Persson M (2003) Biogas as transportation fuel. Swedish Gas Centre, Malmoe
5. Margareta P (2003) Evaluation of upgrading techniques for biogas, Swedish Gas Centre, Malmoe
6. Stern SA, Krishnakumar B, Charati SG, Amato WS, Friedman AA, Fuess DJ (1998) Performance of a bench-scale membrane pilot plant for the upgrading of biogas in a wastewater treatment plant. J Membr Sci 151:63–74
7. Weiland P (2003) Production and energetic use of biogas from energy crops and wastes in Germany. Biochem Biotechnol 109:263–274
8. Kimura Y, Yasui S, Hinata T, Noguchi N, Tsukamoto T, Imai T, Kanai M, Sumie Z, Okamoto E, Matsuda Z (2010) Development of a biogas purifier for rural areas in Japan, XVIIth CIGR, Paper-ID:CSBE100492
9. Nagai K (2007) Gas separation, permeation and barrier membranes. CMC Publication, Vellore. ISBN:978-4-7813-0029-0
10. Harasimowicz M, Orluk P, Zakrzewska-Trznadel G, Chmielewski AG (2007) Application of polyimide membranes for biogas purification and enrichment. J Hazard Mater 144:698–702
11. Hagen M, Polman E (2001) Report Swedish Gas Center, adding gas from biomass to the gas grid. Contract No: XVII/4.1030/Z/99–412, ISSN 1102–7371
12. Air products and chemicals, Inc. http://www.airproducts.com
13. High Pressure Gas Safety Act 1951 Ministry of Economy, Trade and Industry, Japan. http://www.nisa.meti.go.jp/english/resources/legislativeframework/files/HighPressureGas.pdf
14. Umetsu K, Takahata H, Kawamoto T (1990) Effect of temperature on mesopilic anaerobic digestion of dairy cow slurry. Res Bull Obihiro Univ 17:401–408
15. Kimura Y, Takahashi K, Umetsu N, Amano T, Ozaki T, Yasui S, Kasuya H (2002) Farm scale biogas production system of dairy manure slurry using unused tower silos in cold region. AAAP 10:131
16. IPCC (International Panel on Climate Changes) (2007) IPCC Fourth Assessment Report 212

Part VI
Novel Pretreatment Techniques

Chapter 14
Status and Perspective of Organic Solvent Based Pretreatment of Lignocellulosic Biomass for Enzymatic Saccharification

Xiaofei Tian, Zhen Fang and Charles (Chunbao) Xu

Abstract Enzymatic saccharification of lignocellulosic biomass encounters many prohibitive factors which make it difficult to be developed on an industrial scale. Pretreatment has been found essentially effective for increasing the susceptibility of substrates to the enzyme, for example, by removing the lignin barrier and breaking down the crystal structure of cellulose in the raw materials. As a green and efficient technique, pretreatment of lignocellulosic biomass employing organic solvents and organic electrolyte solution (OES) is introduced in this chapter. Future prospects and recommended research work for developing these technologies for practical application, as well as coupling production of high-value bio-products from lignocellulosic biomass, are also discussed.

Keywords Aqueous organic solvent · Delignification · Decrystallized cellulose · Ionic liquids · Selective precipitation · Hydrolysis yield

14.1 Introduction

In biorefineries of lignocellulosic biomass, especially for cellulosic ethanol production, there are three key steps in the bioconversion process: (1) lignocellulose pretreatment, to separate the lignin and hemicellulose components from cellulose in the biomass and destroy the recalcitrant cellulosic structure to reactive intermediates; and (2) enzymatic hydrolysis, saccharification of the cellulose and hemicellulose to fermentable sugars (e.g., glucose and xylose) by cellulase-catalyzed hydrolysis; and

---

Z. Fang (✉) · X. Tian
Biomass Group, Chinese Academy of Sciences, Xishuangbanna Tropical Botanical Garden, 88 Xuefulu, Kunming, Yunnan province, 650223, China
e-mail: zhenfang@xtbg.ac.cn

X. Tian · C. (Chunbao) Xu
The Institute for Chemicals and Fuels from Alternative Resources, Faculty of Engineering, Western University, London, Ontario, N6A 5B9, Canada

Z. Fang (ed.), *Pretreatment Techniques for Biofuels and Biorefineries*, Green Energy and Technology,
DOI 10.1007/978-3-642-32735-3_14,

(3) fermentation, to produce bio-ethanol or other bio-based chemicals (e.g., lactic acid and succinic acid) [1–3].

Effect of the nature of lignocellulosic substrate on the enzymatic attachment and activity of cellulases are two of the critical factors that influence the entire rate and yield during the enzymatic hydrolysis process, and also the big challenges that have to be overcome for the commercial production of lignocellulosic bio-ethanol [4–6]. Consequently, pretreatment is required prior to the hydrolysis to remove the recalcitrance, such as (1) disruption of the carbohydrate–lignin shield by reducing lignin and hemicellulose contents [6–9]; (2) concomitantly decreasing the crystallinity of the cellulosic structure [6, 7]; (3) extension of enzymatic accessible surface area by particle cracking and increasing the volume of micro pores [5]; (4) loosening of the structure of lignin or (and) cellulose by cleavage of beta-1-O-4-aryl-ether bond or (and) beta-1-O-4-glucosidic bond [8]. Since pretreatment accounts for the largest proportion (around 50 %) of bio-ethanol production cost excluding the enzyme and raw material costs [5,9–11], it has great potential for efficiency improvement and cost reduction through research and development [8]. Plenty of work on pretreatment has been conducted and published during the last 10 years. The reported pretreatment methods can be classified into the following categories: (1) physical treatments (i.e., milling, grinding, high energy radiation, etc), (2) chemical treatments (e.g., soaking and boiling by diluted acid or alkalis, ozone oxidation, etc.), (3) physiochemical treatments (steam explosion, ammonia fiber explosion, ammonia recycled percolation [8], pyrolysis, hydrothermal and organosolv methods), and (4) biological (white rot fungi) treatments [12].

The nomenclature of *organosolv pretreatment* is derived from organosolv, which is a pulping technique that uses an organic solvent to solubilize lignin and hemicellulose and was invented by Theodore Kleinert [13]. Organosolv pretreatment is generally classified as a physiochemical technique, which employs non-aqueous or aqueous organic solvents, such as various alcohols, acetone, glycerol, dioxane, ethylene glycol, triethylene glycol, and phenol [14], with (or without) catalysts to remove parts of noncellulosic components (mainly lignin), and (or) convert the form of cell walls, as well as structures of natural cellulose, providing effective fractions of lignocellulosic biomass for the subsequent enzymatic saccharification step. Though organic acid and organic peracid were used to pretreat biomass as well [15], they differ from the organosolv pretreatment of interest here due to their dissimilar fundamentals.

This chapter presents an overview of the status of organosolv pretreatment techniques and their corresponding fundamentals, as well as further processing of the organosolv by-products. In addition, a newly developed technique using organic electrolyte solution (OES) composed of organic solvent and cellulose-soluble electrolyte solvent in pretreatment is also introduced. Future prospects and recommended research work on developing these technologies for practical application, as well as coupling production of high-value bio-products from lignocellulosic biomass are also discussed.

14.2 Organosolv Pretreatment

14.2.1 Advantages Associated with Organosolv Pretreatment

In order to make the saccharification of lignocellulosic materials economically feasible, pretreatment is needed to: (1) maximize the enzymatic susceptibility; (2) minimize the carbohydrate loss; (3) maximize the recovery of lignin and hemicellulose as valuable by-products; (4) minimize the use of energy, chemicals, and capital equipment; (5) be capable of being scaled up to industrial sizes; (6) avoid impregnating the biomass with toxic chemicals that would inhibit the saccharifying enzymes or microbes in fermentation of the sugars [14]; (7) have less and easy-operating procedures; and (8) be enviromentally friendly. As shown in Table 14.1, none of the pretreatment techniques by far can provide all of the above-mentioned desired outcomes on all of lignocellulosic materials [6, 7, 16, 17]. However, organosolv pretreatment is one of the compromised techniques that offers potential for meeting these criteria owing to its specific effects of thermal infiltration and extraction on biomass to modify its components and structure.

From the literature, the fundamentals of enzymatic hydrolysis improvement associated with organosolv pretreatment are given as follows.

14.2.2 The Fundamentals of Organosolv Pretreatment

The enzymatic saccharification of lignocellulosic biomass is affected by the cellulolytic enzymes, their interactions and factors related to lignocellulosic biomass features [16]. The biomass features influenced by organoslv pretreatment have been reported, including the crystallinity of cellulose, its degree of polymerization (DP), lignin content/structure, lignin–carbohydrate complexes, hemicellulose content, surface area, and pore size [14,18–24]. Hereby, they could be normally classified into three categories: chemical composition content, physical distribution of constituents within the biomass matrix, and intrinsic properties of cellulose.

14.2.2.1 Chemical Compositions

Lignin Content

Lignin content is an important recalcitrant factor to enzymatic hydrolysis of lignocellulosic biomass, evidenced by the obvious enhancement of enzymatic hydrolysis yield and rate after removal of lignin [25–27]. The negative effect of lignin in the biomass against enzymatic-catalyzing reaction is caused by the enzyme adsorptive loss to lignin, enzyme deactivation by lignin [28, 29], and steric hindrance between the enzyme and the substrate [30, 31]. To reduce or remove the lignin content from

Table 14.1 Comparison of currently developed pretreatment techniques for enzymatic saccharification. (Reproduced from [16] with permission)

Pretreatment	Compositional features			Advantage	Disadvantage
	Cellulose	Hemicellulose	Lignin		
Ball-milling	Intensive decrystallization	No removal	No removal	Intensive decrystallization	Energy intensive
Steam explosion	Some depolymerization	80–100 % solubilization	Little solubilization, more redistribution	Energy efficient, no recycling cost	Xylan degradation, by-product inhibition
Dilute acid	Some depolymerization	80–100 % solubilization	Little solubilization, more redistribution	Mild condition, high xylose yields	Acid recovery,corrosive, relatively expensive
AFEX	Decrystallization Up to 60 %	solubilization	10–20 % solubilization	Less xylan loss, no inhibitor formation	Ammonia recovery, not effective for high lignin
Sodium hydroxide	Substantial swelling, type I → type II	Substantial solubilization	Substantial solubilization (>50 %)	Effective ester removal	Expensive reagent, alkali recovery
ARP	Less than 5 % depolymerization	~50 % solubilization	~70 % solubilization	Effective delignification	Alkali recovery, relatively expensive
Lime	Little depolymerization	Significant solubilization (to 30 %)	Partial solubilization (~40 %)	Effective lignin & acetyl removal, inexpensive	Less effective due to poor solubility of lime
Ozonolysis	Almost no depolymerization	Little solubilization	Up to 70 % solubilization	Effective delignification, mild condition	Expensive, need more ozone
Organosolvlysis	Significant swelling	Substantial, can be nearly complete	Substantial, can be nearly complete	High xylose yields, effective delignification	Solvent recovery expensive
Biological	20–30 % depolymerization	Up to 80 % depolymerization	~40 % delignification	Low energy requirement, effective delignification	Cellulose loss, slow hydrolysis rate

AFEX: ammonia fiber explosion; ARP: ammonia recycled percolation the lignocellulosic biomass, organic solvents were employed for delignification due to their good solubility for lignin. In the previous studies, organosolv pretreatment could achieve higher delignification yields on most kinds of biomasses than any other methods. The enzymatic digestibility had been consequently more or less enhanced. The relationship between the lignin content in pretreated materials and their enzymatic digestibility was not found to be always proportional [22]. In some cases, significant reduction of lignin (>95 %) can negatively affect cellulose digestibility. In the absence of lignin, decreased enzymatic accessibility was caused by aggregating cellulose microfibrils [32]. Therefore, the process should be optimized in terms of the balance of enzymatic hydrolysis enhancement and delignification.

Hemicellulose Compositions

Hemicellulose is a branched matrix polysaccharide in compound middle lamella together with cellulose and lignin [33]. It forms enzyme-impenetrable cross-links that are the barrier for enzyme-catalyzed deconstruction of cellulose [34]. Because the hemicellulose is amorphous, it is more easily deconstructed by the hydrothermal effect during organosolv pretreatment. Then, the hydrolyzed sugars can be extracted in the aqueous organic solvent. The removal of hemicellulose by organosolv pretreatment could reduce the steric hindrance of enzymes.

Acetyl Group

The enzyme-impenetrable crosslinks were created by covalent bonds between the side chains of branched hemicellulose (e.g., acetyl, uronic acids, and arabinose) and lignin or unbranched hemicellulose that attached to cellulose [34]. Several studies showed that the removal of acetyl groups from side chains of hemicellulose greatly enhanced the cellulose and xylan digestibility [35, 36]. Removal of acetyl groups could cause splitting of the cross-linking structure and thereby accelerate the enzyme recognition to the substrate.

14.2.2.2 Physical Distribution of Constituents within the Lignin–Carbohydrate Complex

Effective releasing of sugars from recalcitrant lignocelluloses was found to depend on increasing substrate accessibility to enzymes rather than high levels of delignification [37].

The substrate accessibility to the enzyme was determined by physical distribution of constituents within the biomass matrix and usually characterized by the maximum adsorption capacity of the recombinant Thioredoxin-Green fluorescent protein-Cellulose binding module (TGC), a non-hydrolytic fusion protein containing a green fluorescence protein and cellulose-binding module [38, 39]. However, it

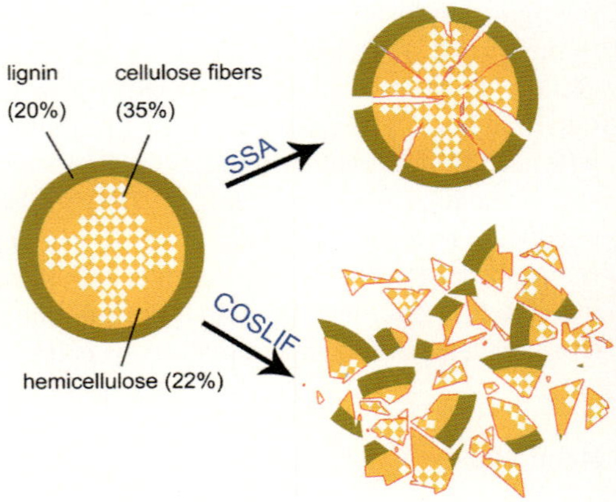

Fig. 14.1 Conceptual images of biomass features after aqueous ammonia soaking (SAA) and cellulose solvent- and organic solvent-based lignocellulose fractionation (COSLIF) pretreatment. Areas of the relative components correspond to their percentage of the microfibril. Cellulose and surfaces susceptible to enzymatic attack are highlighted in red [37]

was not easy to realize TGC protein measurement due to the requirement of specific reagents. In the organosolv pretreatment, the substrate accessibility to the enzyme was indirectly evaluated by some common physical factors of the biomass, such as pore volume, particle size, and surface area [27, 40]. Large pore volume and surface area have a positive effect on the enzymatic hydrolysis [15, 18], whereas each of them has to correlate with the other factors to explain the improvement of cellulose digestibility because they are dependent on each other.

Compared to aqueous ammonia soaking pretreatment, organosolv pretreatment using ethanol as the solvent brought ineffective delignification but a 10.4-fold higher saccharification yield [37]. The organosolv pretreatment fully disrupted the cell wall structure to form component clusters, as shown in Fig. 14.1, which led to increasing of cellulose accessibility [37]. Similarly, Donohoe et al. [41] mentioned that the re-localization of lignin and its movement away from the cellulose microfibril surface, without being removed from the biomass, improved cellulase accessibility. Enzymatic digestibility of biomass with the same delignification may not be equal. Similar findings were demonstrated by the organosolv pretreatment. The changing in the structure of cellulose caused alternate disruption of lignin in the lumen side of the cell wall and in the middle lamella region during the ethanol process. It made the easier accessibility of enzymes and positive influence on enzymatic hydrolysis of *Buddleja davidii*.

14.2.2.3 Cellulose Physical Properties

Crystallinity

Crystallinity represents the relative amount of cellulosic crystalline structure in the cellulose content of feedstock. Higher crystalline causes the lower saccharification rate of cellulose [42–44]. During the organosolv process, the crystallinity structure of feedstocks could be reduced due to the solvent attacking on the cellulose by thermal and infiltration effects. Unfortunately, the effect of organosolv pretreatment on the cellulosic crystallinity cannot be precisely measured in the case of an actual substrate. The cellulosic crystallinity might be masked by the removal of amorphous lignin and hemicellulose, as well as selective hydrolysis of amorphous part of cellulose which are superior to the cellulosic decrystallization [45].

Crystalline Allomorph

Additionally, the most abundant crystalline polymorphs found in higher plants are cellulose Iα and Iβ [34, 46, 47]. They can be converted to allomorphic forms such as cellulose II, III_I, and IV_I by ionic liquids (ILs), alkali, or ammonia-based pretreatment [34,48–50]. Without any relevant loss of crystallinity, these polymorphs have been shown to promote their hydrolysis rates [34, 51, 52]. Accomplished by the thermochemical effect, organosolv pretreatment can also convert cellulose I to other allomorphic forms to enhance the hydrolysis rates. For example, pretreating *B. davidii* by aqueous ethanol resulted in the replacement of cellulose Iα and Iβ by para-crystalline and amorphous cellulose forms [23]. Glycerol can convert ramie cellulose III_I to cellulose IV at 260 °C [53]. Relative to the other factors, the respond of cellulosic allomorphic form transformation to the effect of organosolv pretreatment is less understood and needs more investigation.

Zhu [16] showed that lignin content, acetyl content, and cellulose crystallinity were key factors that determine biomass digestibility. However, it is still not conclusive which the most important factors are to accelerate the biomass digestibility in the organosolv pretreatment, due to the various natural structural features and processing operation conditions. For example, removal of lignin and hemicellulose, reduction in DP, and decrease in the crystalline allomorphs (Iα and Iβ) increased the amenability of the ethanol pretreated *B.davidii* to enzymatic degradation [23]. But the success of ethanol pretreatment on enzymatic hydrolysis of *Loblolly pine* was alternatively contributed to a decrease in cellulose crystallinity, as well as removal of lignin and hemicellulose [54]. Therefore, studying the changes in the structure of cellulose during organosolv pretreatment is still an area of research that remains to be thoroughly investigated under certain conditions [23].

14.2.3 The Current Techniques of Organosolv Pretreatment

14.2.3.1 General Process of Organosolv Pretreatment

A general process of organosolv pretreatment is described in Fig. 14.2. Pure or aqueous organic solvents, such as methanol, ethanol, acetone, ethylene glycol, triethylene glycol, and phenol, were typically used as the working medium. Biomass was thermochemically treated in these working medium by mixing with (or without) the addition of acid, alkaline, or neutral catalysts, such as H_2SO_4, NaOH, or $MgCl_2$, at a

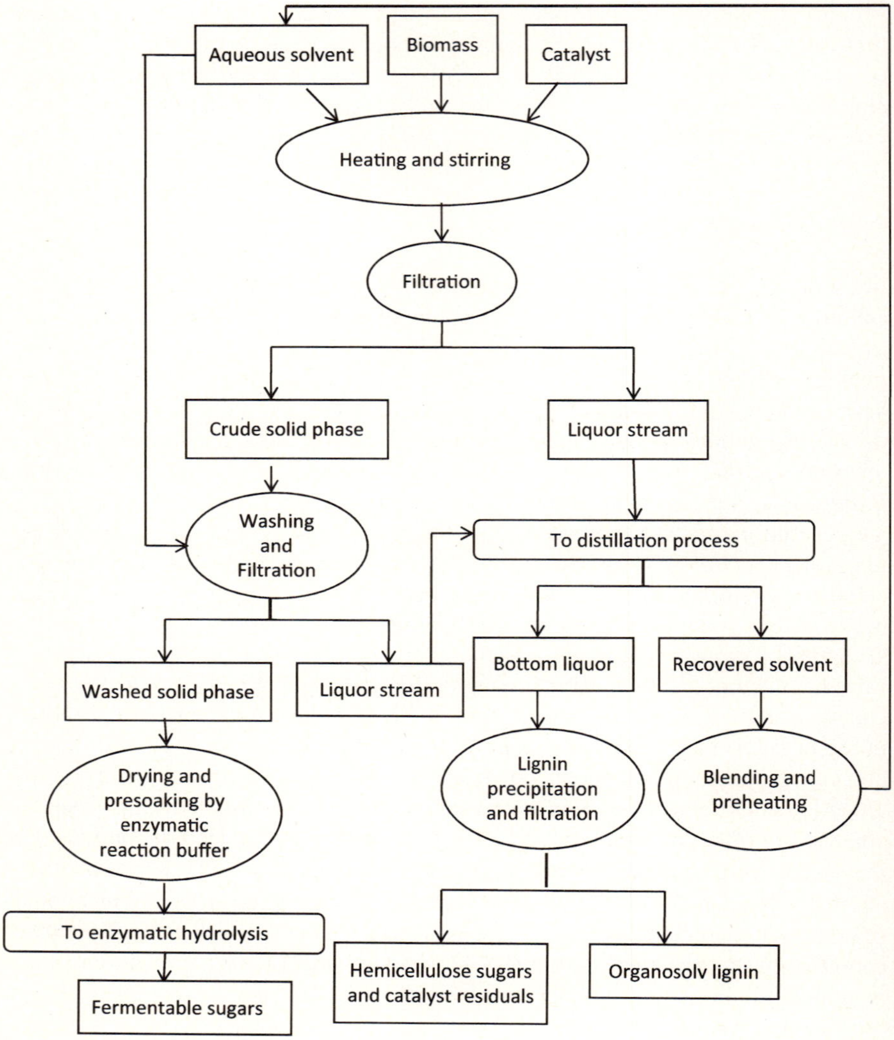

Fig. 14.2 Typical procedure of organosolv pretreatment

relatively lower (<180 °C) or higher temperature (>180 °C) [23, 24, 55, 56]. Organosolv pretreatment also dictated a unique pathway of downstream refinery process: After treating the substrate for several minutes or days, the cellulose-rich solid fraction was separated by filtration and washed by the same solvent and, subsequently, water, and then was ready for hydrolysis. The liquid stream, which contained the solvent, alone with lignin and sugars from hemicellulose, as well as their derived products (e.g., furfural, 5-hydroxymethylfurfural (5-HMF), depolymerized lignin), was recovered by distillation. After distillation, the residual was washed by water to precipitate the organosolv lignin. Chemicals recovered from the water-soluble fraction include xylose, glucose, oligomeric sugars, organic acids, 5-HMF, and other lipophilic extracts. They can be further utilized after separation and concentration [24].

14.2.3.2 Two-Stage Pretreatment

Efficient organosolv pretreatment has been well developed for agricultural residues and hardwoods, rather than softwood biomass because they are considerably more recalcitrant to the pretreatment and enzymatic processes [57, 58]. In order to overcome the problems, organosolv combined with some other approaches have been attempted, such as the two-stage pretreatment approaches as follows:

1. Presoaking/pretreatment of biomass by dilute-acid before the organosolv process. Dilute acid-pretreated sugarcane bagasse was performed prior to an organosolv process using NaOH as a catalyst under the optimized conditions (60 min, 195 °C, using 30 % (v/v) ethanol), where 67.3 % (w/w) of the pretreated solid material was easily converted to glucose in 24 h [59]. The optimized dilute acid presoaking and aqueous ethanol organosolv treatment of *Miscanthus* led to a better recovery of xylans (~20 % (w/w) of dry mass against ~10 % (w/w) without presoaking), enhanced dissolution of lignin in the aqueous ethanol, and increased enzymatic digestibility (98 % cellulose-to-glucose conversion against 80 % conversion without presoaking) [60].
2. Organosolv pretreatment followed by mechanical milling of pretreated substrates. The subsequent mechanical treatment of the aqueous ethanol-treated lodgepole pine by mill refining decreased particle size and crystallinity but increased swelling and fiber delamination of the substrate. However, the hydrolysis process was not accelerated remarkably [61].
3. Organosolv pretreatment and substrate sulfonation of the product. After ethanol treatment, sulfonation of lodgepole pine could reduce none of the specific bindings of lignin to enzymes. Accordingly, it resulted in obvious enhancement of hydrolysis yields even at low enzyme loadings despite less removal of residual lignin. At the same time, enzyme recovery was increased as well [61].

Though some positive results were achieved with the combined method, it was impossible to check the efficiency of the developed methods such as acid presoaking and sulfonation on other kind of softwood. Additionally, alkali reagent, such as NaOH, $NH_3 \cdot H_2O$, or triethanolamine, which has good ability for delignification rather than the improvement of hydrolysis, can be used in the presoaking prior to acid-catalyzed organosolv treatment [23, 62]. Moreover, it was reported that presoaking of straw by the reaction solvent under mild conditions could improve the subsequent organoslv delignification particularly in high solid-to-liquid ratios. This may account for the higher swelling of the material and increased accessibility of solvents to the reaction sites [63]. However, the effects of combined pretreatment on softwood need to be confirmed.

14.2.4 *The Optimization of Effective Factors in Organosolv Pretreatment*

Table 14.2 represents some organosolv pretreatment processes, their effects on the substrate features and their susceptibilities to enzymatic hydrolysis. Although a precise comparison is difficult to make from the data given in this table due to differences in the natures of lignocellulose and the reaction conditions employed, a rough evaluation is possible. By varying the process effective factors, such as the type of solvent (ethanol, acetone, butanol, glycerol, or other high boiling solvents) and aqueous content (50–90 %, w/w), catalyst type (H_2SO_4, NaOH, or $MgCl_2$) and concentration (0.5–1.5 %, w/w), reaction temperature (140–200 °C) and time (10–60 min), ratio of biomass to solvent (1:5–1:15, w/w), the organosolv pretreatment produces a range of substrates with varying enzymatic digestibility of 10–90 %. As shown in the table, the organosolv pretreatment could increase the hydrolysis yields to 1.5–10 times that of the untreated substrate. Most of the previous work defined the delignification, recovery of hemicellulose and cellulose as well as hydrolysis rate and yield of the pretreated materials as the indicators to evaluate the efficiency of the process. To most kinds of biomasses 10 % (w/w) biomass loading, using the aqueous ethanol or aqueous acetone (around 50 %, w/w), with 0.5–1.0 % H_2SO_4 as catalyst, under 160–180 °C for 30 min appear to be the most effective parameters for organosolv pretreatment efficiency.

Reaction conditions were optimized to achieve the highest yield of them. Nevertheless, these reaction factors should not only be optimized to maximize the enzymatic saccharification yield, but also to take into account the energy cost and reduction of the formation of fermentation inhibitors. Using less severe pretreatment conditions for better processing economy, as well as minimizing the generation of fermentation inhibitors (furfural and 5-HMF), have to be taken into consideration [22, 72].

14.2.5 *Further Processing of Hemicellulose and the Derived Lignin*

The economical conversion of lignocellulosic biomass to higher value products requires efficient recovery of cellulose, lignin, and hemicellulose components during the fractionation process. Organosolv pretreatment can meet the requirement due to the selective dissolving ability of the solvent employed in the process. In addition to the cellulose-rich materials, it produces organosolv lignin and hemicellulose sugars that is environmentally friendly.

The hemicellulose sugars recovered from the water-soluble stream can be concentrated for fermentation using special organisms to convert the five-carbon aldose to ethanol or other products [33]. Organosolv lignin is a suitable feedstock for the production of phenolic resins/adhesives [73], antioxidant [74], bio-based polymer

Table 14.2 Comparison of different organosolv pretreatment processes

Species of biomass	Reaction factors					Delignification	Hemicellulose removal	Cellulose recovery	Biomass recovery	Effect on enzymatic hydrolysis			Reference
	Temperature	Time	Solvents	Biomass Loading (w/v)	Catalyst and contents					cellulose-to-glucose conversion	Enzyme loading	Time	
Poplar (52 % cellulose, 23 % hemi-cellulose, 25 % lignin)	160 °C	30 min	25 % 50 % 75 % ethanol/water or butanol/water (w/w)	4 g/40 ml	–	20 %	10–20 %	90 %	NA	<10 % ultimate conversion	4 FPU cellulase and β-glucosidase/g of pretreated material	24 h	[14]
	160 °C	30 min	50 % butanol/water (w/w)	4 g/40 ml	H_2SO_4 (1.0 N)	>90 %	>90 %	60 %	NA	40 % ultimate conversion	4 FPU cellulase and β-glucosidase/g of pretreated material	24 h	[14]
	160 °C	30 min	50 % butanol/water (w/w)	4 g/40 ml	NaOH (1.0 N)	65 %	70 %	90 %	NA	55 % ultimate conversion	4 FPU cellulase and β-glucosidase/g of pretreated material	24 h	[14]
	160 °C	30 min	50 % butanol/water(w/w)	4 g/40 ml	0.5 % $FeCl_3$ (w/v)	>80 %	>80%	90 %	NA	15–20 %	4 FPU cellulase and β-glucosidase/g of pretreated material	24 h	[14]
	160 °C	30 min	Ethanol/butanol	4 g/40 ml	30 % $NH_3.H_2O$ (v/v)	~60 %	NA	~100 %	NA	<10 %	4 FPU cellulase and β-glucosidase/g of pretreated material	24 h	[14]

Table 14.2 (Continued)

Species of biomass	Reaction factors					Deligni-fication	Hemicel-lulose removal	Cellulose recovery	Biomass recovery	Effect on enzymatic hydrolysis			Refer-ence
	Temperature	Time	Solvents	Biomass Loading (w/v)	Catalyst and contents					cellulose-to-glucose conversion	Enzyme loading	Time	
Pinus radiata (45.3 % glucan, 6.4 % xylan, 2.1 % galactan, 1.5 % arabinan, 12.2 % mannan, 26.8 % lignin, 0.20 % ash)	183–197 °C	4–46 min	50 % acetone/water (w/w)	100 g /700 ml	0.9 % H_2SO_4 (w/w, dry wood) pH = 2	20.0–48.2 %	NA	40.9–85.4 %	40.0–54.6 %	38.2–71.8 %	20 FPU cellulase and 40 IU β-glucosidase/g of pretreated material	72 h	[64]
Hybrid poplar substrates (*Populus nigra* and *Prunus maximowiczii*)	155–205 °C	26–94 min	25–75 % ethanol/water (w/w)	20.0 g /140 ml	0.83–1.67 % H_2SO_4 (w/w, dry wood)	NA	NA	NA	NA	NA	NA	NA	[45]
Buddleja davidii (30.2 % lignin, 38.88 %glucose, 21.69 % xylose, 3.04 % mannose, 0.93 % galactose, 0.48 % arabinose)	170–195 °C	40–80 min	50–65 % ethanol/water (w/w)	NA	1.25–2 % H_2SO_4 (w/v)	9–21 %	49%–74%	85 %	NA	60–85 %	20 FPU cellulase and 40 IU β-glucosidase/g of cellulose	72 h	[23]

Table 14.2 (Continued)

Species of biomass	Reaction factors					Deligni-fication	Hemicel-lulose removal	Cellulose recovery	Biomass recovery	Effect on enzymatic hydrolysis			Refer-ence
	Temperature	Time	Solvents	Biomass Loading (w/v)	Catalyst and contents					cellulose-to-glucose conversion	Enzyme loading	Time	
Poplar wood	190–250 °C	-	50 % methanol/ water (w/w)	–	–	NA	NA	NA	35–58 %	50–90 %	1,750 FPU cellulase/g pretreated material	70 h	[65]
Wheat straw	200–240 °C	–	50 % methanol/ water (w/w)	–	–	NA	NA	NA	68–48 %	50–80 %	1,750 FPU cellulase/g pretreated material	70 h	[65]
Douglas-Fir wood chips (43 % cellulose, 23 % hemi-celluloses, 28 % lignin)	181–202 °C	15–90 min	50 % ethanol/ water (w/w)	1 g/7 ml	H_2SO_4 pH = 2.4	NA	NA	NA	NA	>90 %	14 FPU cellulase and 8 IU β-glucosidase/g of cellulose	8 h	[66]
Wheat Straw (34.6 % glucan, 21.5 % xylan, 2.1 % arabinan, 0.5 % galactan, 0.2 % mannan, 16.1 % lignin, 8.5 ash)	190 °C	60 min	0-60 % acetone/ water(w/w)	1 g/14.2 g (w/w)	–	~0-70 %	~50–95 %	89–96 %	NA	NA	2,500 CMC U cellulase and 400 pNPG U β-glucosidase/g of pretreated material	72 h	[67]
	190 °C	0–120 min	50 % acetone/ water(w/w)	1 g/14.2 g (w/w)	–	21–77 %	~5–85 %	89–96 %	NA	31–87 %	2,500 CMC U cellulase and 400 pNPG U β-glucosidase/g of pretreated material	72 h	[67]

Table 14.2 (Continued)

Species of biomass	Reaction factors					Deligni-fication	Hemicel-lulose removal	Cellulose recovery	Biomass recovery	Effect on enzymatic hydrolysis			Refer-ence
	Temperature	Time	Solvents	Biomass Loading (w/v)	Catalyst and contents					cellulose-to-glucose conversion	Enzyme loading	Time	
	160–205 °C	60 min	50 % acetone/ water(w/w)	1 g/14.2 g (w/w)	–	11–79 %	~30–80 %	89–96 %	NA	31–87 %	2,500 CMC U cellulase and 400 pNPG U glucosidase/g of pretreated material	72 h	[67]
	205 °C	60 min	50 % acetone/ water(w/w)	1 g/14.2 g (w/w)	–	79 %	82 %	93 %	NA	87 %	2,500 CMC U cellulase and 400 pNPG U β-glucosidase/g of pretreated material	72 h	[67]
Loblolly pine	170 °C	60 min	65 % ethanol/ water	100 g/ /700 ml	1.1 % H_2SO_4/ (w/w, dry wood)	61.2 %	96.4	79.3	NA	65 %	8 FPU cellulose and 16 IU β-glucosidase/g of cellulose	80 h	[54]
Pinus rigida	180 °C	50 min	50 % ethanol/ water (v/v)	20 g 200 ml	1 % H_2SO_4	9.10 %		77.80 %	NA	95 %	60 FPU cellulase 64 pNPGU β-glucosidase/of pretreated material	72 h	[68]
	210 °C	60 min	50 % ethanol/ water (v/v)	20 g/ 200 ml	1 % $MgCl_2$ (w/v)	1.50 %	–	~100 %	NA	85 %	60 FPU cellulase 64 pNPGU β-glucosidase/of pretreated material	72 h	[68]

Table 14.2 (Continued)

Species of biomass	Reaction factors					Delignification	Hemicellulose removal	Cellulose recovery	Biomass recovery	Effect on enzymatic hydrolysis			Reference
	Temperature	Time	Solvents	Biomass Loading (w/v)	Catalyst and contents					cellulose-to-glucose conversion	Enzyme loading	Time	
	190 °C	70 min	50 % ethanol/water (v/v)	20 g/200 ml	2 % NaOH (w/v)	55.20 %	–	80 %	NA	75 %	60 FPU cellulase 64 pNPGU β-glucosidase/of pretreated material	72 h	[68]
Pinus contorta	170 °C	60 min	65 % ethanol/water	200 g/2 L	1.1 % H_2SO_4	79 %	~90 %	79 %	NA	93–97 %	20 FPU cellulase 40 IU β-glucosidase/g ocellulose	48 h	[69]
Beech	184 °C	~100 min	37.5 % cyclohexan/water (v/v)	4 g/40 ml	–	95.00 %	NA	NA	NA	61 %	420,000 FPU cellulase/g of pretreated material	30 h	[70]
Akamatsu pine	184 °C	~100 min	37.5 % cyclohexan/water (v/v)	4 g/40 ml	–	92.00 %	NA	NA	NA	53 %	420,000 FPU cellulase/g of pretreated material	30 h	[70]
Bagasse	184 °C	~100 min	37.5 %/cyclohexan water (v/v)	4 g/40 ml	–	78 %	NA	NA	NA	69 %	420,000 FPU cellulase/g of pretreated material	30 h	[70]
sugarcane bagasse (49 % cellulose, 15.8 % hemicellulose, 27.2 % lignin, 5.6 %	175 °C	60–90 min	50 % ethanol/water (v/v)	500 g/2.5 L	1.5 % NaOH (w/w)	NA	NA	NA	NA	9.3–13.9 g glucose per 100 g substrate	15–25 FPU/g biomass, xylanase 0–300 UI/g biomass β-glucosidase 100–250 IU/g biomass	24 h	[71]

Table 14.2 (Continued)

Species of biomass	Reaction factors					Deligni-fication	Hemicel-lulose removal	Cellulose recovery	Biomass recovery	Effect on enzymatic hydrolysis			Refer-ence
	Temperature	Time	Solvents	Biomass Loading (w/v)	Catalyst and contents					cellulose-to-glucose conversion	Enzyme loading	Time	
extractives, 3.6 % ashes)													
	175 °C	60–90 min	50 % ethanol/ water (v/v)	500 g/ 2.5 L	1.25 % NaOH (% w/w, dry matter)	NA	NA	NA	NA	13.1–16.9 g glucose per 100 g substrate	15–25 FPU/g biomass, xylanase 0–300 UI/g biomass, β-glucosidase 100–250 IU/g biomass	24 h	[71]
	175 °C	60–90 min	50 % ethanol/ water (v/v)	500 g/ 2.5 L	1.5 % H_2SO_4	NA	NA	NA	NA	8.0–23.9 g glucose per 100 g substrate	15–25 FPU/g biomass, xylanase 0–300 UI/g biomass, β-glucosidase 100–250 IU/g biomass	24 h	[71]
	175 °C	60 min	50 % ethanol/ water (v/v)	500 g/ 2.5 L	1.25 % H_2SO_4	NA	NA	NA	NA	25.1 g per 100 g substrate	15 FPU cellulase and 300 UI xylanase enzyme/g of pretreated material	24 h	[71]

("–" blank; "NA" not available)

composites [75], and even hydrocarbon products for blending with gasoline [76], owing to its unique high purity, low molecular weight, and abundance of reactive groups [21, 77].

However, the water-insoluble property of organosolv lignin may limit its applications [78]. Therefore, conversion of the lignin dissolved in the organic solvents by catalyst or organic-solvent-stable enzymes may be a potential technological approach to resolve the problem [79–81].

14.2.6 Summary and Perspective

As described above, the flexibility and good generality combined with the ease of recovering the lignin and hemicellulose sugar streams make organosolv pretreatment one of the most efficient pretreatment techniques. Its fundamentals and effective operation factors for improvement of the enzymatic hydrolysis have been well demonstrated in this literature. However, organosolv pretreatment still possesses several disadvantages [78] as below:

1. A higher cost that is associated with the handling and recovery of the organic solvent.
2. Pretreatment efficiency is still relatively low for softwood and similar feedstocks.
3. Its lignin product tends to be less water-soluble, which may limit its use in some applications.
4. Currently, there are few organosolv pretreatment operations on a commercial scale, and there are fewer commercial sources of organosolv lignin for further exploitation.

To resolve the above problems, further develop organosolv pretreatment technology, and consequently benefit the economics of the entire cellulosic bio-fuel production, we present the following recommendations:

1. Optimizing the effective factors of organosolv pretreatment on certain biomass materials, such as reaction time, temperature, catalyst concentration, biomass-to-solvent ratio, against the delignification and enhancement of sacharifaction by surface response design or orthogonal design.
2. Advancing the pretreatment process by (i) using cheaper and renewable medium; (ii) using highly efficient microwave-irradiation and ultrasonic techniques to replace the common thermo heating and mechanical stirring methods [82]; (iii) employing two-phase heterogeneous medium; (iv) performing concentration operation with preheated aqueous organic fluids [65]; and (v) combining organosolv pretreatment with other methods to achieve higher delignification and hydrolysis efficiencies, especially for softwoods.
3. Improving the fundamental knowledge on the pretreatment technologies especially for some recalcitrant biomasses (such as softwood).
4. Investigating new processes and products from the organosolv lignin and hemicellulose sugars derived from the organosolv process.

5. Conducting economic evaluation of the entire organosolv pretreatment.
6. Demonstrating organosolv-pretreatment technology on a pilot scale to obtain engineering data for industrial scale processing and sufficient refined products for characterization and exploitation.

14.3 Pretreatment with Organic Electrolyte Solution

14.3.1 Introduction

OES, short for organic electrolyte solution, is defined as non-aqueous or aprotic solvent solution of electrolytes, which usually contains fraction of polar organic solvents and free ions.

Organic solvents mixed with certain lignocellulose-dissolving solutions, such as room temperature ILs [83–86] and concentrated phosphoric acid [87], form new types of homogeneous solutions that can be used to (1) swell and (or) dissolve the component of cellulose and (or) lignin, as a pure lignocellulose-dissolving solution does, and (2) precipitate the cellulose and lignin selectively. These combinations include dimethyl sulfoxide (DMSO) + paraformaldehyde [88, 89], DMSO + [BMIM]Cl [90], DMSO + tetra-*n*-butylammonium fluoride (TBAF) [91, 92], *N,N*-dimethylformamide (DMAC) + 1-allyl-3-methylimidazolium chloride ([AMIM]Cl) [93], (acetone, pyridine, or hexamethylphosphoramide) + ILs [94], 3-dimethyl-2-imidazolidinone + 1-ethyl-3-methylimidazolium acetate ([EMIM]AcO) or 1-butyl-3-methylimidazolium chloride ([BMIM]Cl) [95], DMAC + LiCl [96], and H_3PO_4 + acetone [97], etc. The addition of common lignocellulose-dissolving solution in organic solvents allows OES to have the equivalent abilities to destroy the crystallinity of natural cellulose (acting as a cellulose solvent) [42]. Meanwhile, with the existence of organic solvents, some OES can remove the lignin component from the lignocellulosic homogeneous solutions by precipitating cellulose and hemicellulose (acting as a precipitating agent) [97]. In terms of the selectivity of dissolving abilities, OES-based pretreatment processes can be divided into two categories: OES-dissolving pretreatment and OES-depositing pretreatment.

14.3.2 OES-Dissolving Pretreatment

Tian et al. [42] designed a simple OES system composed of [AMIM]Cl and DMSO, to treat microcrystalline cellulose for enzymatic saccharification. [AMIM]Cl is one of the most effective ILs for dissolving and pretreating wood chips [86]. Compared to [BMIM]Cl, it has a lower melting point. The dissolution mechanism of cellulose in [AMIM]Cl was proposed as: The free Cl- anions associated with cellulose hydroxyl protons and the free cations combined with the cellulose hydroxyl oxygen to form electrovalent bonds, leading to disruption of hydrogen bonding in the cellulose and its consequent dissolution [98]. When an anti solvent, such as water, containing large

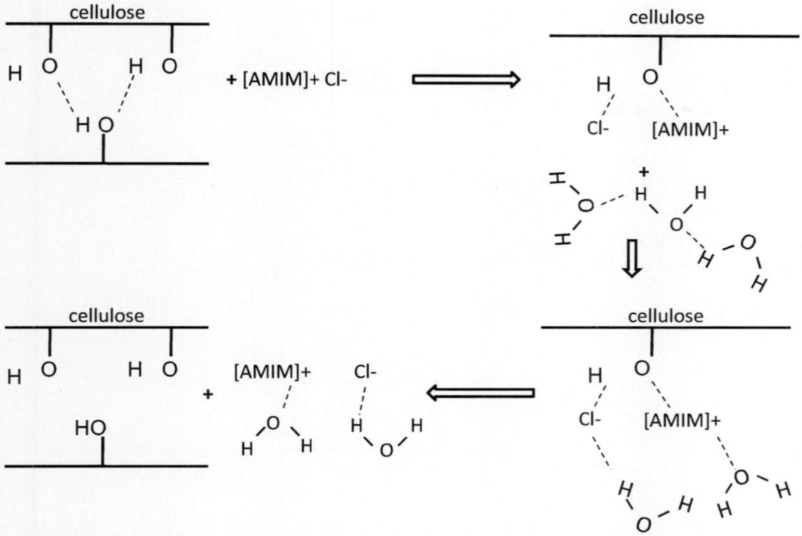

Fig. 14.3 Possible mechanism of cellulose dissolution in [AMIM]Cl and regenerated by water as the regeneration reagent

quantities of hydrogen bonds was homogeneously mixed with [AMIM]Cl–cellulose solution to remove the ILs, the electrovalent bonds were replaced oppositely by the reforming of hydrogen bonds among cellulose chains. However, the structures consisting of cellulose chains and regenerated hydrogen bonds were never arranged as regularly as the crystalline cellulose prior to the pretreatment (see Fig. 14.3). DMSO is an important polar aprotic solvent that dissolves both polar and non-polar compounds and is miscible in a wide range of organic solvents [99]. Therefore, despite of the relatively lower price, it is an important and efficient co-solvent for cellulose dissolution and pretreatment.

In the OES-dissolving pretreatment study, 5 % microcrystalline cellulose was deposited in each of these OES which had different molar fractions of [AMIM]Cl (i.e., χ [AMIM]Cl = 0.1-0.9). Followed by being kept at 110 °C for 1 h under a continuous stirring, the regenerated cellulose was precipitated by water and enzymatically hydrolyzed for 72 h.

Results revealed that the microcrystalline cellulose was rapidly dissolved in the OES (when χ [AMIM]Cl $\geq$ 0.2) within 10 min. During the hydrolysis, with the increase of OES from 0.1 to 0.9, both hydrolysis yield and initial hydrolysis rate of the regenerated cellulose increased gradually. After 72 h, the glucose yield of cellulose treated by OES (at χ [AMIM]Cl = 0.7) was 54.1 %, which was 7.2 times that of the untreated cellulose, and was only slightly lower than the value (59.6 %) obtained by using pure [AMIM]Cl. Characterization of the regenerated cellulose samples was conducted subsequently. With increasing molar fractions of [AMIM]Cl, the crystallinity index (CI) of cellulose I decreased from 0.834 to 0.319, whereas the CI of cellulose II stayed at around 0.284. Meanwhile, the average specific surface area

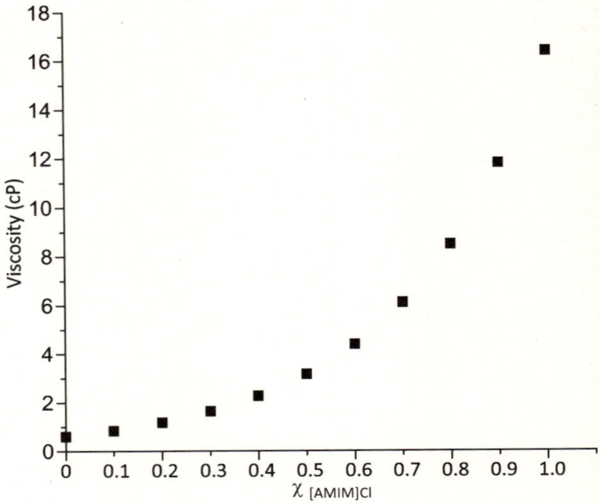

Fig. 14.4 Variable viscosities of OES against the increasing molar fractions of [AMIM]Cl at 110 °C (Calculated according to the Grunberg-Nissan mixing law. Viscosities of DMSO and [AMIM]Cl at 110 °C were caculated according to Arrhenius model and VFT equation, respectively.) [42]

and degree of polymerization remained without significant change, being 4.174 m^2/g and 137.2, respectively.

Although DMSO has no positive effect on promoting cellulose dissolution for being unable to donate cations and anions [42, 90], the OES mixed with DMSO and [AMIM]Cl has positive effect on decrystallinity of cellulose I, which may account for the higher hydrolysis yield and rate [18].

The unique advantages of employing OES as a cellulose solvent in the pretreatment can be summarized below [42, 85]:

1. Replacing a portion of ILs by less expensive OES, which can lower the processing cost.
2. Shortening the dissolution time to form cellulose homogenerous solutions due to its rapid diffusion efficiency.
3. Being more practical and easier for large-scale operations (stirring and pipeline transportation) due to the reduced viscosity of the OES system. For example, the viscosity of OES with molar fraction of [AMIM]Cl = 0.7, is only 37.28 % of the pure IL at 110 °C (see Fig. 14.4).
4. Enabling a higher cellulose recovery rate (95.37 ± 1.41 %) opposite to significant decomposition of polysaccharides in other chemical pretreatment methods [100].
5. Proving to be a simple but effective process to prepare cellulose with controlled CI than some other methods [101, 102]. The CI of cellulose I in the regenerated samples has a strong negative linear correlation against the molar fraction of IL in OES (i.e., with a correlation coefficient of 0.98).

Referring to the advantages, employing OES as a cellulose solvent has a bright perspective for efficient pretreatment of lignocellulosic biomass. However, recycling of OES should be taken into account to remove the barriers to large-scale application. It can be achieved by using commercial distillation technology to separate water from ILs due to their lower vapor pressures. Alternately, using aqueous ethanol or aqueous acetone instead of pure water as an antisolvent would reduce the temperature and vacuum requirement in the distillation. Otherwise, some new technologies, such as nanofiltration, reverse osmosis, and pervaporation [103], three-phase system precipitation [104], and supercritical CO_2 extraction [103, 105] may have potential applications in recycling of OES.

Moreover, investigation of OES on the pretreatment of various lignocellulosic biomass, such as corn stover, switchgrass, poplar, and pine, should be performed to determine the chemical components isolation, cellulose decrystallization, and improvement of enzymatic saccharification against the solvent constituents as well as their molar ratio in OES. Physical features of samples, such as moisture, particle size, and homogeneity, which have been proposed to affect the cellulose-dissolving ability in ILs [106], should also be considered and optimized in the OES system.

14.3.3 OES-Precipitating Pretreatment

Many cellulose solvents were employed to treat biomass and accelerate its enzymatic accessibility and digestibility by disrupting linkages among cellulose, hemicellulose, and lignin component as well as breaking up the hydrogen-bond linkages in the orderly crystalline cellulose to amorphous forms [84,85,107–111]. But this method is still challenging in dealing with real lignocelluloses due to the low efficiency in removal of lignin and hemicellulose from cellulose, resulting in lower cellulose digestibility and slower hydrolysis [42]. It was revealed that cellulose, hemicellulose, and lignin have different solubilities in OES composed of aqueous acetone or ethanol with concentrated phosphoric acid or ILs [89,97,112–114]. Utilization of this lignin-soluble OES to deposit cellulose and/or hemicellulose from the lignocellulosic biomass solutions can be a practical approach to resolve the problem.

Zhang et al. [97] developed a lignocellulose pretreatment technique to fractionate lignocelluloses to amorphous cellulose, hemicellulose, lignin, and acetic acid. It is featured by using 8 ml cellulose solvent (pre-calibrated concentrated phosphoric acid 83–85.9 %) to dissolve 1 g biomass under a modest reaction conditions (50 °C and atmospheric pressure). After the reaction, OES was immediately generated by the addition of 20 ml organic solvent (i.e., acetone) into the biomass–phosphoric acid solutions and then well mixed. Coupled with the formation of OES, the cellulose and hemicellulose were rapidly precipitated. Followed by subsequently centrifuging, the OES supernatants were collected for separation of lignin, acetic acid, and acetone. With the removal of acetone from the OES by simple evaporation as well as acetic acid, the low molecular lignin was precipitated from aqueous phosphoric acid. In this way, partial lignin was removed from the biomass combined with the decrystallization of cellulose. Consequently, the regenerated amorphous cellulose without lignin was hydrolyzed efficiently. The regenerated corn stover, switchgrass, and poplar, were hydrolyzed, respectively, where a high-hydrolysis efficiency ~94 % and ~96– ~97 % was achieved at the 12th and 24th hour with an enzyme loadings of

15 FPU cellulase and 60 IU beta-glucosidase per gram of glucan, respectively. The data of hydrolysis rates and digestibility were the highest in the literature [2, 8, 115]. Alternatively, although this method seems not to be efficient for treating Douglas Fir (∼73 % digestibility at 12 h and ∼75 % at 24 h), it was still ∼1.7 times of the sample pretreated by SO_2 steam explosion [97].

Similar study of OES-precipitating pretreatment for bamboos was conducted using 85 % (w/v) concentrated phosphoric acid as a cellulose solvent and 95 % (v/v) ethanol as an organic solvent [113]. After dissolving by concentrated phosphoric acid and followed by precipitation with simultaneous formation of OES, the glucan recovery yield of the sample was 93.9 % and the delignification yield was 15.3 %. The elevation of hydrolysis yield was due to the increase in cellulose accessibility to cellulase from 0.27 to 9.14 m^2 per gram of biomass. Glucan digestibility attained 88.2 % at the cellulase loading of 1 glucan in 72 h. The overall glucose and xylose yields were 86.0 % and 82.6 %, respectively.

The advantages of OES-precipitating pretreatment by employing a non-volatile cellulose solvent and a highly volatile organic solvent are:

1. Decrystallized cellulose and hemicellulose can be separated efficiently from dissolved biomass by precipitation because they have poor solubility in the OES mixture that is able to partially dissolve lignin.
2. Acetone-soluble lignin can be easily recovered after adding water or evaporation of organic solvent from the OES because it is insoluble either in water or specific cellulose solutions, such as phosphoric acid [97].
3. Organic solvents such as acetone and ethanol can be easily recycled and reused by fractional distillation due to their low boiling point against non-volatile ability of the specific cellulose solutions, such as ILs or concentrated phosphoric acids.

14.3.4 Further Works and Perspective

Further work to improve the OES-precipitating pretreatment may be focused on the fractionation of softwood by new kinds of OES because the substrate applicability of H_3PO_4 + organic solvent precipitating technology is by far limited to hardwood, herbaceous, and bamboo plants [97, 112, 114].

[AMIM]Cl is the only IL that has broad solubility to dissolve both the softwood and hardwood completely among the 96 ILs [85]. It can compete with the lignocellulose components for hydrogen bonding, attributed to its anions, especially chloride anions [85, 101]. However, the π-electrons which are exhibited within imidazolium ring and on side-chain of [AMIM]Cl cation, could have π-π interactions with the aromatic compounds of lignin [86, 101]. It reveals that [AMIM]Cl could take a stronger and more efficient effect on the lignin structures in softwood than other dissolving solvents. As a result, [AMIM]Cl could be employed to dissolve the softwood first, and then a bicomponent or tricomponent OES is homogeneously formed to deposit cellulose and hemicellulose by adding ethanol, or acetone and DMSO, or acetone and PEG mixtures. After filtrations, lignin would be separated from the OES by water and further modified to other products.

Therefore, OES is a green and efficient solvent to be employed in the pretreatment producing a high-reactivity lignocellulosic substrate suitable for enzymatic saccharification as well as high-quality lignin polymers for materials. OES has many advantages such as lower viscosity, rapid and selective solubility, generating less by-products, and recyclable. To meet the requirement of high biomass productivity for biofuels and biorefinery, further research with respects to the OES pretreatment should be focused on:

1. Designing new bicomponent or tricomponent OES which has broad applicability on different biomasses.
2. Developing efficient OES recycling techniques.
3. Conducting continuous flow process by combining unit operations of dissolving, precipitation and solvent recycling.
4. Optimizing the process and evaluating the economics of the entire process.

Acknowledgments This work was funded by Chinese Academy of Sciences (BairenJihua and Knowledge innovation key project (KSCX2-YW-G-075)), Yunnan Provincial Government (Baiming Haiwai Gaocengci Rencai Jihua and Provincial Natural Science Foundation), and China National Natural Science Foundation (No: 21076220). One of the authors (C. XU) would like to acknowledge the funding from Natural Science and Engineering Research Council of Canada (NSERC) through the Discovery Grant.

References

1. Ragauskas AJ, Williams CK, Davison BH, Britovsek G, Cairney J, Eckert CA, Frederick WJ, Hallett JP, Leak DJ, Liotta CL, Mielenz JR, Murphy R, Templer R, Tschaplinski T (2006) The path forward for biofuels and biomaterials. Science 311(5760):484–489. doi:10.1126/science.1114736
2. Wyman CE, Dale BE, Elander RT, Holtzapple M, Ladisch MR, Lee YY (2005) Coordinated development of leading biomass pretreatment technologies. Bioresour Technol 96(18):1959–1966. doi:10.1016/j.biortech.2005.01.010
3. Zhang YHP, Himmel ME, Mielenz JR (2006) Outlook for cellulase improvement: screening and selection strategies. Biotechnol Adv 24(5):452–481. doi:10.1016/j.biotechadv.2006.03.003
4. Guragain YN, De Coninck J, Husson F, Durand A, Rakshit SK (2011) Comparison of some new pretreatment methods for second generation bioethanol production from wheat straw and water hyacinth. Bioresour Technol 102(6):4416–4424. doi:10.1016/j.biortech.2010.11.125
5. Chandra R, Bura R, Mabee W, Berlin A, Pan X, Saddler J, Olsson L (2007) Substrate pretreatment: the key to effective enzymatic hydrolysis of lignocellulosics? Biofuels: Advances in biochemical engineering/biotechnology, vol 108. Springer, Heidelberg, pp 108:67–93. doi:10.1007/10_2007_064
6. Sun Y, Cheng JY (2002) Hydrolysis of lignocellulosic materials for ethanol production: a review. Bioresour Technol 83(1):1–11. doi:10.1016/S0960-8524(01)00212-7
7. Fan L, Lee Y-H, Gharpuray M (1982) The nature of lignocellulosics and their pretreatments for enzymatic hydrolysis. Microbial Reactions: Advances in biochemical engineering/biotechnology, vol 23. Springer, Heidelberg. pp 157–187. doi:10.1007/3540116982_4

8. Mosier N, Wyman C, Dale B, Elander R, Lee YY, Holtzapple M, Ladisch M (2005) Features of promising technologies for pretreatment of lignocellulosic biomass. Bioresour Technol 96(6):673–686. doi:10.1016/j.biortech.2004.06.025
9. Yang B, Wyman CE (2008) Pretreatment: the key to unlocking low-cost cellulosic ethanol. Biofuels, Bioprod Biorefin 2(1):26–40. doi:10.1002/bbb.49
10. Kumar D, Murthy G (2011) Impact of pretreatment and downstream processing technologies on economics and energy in cellulosic ethanol production. Biotechnol Biofuels 4(27). doi:10.1186/1754-6834-4-27
11. Lynd L, Elamder R, Wyman C (1996) Likely features and costs of mature biomass ethanol technology. Appl Biochem Biotechnol 57–58(1):741–761. doi:10.1007/bf02941755
12. Tian X, Fang Z, Guo F (2011) Impact and prospective of fungal pre-treatment of lignocellulosic biomass for enzymatic hydrolysis. Biofuels, Bioprod Biorefin: 6(3):335–350. doi:10.1002/bbb.346
13. Kleinert TN (1971) Organosolv pulping and recovery process. US Patent 3585104.
14. Holtzapple MT, Humphrey AE (1984) The effect of organosolv pretreatment on the enzymatic-hydrolysis of poplar. Biotechnol Bioeng 26(7):670–676. doi:10.1002/bit.260260706
15. Zhao X, Cheng K, Liu D (2009) Organosolv pretreatment of lignocellulosic biomass for enzymatic hydrolysis. Appl Microbiol Biotechnol 82(5):815–827. doi:10.1007/s00253-009-1883-1
16. Zhu L (2005) Fundamental study of structural features affecting enzymatic hydrolysis of lignocellulosic biomass. Ph.D. Dissertation, Texas A&M University, College Station, Texas, p 306
17. Lynd LR, Weimer PJ, van Zyl WH, Pretorius IS (2002) Microbial cellulose utilization: fundamentals and biotechnology. Microbiol Mol Biol Rev 66(3):506–577. doi: 10.1128/MMBR.66.3.506-577.2002
18. Hendriks ATWM, Zeeman G (2009) Pretreatments to enhance the digestibility of lignocellulosic biomass. Bioresour Technol 100(1):10–18. doi:10.1016/j.biortech.2008.05.027
19. Kumar R, Mago G, Balan V, Wyman CE (2009) Physical and chemical characterizations of corn stover and poplar solids resulting from leading pretreatment technologies. Bioresour Technol 100(17):3948–3962. doi:10.1016/j.biortech.2009.01.075
20. Zhu L, O' Dwyer JP, Chang VS, Granda CB, Holtzapple MT (2008) Structural features affecting biomass enzymatic digestibility. Bioresour Technol 99(9):3817–3828. doi:10.1016/j.biortech.2007.07.033
21. Pan X, Arato C, Gilkes N, Gregg D, Mabee W, Pye K, Xiao Z, Zhang X, Saddler J (2005) Biorefining of softwoods using ethanol organosolv pulping: preliminary evaluation of process streams for manufacture of fuel-grade ethanol and co-products. Biotechnol Bioeng 90(4):473–481. doi:10.1002/bit.20453
22. Hallac BB, Ray M, Murphy RJ, Ragauskas AJ (2010) Correlation between anatomical characteristics of ethanol organosolv pretreated *Buddleja davidii* and its enzymatic conversion to glucose. Biotechnol Bioeng 107(5):795–801. doi:10.1002/bit.22884
23. Hallac BB, Sannigrahi P, Pu Y, Ray M, Murphy RJ, Ragauskas AJ (2010) Effect of ethanol organosolv pretreatment on enzymatic hydrolysis of *Buddleja davidii* stem biomass. Ind Eng Chem Res 49(4):1467–1472. doi:10.1021/ie900683q
24. Del Rio LF, Chandra RP, Saddler JN (2010) The effect of varying organosolv pretreatment chemicals on the physicochemical properties and cellulolytic hydrolysis of mountain pine beetle-killed Lodgepole pine. Appl Biochem Biotechnol 161(1–8):1–21. doi:10.1007/s12010-009-8786-6
25. Chang V, Holtzapple M (2000) Fundamental factors affecting biomass enzymatic reactivity. Appl Biochem Biotechnol 84–86(1):5–37. doi:10.1385/abab:84-86:1-9:5
26. Draude KM, Kurniawan CB, Duff SJB (2001) Effect of oxygen delignification on the rate and extent of enzymatic hydrolysis of lignocellulosic material. Bioresour Technol 79(2):113–120. doi:10.1016/S0960-8524(01)00055-4
27. Thompson DN, Chen HC, Grethlein HE (1992) Comparison of pretreatment methods on the basis of available surface area. Bioresour Technol 39(2):155–163. doi:10.1016/0960-8524(92)90135-K

28. Berlin A, Gilkes N, Kurabi A, Bura R, Tu M, Kilburn D, Saddler J (2005) Weak lignin-binding enzymes. Appl Biochem Biotechnol 121(1):163–170. doi:10.1385/abab:121:1-3:0163
29. Gunjikar TP, Sawant SB, Joshi JB (2001) Shear deactivation of cellulase, exoglucanase, endoglucanase, and β-Glucosidase in a mechanically agitated reactor. Biotechnol Progr 17(6):1166–1168. doi:10.1021/bp010114u
30. Mooney CA, Mansfield SD, Touhy MG, Saddler JN (1998) The effect of initial pore volume and lignin content on the enzymatic hydrolysis of softwoods. Bioresour Technol 64(2):113–119. doi:10.1016/S0960-8524(97)00181-8
31. Meunier-Goddik L, Penner MH (1998) Enzyme-catalyzed saccharification of model celluloses in the presence of lignacious residues. J Agr Food Chem 47(1):346–351. doi:10.1021/jf980407b
32. Ishizawa C, Jeoh T, Adney W, Himmel M, Johnson D, Davis M (2009) Can delignification decrease cellulose digestibility in acid pretreated corn stover? Cellulose 16(4):677–686. doi:10.1007/s10570-009-9313-1
33. Zhang XM, Meng LY, Xu F, Sun RC (2011) Pretreatment of partially delignified hybrid poplar for biofuels production: characterization of organosolv hemicelluloses. Ind Crops Prod 33(2):310–316. doi:10.1016/j.indcrop.2010.11.016
34. Chundawat SPS, Beckham GT, Himmel ME, Dale BE (2011) Deconstruction of lignocellulosic biomass to fuels and chemicals. In: Prausnitz JM (ed) Annual review of chemical and biomolecular engineering, vol 2. Annual review of chemical and biomolecular engineering. Annual Reviews, Palo Alto, pp 121–145. doi:10.1146/annurev-chembioeng-061010-114205
35. Kong F, Engler C, Soltes E (1992) Effects of cell-wall acetate, xylan backbone, and lignin on enzymatic hydrolysis of aspen wood. Appl Biochem Biotechnol 34–35(1):23–35. doi:10.1007/bf02920531
36. Mitchell DJ, Grohmann K, Himmel ME, Dale BE, Schroeder HA (1990) Effect of the degree of acetylation on the enzymatic digestion of acetylated xylans. J Wood Chem Technol 10(1):111–121. doi:10.1080/02773819008050230
37. Rollin JA, Zhu Z, Sathitsuksanoh N, Zhang YHP (2010) Increasing cellulose accessibility is more important than removing lignin: a comparison of cellulose solvent-based lignocellulose fractionation and soaking in aqueous ammonia. Biotechnol Bioeng 108(1):22–30. doi:10.1002/bit.22919
38. Hong J, Ye X, Zhang YHP (2007) Quantitative determination of cellulose accessibility to cellulase based on adsorption of a non-hydrolytic fusion protein containing CBM and GFP with its applications. Langmuir 23(25):12535–12540. doi:10.1021/la7025686
39. Zhu Z, Sathitsuksanoh N, Vinzant T, Schell DJ, McMillan JD, Zhang YHP (2009) Comparative study of corn stover pretreated by dilute acid and cellulose solvent-based lignocellulose fractionation: enzymatic hydrolysis, supramolecular structure, and substrate accessibility. Biotechnol Bioeng 103(4):715–724. doi:10.1002/bit.22307
40. Koo BW, Min BC, Gwak KS, Lee SM, Choi JW, Yeo H, Choi IG (2012) Structural changes in lignin during organosolv pretreatment of *Liriodendron tulipifera* and the effect on enzymatic hydrolysis. Biomass Bioenerg 42:24–32. doi:10.1016/j.biombioe.2012.03.012
41. Donohoe BS, Decker SR, Tucker MP, Himmel ME, Vinzant TB (2008) Visualizing lignin coalescence and migration through maize cell walls following thermochemical pretreatment. Biotechnol Bioeng 101(5):913–925. doi:10.1002/bit.21959
42. Tian X, Fang Z, Jiang D, Sun X (2011) Pretreatment of microcrystalline cellulose in organic electrolyte solutions for enzymatic hydrolysis. Biotechnol Biofuels 4:53. doi:10.1186/1754-6834-4-53
43. Kuo CH, Lee CK (2009) Enhancement of enzymatic saccharification of cellulose by cellulose dissolution pretreatments. Carbohydr Polym 77(1):41–46. doi:10.1016/j.carbpol.2008.12.003
44. Zhou W, Xu Y, Schüttler H-B (2010) Cellulose hydrolysis in evolving substrate morphologies III: time-scale analysis. Biotechnol Bioeng 107(2):224–234. doi:10.1002/bit.22814
45. Pan X, Xie D, Kang KY, Yoon SL, Saddler JN (2007) Effect of organosolv ethanol pretreatment variables on physical characteristics of hybrid poplar substrates. Appl Biochem Biotechnol 137:367–377. doi:10.1007/s12010-007-9065-z

46. Hayashi N, Sugiyama J, Okano T, Ishihara M (1997) Selective degradation of the cellulose I alpha component in Cladophora cellulose with *Trichoderma viride* cellulase. Carbohydr Res 305(1):109–116. doi:10.1016/S0008-6215(97)00281-4
47. Baker AA, Helbert W, Sugiyama J, Miles MJ (2000) New insight into cellulose structure by atomic force microscopy shows the I alpha crystal phase at near-atomic resolution. Biophys J 79(2):1139–1145. doi:10.1016/S0006-3495(00)76367-3
48. Zugenmaier P (2001) Conformation and packing of various crystalline cellulose fibers. Prog Polym Sci 26(9):1341–1417. doi:10.1016/S0079-6700(01)00019-3
49. Weimer PJ, French AD, Calamari TA (1991) Differential fermentation of cellulose allomorphs by ruminal cellulolytic bacteria. Appl Environ Microb 57(11):3101–3106
50. Wada M, Chanzy H, Nishiyama Y, Langan P (2004) Cellulose IIII crystal structure and hydrogen bonding by synchrotron X-ray and neutron fiber diffraction. Macromolecules 37(23):8548–8555. doi:10.1021/ma0485585
51. Wada M, Ike M, Tokuyasu K (2010) Enzymatic hydrolysis of cellulose I is greatly accelerated via its conversion to the cellulose II hydrate form. Polym Degrad Stab 95(4):543–548. doi:10.1016/j.polymdegradstab.2009.12.014
52. Igarashi K, Wada M, Samejima M (2007) Activation of crystalline cellulose to cellulose IIII results in efficient hydrolysis by cellobiohydrolase. FEBS J 274(7):1785–1792. doi:10.1111/j.1742-4658.2007.05727.x
53. Gardiner ES, Sarko A (1985) Packing analysis of carbohydrates and polysaccharides. 16. The crystal structures of celluloses IVI and IVII. Can J Chem 63(1):173–180. doi:10.1139/v85-027
54. Sannigrahi P, Miller SJ, Ragauskas AJ (2010) Effects of organosolv pretreatment and enzymatic hydrolysis on cellulose structure and crystallinity in Loblolly pine. Carbohydr Res 345(7):965–970. doi:10.1016/j.carres.2010.02.010
55. Chum HL, Johnson DK, Black SK (1990) Organosolv pretreatment for enzymic hydrolysis of poplars. 2. Catalyst effects and the combined severity parameter. Ind Eng Chem Res 29(2):156–162. doi:10.1021/ie00098a003
56. Parajo JC, Alonso JL, Santos V (1995) Kinetics of catalyzed organosolv processing of pine wood. Ind Eng Chem Res 34(12):4333–4342. doi:10.1021/ie00039a025
57. Galbe M, Zacchi G (2002) A review of the production of ethanol from softwood. Appl Microbiol Biotechnol 59(6):618–628. doi:1007/s00253-002-1058-9
58. Varnai A, Siika-aho M, Viikari L (2010) Restriction of the enzymatic hydrolysis of steam-pretreated spruce by lignin and hemicellulose. Enzyme Microb Technol 46(3–4):185–193. doi:10.1016/j.enzmictec.2009.12.013
59. Mesa L, Gonzalez E, Cara C, Gonzalez M, Castro E, Mussatto SI (2011) The effect of organosolv pretreatment variables on enzymatic hydrolysis of sugarcane bagasse. Chem Eng J 168(3):1157–1162. doi:10.1016/j.cej.2011.02.003
60. Brosse N, Sannigrahi P, Ragauskas A (2009) Pretreatment of Miscanthus x giganteus using the ethanol organosolv process for ethanol production. Ind Eng Chem Res 48(18):8328–8334. doi:10.1021/ie9006672
61. Del Rio LF, Chandra RP, Saddler JN (2011) The effects of increasing swelling and anionic charges on the enzymatic hydrolysis of organosolv-pretreated softwoods at low enzyme loadings. Biotechnol Bioeng 108(7):1549–1558. doi:10.1002/bit.23090
62. Bruce A, Diner JF (2009) Organic solvent pretreatment of biomass to enhance enzymatic saccharification. US Patent: US 2010/0159520 A1
63. Ghose TK, Pannir Selvam PV, Ghosh P (1983) Catalytic solvent delignification of agricultural residues: organic catalysts. Biotechnol Bioeng 25(11):2577–2590. doi:10.1002/bit.260251108
64. Araque E, Parra C, Freer J, Contreras D, Rodriguez J, Mendonca R, Baeza J (2008) Evaluation of organosolv pretreatment for the conversion of pinus radiata D. Don to ethanol. Enzyme Microb Technol 43(2):214–219. doi:10.1016/j.enzmictec.2007.08.006
65. Bonn G, Hormeyer HF, Bobleter O (1987) Hydrothermal and organosolv pretreatments of poplar wood and wheat straw for saccharification by a *Trichoderma viride* cellulase. Wood Sci Technol 21(2):179–185. doi:10.1007/BF00376197

66. Mabee W, Gregg D, Arato C, Berlin A, Bura R, Gilkes N, Mirochnik O, Pan X, Kendall Pye E, Saddler J (2006) Updates on softwood-to-ethanol process development. Appl Biochem Biotechnol 129(1):55–70. doi:10.1385/abab:129:1:55
67. Huijgen WJJ, Reith JH, den Uil H (2010) Pretreatment and fractionation of wheat straw by an acetone-based organosolv process. Ind Eng Chem Res 49(20):10132–10140. doi:10.1021/ie101247w
68. Park N, Kim HY, Koo BW, Yeo H, Choi IG (2010) Organosolv pretreatment with various catalysts for enhancing enzymatic hydrolysis of pitch pine (*Pinus rigida*). Bioresour Technol 101(18):7046–7053. doi:10.1016/j.biortech.2010.04.020
69. Pan XJ, Xie D, Yu RW, Lam D, Saddler JN (2007) Pretreatment of lodgepole pine killed by mountain pine beetle using the ethanol organosolv process: fractionation and process optimization. Ind Eng Chem Res 46(8):2609–2617. doi:10.1021/ie0615761
70. Ooshima H, Kurakake M, Kato J, Harano Y (1992) Pretreatment of wood by uct-solvent for the enzymatic-hydrolysis. Appl Biochem Biotechnol 37(2):165–176. doi:10.1007/bf02921668
71. Mesa L, Gonzalez E, Cara C, Ruiz E, Castro E, Mussatto SI (2010) An approach to optimization of enzymatic hydrolysis from sugarcane bagasse based on organosolv pretreatment. J Chem Technol Biotechnol 85(8):1092–1098. doi:10.1002/jctb.2404
72. Palmqvist E, Hahn-Hagerdal B (2000) Fermentation of lignocellulosic hydrolysates. II: Inhibitors and mechanisms of inhibition. Bioresour Technol 74(1):25–33. doi:10.1016/S0960-8524(99)00161-3
73. Wang M, Leitch M, Xu C (2009) Synthesis of pheno-formaldehyde resol resins using organosolv pine lignins. Eur Polym J 45(12):3380–3388. doi:10.1016/j.eurpolymj.2009.10.003
74. Pan X, Kadla JF, Ehara K, Gilkes N, Saddler JN (2006) Organosolv ethanol lignin from hybrid poplar as a radical scavenger: relationship between lignin structure, extraction conditions, and antioxidant activity. J Agric Food Chem 54(16):5806–5813. doi:10.1021/jf0605392
75. Ramires EC, Megiatto JD, Gardrat C, Castellan A, Frollini E (2010) Valorization of an industrial organosolv–sugarcane bagasse lignin: characterization and use as a matrix in bio-based composites reinforced with sisal fibers. Biotechnol Bioeng 107(4):612–621. doi:10.1002/bit.22847
76. Piskorz J, Majerski P, Radlein D, Scott DS (1989) Conversion of lignins to hydrocarbon fuels. Energy Fuels 3(6):723–726. doi:10.1021/ef00018a011
77. Hu G, Cateto C, Pu Y, Samuel R, Ragauskas AJ (2011) Structural characterization of switchgrass lignin after ethanol organosolv pretreatment. Energy Fuels 26(1):740–745. doi:10.1021/ef201477p
78. Holladay JE, Bozell JJ, White JF, Johnson D (2004) Top value-added chemicals from biomass volume II—results of screening for potential candidates from biorefinery lignin. Prepared for the U.S. Department of Energy under Contract DE-AC05-76RL01830, pp 1–79
79. Milstein O, Nicklas B, Hüttermann A (1989) Oxidation of aromatic compounds in organic solvents with laccase from *Trametes versicolor*. Appl Microb Biotechnol 31(1):70–74. doi:10.1007/bf00252530
80. Sena-Martins G, Almeida-Vara E, Duarte JC (2008) Eco-friendly new products from enzymatically modified industrial lignins. Ind Crops Prod 27(2):189–195. doi:10.1016/j.indcrop.2007.07.016
81. Vishtal A, Kraslawski A (2011) Challenges of lignins. BioResources 6(3):3547–3568
82. Liu J, Takada R, Karita S, Watanabe T, Honda Y, Watanabe T (2010) Microwave-assisted pretreatment of recalcitrant softwood in aqueous glycerol. Bioresour Technol 101(23):9355–9360. doi:10.1016/j.biortech.2010.07.023
83. Swatloski RP, Spear SK, Holbrey JD, Rogers RD (2002) Dissolution of cellose with ionic liquids. J Am Chem Soc 124(18):4974–4975. doi:10.1021/ja025790m
84. Phillips DM, Drummy LF, Conrady DG, Fox DM, Naik RR, Stone MO, Trulove PC, De Long HC, Mantz RA (2004) Dissolution and regeneration of *Bombyx mori* silk fibroin using ionic liquids. J Am Chem Soc 126(44):14350–14351. doi:10.1021/ja046079f
85. Dadi AP, Varanasi S, Schall CA (2006) Enhancement of cellulose saccharification kinetics using an ionic liquid pretreatment step. Biotechnol Bioeng 95(5):904–910. doi:10.1002/bit.21047

86. Zavrel M, Bross D, Funke M, Buchs J, Spiess AC (2009) High-throughput screening for ionic liquids dissolving (ligno-)cellulose. Bioresour Technol 100(9):2580–2587. doi:10.1016/j.biortech.2008.11.052
87. Zhang YHP, Cui JB, Lynd LR, Kuang LR (2006) A transition from cellulose swelling to cellulose dissolution by o-phosphoric acid: evidence from enzymatic hydrolysis and supramolecular structure. Biomacromolecules 7(2):644–648. doi:10.1021/bm050799c.
88. Chanzy H, Peguy A, Chaunis S, Monzie P (1980) Oriented cellulose films and fibers from a mesophase system. J Polym Sci 18(5):1137–1144. doi:10.1002/pol.1980.180180517
89. MacDDM (1977) The spinning of unconventional cellulose solutions. In: Solvent spun rayon, modified cellulose fibers and derivatives. ACS Symp Ser, 58:25–39. doi:10.1021/bk-1977-0058.ch003
90. Fort DA, Remsing RC, Swatloski RP, Moyna P, Moyna G, Rogers RD (2007) Can ionic liquids dissolve wood? Processing and analysis of lignocellulosic materials with 1-*n*-butyl-3-methylimidazolium chloride. Green Chem 9(1):63–69. doi:10.1039/b607614a
91. Åsa Östlund DL, Lars N, Krister H, Magnus N (2009) Dissolution and gelation of cellulose in TBAF/DMSO solutions: the roles of fluoride ions and water. Biomacromolecules 10(9):2401–2407. doi:10.1021/bm900667q
92. Thomas H, Andreas K (2005) Solvents applied in the field of cellulose chemistry: a mini review. Polímeros—Ciê ncia e Tecnologia 5(2):84–90
93. Sui XF, Yuan JY, Yuan WZ, Zhou M (2008) Preparation of cellulose nanofibers/nanoparticles via electrospray. Chem Lett 37(1):114–115. doi:10.1246/cl.2008.114
94. Luo MK, Neogi AN, West H (2009) Dissolution of cellulose in mixed solvent systems. USPTO Patent application no.20090088564
95. Rinaldi R (2011) Instantaneous dissolution of cellulose in organic electrolyte solutions. Chem Commun 47(1):511–513. doi:10.1039/c0cc02421j
96. Ishii D, Tatsumi D, Matsumoto T, Murata K, Hayashi H, Yoshitani H (2006) Investigation of the structure of cellulose in LiCl/DMAc solution and its gelation behavior by small-angle X-ray scattering measurements. Macromol Biosci 6(4):293–300. doi:10.1002/mabi.200500231
97. Zhang YHP, Ding SY, Mielenz JR, Cui JB, Elander RT, Laser M, Himmel ME, McMillan JR, Lynd LR (2007) Fractionating recalcitrant lignocellulose at modest reaction conditions. Biotechnol Bioeng 97(2):214–223. doi:10.1002/bit.21386
98. Zhang H, Wu J, Zhang J, He JS (2005) 1-Allyl-3-methylimidazolium chloride room temperature ionic liquid: a new and powerful nonderivatizing solvent for cellulose. Macromolecules 38(20):8272–8277. doi:10.1021/ma0505676
99. Novak KM (ed) (2002) Drug facts and comparisons, 56th edn. Wolters Kluwer Health, St. Louis, p 619
100. Mora-Pale M, Meli L, Doherty TV, Linhardt RJ, Dordick JS (2011) Room temperature ionic liquids as emerging solvents for the pretreatment of lignocellulosic biomass. Biotechnol Bioeng 108(6):1229–1245. doi:10.1002/bit.23108
101. Kilpelainen I, Xie H, King A, Granstrom M, Heikkinen S, Argyropoulos DS (2007) Dissolution of wood in ionic liquids. J Agric Food Chem 55(22):9142–9148. doi:10.1021/jf071692e
102. Zhang J, Luo J, Tong D, Zhu L, Dong L, Hu C (2010) The dependence of pyrolysis behavior on the crystal state of cellulose. Carbohydr Polym 79(1):164–169. doi:10.1016/j.carbpol.2009.07.038
103. Haerens K, Van Deuren S, Matthijs E, Van Der Bruggen B (2010) Challenges for recycling ionic liquids by using pressure driven membrane processes. Green Chem 12(12):2182–2188. doi:10.1039/c0gc00406e
104. Shill K, Padmanabhan S, Xin Q, Prausnitz JM, Clark DS, Blanch HW (2010) Ionic liquid pretreatment of cellulosic biomass: enzymatic hydrolysis and ionic liquid recycle. Biotechnol Bioeng 108(3):511–520. doi:10.1002/bit.23014
105. Abu-Eishah SI (2011) Ionic liquids recycling for reuse. In: Handy ST. Rijeka (ed) Ionic liquids—classes and properties. InTech, pp 239–272
106. Yuan T, He J, Xu F, Sun R (2010) A new vision in the research of biomass resources: complete-lignocellulose-dissolution system. Prog Chem 22(0203):472–481

107. Li Q, He YC, Xian M, Jun G, Xu X, Yang JM, Li LZ (2009) Improving enzymatic hydrolysis of wheat straw using ionic liquid 1-ethyl-3-methyl imidazolium diethyl phosphate pretreatment. Bioresour Technol 100(14):3570–3575. doi:10.1016/j.biortech.2009.02.040
108. Li Q, Jiang X, He Y, Li L, Xian M, Yang J (2010) Evaluation of the biocompatibile ionic liquid 1-methyl-3-methylimidazolium dimethylphosphite pretreatment of corn cob for improved saccharification. Appl Microbiol Biotechnol 87(1):117–126. doi:10.1007/s00253-010-2484-8
109. Shafiei M, Karimi K, Taherzadeh MJ (2010) Pretreatment of spruce and oak by *N*-methylmorpholine-*N*-oxide (NMMO) for efficient conversion of their cellulose to ethanol. Bioresour Technol 101(13):4914–4918. doi:10.1016/j.biortech.2009.08.100
110. Kuo CH, Lee CK (2009) Enhanced enzymatic hydrolysis of sugarcane bagasse by *N*-methylmorpholine-*N*-oxide pretreatment. Bioresour Technol 100(2):866–871. doi: 10.1016/j.biortech.2008.07.001
111. Hamilton TJ, Dale BE, Ladisch MR, Tsao GT (1984) Effect of ferric tartrate/sodium hydroxide solvent pretreatment on enzyme hydrolysis of cellulose in corn residue. Biotechnol Bioeng 26(7):781–787. doi:10.1002/bit.260260724
112. Sathitsuksanoh N, Zhu Z, Ho TJ, Bai MD, Zhang YHP (2009) Bamboo saccharification through cellulose solvent-based biomass pretreatment followed by enzymatic hydrolysis at ultra-low cellulase loadings. Bioresour Technol 101(13):4926–4929. doi:10.1016/j.biortech.2009.09.081
113. Moxley G, Zhu Z, Zhang YHP (2008) Efficient sugar release by the cellulose solvent-based lignocellulose fractionation technology and enzymatic cellulose hydrolysis. J Agric Food Chem 56(17):7885–7890. doi:10.1021/jf801303f
114. Sathitsuksanoh N, Zhu Z, Templeton N, Rollin JA, Harvey SP, Zhang YHP (2009) Saccharification of a potential bioenergy crop, *Phragmites australis* (common reed), by lignocellulose fractionation followed by enzymatic hydrolysis at decreased cellulase loadings. Ind Eng Chem Res 48(13):6441–6447. doi:10.1021/ie900291s
115. Wyman CE, Dale BE, Elander RT, Holtzapple M, Ladisch MR, Lee YY (2005) Comparative sugar recovery data from laboratory scale application of leading pretreatment technologies to corn stover. Bioresour Technol 96(18):2026–2032. doi: 10.1016/j.biortech.2005.01.018

Chapter 15
Solid- and Nano-Catalysts Pretreatment and Hydrolysis Techniques

Guo Feng and Zhen Fang

Abstract Conversion methods for lignocellulosic materials are essential for the production of biofuels and bio-chemicals via sugar platform biorefineries. Sugars are commonly derived by a two-step process involving liquid acid pretreatment and subsequent enzymatic hydrolysis. Recent advances in biomass pretreatment and hydrolysis with solid acid catalysts have shown that catalytic method can be a promising replacement for liquid acid hydrolysis. This chapter covers works concerning the pretreatment and hydrolysis of lignocellulosic materials using solid- and nanocatalysts. First, biomass pretreatment is introduced briefly. The properties and synthesis of solid acid catalysts, such as acid site density, acid strength, structure of supports, acid site distribution, and tolerance to water are introduced in detail. Influences of reaction conditions on hydrolysis efficiency are also summarized. Moreover, the chapter discusses obstacles in the applications of solid acid catalysts for biomass pretreatment and hydrolysis. Suggestions are given for promoting catalytic efficiency, recycling, and regeneration of solid acid catalysts. Finally, nanosized solid catalysts are introduced and discussed that can promote biomass pretreatment and hydrolysis.

Keywords Biomass · Pretreatment · Cellulose · Solid catalysts · Magnetic nanocatalysts · Hydrolysis · Glucose

15.1 Introduction

Lignocellulosic biomass is mostly composed of cellulose, hemicellulose, lignin, and some extractives [1]. Cellulose, a straight chain biopolymer that is insoluble in water, is composed of cellobiose unit formed from D-glucose via β-1, 4 glycosidic bonds. With annual net photosynthesis yield of approximately 720 billion tonnes, cellulose is the world's largest organic raw material resource, and the most potential alternative for petroleum resource [2]. Cellulosic biomass, such as agricultural and forestry residues, is often difficult to be hydrolyzed directly by enzymatic hydrolysis due to the crystalline form of cellulose and the hydrogen bonds involved as well as lignin that binds the cellulose [3]. These are the main hurdles that significantly

Z. Fang (✉) · F. Guo
Biomass Group, Chinese Academy of Sciences, Xishua-ngbanna Tropical Botanical Garden, 88 Xuefu Road, Kunm-ing, Yunnan province 650223, China
e-mail: zhenfang@xtbg.ac.cn

Z. Fang (ed.), *Pretreatment Techniques for Biofuels and Biorefineries*, Green Energy and Technology,
DOI 10.1007/978-3-642-32735-3_15,

hinder their industrial applications through inexpensive and less energy-intensive biorefinery processes [4].

Hydrolysis methods for biomass are essential for the production of biofuels and bio-chemicals via sugar-platform biorefineries. Liquid acid pretreatment followed by enzymatic hydrolysis and fermentation is one of the most common processes for cellulosic ethanol production [5]. The major issues in processes with liquid acids are: (1) the pretreatment leading to secondary degradation products that inhibit subsequent hydrolysis and fermentation; and (2) the recovery and recycle of the used acid solutions. Another method is direct chemical catalytic hydrolysis that is primarily conducted over catalysts with proper acidity/basicity, such as inorganic or organic acids and various solid acids [6]. Enzymatic hydrolysis of cellulosic biomass has got much attention because it produces better yields than acid-catalyzed hydrolysis. However, it is a very low process and results in difficulty in the production of concentrated saccharide solution. When a large amount of cellulase was used, cost was also increased with it. Recent advances in biomass hydrolysis with solid acid catalysts have shown that they can be promising replacements for harsh liquid acids for pretreatment and hydrolysis. Solid acid catalysts have many advantages over liquid catalysts regarding activity, selectivity, catalyst life, and ease in recovery and reuse [7–10]. However, it is a challenge to develop hydrothermal catalytic hydrolysis processes with solid acid catalysts.

Innovation and breakthrough in hydrolysis are keys to the commercialization of solid acid catalysis. The development of novel highly acidic solid catalysts with nanometer size and special characteristics (e.g., magnetic properties) is a key issue for the effective hydrolysis of biomass. Several reviews concerning the conversion of hemicellulose and lignin have been reported [11–14]. Moreover, several comprehensive reviews were published recently which covered the details of enzymatic hydrolysis approaches [1, 15–18]. This chapter covers works concerning the pretreatment and hydrolysis of lignocellulosic materials using solid- and nano-catalysts. Moreover, the article discusses obstacles in the applications of solid acid catalysts for cellulose hydrolysis.

15.2 Biomass Pretreatment

Pretreatment is the first step in the conversion of lignocellulosic biomass to biofuels and chemicals. The purpose of pretreatment is to break down the lignin that binds cellulose and to destroy the crystalline structure of cellulose and increase its surface area so that fragments become accessible to chemical or enzyme active sites [1, 13]. Pretreatment is the most expensive step in the production of cellulosic ethanol by enzymatic hydrolysis and fermentation. The conventional methods of pretreatment are physical, chemical, physical-chemical, and biological pretreatments. Typical physical pretreatment includes chipping, grinding, milling, and thermal methods. Mais et al. [19] pretreated cellulose by milling, and subsequently obtained up to 100 % hydrolysis yield with a relatively low enzyme loading. However, ball-milling consumes significant energy. It is widely recognized that chemical pretreatment is more economic.

15.2.1 Chemical Pretreatment

Chemical pretreatment is a widely used method. It effectively removes and recovers most of hemicellulose portion as soluble sugars, and disrupts lignin to be partially dissolved in aqueous acidic/basic solution (e.g., H_2O_2 and ammonia) [1]. Chemical pretreatment with acids has proven to be effective for breaking down hydrogen bonds leading to intra-crystalline cellulose swelling. Acid pretreatment is a process in which hydronium ions break down or attack inter- and intra-molecular bonds among cellulose, hemicellulose, and lignin. It increases the porosity of substrate and accessibility of cellulose to enzymes for subsequent hydrolysis. During acid pretreatment process, very little cellulose is hydrolyzed. However, the process is usually accompanied by further degradation of monomers, and it has drawbacks such as equipment corrosion and issues in the recovery and recycle of the acid.

Alkali (e.g., NaOH, KOH) solution is a swelling agent for both crystalline and amorphous celluloses that can destroy the linkages between lignin and carbohydrates by saponification of intermolecular ester bonds [20]. Kumar et al. [21] reported that NaOH pretreatment increased hardwood digestibility from 14 % to 55 % by reducing lignin content from 24–55 % to 20 %. Some chemical agents, such as peroxides and acidic alcohol solutions, have advantages for dissolving lignin and loosening hemicellulose from insoluble crystalline cellulose. Pan et al. [22] pretreated woody biomass with an organic solvent (1.25 % H_2SO_4 and 50 % ethanol) under conditions of 180 °C for 60 min, about 75 % lignin was removed from the substrate.

Ionic liquids (ILs) are efficient for the pretreatment and hydrolysis of lignocellulosic materials, and can dissolve biomass and overcome many of the physical and biochemical barriers for hydrolysis at ambient conditions. Many ILs have been shown to be effective solvents for cellulose. After IL pretreatment, glucose yield from subsequent cellulose hydrolysis is greatly increased with 97 % enzymatic glucose conversion rate being reported [23].

15.2.2 Pretreatment with Solid Acid Catalysts

Solid acids are widely used for the study of biomass hydrolysis that will be introduced in Sect. 15.3 in detail. The hydrolysis mechanistic route of a water-soluble polysaccharide consists of the following steps: (1) The soluble polysaccharide diffuses onto the surface acidic sites of a solid acid; (2) β-1,4 glycosidic bonds access to the acidic sites; and (3) These bonds cleave randomly, and the polysaccharide is hydrolyzed to mono-sugars. Therefore, ideally, cellulosic materials need to be dissolved in a solvent and converted into short-sugar chains to make full use of the acid sites.

Liquid acid has some important limitations including the corrosion of reactor, recycle, and neutralization of the acid for microbial fermentation. Formation of degraded products and release of fermentation inhibitors are other characteristics of liquid acid pretreatment.

Unlike hydrolysis, in the pretreatment process, few mono-sugars are formed. Therefore, it performs under milder conditions, such as lower temperature, short processing time, and weaker acidity that can be selectively controlled. After pretreatment, biomass can be further hydrolyzed by either enzymatic or chemical method. Solid acid catalysts mediated pretreatment is superior in the destruction of biomass structure as compared with traditional liquid acid pretreatment because it achieves low reaction rate, and fewer side-reactions occur. After pre-hydrolysis of cellulose, the formed oligosaccharides can extend into the vicinity of Brønsted or Lewis acid sites where catalytic hydrolysis occurs. Pre-converting cellulose into oligomers is important to the subsequent hydrolysis with solid acid catalysts.

Solid acid pretreatment is a novel method for biomass pretreatment. Although little work is reported, it can be considered as the initial stage or lesser degree of hydrolysis, and is easily controlled by changing the parameters of solid acid hydrolysis conditions. With the development and application of solid acids in biomass hydrolysis, they will find applications in pretreatment. So, hydrolysis is focused on the following sections.

15.3 Hydrolysis with Solid Acid Catalysts

Chemical catalytic systems with liquid acids have the difficulty in separating the homogeneous catalysts from product solutions. Solid acid catalysts have many advantages over liquid catalysts. They are widely studied as direct replacements for liquid acids to reduce pollutants and operating costs. A solid acid catalyst is defined as solid that can donate protons (Brønsted, B acid) or accept electrons during reactions (Lewis, L acid). The catalytic function for a solid acid catalyst is derived from its acidic centers, existing mainly on its surface. Accordingly, solid acids with B acid sites can catalyze biomass hydrolysis. Because L acid catalysis of cellulose conversion was reviewed [24], this section focuses on B acid catalysis.

The mechanistic route of cellulose hydrolysis by solid acid catalysts consists of the following steps: (1) crystalline cellulose dissolves in some chemical agents, such as ILs; (2) the soluble polysaccharide diffuses onto the surface of the solid acid catalysts; (3) the polysaccharide undergoes hydrolysis over the acid sites; (4) some hydrolysis oligosaccharides diffuse into the internal pores of solid acid catalysts; (5) the poly-oligosaccharides undergo hydrolysis over the surface and internal acid sites; (6) the hydrolysis products, mainly glucose, diffuse into the reaction medium. Therefore, the properties of solid B acid catalysts, such as acid site density, acid strength, structure of supports, acid site distribution, and tolerance to water, have great influence on their activities and selectivities. Under ordinary circumstances, acid strengths and catalytic activities of solid acid catalysts decrease in the presence of water. Moreover, most solid acids do not function effectively for cellulose hydrolysis because the surfaces of these solids do not have strong acid sites or cannot allow their close contact to β-1,4-glucans. Therefore, it is a challenge to develop hydrothermal catalytic hydrolysis processes with solid acid catalysts. It is important to acquaint with the properties of solid B acid catalysts considerably.

15.3.1 Preparation of Solid Acids

Co-precipitation or filling supports with an aqueous solution of active precursor is a conventional method for preparing solid acid catalysts. Metal oxides are widely used as catalyst supports because of their thermal and mechanical stability, high specific surface area, and large pore size (15 nm) and pore volume (>0.2 mL/g) [25, 26]. Because solid acids function the same as $[H^+]$ for cellulose hydrolysis, sulfonated metal oxides, such as SO_4^{2-}/Al_2O_3, SO_4^{2-}/TiO_2, SO_4^{2-}/ZrO_2, SO_4^{2-}/SnO_2, and SO_4^{2-}/V_2O_5, can supply many acidic species. Such solid acids are usually prepared by impregnating the hydroxides from ammonia precipitation of corresponding metal salt solutions with aqueous sulfuric acids followed by calcination. One limitation of these types of solid catalysts is that the acidic sites are leached under hydrolytic conditions. It is difficult to control the catalyst particle size and shape, therefore, novel synthesis technology should integrate with conventional methods to resolve these issues. Fang et al. [27] have successfully prepared activated hydrotalcite nanoparticles by co-precipitation of $Mg(NO_3)_2 \cdot 6H_2O$ and $Al(NO_3)_3 \cdot 9H_2O$ in urea solution and subsequent with microwave-hydrothermal treatment. The particles were activated with $Ca(OH)_2$ and used to hydrolyze cellulose. X-ray diffraction (XRD) pattern indicated that it had layered and well-crystallized structures with characteristic and symmetric reflections.

Carbonaceous solid acid catalysts are known to have one of the highest catalytic activities for cellulose hydrolysis. These catalysts were typically prepared from carbohydrates by carbonizing at 400 °C under N_2 and then sulfonating at 150 °C [28]. Glucose, sucrose, cellulose, lignin, and activated-carbon can be used as raw materials for their preparation [28–31]. The carbon in the catalysts is in amorphous forms consisting of polycyclic aromatic carbon sheets. All S-atoms in the catalysts are in $-SO_3H$ groups, which are the active sites. Carboxylic acid species, –COOH, generally provide more active sites than Nafion NR50 and Amberlyset-15 which could not help to hydrolyze cellulose into glucose. Pang et al. [29] reported that high glucose yield of up to 74.5 % with 94.4 % selectivity was obtained at 150 °C and 24 h. Lignin is the second-most abundant natural organic material after cellulose, and the richest aromatic organic biopolymer. It has high carbon content and should be usable as a precursor for activated carbon. Pua et al. [30] prepared a solid acid catalyst from Kraft lignin by treatment with phosphoric acid, pyrolysis, and sulfuric acid, and subsequently it was successfully used as catalyst to synthesize biodiesel from high acid value *Jatropha* oil. It is speculated that lignin derived carbonaceous catalyst is more advantageous for cellulose hydrolysis.

Homogeneous catalysis by heteropoly acids (HPAs) is in principle similar to sulfuric acid in that $[H^+]$ leaches into solution and interacts with the oxygen atoms in the glycosidic bonds of cellulose. However, recovery of the homogeneous catalysts is problematic. Cellulose hydrolysis using solid HPAs was reported by Tian et al. [32]. Several types of acidic cesium salts, $Cs_xH_{3-x}PW_{12}O_{40}$ ($X = [1–3]$), were prepared. The salt $Cs_1H_2PW_{12}O_{40}$ was found to give the highest glucose yield (30 %) at 160 °C for 6 h reaction time. $Cs_xH_{3-x}PW_{12}O_{40}$ catalysts were prepared by adding dropwise the required amount of aqueous solution of cesium carbonate to aqueous solution of $H_3PW_{12}O_{40}$ with cesium content ranging from 1 to 3 at room temperature with stirring. After the resultant milky suspension was aged at room temperature overnight, the solution was slowly heated at 50 °C to obtain white solid powers. It was found that $Cs_1H_2PW_{12}O_{40}$, with strong protonic acid sites, showed the best catalytic performance in terms of the conversion of cellulose and the yield for glucose. $Cs_{2.2}H_{0.8}PW_{12}O_{40}$ showed the highest selectivity in terms of glucose, which is due to its micro-porous structure.

For industrial applications, low cost catalysts with good performance are required. The catalyst cost can be reduced by selecting cheap materials and simple preparation process. Relatively speaking, carbonaceous solid acid catalysts are considered as the cheapest catalyst, since they were obtained from biomass (such as glucose, cellulose, lignin, wood et al.) by a simple process of carbonization and sulfonation.

15.3.2 Acid Site Density

Some supported solid acid catalysts (e.g., sulfonated carbon based solid acid, sulfonated metal oxides and sulfonated activated-carbon) showed good activity in cellulose hydrolysis [24, 33, 34]. Supported solid acid catalysts are promising for the depolymerization of cellulose in water, since they have substantial surface acidic species (e.g., 1.5 mmol/g; sulfonated activated-carbon) as compared with zeolites (e.g., 0.3 mmol/g; HSM-5B) and transition-metal oxides (e.g., 0.3 mmol/g; Nb_3W_7 oxide) [35, 36], and specific functional groups. The active species of protons $[H^+]$ in such catalysts are more accessible to the β-1,4-glucans in cellulose than L acid sites [37].

Solid acid catalysts, especially strong acids such as sulfated metal oxides (e.g., SO_4^{2-}/Al_2O_3, SO_4^{2-}/TiO_2, SO_4^{2-}/ZrO_2, SO_4^{2-}/SnO_2 and SO_4^{2-}/V_2O_5), are being studied extensively in an effort to replace liquid acid catalysts. Sulfated metal oxides are solid acid catalysts with the combination of B and L acids in a certain form. Formation of proton acid centers is related to the adsorption of hydroxyl groups or H_2O on SO_4^{2-}. However, most acid sites are mainly formed by coordination adsorption of SO_4^{2-} on the surface of metal oxides. The coordination adsorption makes strong migration of the electron cloud in metal-oxygen bond, leading to strengthening the L acid center. Many studies have proposed that water, present in biomass or produced as a reaction product, converts L acid sites to B acid sites. Therefore, in the hydrolysis of cellulose into glucose, B acid sites play an important role.

The relative activity of B acid sites depends on many factors, such as the structure of supports, the nature of reactions and the polarity of reaction media. Supported acid catalysts are the most extensively studied ones for organic synthesis [38]. As the same kind of catalysts, the number of strong B acid sites is correlated with the catalytic activity. Zhang et al. [39] studied the skeletal isomerization of *n*-butane and found that the catalytic activity of $H_4SiW_{12}O_{40}/SiO_2$ reached a maximum when the loading amount of $H_4SiW_{12}O_{40}$ was 50 wt%. The density of strong B acid sites increased with loading of $H_4SiW_{12}O_{40}$ on SiO_2 (10–50 wt%). It was concluded that the direct interaction of $H_4SiW_{12}O_{40}$ with the surface of SiO_2 caused strong B sites to form in the second-layer of $H_4SiW_{12}O_{40}$ exposed on the surface. However, in the previous study about the effect of support identity on B acid site density [40], it was found that the reactivity of B acid sites was not affected by the identity of supports. The turnover rate increased with the rising of density of acid sites on all supports.

The influence of supports on activity of B acid sites in cellulose hydrolysis still remains undiscovered. Sulfonic group functionalized magnetic SBA-15 catalyst (Fe_3O_4-SBA-SO_3H) [41], which gave high glucose yield (98 %) in cellobiose conversion, can be recovered for reuse by an external magnetic field. Fe_3O_4-SBA-SO_3H not only provides good access of reactants to the $-SO_3H$ groups, but also has functional characteristics that allow it to be separated and regenerated. Sulfated ZrO_2 [31] has similar active acid sites (sulfonic acid group: 1.2 vs. 1.09 mmol/g) but lower glucose yield (14 % vs. 98 %) as compared with Fe_3O_4-SBA-SO_3H. It still could not prove whether their activities were influenced by effect of supports on activity of B acid sites or by access of reactants to $-SO_3H$ groups. Suganuma et al. [42] reported that carbon materials can incorporate large amounts of hydrophilic molecules into the carbon bulk, due to the high density of the hydrophilic functional groups bound to the flexible carbon sheets. Compared with niobic acid, H-mordenite, Nafion, and Amberlyst-15 resins, the carbon catalyst had the highest catalytic activity because it can adsorb β-1,4-glucans, which are not adsorbed by the other four solid acids.

The activity of solid acid catalysts for cellulose hydrolysis was not only related to the density of B acid sites even for the same support. Onda et al. [31] studied the hydrolysis of cellulose into glucose using sulfonated activated-carbon as catalyst (acid site density of 0.58 mmol/g), and a glucose yield of 41.4 % was achieved at 150 °C for 24 h. They reported a similar glucose yield of 40 % under the same reaction conditions using the same sulfonated activated-carbon with 1.25 mmol/g acid site density [43]. The high catalytic activity of the sulfonated carbonaceous material was attributed to (1) its ability to absorb the β-1,4-glucans, (2) its large effective surface area in water, and (3) the presence of $-SO_3H$ groups that are tolerant to hydrolysis [44]. Vigier and Jérôme [33] found that an increase in loading of sulfonic sites (i.e., the proton concentration) on poly(tetrafluoroethylene-*co*-perfluorovinyl ether)-*graft*-polystyrenesulfonic acid (PFA-*g*-PSSA) membrane surface from 28 % to 63 % enhanced the catalytic activity from 7.5×10^{-3} to 27.5×10^{-3} min^{-1}. It was suggested that this increase in activity might be ascribed to the rise of the amount of accessible sulfonic sites and the catalyst hydrophilicity, thus improving the polymer chains mobility and therefore accessibility of the catalytic sites [24]. Similar phenomenon was obtained by Zhang and Zhao [45] who used different

H-form zeolite catalysts (i.e., HZSM-5a, HZSM-5b and H-beta) for cellulose hydrolysis with microwave-heating at 240 W. HZSM zeolites had a higher glucose yield (35 %) than that of H-beta zeolite (30 %) because of their higher acidity and more reactive sites. Therefore, disregarding the accessibility of catalytic sites to β-1,4-glucans in cellulose, hydrolysis yield is positively related to acid site density.

15.3.3 Strength of Solid Acid Sites

The concentration and strength of acid sites in solid acid catalysts are usually measured by amine titration, infrared spectroscopy (IR), temperature-programmed desorption (TPD), solid-state nuclear magnetic resonance (NMR) spectroscopy and thermal analysis [46]. The traditional amine titration method can be used for quantitative analysis of acid amount. However, it could not accurately analyze the types of acid and the distribution of acidic centers on the surface of solid acid catalysts. The analysis of total acid amount and strength of solid acid catalysts using TPD method is very accurate, except for distinguishing acid types. As a supplementary method, IR spectroscopy can be used well to distinguish B- and L acids but not suitable for exact quantitative analysis of acid amounts. Therefore, the combination of TPD method and IR spectroscopy is more accurate to get acidic characteristics of solid acid catalysts. Currently-developed solid-state NMR spectroscopy can realize qualitative and quantitative analysis of acidic center distribution, acid amount, and acid strength, but the analysis is complex and costly.

The acidic strength and acid site density are determined by the preparation process conditions, such as calcination temperature, crystal form of oxides, and sulfonation process. The main functions of calcination are as follows: (1) change amorphous oxides into crystals; and (2) promote solid phase reaction between sulfuric acid and metal oxides, forming acid structure of strong bonding of sulfuric acid with metal oxides. An appropriate calcining temperature is conducive to have good porous structure, many acid sites and high acid strength. Loss of sulfur species, decrease of specific surface and crystal transformation will occur if calcination temperature is too high. On the contrary, the ideal acid structure, crystal and pore structure will not be obtained if calcination temperature is too low. Pang et al. [29] found that the higher the sulfonation temperature, the higher the acid density of the sulfonated carbon. The active carbon sulfonated at 300 °C (AC-SO_3H-300) showed an acid density of 2.19 mmol/g, which was 15-fold that of the untreated active carbon. However, a higher temperature (>250 °C) caused the decomposition of sulfonate. The highest glucose yield of 74.5 % was achieved by AC-SO_3H sulfonated at 250 °C, which possessed the highest density of –SO_3H groups.

Mesoporous metal oxides with high catalytic efficiencies have the following special structure properties: (1) high specific surface area, (2) adjustable pore size, and (3) enhanced thermal stability [47]. However, it has been shown that highly crystalline pore walls and high mesoporous order cannot always be achieved for the same material. High-temperature heat treatment helps to increase the crystallinity of pore walls, but leads to the collapse of mesoporous structure. Many reports on sulfated metal oxides showed that the amorphous metal oxides are needed for preparation of solid superacids except for γ-Al_2O_3 [47].

During the preparation process of metal oxides, the pH should be controlled by adding precipitation agent to obtain the precipitates of metal hydroxides. Ammonia and urea are the most used precipitation agents. The precipitation process using ammonia is more slowly, thus improving crystal growth efficiency. B acid sites (proton donors) can be generated from highly polarized hydroxyl groups. They can also be formed on oxide-based catalysts via proton balance of a net negative charge introduced by substituting cations with a lower valence charge [48]. Sulfide type should be considered as the main factor in acid strength, acid type, and catalytic activity. Commonly-used sulfides include H_2S, SO_2, SO_3, CS_2, H_2SO_4, $(NH_4)_2SO_4$ and benzenesulfonic acid, but only high-valence sulfides perform acid action. In order to obtain enough strong superacid centers, the optimum calcination temperature should be slightly lower than the decomposition temperature of acid sites. Besides sulfated single metal oxides, manipulation of sulfate content and activation temperature also provides the means for controlling the strength of surface B acid sites in the sulfated mixed metal oxides [49].

Certainly, not only sulfides are used as active sites. Metal oxide supported Pt or Ru showed highly-active for converting cellulose to sugar alcohols with 31 % yield being reported [50]. They were active for catalyzing cellulose conversion and glucose isomerization, simultaneously with 88 % selectivity of glucose being obtained. It should be noted that they presented L acid catalysis.

15.3.4 Modification of Solid Acid Catalysts

Most solid acid catalysts presented excellent catalytic active in the first run. However, a considerable loss in catalytic activity occurred when the catalyst was recycled for several times [51–53]. Hegner et al. [51] reported that glucose yield from the hydrolysis of cellulose with $FeCl_3$/silica was 11 %, while for the second and third cycles, it reduced to 8 % and 7 %, respectively. Dhepe et al. [52] studied the hydrolysis of sucrose using water-tolerant sulfonated mesoporous silicas. The reactions were tested at 80 °C for 24 h up to three recycles, and no decrease in activity was found. However, the catalysts were not tested for cellulose hydrolysis which needs a higher reaction temperature (~150 °C). Up to now, an amorphous carbon bearing –SO_3H, –COOH and –OH groups showed the best recyclability among the reported results [42]. No decrease in activity was observed even after 25 cycles of the catalyst (total reaction time, 150 h). Although such sulfonated carbons are highly efficient for cellulose hydrolysis, there are needs for their improvement in the areas of separation and recovery from un-hydrolyzed cellulose residues.

A modification technology to increase the surface area, mechanical strength, and stability of catalyst supports and improve the acid density, strength, and recyclability of acid sites is required. The modification usually starts with discovering the cause of catalyst deactivation and seeking cheap raw materials to reduce the cost of catalyst preparation. Reusability is a unique character of solid catalysts as distinguished from liquid catalysts, also is the key to reduce the cost of catalytic process. Metal oxides, zeolites, cation-exchange resins, and carbonaceous solid are common supports of solid acids. Herein, modification of catalyst supports is discussed in detail.

15.3.4.1 Metal Oxides

Metal oxides are usually used to synthesize solid superacids as introduced in Sect. 15.3.1. Active sites supported on single metal oxides performed high catalytic activity in many organic reactions [54–56]. Sulfonated metal oxides, such as SO_4^{2-}/Al_2O_3, SO_4^{2-}/TiO_2, SO_4^{2-}/ZrO_2, and SO_4^{2-}/V_2O_5 [57, 58], can supply much acid species, which function the same as $[H^+]$ in sulfuric acid for cellulose hydrolysis. Such acid solid catalysts are widely used in biodiesel synthesis, but it is generally very difficult to retain a strong B acid site as sulfonic acid in the framework. Jitputti et al. [57] studied the transesterification of crude oil by different solid acids, such as SO_4^{2-}/SnO_2, SO_4^{2-}/ZrO_2, and resulted in the production of over 90 % fatty acid methyl esters (FAMEs; biodiesel). It was found that the spent SO_4^{2-}/ZrO_2 could not be directly reused for the transesterification. Such solid acid catalysts cannot be used for cellulose hydrolysis at the current structure and composition. Kulkarni and Muggli [59], reported that an apparent increase in B acidity was found upon treating SO_4^{2-}/TiO_2 with H_2O, which proved that H_2O displaced isopropylamine from approximately one-third to one-half of the B acid sites.

In order to improve the catalytic efficiency and stability of sulfonated metal oxides, some modifications are usually proposed as below: (1) Introducing other metals or metal oxides. Metals that promote the catalytic activity include aluminum, iron, and manganese, and platinum can also increase the stability of solid catalysts [60, 61]. (2) Introducing lanthanum. Lanthanun was widely used to improve the stability of active sites [62]. (3) Synthesizing nanoparticles. Nano-catalysts have the advantage of supplying larger surface area, good stability, and high effect active, which will be introduced in detail in Sect. 15.4.

15.3.4.2 Zeolites

Zeolites are widely used as catalysts in organic synthesis, as they are non-toxic and non-corrosive, and easy to recover for reuse. They can be synthesized with different crystal structures and definitive pore size, and adjusted acid centers to have some important catalytic properties. The number of B acid sites in H-form zeolites is related to the atomic ratio of Al/Si. Acidic zeolites have been successfully used for conversion of sucrose into mono-sugars at mild conditions. Zhang and Zhao [45] performed cellulose hydrolysis with H-zeolite in IL solvents heating at 100 °C by oil bath, and a glucose yield of 2.1 % was achieved after 10 h. After cellulose was milled, a glucose yield of 12 % was obtained using H-ZSM-5 catalyst, but the catalytic efficiency is still lower than that of other solid acid catalysts. For cellulose hydrolysis with an H-zeolite catalyst, cellulosic materials need to be dissolved in a solvent, and be converted into short-sugar chains to make full use of B acid sites in internal channels of the zeolite. Four modifications were proposed to improve the catalysts [63–68]: loading of cations, synthesis of zeolites with super-large pores, loading of super acids, and synthesis of composite zeolites. Herein, other kinds of modification technologies are proposed further:

1. *Synthesis of nanosized zeolites* Confined space synthesis is a novel method in zeolite preparation. It involves the crystallization of zeolite inside the pore system of an inert mesoporous matrix [69]. By using this technology, Schmidt et al. [69] synthesized ZSM-5 with Al/Si ratios of 0, 0.02 and 0.01 which have controlled-average crystal size in the range 20–75 nm. Nanosized zeolite suspension has characteristics of a fluid solution to provide more active sites per gram. By means of adjusting the particle size, some desired properties can be obtained for nanoparticles. However, recycling of nanoparticles can be seen as a research barrier due to the adsorption, agglomeration, and viscous effects of the reaction mixture.
2. *Introduction of lanthanum* La as an effective element to adjust acidity of zeolites can maintain a certain amount of acid strength, and significantly improve the water-tolerance and thermal-stability of the zeolites. Yu et al. [70] studied the influence of lanthanum and cerium cations on the stability of Y zeolite. It was confirmed that the stability of Y zeolite was enhanced by the introduction of La. A strong interaction between the rare earth species and Y zeolite clusters was also found. Their small pore diameters limited the accessibility of acid sites to β-1,4-glucans in cellulose, leading to a poor performance in cellulose hydrolysis. Therefore, synthesis of zeolites by combination of La modification and aperture increase maybe a good solution. The tolerance to hot-water is also an important factor in catalytic applications of zeolites. Therefore, the combination of above two proposed technologies is necessary to synthesize stable zeolites for cellulose hydrolysis.

15.3.4.3 Cation-Exchange Resins

Cation-exchange resins have been used commercially as solid acid catalysts in esterification, alkylation, hydration/dehydration, isomerization, oligomerization, and condensation reaction [71]. Amberlyst® 15 and Nafion® resins are the mostly applied catalysts in organic catalysis. These resins are both chemically and thermally stable (up to 280 °C). Amberlyst-15 is an effective catalyst for the selective conversion of cellulose to glucose. More than 25 % glucose yield was obtained under reaction conditions of 150 °C for 24 h with Amberlyst® 15 (50 mg) from milled cellulose (45 mg), and distilled water (5.0 mL) [72]. Nafion® NR50 has similar acidic character but better thermal stability as compared with Amberlyst-15. However, such resins have obvious drawbacks such as low surface area and activity in both aqueous solvent and gas phases [72]. Kim et al. [73] proposed a strategy to enhance the overall process efficiency by combining of 1-*n*-butyl-3-methylimidazolium chloride {[BMIM][Cl]} and Nafion® NR50, which resulted in 35 % glucose yield.

These resin catalysts have high surface area (up to 539 m^2/g) and controllable hydrophobicity, but their acidic concentration is still blocked by the limitation of sulfonation in phenolic rings due to the strong steric hindrance [74]. The hydrophobic feature of resin catalysts limited their efficiency in cellulose hydrolysis. Therefore, special solvent like, ILs is needed to dissolve cellulose. Qi et al. [75] using a strong acidic cation-exchange resin to achieve high glucose yield of 83 % from cellulose hydrolysis in 1-ethyl-3-methyl imidazolium chloride [EMIM][Cl] with gradual addition of water.

Many efforts focused on the design and synthesis of nanosized resins. Via an in-situ sol–gel route [76], Nafion/SiO_2 with high specific surface area was synthesized, and further exposure of more acid sites is obtained. The feasibility of using Lewatit FO36 nano ion-exchange resin was investigated as adsorption for removal Cr (VI) from aqueous solutions using batch technique under various conditions [77]. Nanosized resins have advantages of allowing the β-1,4 glycosidic bonds in cellulose to become more accessible to catalytic sites. The other method to improve catalytic efficiency is the modification of cation-exchange resins using metallic salt. Chen et al. [78] demonstrated that gallium sulfate modified strong acid cation ion-exchange resin was highly active for the synthesis of butyl lactate with conversion rate of 92 % in 70 min. Moreover, a cation-exchange resin modified by incorporation of magnetic components (e.g., iron, cobalt, and nickel) can be used to separate resin catalysts easily.

15.3.4.4 Carbonaceous Support

Of all the solid acid supports, carbonaceous support seems to be the most effective. It was reported that $-SO_3H$ groups are linked to amorphous structure in the form of $C-O-SO_3H$ [31]. As mentioned in Sect. 15.3.1, a carbon material can incorporate large amounts of hydrophilic molecules, which provide good access to β-1,4 glucans by $-SO_3H$ groups, giving high catalytic performance for hydrolysis [42]. The incorporation results in the decease of activation energy for the hydrolysis of cellulose.

Suganuma et al. [42] reported that carbonaceous solid acids obtained at mild carbonization temperatures (ca. 450 °C) exhibited high catalytic activity, because $-SO_3H$ groups were bonded to carbon sheets with poor cross-linking, which made it easy for access to reactants by the $-SO_3H$ groups. However, when carbonized at higher temperature (ca. 550 °C), most of the $-SO_3H$ groups bonded to the carbon sheets were not located on surface, which resulted in a poor catalytic performance. Onda et al. [31] used sulfonated activated-carbon catalyst to selectively hydrolyze cellulose at 150 °C for 24 h, and obtained 40.5 % glucose yield with 90 % glucose selectivity. Only few amount of SO_4^{2-} was leached (<0.03 mmol/L) after the reaction.

Carbonaceous solid acid catalysts ground to nanosize (10–100 nm) could achieve high catalytic activity. Vyver et al. [79] proposed an effective conversion technique for cellulose hydrolysis by using a sulfonated silica/carbon nano-composite as catalyst. Glucose formation was faster on the nano-composite as compared with ion-exchange resins (glucose yield of 50 % vs. 29 %) under reaction conditions of 150 °C for 24 h with 0.05 g of ball-milling cellulose, 0.05 g of catalysts, and 5 mL of water. The catalyst was superior in the formation rate of glucose (4.6 μmol h^{-1}), and turnover frequency (0.37 h^{-1}) as compared with typical sugar catalysts (formation rate of glucose, 0.93 h^{-1}; turnover frequency, 0.06 h^{-1}). However, separation and recovery of it from un-hydrolyzed cellulose residues are needed for further study. Lai et al. [41] developed a process for recovering carbonaceous solid catalysts by using a paramagnetic solid acid (Fe_3O_4–SBA–SO_3H). When microcrystalline cellulose was pretreated with [BMIM][Cl], glucose yield reached 52 % in 3 h. The incorporation of paramagnetic nanoparticles into the carbonaceous carriers not only provides good access of reactants to the –SO_3H groups, but also has functional characteristics that allow it to be separated and regenerated.

Carbonaceous solid acid catalysts are considered as the most promising catalyst for cellulose hydrolysis, since they provide good access of reactants to the acidic sites of $-SO_3H$ groups. High glucose yields of up to 75 % with 80 % selectivity have been achieved at 150 °C for 24 h with carbonaceous solid acid catalysts. However, separation of carbonaceous solid acid catalysts from un-hydrolyzed cellulose residues after hydrolysis needs further research since these catalysts have similar physical and chemical properties to the residues. Use of functionalized carbonaceous solid acid catalysts that contain paramagnetic groups is one method to improve carbonaceous solid acid catalysts' separation and reuse.

15.3.5 *Influence of Reaction Conditions*

Heterogeneous catalysts promote reactions at the active sites on their surfaces, so the accessibility of cellulose to internal acidic sites and the dispersion limitation are important limiting factor for the catalytic activity. Heterogeneous catalysts can offer technical advantages related to stability, separation, handling, recycling of catalysts, and reactor design. However, none of the catalysts (metal oxides, zeolites, supported acids, and cation-exchange resins) are highly active and selective under mild conditions [80]. Reaction conditions such as reaction temperature, catalyst loading, and reaction medium are required to be optimized for cellulose hydrolysis with solid catalysts to achieve practical applications.

15.3.5.1 Temperature

Hydrolysis of cellulose is highly dependent upon reaction temperature. Dilute sulfuric acid exhibited highly catalytic performance for cellulose hydrolysis at temperatures above 180–240 °C, but efficient hydrolysis did not occur at 100 °C [81]. When supercritical water was used as a medium for hydrolysis, harsh conditions such as high temperature (e.g., 400 °C), and high pressure (e.g., 30 MPa) were required

Table 15.1 Comparison of loading of solid acid catalysts for cellulose hydrolysis to glucose

Catalyst	Amount	Cellulose amount	Reaction temp. (°C)	Reaction time (h)	Glucose yield (%)	Ref.
Nafion-50	0.1 g	0.2 g	160	4	35	[35]
$FeCl_3$/silica*	0.47 g	2.0 g	190	24	9	[35]
Amberlyst 15	50 mg	45 mg	150	24	25	[43]
Fe_3O_4–SBA–SO_3H	1.5 g	1.0 g	150	3	50	[41]
Cs-HPA	0.21 g	0.1 g	170	8	39	[84]

* carried out in [BMIM][Cl].

[82, 83]. Most solid acid catalysts have higher activation energy for cellulose conversion into glucose than that for sulfuric acid (170 kJ/mol) under optimal conditions except carbonaceous solid acid catalysts (110 kJ/mol) [41].

Most experiments were conducted at 120–190 °C [1]. Lai et al. [41] performed cellulose hydrolysis with sulfonic group functionalized magnetic SBA-15 catalyst (Fe_3O_4–SBA–SO_3H) in [BMIM][Cl] at 120 °C, and a 50 % glucose yield was obtained in 3 h. The low reaction temperature and short reaction time are attributed to the high solubility of [BMIM][Cl] to dissolve cellulose by disrupting hydrogen bonds among the molecules. Suganuma et al. [42] demonstrated that the formation rate of glucose and water-soluble β-1,4-glucans on carbonaceous acid catalysts increased exponentially with temperature from 60 to 120 °C. When temperature increased from 60–90 °C to 90–120 °C, the apparent activation energy decreased from 128 to 44 kJ/mol. At higher temperature (180 °C), the yield of glucose markedly decreased from 43 % to 3 % due to the formation of side-products, such as levoglucosan, cellobiose, maltose, levulinic, and formic acids [41, 42].

15.3.5.2 Catalyst Loading

Most solid acids do not function effectively for cellulose hydrolysis because the surfaces of these solids do not have strong acid sites or the acid sites are hard to contact β-1,4-glucans. The type and amount of catalysts have a great impact on the conversion of cellulose. High catalytic acidity leads to higher reaction rate, so the most convenient method to improve hydrolysis efficiency is to load more catalyst to provide more acid sites access to β-1,4-glucans in cellulose. The adequate catalyst loading is necessary to obtain the maximum conversion rate of cellulose to glucose. Table 15.1 summarizes hydrolysis results over some typical solid acid catalysts in water (except for $FeCl_3$/silica). It can be seen that, catalyst weight is higher than that of cellulose in most of cases. Fe_3O_4–SBA–SO_3H is the most active one, and that is related to its high amount of B acid sites. Catalyst amount is closely related to reaction time and hydrolysis yield. It was found that less catalyst requires longer reaction time to achieve high conversion rate [4]. The more catalyst, the higher concentration of acidic sites, resulting in [H^+] attacking on β-1,4-glucans to produce more glucose.

15.3.5.3 Reaction Medium

Water is the mostly used medium for cellulose hydrolysis. It was also shown that catalyst activity and selectivity are closely determined by the amount of water. When the amount of water is close to the weight of solid catalyst, a maximum yield of glucose is obtained. However, a lower amount of water and longer reaction time (e.g., 24 h) mainly drive the reaction towards the formation of water–soluble β-1,4-glucans. ILs are efficient for the pretreatment and hydrolysis of lignocellulosic materials, and can dissolve biomass and overcome many of the physical and biochemical barriers for hydrolysis at ambient conditions. [BMIM][Cl] and 3-allyl-1-methylimidazolium chloride {[AMIM][Cl]} can dissolve 10 wt% cellulose at 50–100 °C [85]. Wang et al. [86] conducted a study on the extraction of cellulose from wood chips with [AMIM][Cl] and found that 62 wt% cellulose dissolved in [AMIM][Cl] under mild conditions. So far, water, organic solvents, ILs, and their mixtures are widely used as reaction media.

Zhang et al. [87] demonstrated that high yield of total–reducing–sugars (TRS) (60 %) was achieved when depolymerization of chitosan was performed in ILs in the presence of mineral acids. This might be due to the fact that the reaction medium 1-butyl-3-methylimidazolium bromine {[BMIM][Br]} reinforced the acidity of the mineral acids. A physical barrier for hydrolysis disappeared through the formation of a solution with [BMIM][Br]. With good solubility to dissolve cellulose, IL is a good medium for directly catalytic hydrolysis of cellulose with high efficiency. ILs can promote the dissolution and dispersion of cellulose molecules, leading to the complete mixture between cellulose and acidic sites in a homogeneous phase. Relatively, small cations are often efficient in dissolving cellulose. In another work [88], when hydrolysis reaction was carried out in [BMIM][Cl] in the presence of 7 wt% HCl at 100 °C under atmospheric pressure for 60 min, TRS yield was 66 %, 74 %, 81 %, and 68 % for the hydrolysis of corn stalk, rice straw, pine wood, and bagasse, respectively. The high TRS yield is due to the dissociated Cl^{-1}/Br^{-1} and the electron-rich aromatic system of [$BMIM^{+}$] weakening the glycosidic linkage to facilitate hydrolysis.

Water addition was found to have a significant impact on the degree of cellulose hydrolysis, because water acts both as a reactant (for producing monosaccharides) and an inhibitor [for producing 5-hydroxymethylfurfural (5-HMF)] in the overall cellulosic conversion. As mentioned above, Qi et al. [75] developed an effective conversion technique for transforming cellulose into 5-HMF via a two-step process. In the first step, high glucose yield (83 %) was obtained from the cellulose hydrolysis by a strong acidic cation-exchange resin in [EMIM][Cl] with gradual addition of water. In their study, cellulose firstly dissolved in [EMIM][Cl]. Then a certain amount of water and the cation-exchange resin was added. Hydrolysis reaction started when the system was heated to 110 °C. Compared with one-time addition of water, glucose yield was improved by adding water to [EMIM][Cl] system during reaction. In the hydrolysis step, increasing the amount of water could increase the yield of monosaccharides, and the maximum yield of 25 % (monosaccharides + 5-HMF) was obtained when H_2O/cellulose molar ratio was 10 [89]. However, excessive water addition caused cellulose to precipitate from IL.

15.3.6 Recycling and Regeneration

Many solid acid catalysts, such as cation-exchange resins [76], carbonaceous solid acids [79], and solid HPA catalysts [32], have shown high activity and selectivity for cellulose hydrolysis. Up to now, amorphous carbon bearing $-SO_3H$ showed the best recyclability among the reported results, and no decrease in activity was observed even after 25 cycles [42]. However, all these solid acid catalysts still suffered deactivation after a period of time, so the operation life of the catalysts still cannot meet the requirement for industrial applications. Catalyst deactivation is a problem that must be solved in the hydrolysis reaction. Fundamental understanding of the deactivation mechanisms during cellulose hydrolysis is the key to extending catalyst lifetime. The main deactivation mechanisms of solid acid catalysts are: (1) leaching of surface acid sites, (2) carbon deposition on the catalyst and poisoning by toxic substance, and (3) surface reconstruction.

Leaching of acid sites, especially sulfonic or sulfuric acid species, limits the reusability of these families of catalysts. It has been reported that all the reactants and products could cause leaching of such active species (especially H_2O) even at low temperatures (e.g., 100 °C) [90]. The leaching seems unavoidable under these operation conditions. Therefore, such kinds of solid acid catalysts are not proposed to be used in reactions in aqueous solutions. Moreover, it is rather complicated to regenerate them by a pickling technique. It is well known that Al, Fe, Ni, and Pt are effective promoters to increase the stability and activity of sulfated metal oxide catalysts [91]. The promoters improve sulfate contents and acidities of solid acid catalysts.

Carbon deposition represents a significant effect on the recyclability of a viable catalyst for cellulose hydrolysis. Shi et al. [92] prepared a series of Al-promoted $SO_4^{2-}/ZrO_2/SBA\text{-}15$ catalysts and investigated the deactivation and regeneration capacities of the catalysts during the dehydration of xylose. It was found that when the catalysts were reused without regeneration, the yield of furfural decreased from 52.7 % to 19.1 %. Based on the characterization of the catalysts, the accumulation of byproducts was the main reason for the deactivation. Regeneration with H_2O_2 can completely recover the catalytic activity of the deactivated catalysts. After first regeneration, the catalytic activity recovered completely. The corresponding xylose conversion rate and furfural selectivity were 98.6 % and 53.5 %, respectively, very close to those with the fresh catalysts (98.7 % and 53.4 %, respectively). During the hydrolysis of cellulose, the produced glucose was polymerized into carbonaceous polymers under certain hydrothermal conditions [93]. The carbonaceous polymers were further deposited on the surface of solid acids, which decreased the catalytic efficiency. The deactivated catalysts can be easily regenerated by calcination to remove the deposited coke. However, some catalysts (e.g., carbonaceous solid acid catalysts) cannot be treated by calcination at high temperature because both deposits and catalysts themselves were combusted. As mentioned in Sect. 15.3.4.4, recovery of carbonaceous solid catalysts by incorporating paramagnetic compounds may be feasible.

Carbon deposition not only covers the active sites, but also binds to catalyst surface inducing a surface reconstruction and affecting activity. Especially for metal oxides, the diffusion of carbon into metal results in the formation of bulk metal carbide, causing the loss in both activity and selectivity [94]. The oxidative treatment was usually applied to regenerate used catalysts by removing carbonaceous phases. Saib et al. [95] studied the deactivation and regeneration of cobalt Fischer–Tropsch synthesis catalysts. The spent catalysts recovered their activity completely by oxidative regeneration.

15.4 Hydrolysis with Nano-Catalysts

Solid acid catalysts are promising for the conversion of cellulosic materials into soluble sugars and have the characteristics that they are environment-friendly and recoverable. However, the existing catalytic systems showed low efficiency, leading to high-energy consumption and generation of by-products. Nano-catalysts can be used to improve many aspects of solid acid catalysts. Nano-catalysts are defined as solid catalysts with particle size in the nanometer level (usually $\sim$100 nm), and mainly used in acid-catalyzed organic reactions, such as hydrogenation, oxidation, alkylation, transesterification, condensation, and polymerization. Nano-catalysts can be used for most of the reactions catalyzed by traditional solid acid catalysts. Compared with traditional solid acid catalysts, nano-catalysts usually showed improved structural and textural properties in terms of high active site loading, small crystalline size, high surface area and pore volume, and high catalytic activity and selectivity. Moreover, most traditional solid acid catalysts can be used for synthesizing nano-catalysts by supporting acidic sites on nano-carriers or directly preparing nano-scale particles. Up to now, nano oxides are mostly studied supports.

15.4.1 Preparation of Nano-Catalysts

Recently, novel synthesis methods for nano-catalysts and nanostructures have been widely reported. The synthesis of silica-based particles focuses on nanostructures of nano-ropes, nano-tubes, and paintbrushes [96]. Hydrothermal treatment with acid can be used to improve the order and stability of the nanostructures after synthesized with an aqueous ammonia solution. By sol–gel synthesis, silica nano-tubes can be bundled to form nanostructures known as paintbrushes. The sol–gel processing of nanoparticles is commonly performed by a so-called semi-alkoxide route in which inorganic compounds like hydroxides, acetates, carbonates, and chlorides, are used as sources to alkaline earth ions [97]. Nanosized transition metal components are usually synthesized using alkoxide and semi-alkoxide routes.

Hybrid nanocomposites will find applications in catalysts. Hybrid (inorganic–organic) nano objects and higher level nanostructured networks were obtained by novel equilibrium and non-equilibrium self-assembly approaches [98]. By controlling over the morphology of hybrid materials, controllable particle size and size distribution were achieved. Dong et al. [99] proposed a feasible and effective self-assembly method to synthesize different scale coordination polymers in highly dilutes solution. The nano and microscale particles gave better catalytic conversion rate (73 %) and selectivity in the hydroxylation of phenols than the bulk crystals.

Reversed micelle technique has been reported on micellular HPA [84]. The synthetic method was based on the reaction between the components dissolved in the lyophilic media and the reversed micelles. Cellulose was hydrolyzed for three continuous repeated runs under the same reaction conditions, and complete hydrolysis was achieved. The highest glucose yield reached with the micellular HPA was 60 % with 85 % selectivity for the three continuous runs. The catalyst was separated from the reaction mixtures by centrifugation.

Supercritical anti-solvent precipitation synthesis was recently used for synthesizing MnO_x–CeO_2 hollow spheres. As reported by Jiang et al. [100], a mixed solution of manganese acetylacetonate, and cerium acetylacetonate in methanol was injected into the precipitator. As the solution droplets contacting with supercritical carbon dioxide, the nanoparticles were precipitated. Then, the supercritical CO_2 was allowed to flow to remove the residual methanol. The system was further depressurized to atmospheric pressure, and the generated nanoparticles were collected. Finally, they were calcined in a muffle furnace, and the MnO_x–CeO_2 hollow nano-spheres with an average diameter of about 50 nm and a wall thickness of 10–20 nm were obtained.

Magnetic nano-catalysts with ordered or disordered array of nano-crystallites have attracted great attention recently because of their wide application potentials. However, it is still a big challenge to tune the magnetic nano-crystallites into three-dimensional regular aggregates with varied nanostructures. By ultrasonic-chemical precipitation synthesis, magnetite particles with 15 nm average diameter were obtained [101]. Under ultrasonic agitation, Fe_3O_4 precipitates were produced immediately by adding sodium hydroxide into mixture of $FeSO_4$ and $FeCl_3$ with Fe^{3+} and Fe^{2+} molar ratio of 1.5:1. Moreover, $C_{12}H_{25}OSO_3Na$ could be added as surface active agent, assisting to obtain Fe_3O_4 nanoparticles with homogenous size and shape distribution.

The development of highly acidic nano-solid catalysts that have special characteristics (e.g., paramagnetic properties) is an interesting area for developing practical systems for biomass hydrolysis. Through the combination of novel nano-catalysts preparation techniques, it is expected that chemical processes based on the hydrolysis of cellulose will be developed rapidly.

15.4.2 Supports of Nano-Catalysts

15.4.2.1 Nano-Silica

In recent years, silica-supported acid catalysts are gaining considerable attention because of their high activities due to large surface area, and high selectivities. Silica-supported acid catalysts have many advantages over liquid catalysts such as high mechanical and thermal stabilities, easy handling, low toxicity, easy separation, and reusability of the catalysts, which make them promising for both academic study and industrial application [102].

In a typical preparation process, first, silica was calcined at high temperature (e.g., 400 °C) and used as support. Silica-supported acid catalyst was then prepared by impregnating the silica support with calculated amount of an aqueous solution of acid functional groups to achieve the required loading. Ghanbaripour et al. [103] used nano-silica-supported $H_3PW_{12}O_{40}$ to catalyze the self-condensation of acetophenones. The products (1,3,5-triarylbenzenes) were obtained with high yield (80–93 %) in very short reaction time (18–30 min). Moreover, the catalyst can be used at least five consecutive times without significant loss in product yield.

Although nano-silica-supported acid catalysts have enormous potential application in cellulose hydrolysis, their improvement in the catalytic stability is needed. In most cases, nano-solid acid catalysts were prepared under mild conditions. As a result, surface passivation occurred, leading to the high density of surface active sites and high specific surface area [104]. However, large specific surface area and surface tension result in an absorption/aggregation process and affect the mass transfer. The combination of ILs and nano-solid acids can be used to resolve these issues, because ILs can help to dissolve cellulose and hemicellulose completely and facilitate catalysts to approach the β-1,4 glycosidic bonds in cellulose.

Unfortunately, some Si–O–Si links were easily ruptured by H_2O under moderate conditions and caused the formation of Si–OH groups [105]. Wang et al. [106] proposed an effective method to improve the surface chemical reactivity of a magnetic SiO_2 catalyst support. More Ti–OH groups were bonded on the silica surface due to the formation of Si–O–Ti. Therefore, the incorporation of metal oxides such as TiO_2, Fe_2O_3, and Al_2O_3 can greatly enhance the stability of nano-silica supported acid catalysts.

15.4.2.2 Nano-Ferric Oxide

Although nano-catalysts are highly efficient for cellulose hydrolysis, there are needs for their improvement in separation and recovery from un-hydrolyzed cellulose residues. As one possible method, ferric oxide could be introduced into supports to provide an efficient way for catalyst recovery. The fundamental prerequisite is the incorporation of paramagnetic nanoparticles into the carriers. Thus, ferric oxide supported catalysts not only provide good access of reactants to the active groups, but also have functional characteristics that allow them to be recovered and regenerated. As reported by Lai et al. [41], the paramagnetic nanoparticles (Fe_3O_4-SBA-SO_3H) can be easily recovered for reuse by magnetic filtration.

Ferric oxide, with advantages of magnetism, innocuity, high-temperature stability, and anti-aging properties has been widely used in the preparation of catalysts. It was reported that the activity and selectivity of nanoparticle-supported catalytic species were found to be comparable to their parent catalysts in solution or their counterparts immobilized on the solid-phase [107]. There are many techniques for preparing nano-ferric oxide, classified as wet method (oxidation precipitation, hydrothermal synthesis, gel–sol method and colloid chemical method), and dry method (thermal decomposition, chemical vapor deposition, low-energy deuterium-plasma exposure and laser pyrolysis) [108–114]. Wet method has drawn much attention due to its advantages of direct use, simple operation, and adjustable particle properties. By colloid chemical method, super-fine, uniform and global Fe_2O_3 particles could be obtained. However, plenty of organic agents were used, leading to pollution in the environment.

Incorporation of different methods, such as self-assembly technology derived from combination and reformation of reactant particles (molecules, atoms and ion), has been used in the preparation of nano-catalysts. However, it is difficult to resolve issues of poor dispersion and easy aggregation induced by calcination process under high temperature. Some of the important alternative routes may be the use of microwave and ultrasound as alternative energy sources for the synthesis [115]. Microwaves expedite the chemical transformation due to the selective adsorption of microwave energy by polar compounds or intermediates. The use of microwave irradiation as a heating source for catalyst synthesis is a rapidly growing research area. Under high-frequency electro-magnetic field, the produced inner heat source is helpful for the dissolution and recrystallization of $Fe(OH)_3$ gel. Sonication is the act of applying ultrasonic energy to agitate particles in a sample for various purposes. The use of ultrasonic irradiation during the synthetic procedure has been reported to help obtain smaller homogeneous nanoparticles and lead to an increase in surface area [116]. Increasing ultrasound power results in the growth of particle size due to the rise of energy in the micro-environment of localized super-saturation. It is expected that the nano-catalysts prepared by microwave/ultrasound irradiation are more stable in the microwave/ultrasound assisted cellulose hydrolysis.

15.4.2.3 Nano-Alumina

Alumina is now the most widely used catalyst support in the industry. The rapid development of nano-alumina in functionalization, controllable preparation, and conformation diversification has brought strong impact on the catalytic areas of traditional alumina. Nano-alumina is usually prepared by physical and chemical methods. Physical technology includes high-performance grinding, ball milling, vibration grinding, and ultrasonic methods. Nano-alumina prepared by chemical method involves the chemical reactions between ions and molecules, accompanied by the growth of nuclei. By using chemical method, the size and particle distribution could be effectively controlled to improve the structural and textural properties.

Nano-alumina catalysts have advantages of high selectivity, small particle size, and controllable size and aperture distribution. Boz et al. [117] prepared KF-impregnated nano-γ-Al_2O_3 as heterogeneous catalysts for the transesterification of vegetable oil with methanol, and methyl ester yield of 98 % was obtained with 15 wt% KF loading. The relatively high basicity of the catalyst surface (1.68 mmol/g) and the high BET surface area (about 12 m^2/g) are considered as the main reason for the high conversion rate. Nano-alumina could improve the thermal performance and acidity of the incorporated catalysts [118].

However, the nano-alumina crystal grain size had wide distribution and poor preparation repeatability. Furthermore, aggregation of nano-alumina affected its application in catalysis fields. Under high preparation temperature, the activity of nano-alumina-supported catalysts decreased due to the decline of surface area and the contraction of volume. There are several new preparation technologies to improve the catalyst properties, such as supercritical anti-solvent precipitation, ultrasonic-chemical precipitation, and combustion synthesis [119].

15.4.3 Magnetic Nano-Catalysts

Although some desired properties can be obtained by means of adjusting the particle size, recycling of nanoparticle materials can be seen as an application barrier due to adsorption and viscous effects of reaction mixture. Carbonaceous solid acids ground to nanosize (10–100 nm) had high catalytic activity for the cellulose that was similarly grounded with product selectivity higher than 90 % [120]. It is interesting to note that in some cases, the recycled catalysts exhibited no deactivation in the catalytic hydrolysis of fresh non-treated cellulose. In our research, the mixed metal oxide, Zn–Ca–Fe oxide, exhibited moderately good catalytic activity for hydrolyzing crystalline cellulose [121]. Cellulose conversion rate and glucose selectivity was 42.6 % and 69.2 %, respectively.

By introducing magnetic nano-components into nano-catalysts, it may be possible to separate the catalysts with an external magnetic field [122]. Magnetic supported nano-catalysts are highly efficient and environment-friendly. They have double functions of magnetism and acidity. They are obtained according to the following steps: (1) preparation of magnetic nanoparticles (magnetic core); (2) coupling anchor points on the surface of the nanoparticles; and (3) coordination of active sites to the anchor points [123]. By additional magnetic field, magnetic nano-catalysts disperse into liquid phase equably, avoiding the aggregation of nanoparticles and increasing the contact area between reactants and catalysts. In our work, hydrotalcite nanoparticles were synthesized and activated, and used for catalytic cellulose hydrolysis. Cellulose conversion rate and glucose selectivity of 46.6 % and 85.3 %, respectively, were obtained and remained stable for four cycles [27].

However, hetero-junctions between different active oxide and magnetic core lead to the decrease of activity as compared with single phase active oxide. Beydoun et al. [124] synthesized magnetic photocatalysts by coating TiO_2 particles on Fe_3O_4 magnetic nanoparticles. It was found that the photo–activity of titania-coated magnetite decreased. Moreover, nanometer-scale apertures are not large enough to allow the transport of sub-micron scale cellulose. Increasing the aperture in nanoparticles would facilitate mass-transfer of oligosaccharides to catalytic sites. Techniques of increasing aperture include change of synthesis conditions, addition of pore-formers and change of templates. The external surface hydrolysis reactions result in a non-shape-selective reaction as well as deposition of un-hydrolyzed cellulose, leading to short lifetime for the catalysts. These issues can be avoided by adopting other treatment methods such as ultrasound. To use the internal acid sites effectively, cellulose or its constituents can be dissolved in organic solvents or ILs for further reactions. It is expected that acid-functionalized magnetic nano-catalysts are promising materials for the hydrolysis of biomass.

15.5 Summary

Solid acid catalysts, which have favorable characteristics such as efficient activity, high selectivity, long catalyst life, and ease in recovery and reuse have great potential for efficiently transforming lignocellulosic biomass into biofuels and chemicals, and can replace many conventional liquid acids for hydrolysis and pretreatment. Besides specific surface area, pore size, and pore volume, the active site concentration and acidic type are important factors for solid acid performance. Solid acid catalysts being considered for biomass hydrolysis should have a large number of B acid sites, a good affinity for the reactant substrates, and good thermal stability. Catalyst composition, porosity, and stability in the presence of water are other important properties for solid acids in biomass hydrolysis. A good solid catalyst with sufficient catalytic activity combined with appropriate reactor design should make it possible to realize biomass hydrolysis on a practical scale. The development of highly acidic solid catalysts with nanometer size that have special characteristics (e.g., paramagnetic properties) is an interesting area of research for developing practical systems for biomass hydrolysis. In the near future, through the combination of green solvents, nanoparticle techniques, and functional solid acid catalysts, it can be expected that chemical processes based on the catalysis of biomass will begin to replace petroleum-based processes to reach a sustainable economy.

References

1. Sun Y, Cheng J (2002) Hydrolysis of lignocellulosic materials for ethanol production: a review. Bioresource Technol 83:1–11
2. Van de Vyver S, Geboers J, Jacobs PA, Sels BF (2011) Recent advances in the catalytic conversion of cellulose. Chem Cat Chem 3:82–94
3. Bellesia G, Asztalos A, Shen T, Langan P, Redondo A, Gnanakaran S (2010) In silico studies of crystalline cellulose and its degradation by enzymes. Acta Cryst D 66:1184–1188
4. Zakrzewska ME, Bogel-Łukasik E, Bogel-Łukasik R (2011) Ionic liquid-mediated formation of 5-hydroxymethylfurfural-a promising biomass-derived building block. Chem Rev 111:397–417
5. Mussatto SI, Roberto IC (2004) Alternatives for detoxification of diluted-acid lignocellulosic hydrolyzates for use in fermentative processes: a review. Bioresource Technol 93:1–10
6. Zhou CH, Xia X, Lin CX, Tong DS, Beltramini J (2011) Catalytic conversion of lignocellulosic biomass to fine chemicals and fuels. Chem Soc Rev 40:5588–5617

7. Takagaki A, Tagusagawa C, Hayashi S, Hara M, Domen K (2010) Nanosheets as highly active solid acid catalysts for green chemical syntheses. Energy Environ Sci 3:82–93
8. Reddy BM, Benjaram M, Meghshyam K (2008) Promoted zirconia solid acid catalysts for organic synthesis. Curr Org Chem 12:118–140
9. Jacobson K, Gopinath R, Meher LC, Dalai AK (2008) Solid acid catalyzed biodiesel production from waste cooking oil. Appl Catal B—Env 85:86–91
10. Helwani Z, Othman MR, Aziz N, Kim J, Fernando WJN (2009) Solid heterogeneous catalysts for transesterification of triglycerides with methanol: a review. Appl Catal A—Gen 363:1–10
11. Gírio FM, Fonseca C, Carvalheiro F, Duarte LC, Marques S, Bogel-Łukasik R (2010) Hemicelluloses for fuel ethanol: a review. Bioresource Technol 101:4775–4800
12. Pandey MP, Kim CS (2011) Lignin depolymerization and conversion: a review of thermochemical methods. Chem Eng Technol 34:29–41
13. Sannigrahi P, Pu Y, Ragauskas A (2010) Cellulosic biorefineries—unleashing lignin opportunities. Curr Opin Env Sust 2:383–393
14. Doherty WOS, Mousavioun P, Fellows CM (2011) Value-adding to cellulosic ethanol: lignin polymers. Ind Crop Prod 33:259–276
15. Mandels M, Hontz L, Nystrom J (1974) Enzymatic hydrolysis of waste cellulose. Biotechnol Bioeng 16:1471–1493
16. Galbe M, Zacchi G (2007) Pretreatment of lignocellulosic materials for efficient bioethanol production. Adv Biochem Eng/Biotechnol 108:41–65
17. Tengerdy RP, Szakacs G (2003) Bioconversion of lignocellulose in solid substrate fermentation. Biochem Eng J 13:169–179
18. Hosseini SA, Shah N (2011) Modelling enzymatic hydrolysis of cellulose part I: population balance modelling of hydrolysis by endoglucanase. Biomass Bioenergy 35:3841–3848
19. Mais U, Esteghlalian AR, Saddler JN, Mansfield SD (2002) Enhancing the enzymatic hydrolysis of cellulosic materials using simultaneous ball milling. Appl Biochem Biotechnol 98:815–832
20. Mosier N, Wyman C, Dale B, Elander R, Lee YY, Holtzapple M, Ladisch M (2005) Features of promising technologies for pretreatment of lignocellulosic biomass. Bioresource Technol 96:673–686
21. Kumar P, Barrett DM, Delwiche MJ, Stroeve P (2009) Methods for pretreatment of lignocellulosic biomass for efficient hydrolysis and biofuel production. Ind Eng Chem Res 48:3713–3729
22. Pan XJ, Gilkes N, Kadla J, Pye K, Saka S, Gregg D, Ehara K, Xie D, Lam D, Saddler J (2006) Bioconversion of hybrid poplar to ethanol and co-products using an organosolv fractionation process: optimization of process yields. Biotechnol Bioeng 94:851–861
23. Nguyen TAD, Kim KR, Han SJ, Cho HY, Kim JW, Park SM, Park JC, Sim SJ (2010) Pretreatment of rice straw with ammonia and ionic liquid for lignocellulose conversion to fermentable sugars. Bioresource Technol 101:7432–7438
24. Shimizu K, Satsum A (2011) Toward a rational control of solid acid catalysis for green synthesis and biomass conversion. Energy Environ Sci 4:3140–53
25. Yang Z, Niu L, Ma Z, Ma H, Lei Z (2011) Fabrication of highly active Sn/W mixed transition-metal oxides as solid acid catalysts. Transition Met Chem 36:269–274
26. Delaney P, McManamon C, Hanrahan P, Copley MP, Holmes JD, Morris MA (2011) Development of chemically engineered porous metal oxides for phosphate removal. J Hazard Mater 185:382–391
27. Fang Z, Zhang F, Zeng HY, Guo F (2011) Production of glucose by hydrolysis of cellulose at 423 K in the presence of activated hydrotalcite nanoparticles. Bioresource Technol 102:8017–8021
28. Toda M, Takagaki A, Okamura M, Kondo JN, Hayashi S, Domen K, Hara M (2005) Biodiesel made with a sugar catalyst. Nature 438:178
29. Pang JF, Wang AQ, Zheng MY, Zhang T (2010) Hydrolysis of cellulose into glucose over carbons sulfonated at elevated temperatures. Chem Commun 46:6935–6937

30. Pua FL, Fang Z, Zakaria S, Guo F, Chia CH (2011) Direct production of biodiesel from high-acid value Jatropha oil with solid acid catalyst derived from lignin. Biotechnol Biofuel 4:56
31. Onda A, Ochi T, Yanagisawa K (2009) Hydrolysis of cellulose selectively into glucose over sulfonated activated-carbon catalyst under hydrothermal conditions. Top Catal 52:801–807
32. Tian J, Fan CY, Cheng MX, Wang XH (2011) Hydrolysis of cellulose over $Cs_xH_{3-x}PW_{12}O_{40}$ ($X = 1$–3) heteropoly acid catalysts. Chem Eng Technol 34:482–6
33. Vigier KDO, Jérôme F (2010) Heterogeneously-catalyzed conversion of carbohydrates. Top Curr Chem 295:63–92
34. Degirmenci V, Uner D, Cinlar B, Shanks BH, Yilmaz A, Santen RA, Hensen EJM (2011) Sulfated zirconia modified SBA-15 catalysts for cellobiose hydrolysis. Catal Lett 141:33–42
35. Takagaki A, Tagusagawa C, Domen K (2008) Glucose production from saccharides using layered transition metal oxide and exfoliated nanosheets as a water-tolerant solid acid catalyst. Chem Commun 42:5363–5365
36. Rinaldi R, Palkovits R, Schüth F (2008) Depolymerization of cellulose using solid catalysts in ionic liquids. Angew Chem 120:8167–8170
37. Massam J, Brown DR (1995) The roles of Brønsted and Lewis surface acid sites in acid-treated montmorillonite supported $ZnCl_2$ alkylation catalysts. Catal Lett 35:335–343
38. Mizuno N, Misono M (1998) Heterogenous catalysis. Chem Rev 98:199–217
39. Zhang J, Kanno M, Wang Y, Nishii H, Miura Y, Kamiya Y (2011) Changes in surface acidity of silica-supported dodecatungstosilicic acid in relation to the loading amount. J Phys Chem C 115:14762–14769
40. Macht J, Baertsch CD, May-Lozano M, Soled SL, Wang Y, Iglesia E (2004) Support effects on Brønsted acid site densities and alcohol dehydration turnover rates on tungsten oxide domains. J Catal 227:479–491
41. Lai DM, Deng L, Guo QX, Fu Y (2011) Hydrolysis of biomass by magnetic solid acid. Energy Environ Sci 4:3552–3557
42. Suganuma S, Nakajima K, Kitano M, Yamaguchi D, Kato H, Hayashi S, Hara M (2008) Hydrolysis of cellulose by amorphous carbon bearing SO_3H, COOH, and OH groups. J Am Chem Soc 130:12787–12793
43. Onda A, Ochi T, Yanagisawa K (2008) Selective hydrolysis of cellulose into glucose over solid acid catalysts. Green Chem 10:1033–1037
44. Nasef MM, Saidi H, Senna MM (2005) Hydrolysis of sucrose by radiation grafted sulfonic acid membranes. Chem Eng J 108:13–17
45. Zhang Z, Zhao ZK (2009) Solid acid and microwave-assisted hydrolysis of cellulose in ionic liquid. Carbohyd Res 344:2069–2072
46. Veefkind VA, Smidt ML, Lercher JA (2000) On the role of strength and location of Brønsted acid sites for ethylamine synthesis on mordenite catalysts. Appl Catal A—Gen 194–195:319–332
47. Tagusagawa C, Takagaki A, Iguchi A, Takanabe K, Kondo J, Ebitani K (2010) Highly active mesoporous Nb-W oxide solid-acid catalyst. Angew Chem Int Ed 49:1128–1132
48. Barton DG, Shtein M, Wilson RD, Soled SL, Iglesia E (1999) Structure and electronic properties of solid acids based on tungsten oxide nanostructures. J Phys Chem B 103:630–640
49. Miller JB, Ko EI (1996) A Brønsted acid strength hierarchy for zirconia-silica-sulfate aerogels. Chem Eng J Biochem Eng J 64:273–281
50. Fukuoka A, Dhepe PL (2006) Catalytic conversion of cellulose into sugar alcohols. Angew Chem Int Ed 45:5161–5263
51. Hegner J, Pereira KC, DeBoef B, Lucht BL (2010) Conversion of cellulose to glucose and levulinic acid via solid-supported acid catalysis. Tetrahedron Lett 55:2356–2358
52. Dhepe PL, Ohashi M, Inagaki S, Ichikawa M, Fukuoka A (2005) Hydrolysis of sugars catalyzed by water-tolerant sulfonated mesoporous silicas. Catal Lett 102:163–169
53. Akiyama G, Matsuda R, Sato H (2011) Cellulose hydrolysis by a new porous coordination polymer decorated with sulfonic acid functional groups. Adv Mater 23:3294–3297

54. Reddy BM, Khan A (2005) Recent advances on TiO_2–ZrO_2 mixed oxides as catalysts and catalyst supports. Catal Rev 47:257–296
55. Sinhamahapatra A, Sutradhar N, Ghosh M, Bajaj HC, Panda AB (2011) Mesoporous sulfated zirconia mediated acetalization reactions. Appl Catal A—Gen 402:87–93
56. Ropero-Vega JL, Aldana-Pérez A, Gómez R (2010) Sulfated titania [$TiO_2/SO_4{}^{2-}$]: a very active solid acid catalyst for the esterification of free fatty acids with ethanol. Appl Catal A—Gen 379:24–29
57. Jitputti J, Kitiyanan B, Rangsunvigit P, Bunyakiat K, Attanatho L, Jenvanitpanjakul P (2006) Transesterification of crude palm kernel oil and crude coconut oil by different solid catalysts. Chem Eng J 116: 61–66
58. Peng BX, Shu Q, Wang JF, Wang GR, Wang DZ, Han MH (2008) Biodiesel production from waste oil feedstocks by solid acid catalysis. Process Saf Environ 8:441–447
59. Kulkarni AP, Muggli DS (2006) The effect of water on the acidity of TiO_2 and sulfated titania. Appl Catal A—Gen 302:274–282
60. Yori YC, Damato MA, Darea J, Parera JM (1995) Isomerization of *n*-Butane on $Pt/SO_4{}^{2-}ZrO_2$ and mechanical mixtures of $Pt/Al_2O_3 + SO_4{}^{2-}ZrO_2$. J Catal 153:218–223
61. Liu J, Zhao J, Miao C, Yue Y, Hua W, Gao Z (2011) Liquid-phase-pinene isomerization over Fe-doped sulfated zirconia prepared by a hydrothermal treatment-assisted process. Chinese J Chem 29:1095–1100
62. Reddy BM, Sreekanth PM, Reddy VR (2011) Modified zirconia solid acid catalysts for organic synthesis and transformations. J Mole Catal A—Chem 225:71–78
63. Huang J, Jiang Y, Marthala VRR, Ooi YS, Weitkamp J, Hunger M (2007) Concentration and acid strength of hydroxyl groups in zeolites La, Na-X and La, Na-Y with different lanthanum exchange degrees studied by solid-state NMR spectroscopy. Micropor Mesopor Mat 104:129–136
64. Jia CJ, Liu Y, Schmidt W, Lu AH, Schüth F (2010) Small-sized HZSM-5 zeolite as highly active catalyst for gas phase dehydration of glycerol to acrolein. J Catal 269:71–79
65. Gagea BC, Lorgouilloux Y, Altintas Y, Jacobs PA, Martens JA (2009) Bifunctional conversion of n-decane over HPW heteropoly acid incorporated into SBA-15 during synthesis. J Catal 265:99–108
66. Xie S, Liu K, Liu S, Liu Y, Zhang W, Xu L (2010) The crystallization process of MCM-49/ZSM-35 composite zeolites in a mixed-amine system. Chinese J Catal 31:1071–1076
67. Prokešová P, Mintova S, Čejka J, Bein T (2003) Preparation of nanosized micro/mesoporous composites via simultaneous synthesis of Beta/MCM-48 phases. Micropor Mesopor Mat 64:165–174
68. Schmidt H, Koch D, Grathwohl G, Colombo P (2001) Micro-/macroporous ceramics from preceramic precursors. J Am Ceram Soc 84:2252–2255
69. Schmidt I, Madsen C, Jacobsen CJH (2000) Confined space synthesis. A novel route to nanosized zeolites. Inorg Chem 39:2279–2283
70. Yu S, Tian H, Dai Z, Long J (2010) Mechanism of the influence of lanthanum and cerium on the stability of Y zeolite. Chinese J Catal 31:1263–1270
71. Harmer MA, Sun Q (2001) Solid acid catalysis using ion-exchange resins. Appl Catal A—Gen 221:45–62
72. Harmer MA, Farneth WE, Sun Q (1996) High surface area Nafion® resin/silica nanocomposites: a new class of solid acid catalyst. J Am Chem Soc 18:7708–7715
73. Kim SJ, Dwiatmoko AA, Choi JW (2010) Cellulose pretreatment with 1-n-butyl-3-methylimidazolium chloride for solid acid-catalyzed hydrolysis. Bioresource Technol 101:8273–8279
74. Chen F, Meng X, Xiao FS (2011) Mesoporous solid acid catalysts. Catal Surv Asia 15:37–48
75. Qi XH, Watanabe M, Aida TM (2011) Catalytic conversion of cellulose into 5-hydroxymethylfurfural in high yields via a two-step process. Cellulose 18:1327–1333
76. Ke CC, Li XJ, Shen Q (2011) Investigation on sulfuric acid sulfonation of in-situ sol–gel derived Nafion/SiO_2 composite membrane. Int J Hydrogen Energy 36:3606–3613

77. Rafati L, Mahvi AH, Asgari AR (2010) Removal of chromium (VI) from aqueous solutions using Lewatit FO36 nano ion exchange resin. Int J Environ Sci Technol 7:147–156
78. Chen DY, Zhang X, He JY (2008) Synthesis of butyl lactate catalyzed by modified strong acid cation ion exchange resin by gallium sulfate. Sci Technol Food Ind 29:191–192
79. Vyver SV, Peng L, Geboers J, Schepers H, Clippel F, Gommes CJ, Goderis B, Jacobs PA, Sels BF (2010) Sulfonated silica/carbon nanocomposites as novel catalysts for hydrolysis of cellulose to glucose. Green Chem 12:1560–1563
80. Dai WL, Luo SL, Yin SF, Au CT (2009) The direct transformation of carbon dioxide to organic carbonates over heterogeneous catalysts. Appl Catal A—Gen 366:2–12
81. Lee YY, Iyerl P, Torget RW (1999) Dilute-acid hydrolysis of lignocellulosic biomass. Adv Biochem Eng Biotechnol 65:93–115
82. Sasaki M, Fang Z, Fukushima Y, Adschiri T, Arai K (2000) Dissolution and hydrolysis of cellulose in subcritical and supercritical water. Ind Eng Chem Res 39:2883–2890
83. Fang Z, Fang C (2008) Complete dissolution and hydrolysis of wood in hot water. AIChE J 54:2751–2758
84. Cheng MX, Shi T, Guan HY, Wang ST, Wang XH, Jiang ZJ (2011) Clean production of glucose from polysaccharides using a micellar heteropolyacid as a heterogeneous catalyst. Appl Catal B-Env 107:104–109
85. Zakrzewska ME, Bogel-yukasik E, Bogel-yukasik R (2010) Solubility of carbohydrates in ionic liquids. Energy Fuel 24:737–745
86. Wang XJ, Li HQ, Cao Y, Tang Q (2011) Cellulose extraction from wood chip in an ionic liquid 1-allyl-3-methylimidazolium chloride (AmimCl). Bioresource Technol 102:7959–7965
87. Zhang ZH, Li CZ, Wang Q, Zhao ZK (2009) Efficient hydrolysis of chitosan in ionic liquids. Carbohyd Polym 78:685–689
88. Li CZ, Wang Q, Zhao ZK (2008) Acid in ionic liquid: An efficient system for hydrolysis of lignocellulose. Green Chem 10:177–182
89. Hsu WH, Lee YY, Peng WH, Wu KXW (2011) Cellulosic conversion in ionic liquids (ILs): Effects of H_2O/cellulose molar ratios, temperatures, times, and different ILs on the production of monosaccharides and 5-hydroxymethylfurfural (HMF). Catal Today 174:65–69
90. Alba-Rubio AC, Vila F, Martín Alonso D, Ojeda M, Mariscal R, López Granados M (2010) Deactivation of organosulfonic acid functionalized silica catalysts during biodiesel synthesis. Appl Catal B—Env 95:279–287
91. Dias AS, Lima S, Pillinger M, Valente AA (2007) Modified versions of sulfated zirconia as catalysts for the conversion of xylose to furfural. Catal Lett 114:151–160
92. Shi XJ, Wu YL, Li PP, Yi H, Yang M, Wang G (2011) Catalytic conversion of xylose to furfural over the solid acid $SO_4^{2-}/ZrO_2–Al_2O_3$/SBA-15 catalysts. Carbohydr Res 346:480–487
93. Wen Z, Ci S, Li J (2009) Pt nanoparticles inserting in carbon nanotube arrays: nanocomposites for glucose biosensors. J Phys Chem C 113:13482–13487
94. Moodley DJ, Loosdrecht J, Saib AM, Overett M, Datye A, Niemantsverdriet J (2009) Carbon deposition as a deactivation mechanism of cobalt-based Fischer–Tropsch synthesis catalysts under realistic conditions. Appl Catal A—Gen 354:102–110
95. Saib AM, Moodley DJ, Ciobîcă IM, Hauman MM, Sigwebela BH, Weststrate CJ, Niemantsverdriet JW, Van De Loosdrecht J (2010) Fundamental understanding of deactivation and regeneration of cobalt Fischer–Tropsch synthesis catalysts. Catal Today 154:271–282
96. Gai PL, Roper R, White MG (2002) Recent advances in nanocatalysis research. Curr Opin Solid St M 6:401–406
97. Veith M, Mathur S, Lecerf N, Huch V, Decker T, Beck HP, Eiser W, Haberkorn R (2000) Sol–gel synthesis of nano-scaled $BaTiO_3$, $BaZrO_3$ and $BaTi_{0:5}Zr_{0:5}O_3$ oxides via single-source alkoxide precursors and semi-alkoxide routes. J Sol–Gel Sci Technol 15:145–158
98. Pietsch T, Metwalli E, Roth SV, Gebhardt R, Gindy N, Müller-Buschbaum P, Fahmi A (2009) Directing the self-assembly of mesostructured hybrid materials: Effect of polymer concentration and solvent type. Macromol Chem Phys 210:864–878
99. Dong H, Tao W, Bi J (2011) Self-assembly of copper and cobalt complexes with hierarchical size and catalytic properties for hydroxylation of phenol. Nanoscale Res Lett 6:484

100. Jiang D, Zhang M, Jiang H (2011) Preparation and formation mechanism of nano-sized MnO_x-CeO_2 hollow spheres via a supercritical anti-solvent technique. Mater Lett 65:1222–1225
101. Wu S, Sun A, Zhai F, Wang J, Xu W, Zhang Q, Volinsky AA (2011) Fe_3O_4 magnetic nanoparticles synthesis from tailings by ultrasonic chemical co-precipitation. Mater Lett 65:1882–1884
102. Khan AT, Choudhury LH, Ghosh S (2006) Silica supported perchloric acid ($HClO_4$-SiO_2): A highly efficient and reusable catalyst for geminal diacylation of aldehydes under solvent-free conditions. J Mol Catal A—Chem 255:230–235
103. Ghanbaripour R, Mohammadpoor-Baltork I, Moghadam M, Khosropour AR, Tangestaninejad S, Mirkhani V (2012) Microwave-promoted efficient conversion of acetophenones to 1,3,5-triarylbenzenes catalyzed by $H_3PW_{12}O_{40}$ and nano-silica supported $H_3PW_{12}O_{40}$ as reusable catalysts. Polyhedron 31:721–728
104. Zhao Y, Xue M, Cao M, Shen J (2011) A highly loaded and dispersed Ni_2P/SiO_2 catalyst for the hydrotreating reactions. Appl Catal B—Env 104:229–233
105. Guo F, Wei NN, Xiu ZL, Liang ZX (2010) Transesterification mechanism of soybean oil to biodiesel catalyzed by calcined sodium silicate. Fuel 93:468–472
106. Wang L, Wang H, Wang A, Liu M (2009) Surface modification of a magnetic SiO_2 support and immobilization of a nano-TiO_2 photocatalyst on it. Chinese J Catal 30:939–944
107. Fan J, Gao Y (2006) Nanoparticle-supported catalysts and catalytic reactions—a mini-review. J Exp Nanosci 4:457–475
108. Burns JM, Craig PS, Shaw TJ, Ferry JL (2011) Short-term Fe cycling during Fe(II) oxidation: Exploring joint oxidation and precipitation with a combinatorial system. Environ Sci Technol 45:2663–2669
109. Lee SH, Sonf Y, Hosein ID, Liddell CM (2009) Magnetically responsive and hollow colloids from nonspherical core–shell particles of peanut-like shape. J Mater Chem 19:350–355
110. Yin H, Ma Z, Chi M, Dai S (2011) Heterostructured catalysts prepared by dispersing Au@Fe_2O_3 core–shell structures on supports and their performance in CO oxidation. Catal Today 160:87–95
111. Wen Y, Lei F (2011) Preparation and catalytic activity of Fe_2O_3/CNT to thermal decomposition of ammonium perchilorate. J Adv Mater Res 396–398:837–840
112. Zhang X, Lei L (2008) Preparation of photocatalytic $Fe_2O_3-TiO_2$ coatings in one step by metal organic chemical vapor deposition. Appl Surf Sci 254:2406–2412
113. Matsunami N, Sogawa T, Sakuma Y, Ohno N, Tokitani M, Masuzaki S (2011) Deuterium retention in Fe_2O_3 under low-energy deuterium-plasma exposure. Phys Scr 145:1–4
114. Morjan I, Alexandrescu R, Dumitrache F, Birjega R, Fleaca C, Soare I, Luculescu CR, Filoti G, Kuncer V, Vekas L, Popa NC, Prodan G, Ciupina V (2010) Iron oxide-based nanoparticles with different mean sizes obtained by the laser pyrolysis: structural and magnetic properties. J Nanosci Nanotechnol 10:1223–1234
115. Pillai UR, Sahle-Demessie E, Varma RS (2003) Alternative routes for catalyst preparation: use of ultrasound and microwave irradiation for the preparation of vanadium phosphorus oxide catalyst and their activity for hydrocarbon oxidation. Appl Catal A—Gen 252:1–8
116. Guo J, Hou Y, Yang C, Wang Y, He H, Li W (2011) Preparation of Ni−B amorphous alloy catalyst from nickel hydrazine complex with ultrasonic assistance. Catal Commun 16:86–89
117. Boz N, Degirmenbasi N, Kalyon DM (2009) Conversion of biomass to fuel: Transesterification of vegetable oil to biodiesel using KF loaded nano-γ-Al_2O_3 as catalyst. Appl Catal B—Env 89:590–596
118. Farhadi A, Takassi MA, Dayer M (2011) Investigation of the interaction of ethyl acetoacetate with nano alumina particle as Lewis acid in acetonitrile solvent. Front Chem China 6:248–252
119. Norouzbeigi R, Edrissi M (2011) Preparation of nano alumina powder via combustion synthesis: Porous structure optimization via taguchi L16 design. J Am Ceram Soc 94:4052–4058
120. Yamaguchi D, Kitano M, Suganuma S, Nakajima K, Kato H, Hara M (2009) Hydrolysis of cellulose by a solid acid catalyst under optimal reaction conditions. J Phys Chem C 113:3181–3188

121. Zhang F, Deng X, Fang Z, Tian XF, Kozinski JA (2011) Hydrolysis of crystalline cellulose over Zn–Ca–Fe oxide catalyst. Petrochem Technol 40:43–48
122. Nanda KK (2009) Size-dependent melting of nanoparticles: Hundred years of thermodynamic model. Pramana 72:617–628
123. Gill CS, Price BA, Jones CW (2007) Sulfonic acid-functionalized silica-coated magnetic nanoparticle catalysts. J Catal 251:145–152
124. Beydoun D, Amal R (2000) Novel photocatalyst: titania-coated magnetite. Activity and photodissolution. J Phys Chem B 104:4387–4396

Part VII
Pretreatment of Different Types of Biomass

Chapter 16
Pretreatment of Sugarcane Bagasse and Leaves: Unlocking the Treasury of "Green Currency"

Anuj K. Chandel, Ellen C. Giese, Felipe A. F. Antunes, Ivy dos Santos Oliveira and Silvio Silvério da Silva

Abstract Sugarcane residues (bagasse and leaves/trash) are the principal feedstock in Asia, South America, Africa, and other parts of the world. The judicious application of this feedstock into value-added products such as fuel ethanol, xylitol, organic acids, industrial enzymes, etc. may provide a strong economic platform along with clean and safe environment. Pretreatment is an inevitable process to harness the carbohydrate fraction of sugarcane bagasse and leaves into readily available sugars by cellulase-mediated process for the production of house-hold commodities. Several methods (physical, physico-chemical, chemical, and biological) have been adopted for the pretreatment of sugarcane residues. Pretreatment methods with pros and cons are employed either to depolymerize hemicellulosic fraction or lignin degradation to make cellulose more amenable for improved cellulolytic enzymes action. The choice of pretreatment methods depends upon its precise mechanistic action on lignin or hemicelluloses with fewer inhibitory products, minimal sugar loss by increasing the cellulosic surface area for subsequent enzymatic action to obtain desired sugars recovery. Furthermore, economics and environmental impacts are two important considerations for the selection of pretreatment method. This chapter aims to explore a better understanding of multiple pretreatment methodologies applied to the sugarcane residues along with economics and environmental impacts.

Keywords Sugarcane bagasse · Sugarcane leaves · Pretreatment · Enzyme hydrolysis · Bioethanol · Biomass recalcitrance · Fermentable sugars

16.1 Introduction

In recent years, numerous efforts have been considered to harness the commercial potential of sugarcane residues (bagasse and leaves) into value-added products of commercial significance such as ethanol, xylitol, organic acids, industrial enzymes,

---

A. K. Chandel (✉) · S. S. da Silva (✉) · E. C. Giese ·
F. A. F. Antunes · I. dos S. Oliveira
Department of Biotechnology, School of Engineering of Lorena,
University of São Paulo (USP), Lorena 12.602.810, Brazil
e-mail: anuj.kumar.chandel@gmail.com; silvio@debiq.eel.usp.br

Z. Fang (ed.), *Pretreatment Techniques for Biofuels and Biorefineries*,
Green Energy and Technology,
DOI 10.1007/978-3-642-32735-3_16,

Table 16.1 Top ten sugarcane producing countries in the world in 2010–2011 [6]

Country	Production (million metric tons)
Brazil	38.745
India	26.000
China	11.475
Thailand	10.061
United States	7.210
Mexico	5.495
Pakistan	4.400
France	4.275
Australia	3.800
Germany	3.565

and others [1, 2]. The judicious application of sugarcane based feedstock into commercial entities is a sustainable process which may influence the economy at the forefront in sugarcane producing countries [3, 4]. Countries in Asia, Asia pacific, South America, and Africa grow a copious amount of sugarcane, which has a key role in their economy [5]. Table 16.1 presents the major sugarcane producing countries in the world.

Production of sugar from cane juice and ethanol from either juice or cane molasses are two major driving forces of agro-economy in these countries. This impact could be wider if the residues of sugarcane (sugarcane bagasse (SB) and sugarcane leaves(SL/ST)) are being used for the production of household commodities employing biotechnological routes [3, 7]. In 2011, the annual worldwide production of sugarcane residues was recorded around 279 million metric tons [2, 5]. From the data of sugarcane production in the world, it can be estimated that a huge amount of sugarcane residue is generated every year which is totally renewable and could be a promising source of bioenergy and other value-added products generation. Each ton of sugarcane yields about 140 kg of humid bagasse, which is a fibrous residue obtained after crushing sugarcane in sugar producing factories, and is currently used for steam generation in boilers [1, 2, 5]. During the harvesting of sugarcane in agriculture fields, cane leaf residues are removed from the cane stem and left on the fields, causing loss of green energy [2, 8]. SL/ST is generated in huge amount in fields (6–8 tons from one hectare of sugarcane crop) [8]. Sugarcane derived lignocellulosic feedstock (SB and SL/ST) are being used in industries now-a-days for steam or electricity generation. Together, both residues constitute a foreseeable amount of biomass, which can be used for bioethanol production and other bio-products of high economic value (Fig. 16.1).

Apart from bioenergy generation from these residues, products of high economic value such as industrial enzymes, organic acids, food/feed products, amino acids, vitamins or cosmetics can also be produced by microbial fermentation. A lot of research work is underway in this line. However, a full technological road-map is yet to come for the industrial production of these commodity chemicals from SB/SL. SB and SL contain an appreciable amount of carbohydrates in cell wall along with the lignin. Pretreatment is an inevitable process to break down the carbohydrate fraction into simpler sugars making them readily available for fermenting by microorganisms

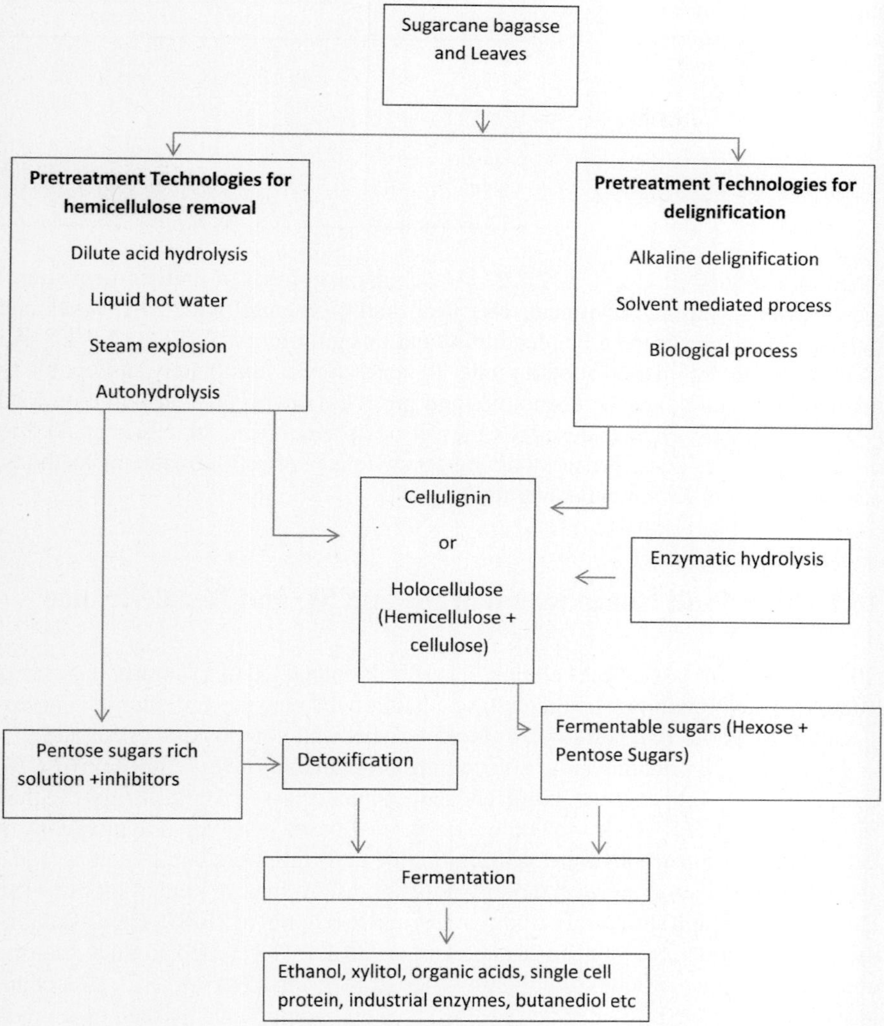

Fig. 16.1 Pretreatment of sugarcane residues (SB and SL) for the production of products of commercial significance

[9–11]. The aim of pretreatment of SB/SL is either to remove lignin or hemicellulose for enhancing the amenability of cellulases toward cellulosic fraction of cell wall. Hemicellulose is generally degraded by weak acid catalyzed process removing its heterogeneity by producing monomeric sugars (xylose, arabinose, mannose, galactose, and glucose) [12]. Lignin is degraded by alkali based reactions or microbial action by selective lignin degraders and thus leaving cellulose and hemicellulose network together, but in less compacted form. The pretreated material is subsequently hydrolyzed by cellulolytic enzymes for sugar recovery toward the fermentation of

Table 16.2 Cell wall composition of sugarcane bagasse and trash on dry matter basis [19]

Composition	Bagasse	Leaves
Glucan	41.4	33.3
Xylan	22.5	18.1
Arabinan	1.3	3.1
Galactan	1.3	1.5
Mannan	3.4	1.5
Lignin	23.6	36.1
Total	*93.5*	*93.6*

other value-added products [1, 2, 12, 13]. In past, a number of pretreatment strategies (physical, physico-chemical, chemical, and biological) have been developed by researchers considering hemicellulose and/or lignin removal [9–11, 14–17]. An ideal pretreatment method should render lignocellulosics completely susceptible to the action of cellulases, be economic, and pose less environmental pollution load [10, 13, 17]. This chapter describes the various pretreatment strategies applied to the sugarcane residues, process parameters, mechanism of pretreatment methods, economic, and environmental aspects.

16.2 Cell Wall Composition of SB and SL and Recalcitrance

SB/SL consists of crystalline cellulose nanofibrils embedded in an amorphous matrix of cross-linked lignin and hemicelluloses that impairs enzyme and microbial accessibility [18]. Table 16.2 summarizes the detail cell wall composition of SB and SL.

In general, SB contains more holocellulose (hemicellulose + cellulose) (67.8 %) than SL (61.7 %). In contrast, lignin and ash content is less in SB (23.6 %, 1 %) than SL (36.1 %, 7.8 %) which limits the biochemical-based conversion applications of latter. Apart from carbohydrates and lignin, the cell wall ingredients such as silica, ash, and extractives along with natural moisture resides in both kind of biomass [8]. Factors such as high structural carbohydrates and less lignin in SB/SL, large availability, almost no food/feed value make it better feedstock for bio-based products than the contemporary agro residues (wheat straw, rice straw, corn stover, etc.) [2]. Holocellulose content in SB/SL (67.8 % and 61.7 %) is of high importance for their bioconversion into various products by microbial fermentation. This content (% dry weight) is fairly comparable with the other lignocellulosic materials such as wheat straw (56.1), corn stover (64.1), switch grass (61.8), and Spruce wood (71.9) [9, 10, 13].

In order to degrade the holocellulosic fraction of plant cell wall, the current pretreatment methodologies are unattractive due to economic concerns. This is basically because of the special arrangement of cross-linked lignin with holocellulose network; biomass has evolved a superb mechanism to protect itself from microbial invasion. This mechanism of natural resistance in plant cell wall is called "biomass recalcitrance" [18]. Accessibility to the carbohydrate fraction of cell wall is a multi-scale phenomenon encompassing several orders of magnitude due to both macroscopic (compositional heterogeneity, mass transfer limitations) and microscopic barriers (holocellulose crystallinity, lignin-holocellulose linkage) [20].

Table 16.3 Summarization of different pretreatment technologies applied to sugarcane feedstock and their mechanistic impact on plant cell wall

Substrate	Type of pretreatment	Pretreatment conditions	Effects observed	References
SB	Hydrogen peroxide (H_2O_2) in alkaline media	H_2O_2 (6 % w/v), 4 h, 20 °C	Hemicellulose sugars recovery [xylose (80.9 %), arabinose (3.8 %), glucose (4.2 %), and uronic acid (3.2 %)]	[21]
SB and cane leaf	AFEX	200 g SB ammonia (2:1 w/w), 30 min, 140 °C	85 % Glucan conversion	[8]
		200 g Cane leaf, ammonia (1:1 w/w), 30 min, 140 °C	95–98 % xylan conversion into sugars	
SB	Ethanol organosolv	175 °C, 500 g (Dry matter), ethanol solution (50 % v/v), 1:5 solid: liquid ratio	Higher glucose yield (20.9 g glucose/100 g SB) was obtained using sulfuric acid (1.25 % v/v) as a catalyst for 60 min	[22]
SB	Ethanol organosolv	In the presence of sodium hydroxide (NaOH), 60 min, 195 °C, using ethanol 30 % (v/v)	Higher glucose yield (80.5 g glucose/100 g acid-pretreated SB or 58.33 g glucose/100 g original SB)	[23]
SB	Ionic liquids (ILs)	IL [Emim] [Ac] (4.0 g), SB (200 mg), 120 min, 60–120 °C	Enhanced SB enzymatic saccharification rate	[24]

Cell wall is a complex and highly arranged structure and thus resists the accessibility of cellulase enzymes. Pretreatment allows its breakdown and increase the amenability of enzymes for the sugars monomers recovery. Vascular bundles are consequently arranged in native SB/SL. Pretreatment disrupts the compactness of cell wall. Scanning electron and atomic force microscopic view of cell wall after dilute acid mediated pretreatment, clearly shows the disorganization of cell wall components which is pivotal for the improved cellulase action on carbohydrate polymer in order to yield simple sugars. From raw SB/SL (highly complex structure) to glucose or other sugars (monomers) production is mutli-sclae phenomenon spanning several orders of magnitude (10^{-9} meters) [18, 20].

16.3 Pretreatment of SB/SL

Pretreatment methods can be divided into four different categories: Physical, physico-chemical, chemical, and biological processes. Each method has its own specificity toward the plant cell wall fractions. Table 16.3 summarizes the effect of different pretreatment technologies applied to the sugarcane feedstock for the recovery of sugars.

16.3.1 Physical Pretreatment

16.3.1.1 Milling

Milling is a mechanical process of pretreatment that breaks down the structure of lignocellulosic materials and reduces the crystallinity of the cellulose [14]. During ball milling, biomass is grounded with the contact of the balls inside a cycle machine to get the uniform particles size [25]. This method can be considered environment-friendly due the absence of added chemicals in this process that produce toxic substances [14]. A disadvantage of milling is the high power required by the machines and consequent high energy costs [14]. Buaban et al. [26] reported that the increase of the time of milling increased the amounts of the sugars (glucose, 89.2 ± 0.7 % and xylose, 77.2 ± 0.9 %) after 4 h of milling.

16.3.1.2 Irradiation

Gamma-rays-mediated pretreatment is irradiation pretreatment which allows the breakdown of beta-1, 4 glycosidic linkages, thus enhancing the surface of area and crystallinity of cellulose [16]. It is a physical pretreatment which increase the surface area, consequently reducing the crystallinity. This method is expensive for large-scale operations with considerable environmental and safety concerns [16].

16.3.1.3 Microwave Irradiation

The use of high-energy radiation such as microwave causes one or more changes in the characteristics of cellulosic biomass, such as an increase in surface area, reduction in the degrees of polymerization, and crystallinity of cellulose and hemicelluloses, and the partial depolymerization of lignin [27]. However, irradiation pretreatment have the disadvantage of high energy consumption, and the methods are slow and expensive. Microwave irradiation process acts under the structural change in cellulose with the occurrence of the lignin and hemicellulose degradation, thus increasing the enzymatic accessibility [14]. To further improve the sugar yield after pretreatment, microwave radiation process was combined with chemicals. Binod et al. [28] tested different microwave pretreatment conditions for SB and reported highest reducing sugar yield (0.83 g/g dry biomass) in the microwave–alkali pretreatment followed by acid pretreatment.

16.3.2 Physico-Chemical

16.3.2.1 Liquid Hot Water (LHW)

Hot-water pretreatment changes the structure of lignocellulose and solubilizes the hemicellulose, thus keeping cellulose amenable for cellulolytic enzymic action [18, 29]. LHW penetrates inside the lignocellulose matrix and degrades hemicellulose into xylose and other accessory sugars with the least generation of inhibitors [17]. The required pressure was applied at high temperatures (160–240 °C) for the cell wall degradation [29]. Laser et al. [30] reported 80 % xylan recovery from SB after hot-water pretreatment (170–230 °C, 1–46 min) which showed 90 % ethanol conversion after simultaneous saccharification and fermentation (SSF).

16.3.2.2 Autohydrolysis

Autohydrolysis is a process that involves the use of steam with or without explosion, based on the selective depolymerization of hemicelluloses from SB/SL [14]. The hydrolysis of acetyl groups generates acetic acid, and hemicelluloses breakdown into its monomeric constituents. Mechanically, a quick reduction of pressure (0.69–4.83 MPa) at high temperature (160–260 °C) provokes an explosive decompression of lignocellulosic material effectively releasing hemicellulosic sugars, preserving the physical–chemical properties of the cellulose [14].

During autohydrolysis, due to depolymerization and repolymerization reactions lignin moieties are redistributed on the fiber surface promoting an enhancement of the pore size and surface area of SB/SL. Hemicellulose is solubilized under high temperature and short residence time (270 °C, 1 min) or lower temperature and longer residence time (190 °C, 10 min) [15, 31]. Dekker and Wallis [32] performed auto hydrolysis of SB at 200 °C for 4 min and observed 90 % solubilization of hemicellulose. The pretreated bagasse enzymatically hydrolysed, revealed 80 % saccharification after 24 h.

16.3.2.3 Steam Explosion

Steam explosion can be considered as one of the most promising techniques of fractionation of SB/SL. In the process, hydrolysis of the hemicellulose is accelerated by the contact of the biomass through steam at high temperature (160–240 °C) and high pressure (between 0.7 and 4.8 MPa) followed by quick decompression [16], which leads to breakdown of biomass [33]. Singh et al. [34] reported the steam explosion of SB which eventually showed the enzymatic hydrolysis efficiency of 100 % after 24 h of incubation by using the cellulases from *Penicillium* sp. Steam explosion disrupts the compactness of fibrils, thus increasing the surface area for better enzymatic action [16].

Sendelius [35] investigated steam explosion in SB at varying temperatures (180, 190, and 205 °C) for different time periods (5 and 10 min) using different impregnating agents (water, 2 % SO_2, and 0.25 g sulfuric acid (H_2SO_4) per 100 g dry matter). These optimization studies showed 80 % theoretical hydrolysis yield at SO_2-impregnation–180 °C–5 min conditions. One of the major advantages of the steam explosion process is the rare or no use of chemicals, eventually reducing the operational costs and minimum production of potential inhibitors [17].

16.3.2.4 CO_2 Explosion

Carbon dioxide (CO_2) explosion is based on the utilization of CO_2 as supercritical fluid leading to lignin removal. CO_2 molecules at high pressure are able to penetrate small pores of SB disrupting the chemical nature of the substrate [36, 37]. Carbonic acid is formed in aqueous CO_2 solution penetrating into the small pores of lignocellulosic material under high pressure. This method is considered environment friendly due to minimum by-products generation [14]. CO_2 in presence of water forms carbonic acid which allows the depolymerization of holocellulose eventually increasing the surface area of substrate. Zheng et al. [38] pretreated SB using CO_2 as supercritical fluid which showed an increase in glucose yield by 50 % after enzymatic hydrolysis.

16.3.2.5 SO_2 Explosion

Addition of sulfur dioxide (SO_2) at elevated temperatures has been described as an alternative approach to enhance the recovery of both cellulose and hemicellulose fractions from SB [14, 29]. Carrasco et al. [39] studied SO_2 explosion mediated pretreatment of SB at temperatures 180–205 °C, with residence times of 5–10 min using SO_2 as a catalyst. Pretreatment at 190 °C for 5 min generated pentose yield 57 %. The pretreated SB showed 87 % hydrolysis at 2 % substrate concentration after enzymatic hydrolysis.

16.3.2.6 Ammonia Fiber Expansion (AFEX)

Ammonia fiber expansion (AFEX) is an attractive method of pretreatment for SB/SL because of several advantages such as economic, fast, and highly efficient. However, handling of ammonia solution for the pretreatment at large scale is a problem due to the environmental concerns [14, 15]. Ammonia acts strongly toward lignin removal with the minimum degradation of hemicelluloses and ameliorating the accessibility of cellulase enzyme action on cellulose and remaining hemicellulose [8, 20]. Ammonia acts on C–O–C linkage in lignin including ether and ester bonds in cellulignin complex [20]. Ammonia after pretreatment can be recycled for further applications. It is important to emphasize that the formation of sugar degradation products are minimized during ammonia pretreatment of SB/SL [8, 40]. This process can improve the enzymatic hydrolysis of lignocellulosic material depending on use and the optimal conditions used in the process [41].

In this method, SB/SL is exposed to liquid aqueous ammonia (1–2 kg of ammonia/kg of dry biomass) under moderate or high temperature (40–140 °C) and pressure (250–300 psi) for a period of time ($\leq$30 min). During AFEX, no significant inhibitory by-products are produced and ammonia can be recycled after pretreatment [15]. Maximum polysaccharide conversion of AFEX pretreated SB and SL by enzymatic hydrolysis using cellulases was almost 85 % while the use of hemicellulases promoted the xylan conversion to 95–98 % levels [8]. Xylanase supplementation also promoted the cellulose conversion into glucose [8, 42].

Ammonia recycle percolation (ARP), modified method of ammonia-mediated pretreatment uses aqueous ammonia (10–15 %) at elevated temperatures (150–170 °C) followed by its recovery and separation of biomass. Similar to steam explosion, this method also promotes the high degree of depolymerization of lignin and cleavage of lignin-carbohydrate linkages, increasing the accessibility of cellulolytic enzymes to carbohydrate skeleton of SB/SL [15, 43]. Ammonia pretreatment in any form (aqueous ammonia pretreatment, soaking in aqueous ammonia, ammonia freeze explosion, ammonia recycling percolation) is generally preferred to low lignin containing lignocellulosic substrates like corn stover than SB/SL [20, 29]. In fact, more research needs to be done for ammonia pretreatment optimization of SB/SL.

16.3.3 Chemical Pretreatments

16.3.3.1 Acid Pretreatment

Among different types of pretreatments, the acid hydrolysis is one of the most commonly used methods for SB/SL. This method is usually employed to solubilize the hemicelluloses fraction of the cell wall which eventually aids the accessibility of cellulolytic enzymes action [12, 29, 44]. In the acid pretreatment, the hemicellulose can be hydrolyzed keeping the biomass in contact with diluted acid or concentrated acid under high temperature and low temperature respectively [10, 45]. Figure 16.3 shows the mechanistic demonstration of acidic pretreatment applied to SB/SL.

The hemicellulose fraction of SB/SL is depolymerized primarily into pentose sugars (xylose and arabinose) and hexose sugars (glucose, galactose, mannose, etc) along with inhibitory compounds [12, 46, 47]. Recently, Moutta et al. [46] reported 56.5 g/L total reducing sugars from the hemicellulosic fraction of SL under the optimized set of conditions (130 °C, 2.9 % w/v H_2SO_4, 1:4 solid:liquid ratio, and 30 min of residence time).

Multiple methods showed their feasibility to eliminate the inhibitors from the hydrolysates prior to fermentation [47, 48]. The hemicellulosic hydrolysate after detoxification can be converted into value-added products such as xylitol, ethanol, lactic acid, etc [2, 49]. The advantage of the dilute acid hydrolysis process is the low operation and energy costs [14, 15, 49]. Concentrated acid hydrolysis of SB/SL leads to the issues such as equipment corrosion and expensive costs of maintenance [29, 45]. The diluted acid pretreated SB/SL is subsequently hydrolysed with cellulase to depolymerize the cellulosics fraction into glucose [44, 37].

16.3.3.2 Alkaline

Alkali pretreatment is employed for the removal of lignin from SB/SL. Alkaline pretreatment cleaves the intermolecular ester bonds between xylan, lignin, and other hemicelluloses present in the SB/SL cell wall contributing to removal of acetyl and uronic acids [27, 29]. The alkaline-based pretreatment methods constitute mainly sodium hydroxide (NAOH), calcium hydroxide (lime), potassium hydroxide, and aqueous ammonia hydroxide [10]. Use of alkali compounds increase the internal surface of cellulose promoting a decrement of its degree of polymerization and crystallinity with less sugar degradation and have been described as more effective on SB/SL that contain low lignin than wood materials [1, 37].

The alkali action is based on the lignin structure disruption by breaking of ester bonds between lignin, hemicellulose, and cellulose, and increasing the hydrolysis efficiency of the carbohydrate portion during cellulase mediated enzyme action [10, 50]. Specific at higher temperatures, the alkali pretreatment can be performed at low temperatures with extended time of reaction and high concentration of the alkali [11, 27].

A disadvantage of alkaline pretreatment is the formation of non-recoverable salts that can be incorporated into the biomass during the pretreatment reactions. Alkali pulping can result in formic and acetic acids, irrecoverable salts, and low-molecular-mass fragments as reaction products [10]. NaOH mediated pretreatment is one of the most common methods applied to pretreatment of SB. Wu et al. [51] evaluated the optimal low temperature NAOH pretreatment for SB hydrolysis in order to convert cellulose into ethanol using separate enzymatic hydrolysis and fermentation (SHF). Almost 98 % of glucan was hydrolyzed to glucose after 72 h using lower loading of commercial enzymes resulting in 88 % of fermentable sugars to ethanol by *Saccharomyces cerevisiae* strain. NAOH pretreatment can be more effective when combined to H_2O_2, a bleaching agent used in the paper and cellulose industry. In this alkaline/oxidative process, hemicellulose is solubilized in the first step, followed by solubilization and oxidation of lignin fraction from lignocellulosic material without formation of secondary products such as furfural and hydroxyl methyl furfural [14]. Some studies indicated that the use combination of both reagents promotes a total conversion of cellulose in glucose units [52]. Figure 16.2 shows the mechanistic demonstration of alkaline pretreatment applied to SB/SL.

Use of lime or calcium hydroxide promotes the removal of acetyl groups from hemicellulose and contributes to enhance the cellulose digestibility with lower cost and less safety requirements compared to sodium and potassium hydroxides despite their slow action on lignin [29]. Lime treatment can remove amorphous substances like lignin and hemicellulose increasing the crystallinity index and is also easily

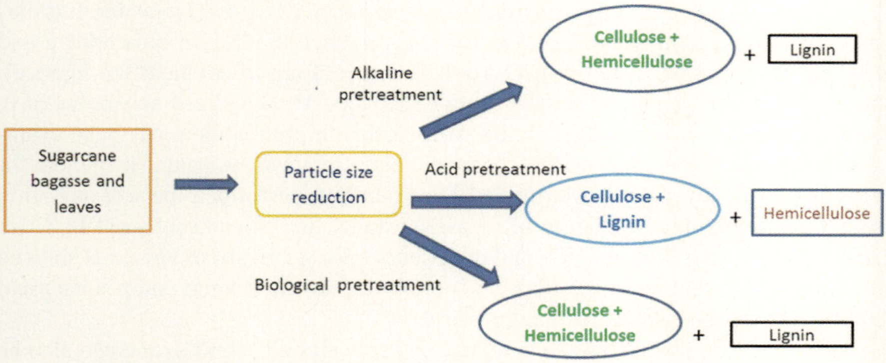

Fig. 16.2 Mechanistic demonstration of various pretreatment methods applied to sugarcane residues (SB and SL)

recovered as calcium carbonate by neutralization with CO_2 and regenerated [15, 53]. A comparison of the yield of total reducing sugars and glucose released from pretreated SB using lime or alkaline H_2O_2 pretreatments after enzymatic hydrolysis was developed. In this case, SB was previously hydrolyzed with H_2SO_4 (3 % v/v), followed by a low temperature alkali pretreatment method. Highest amounts of glucose were released from lime pretreated SB [54]. Lime and alkaline H_2O_2 pretreatments of SB were also compared aiming the evaluation of the potential of biogas production from the residues of second generation bioethanol. Results have shown that the highest methane production was obtained from SB residues pretreated in the presence of peroxide [55]. Rocha et al. [56] pretreated SB with steam explosion (1.3 MPa, 190 °C, 15 min) at pilot scale followed by the alkaline delignification (1.0 % w/v NaOH). Almost 94 % of hemicellulose hydrolysis and 92 % of lignin solubilization was recorded during this strategy.

Combined alkali and microwave treatment of SB have been also described. Microwave treatment (600 W) of SB in the presence of NaOH (1 % w/v) for 4 min followed by enzymatic hydrolysis resulted in 0.665 g reducing sugars/g dry biomass [28]. Xu et al. [57] realized two-stage pretreatments of SB with mild alkali (NaOH 1M) and acidic 1, 4-dioxane in HCl (2M). Alkali treatment resulted in 62.1 % of hemicelluloses (78–82.2 % of xylose) released after 18 h at 40 °C against only 10.6 % under acidic conditions (44.9–46.8 % of xylose). Delignification of SB with alkali (NaOH 10 % w/v; 90 °C; 1.5 h) and per acetic acid (PAA) (15 % w/v; 75 °C; 3 h) in a two-stage process was developed for generation of enzymatic digestible pulp as well as to be used for simultaneous saccharification fermentation (SSF) to ethanol production. The alkali-PAA pulp showed more xylose concentration and higher ethanol conversion when compared to dilute acid pretreated SB [58].

16.3.3.3 Oxidative Pretreatment

Wet oxidation solubilizes hemicellulose in consequence with lignin removal from the lignocellulosic substrates and thus fractionates lignocellulosic structure. Wet oxidation utilizes oxygen as an oxidizer for compounds dissolved in water. It is carried out at high temperature for less time in presence of sodium bicarbonate and oxygen. Sodium carbonate (Na_2CO_3) at high temperature fractionates the lignocellulose producing sugars in addition to by-products. Wet-oxidized substrates have shown increased enzyme digestibility. Wet-oxidation pretreatment has been evaluated for wheat straw, corn stover, sugarcane bagasse, cassava, peanuts, rye, etc for enzymatic hydrolysis [17]. Martín and Thomsen [59] performed the wet oxidation for the fractionation of SB, rice hulls, cassava stalks, and peanut shells at 195 °C for 10 min, with 2 gkg^{-1} of Na_2CO_3 and under either 3 or 12 bar of oxygen. SB showed the enhanced the enzyme digestibility of cellulose showing the maximum sugar yield (670.2 g/kg) at the highest oxygen pressure.

Different oxidative agents such as H_2O_2, per acetic acid, and ozone can also be used with wet oxidation. Azzam [60] performed oxidative pretreatment with H_2O_2 for bagasse pretreatment (2 % H_2O_2, 30 °C, 8 h) which showed improved enzymatic hydrolysis (95 % efficiency). Per acetic acid was found effective for the acetylation of cellulose in bagasse in conjunction with 80 % lignin removal which showed more than 80 % conversion of cellulose of hydrolysis. Structural changes in bagasse after pretreatment with per acetic acid was confirmed by scanning electron microscope, X-ray diffraction, and Fourier transform infra-red spectroscopy [61].

Pretreatment with ozone of lignocellulosic materials, so called ozonolysis, reduce lignin content without the production of any toxic compounds. It is expensive method due to requirement of high amount of ozone. Ozonolysis effectively remove lignin with the partial loss of hemicellulose [14]. It is performed generally at room temperature and produces minimum inhibitory compounds. This method has been less studied for the pretreatment of SB. Neely [62] reported 40 % digestibility of bagasse after ozone pretreatment of 2 h.

16.3.3.4 Ionic Liquids

Ionic liquids (IL) are salts containing large organic cations and small inorganic anions that exist as liquids at relatively low temperatures, and their solvent properties vary according the anion and alkyl constituents of the cation [63]. Both fractions from lignocellulosic materials can be dissolved in the presence of IL by formation of hydrogen bonds between the non-hydrated anions present in IL and the sugar hydroxyl protons portion from carbohydrates or lignin [29]. This non-covalent interaction causes the disruption of the strong biomass linkages minimizing undesirable products formation [63, 64]. IL, such as 1-butyl-3-methylimidazolium chloride [Bmim] [Cl], 1-allyl-3-methylimidazolium chloride [Amim] [Cl], and 1-ethyl-3-methylimidazolium acetate [Emim] [Ac], have been used as alternative solvents for pretreatment of lignocellulosic materials mainly by its chemical inertness and thermal stability [11, 60]. Liu et al. [65] had used the [Bmim] [Cl] IL as reaction medium to modify native SB cellulose generating phthalated cellulosic derivatives. This compound was also evaluated as a pretreatment solvent for SB and demonstrated to increase in 8-fold the cellulose conversion in comparison to untreated material [24]. ILs can easily dissolve and regenerate the cellulose molecule and is also a promising technology forproducing modified cellulose. The combined use of the [Bmim] [Cl] and ultrasound can be used for fractionating SB cellulose [66].

16.3.3.5 Organosolv

Organosolv process involves the use of an organic or aqueous organic solvent mixture (methanol, ethanol, acetone, ethylene glycol, triethylene glycol, tetrahydrofurfuryl alcohol, glycerol, aqueous phenol, aqueous *n*-butanol) and water, with or without addition of an acid (oxalic, salicylic, and acetyl salicylic acid) or base as catalyst agents at high temperature (150–200 °C) for the lignin degradation and partial solubilization of hemicellulose sugars [10]. Solvents used during the pretreatment can be eliminated by evaporation and condensation and recycled to reduce the operational costs of the process. This step allows avoiding the presence of inhibitory compounds, which may reduce the rate of enzymatic hydrolysis [14, 27]. During organosolv, low boiling point alcohols such as methanol, ethanol seems more effective due to low cost and easy recovery of solvents. Lignocellulosic substrates after organosolv pretreatment have high crystallinity and increased surface area for better cellulolytic enzymes action. However, organosolv pretreatment methods are expensive at large scale. Zhao et al. [67] comprehensively reviewed the various organosolv pretreatment methods (alcohol mediated, ethanol, acetone, per acetic acid, per formic acid, etc). They further recommended developing the pretreatment strategies based on continuous processes in reactors with fewer loads of organic solvents with high solid biomass and exploration of the increased applications of by-products generated during the organosolv pretreatment.

Mesa et al. [23] described a combination of a dilute-acid pretreatment followed by the organosolv pretreatment with NaOH (60 min, 195 °C, ethanol 30 % v/v) to fractionate the SB which resulted in 67.3 % (w/w) glucose released.

16.3.4 Biological Pretreatment

The selective delignifying microorganisms usually applied to SB/SL for the degradation of lignin. This process can also be referred as in-situ microbial delignification (ISMD). Generally white-rot fungi are used for the degradation of lignin and hemicelluloses fraction of SB/SL. The myco-SB/SL (SB/SL after fungal growth) can be used subsequently for enzymatic hydrolysis which in turn yields high amount of fermentable sugars [13, 68]. Figure 16.2 shows the mechanistic demonstration of biological pretreatment applied to SB/SL.

This is the low energy and capital intensive process where mild environmental conditions are required [68]. Selective lignin degraders such as *Pycnoporous cinnabarinus*, and *Ceriporiopsis subvermispora* are more useful in ISMD, having affinity toward lignin breakdown than cellulose and hemicelluloses [13]. SB was pretreated with the *C. subvermispora* for 30 days of incubation. The results of chemical analysis and mass component loss showed that *C. subvermispora* was selective to lignin degradation. Pretreated SB after soda/anthraquinone pulping showed improved pulp yields, kappa number, and viscosity pretreated SB [69].

16.4 Enzymatic Hydrolysis of Pretreated SB/SL

Inherent properties of SB/SL cell wall make them resistant to cellulase mediated act for releasing monomeric sugars from hemicelluloses or cellulose. Pretreatment of SB/SL increase the accessible surface area and crystallinity of holocellulose which ameliorates the enzymatic action, in turn yielding optimum sugars recovery [14]. Table 16.4 summarizes the enzymatic hydrolysis profile of SB/SL after various kinds of pretreatment methods employed.

The hydrolysis rate of SB/SL directly depends upon the efficiency of pretreatment method used. Removal of lignin from the substrate determines the enzyme accessibility to the carbohydrate fraction of cell wall. A direct correlation exists between the removal of lignin and hemicelluloses on cellulose saccharification [11, 17]. Dilute H_2SO_4 pretreated SB followed by NAOH pretreatment (1 % NaOH, 60 min, 120 °C) produced 35 g/L sugars (86.2 % hydrolytic efficiency) after 96 h of enzymatic hydrolysis (10 FPU/g; 15 beta-glucosidase IU/g) [70]. Scanning electron microscopic (SEM) analysis also reveals the effect of dilute-acid pretreatment on native SB, alkaline pretreatment on cellulignin followed by enzymatic hydrolysis of cellulose. The un-homogeneity in structure and the disruption of first hemicelluloses followed by lignin and cellulose is clearly evident in SEM analysis (Fig. 16.3).

In the other study, oxalic acid pretreated SB (160 °C, 16 min, 3.5 % w/v Oxalic acid) produced total reducing sugars 56.3 g/g bagasse (92.30 % hydrolytic efficiency) after 120 h of enzymatic hydrolysis (20 FPU/g; 25 beta glycosidase (IU/g)) [71]. Among alkaline pretreatments, hydrated ammonia based pretreatment methods have found great interest recently [8]. Hydrated ammonia precisely act on lignin removal from the SB leaving hemicelluloses and cellulose together but in fragile form which yields appreciable sugar recovery upon enzymatic hydrolysis [8]. In association, our laboratory reveals the maximum sugars recovery (28 g/L) after 96 h of enzymatic hydrolysis of ammonia pretreated SB (20 % ammonia, 24 h, 70 °C). The optimum enzyme loadings (15 FPU/g and 17.5 beta glucosidase IU/g) were used (Chandel et al. Unpublished work).

Other important factors such as substrate concentration, cellulase loading, and end-product inhibition also plays an important role for the hydrolytic efficiency of SB/SL [9, 50]. In order to enhance the surface area of holocellulosic fraction in the cell wall, surfactants like Tween-20 have been used [44, 23]. To overcome the problems of end-product inhibition and process complexities, integrated process configurations such as SSF and consolidated bioprocessing (CBP) have been found successful [14, 50].

16.5 Economic and Environmental Concerns

In current scenario to utilize the lignocellulosic material for green energy, the pretreatment is an unavoidable step in biorefinery with the cost as high as 30 cents/gallon ethanol produced [10]. Approximately, pretreatment alone costs around 30 % to the

Table 16.4 Effect of different pretreatment methods on sugarcane residues for the sugars recovery after enzymatic hydrolysis

Pretreatment type	Enzymatic hydrolysis conditions	Sugars recovery (g/g or g/L)	References
Low alkali (NaOH)	Celluclast 1.5 L and Novozyme188 at 3:1 ratio (v/v) in citrate buffer (50 mM pH 4.8) at 50 °C	98 % Cellulose conversion	[51]
Ethanol organosolv	Celluclast 1.5 L (15 FPU/g substrate) and Novozym 188 (15 IU/g substrate) in citrate buffer (pH 4.8), SB 5 % (w/v), 150 rpm at 50 °C	20.9 g Glucose/100 g SB	[22]
H_2O_2 in alkaline media	Crude enzymatic extract from *Thermoascus aurantiacus*in citrate buffer (0.05 M, pH5.0) at 50 °C	39 % of Xylose, 59 % of xylobiose and 2 % of other xylooligosaccharides	[21]
AFEX	Spezyme CP (33 mg/g glucan), Novozyme 188 (31 mg/g glucan), Multifect xylanase (15 mg/g glucan), sodium citrate buffer (50 mM, pH 4.8), 96 h, 250 rpm, 50 °C	Conversion of xylan to xylose was 10 % higher for cane leaf residue (72 %) when compared with SB (62 %)	[8]
Ethanol organosolv	Celluclast 1.5 L (15 FPU/g substrate) and Novozym 188 (15 IU/g substrate) in citrate buffer (pH 4.8), SB 5 % (w/v), 150 rpm at 50 °C	18.1 g/L Glucose corresponding to 29.1 g glucose/100 g SB	[22]
IL	2 ml Of enzymatic mix containing *Acremonium* cellulase per gram of substrate (15 FPU/g substrate), Optimash^{TM} (0.2 % v/v), BG β-xylosidase in sodium acetate buffer (50 mM pH 5.0), 0.05 g of IL-treated substrates (2.5 % w/v), 48 h at 45 °C	Higher glucose (98.2 %) and xylose (60.7 %) saccharification yields in the presence of [Emim] [Ac]. Treatment with [Mmim] [DMP] resulted in glucose (61.9 %) and xylose (43.9 %)	[24]

total processing cost in the conversion of lignocellulosics into ethanol [29]. Performance of pretreatment methods and the incurred cost on bioconversion process was comprehensively analyzed by Eggeman and Elander [72]. The ideal pretreatment process needs to be highly efficient with imposing low operational and capital cost with less pollution [10, 16, 73]. Recovery of maximum sugars after pretreatment, less chemical load, usage of by-products, faster kinetics, and less process complexity are the important criteria to determine the overall impact of pretreatment methods [11, 16, 17]. All these features determine the cost on downstream processing steps and the trade-off with operational cost, capital cost, and biomass cost [10, 68, 73]. Table 16.5 analyzes the environment and economic impact of various pretreatment

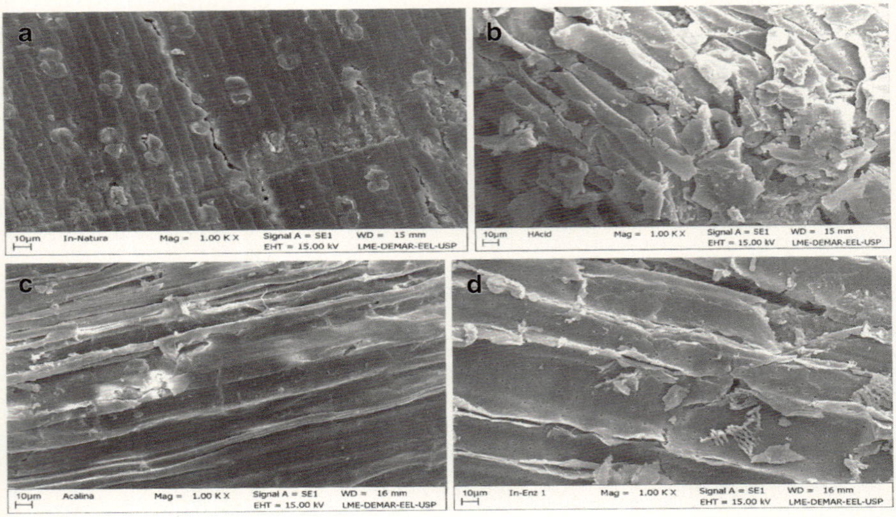

Fig. 16.3 Scanning electron microscopic analysis of SB after various pretreatments: **a** Native SB: Lignin-cellulose-hemicellulose close network in highly organized manner, **b** dilute sulfuric acid pretreated SB: Removal of hemicelluloses, less organized structure, **c** sodium hydroxide pretreatment of acid pretreated SB: Removal of lignin from cellulignin leaving cellulose in disorganized manner, making it amenable for cellulase action, **d** cellulase mediated hydrolysed SB: Coordinated action of cellulase leads the breakdown of cellulose polymer into glucose as monomeric units

methods applied to SB/SL. Critical features like upstream and downstream processing cost, capital investment, chemical recycling, usage of by-products, and waste treatment systems makes the comparison and evaluation of pretreatment methods difficult [74]. Net impact of each pretreatment method concerning the economics and environmental impact shows that chemical-based pretreatment methods have strong impact on economics of overall biomass conversion along with considerable environmental pollution burden (Table 16.5).

Pure chemical based pretreatment methods require significant amount of chemicals for biomass destruction with the significant amount of by-products generation [14, 17]. However, these methods are highly effective toward either lignin removal or hemicellulose degradation from lignocelluloses in short reaction times. Physico-chemical methods (dilute acid hydrolysis, AFEX, steam explosion, etc) have considerable effect on economics and environmental concerns [14, 17]. These methods are more specific toward hemicellulose or lignin degradation leaving cellulignin and holocellulose together but in disorganized manner amenable for better cellulases action. In comparison to alkaline methods, physic-chemical methods need less chemical load for the hemicellulose degradation. Biological, physical and LHW pretreatment methods do not require chemicals and are generally considered as moderate. Biological methods are generally safe but take longer time periods for lignin removal from the substrates.

Table 16.5 Economic and environmental aspects of different pretreatment methods applied to sugarcane residues

Pretreatment	Types	Factors governing economic impact	Factors governing environmental impact	Net impact*
Physical	Milling	Energy and capital intensive	Minimal	+
	Irradiation	Energy intensive	Radiations	+
Physico-chemical	Hot water	Electricity consumption		+
		Partial hemicellulose breakdown	By-products generation	
	Autohydrolysis	Electricity consumption		+
		Partial hemicellulose breakdown	By-products generation	
	Steam explosion	Electricity consumption		++
		Partial hemicellulose breakdown	By-products generation	
		Capital intensive		
	Ammonia fiber expansion	Ammonia consumption	High chemical load	++
		Capital intensive	Ammonia recovery	
		Lignin recovery		
Chemical	Acid pretreatment	Acid consumption	Acid load	++
		Capital intensive	By-products generation	
		By-products generation		
	Alkaline	Capital intensive	High chemical load	+++
		Sugars loss	By-products generation	
		Lignin recovery		
	IL	Capital intensive	High chemical load	+++
		Cost of chemicals used	By-products generation	
		Lignin recovery		
	Organosolv	Capital intensive	High chemical load	+++
		Lignin recovery	By-products generation	
	Oxidative delignification	Capital intensive	High chemical load	+++
		Lignin recovery	By-products generation	
Biological	ISMD	Longer incubation time	Negligible environment pollution	+
		Lignin recovery	Green technology	
		Sugars loss		

*Net impact of pretreatment methods can be defined as the overall impact of each pretreatment technology encompassing economics, environmental assessment, and efficiency to improve sugars recovery after enzymatic hydrolysis; +: moderate, ++: considerable, + + +: strong; *IL*: Ionic liquids, *ISMD*: in-situ microbial delignification

Application of lignin generated during alkaline pretreatment has been found important in many products of commercial significance such as resins, adhesives and coatings. Many industries are aiming forward to commercialize the lignin derived products [73]. Pretreatment like dilute acid hydrolysis, auto-hydrolysis and LHW degrade hemicellulose fraction of cell wall into varieties of sugars (xylose, arabinose, glucose, mannose and galactose) are used for the production of value-added products like D-xylitol, ethanol, lactic acid, single cell protein etc [2].

A detail economic analysis of each pretreatment strategy considering all the involving factors will help to direct research and development efforts in the success of commercialization of bioconversion processes [72, 73]. The renewed interest in sustainable development and environment friendly based practices, biotransformation processes are generally preferred over the conventional chemical conversion process. Unfortunately, most pretreatment protocols involve either strong chemicals or harsh physical conditions except bio-delignification. Pretreatment methods autohydrolysis, LHW, steam explosion do not deal with corrosive chemicals however strong physical parameters (high temperature and pressure) are the matter of concern [10]. Biological pretreatment are the least environmental pollution causing methods but their slow reaction time and loss of significant amount of carbohydrates are the important concerns while selecting them as pretreatment method of choice [13, 16]. Biological pretreatment methods have important benefits in a life cycle context also [68].

16.6 Expert Commentary and Five-year View

Pretreatment is an important process to overcome the recalcitrance of SB/SL. The primary goal of pretreatment is either to remove lignin or hemicellulose making the remaining carbohydrate accessible for enzymatic hydrolysis into simple sugars. Each pretreatment method has its pros and cons. The enzymatic hydrolysis efficiency directly depends upon the effectiveness of pretreatment strategy. In the last five years, several reports have been published describing the pretreatment method applied to sugarcane residues. The choice of pretreatment option depends upon maximum removal of lignin, less generation of inhibitors and high recovery of sugars with minimum enzyme dosage. Dilute acid hydrolysis, steam explosion and NAOH pretreatment are largely studied for SB hydrolysis. AFEX has been studied rationally less to pretreat SB or SL. Pretreatment of lignocellulose in tandem with enzyme cost play a very crucial role in overall economics of biorefinery or other lignocellulose based bioconversion industries. An effective pretreated lignocellulosic material requires less enzyme amount for complete holocellulose degradation. Since enzymes are expensive hence the fewer amounts of enzymes for the maximum degradation of cellulosics will impact the cost economics of lignocellulose based industries. Table 16.4 presents the effect of different pretreatment methods on sugarcane residues for the sugars recovery after enzymatic hydrolysis. It is clearly evident with the table that effective pretreatment of SB/SL require less enzyme loading in order to get maximum depolymerization of cellulose/hemicelluloses into their constituent sugar monomers.

16.7 Future Considerations and Conclusion

Sugarcane residues are generated in huge amount in the world every year and can be referred as "Green economy". To harness its fullest potential, pretreatment is a key to unlock the green economy. Pretreatment specifically acts on either hemicellulose or lignin removal increasing the accessibility of cellulose to cellulases eventually releasing monomeric sugars in ready-to-use form. These sugars are further converted into value-added products by microbial mediated processes. In the past, considerable research progress has been made to develop ideal pretreatment strategy considering SB/SL as raw material. Among the pretreatment methods, alkaline and acid mediated processes have been largely explored for pretreatment of SB/SL. Biological pretreatment has the maximum environmental and economic benefits but their slow reaction rates make them unattractive. Pretreatment methods like auto-hydrolysis, steam explosion and LHW could be effective for the removal of hemicelluloses posing less environment pollution but their efficacy toward SB/SL has limits. The need of hour is to develop the tailor made efficient pretreatment technologies aiming toward the specific, fast, economic and environmental friendly. Greater fundamental knowledge of chemicals and their possible action on SB/SL cell wall with the software-aided approach considering cheminformatics principles is required to choose the best pretreatment strategy for the effective carbohydrate depolymerization into fermentable sugars. Such an initiative will forward the commercialization of bio-based applications into new horizons which ultimately usher the overall economy of sugarcane producing countries.

Acknowledgments We are grateful to the BIOEN/FAPESP, CNPq and CAPES, Brazil for the financial assistance. We are also thankful to Dr. Om V. Singh from University of Pittsburgh, Bradford for the critical reading of chapter and valuable suggestions.

References

1. Pandey A, Soccol CR, Nigam P, Soccol VT (2000) Biotechnological potential of agro-industrial residues. I: Sugarcane bagasse. Bioresour Technol 74:69–80
2. Chandel AK, Silva SS, Carvalho W, Singh OV (2012) Sugarcane bagasse and leaves: foreseeable biomass of biofuel and bio-products. J Chem Technol Biotechnol 87:11–20
3. Dias MOS, Modesto M, Ensinas AV, Nebra SA, Filho RM, Rossel CEV (2009) Production of bioethanol and other bio-based materials from sugarcane bagasse: integration to conventional bioethanol production process. Chem Eng Res Des 87:1206–1216
4. Ojeda K, Avila O, Suarez J, Kafaro V (2010) Evaluation of technological alternatives for process integration of sugarcane bagasse for sustainable biofuels production—Part 1. Chem Eng Res Des 89:270–279
5. Giese EC, Chandel AK, dos Oliveira IS, Silva SS (2012) Prospects for the bioethanol production from sugarcane feedstock: Focus on Brazil. In: Goncalves JF, Correia KD (eds) Sugarcane: Production, Cultivation and Uses. Nova Science Publishers, New York (In Press)

6. Blair Euteneuer (2011) World's top 10 sugar-producing countries in 2010–2011. http://www.bloomberg.com/news/2011–10-06/world-s-top-10-sugar-producing-countriesin-2010–2011*-table-.html*. Accessed 20 June 2012
7. Kim S, Dale EB (2004) Global potential bioethanol production from wasted crops and crop residues. Biomass Bioenergy 26:361–375
8. Krishnan C, Sousa LC, Jin M, Chang L, Dale BE, Balan V (2010) Alkali based AFEX pretreatment for the conversion of sugarcane bagasse and cane leaf residues to ethanol. Biotechnol Bioeng 107:441–450
9. Sun Y, Cheng J (2002) Hydrolysis of lignocellulosic materials for ethanol production: a review. Bioresour Technol 83:1–11
10. Moiser N, Wyman C, Dale B, Elander R, Lee YY, Holtzapple M, Ladisch M (2005) Features of promising technologies for pretreatment of lignocellulosic biomass. Bioresour Technol 96:673–686
11. Hendriks ATWM, Zeeman G (2009) Pretreatments to enhance the digestibility of lignocellulosic biomass. Bioresour Technol 100:10–18
12. Chandel AK, Kapoor RK, Singh AK, Kuhad RC (2007) Detoxification of sugarcane bagasse hydrolysate improves ethanol production by Candida shehatae NCIM 3501. Bioresour Technol 98:1947–1950
13. Chandel AK, Kapoor RK, Narasu ML, Viswadevan V, Kumaran SSG, Ravinder R, Rao LV, Tripathi KK, Lal B, Kuhad RC (2007) Economic evaluation and environmental benefits of biofuel: an Indian perspective. Int J Global Energy Issues 28:357–381
14. Taherzadeh MJ, Karimi K (2008) Pretreatment of lignocellulosic wastes to improve ethanol and biogas production: a review. Int J Mol Sci 9:1621–1651
15. Kumar P, Barrett DM, Delwiche MJ, Stroeve P (2009) Methods for pretreatment of lignocellulosic biomass for efficient hydrolysis and biofuel production. Ind Eng Chem Res 48:3713–3729
16. Agbor VB, Cicek N, Sparling R, Berlin A, Levin DB (2011) Biomass pretreatment: fundamentals toward application. Biotechnol Adv 29:675–685
17. Brodeur G, Yau E, Badal K, Collier J, Ramachandran KB, Ramakrishnan S (2011) Chemical and physicochemical pretreatment of lignocellulosic biomass: a review. Enzyme Res 2011:1–17
18. Himmel ME, Ding S, Johnson D, Adney W, Nimlos M, Brady J, Foust T (2007) Biomass recalcitrance: engineering plants and enzymes for biofuels production. Science 315:804–807
19. Ferreira-Leitao V, Perrone CC, Rodrigues J, Franke APM, Macrelli S, Zacchi G (2010) An approach to the utilization of CO2 as impregnating agent in steam pretreatment of sugarcane bagasse and leaves for ethanol production. Biotechnol Biofuels 3:1–8
20. Chundawat SPS, Donohoe BS, Sousa LDS, Elder T, Agarwal UP, Lu F, Ralph J, Himmel ME, Balan V, Dale BE (2011) Multi-scale visualization and characterization of lignocellulosic plant cell wall deconstruction during thermochemical pretreatment. Energy Environ Sci 4:973–984
21. Brienzo M, Carvalho W, Milagres AMF (2010) Xylooligosaccharides production from alkali-pretreated sugarcane bagasse using xylanases from Thermoascus aurantiacus. Appl Biochem Biotechnol 162:1195–1205
22. Mesa L, González E, Ruiz E, Romero I, Cara C, Felissia F, Castro E (2010) Preliminary evaluation of organosolv pre-treatment of sugar cane bagasse for glucose production: application of 23 experimental design. Appl Energy 87:109–114
23. Mesa L, González E, Cara C, González M, Castro E, Mussatto SI (2011) The effect of organosolv pretreatment variables on enzymatic hydrolysis of sugarcane bagasse. Chem Eng J 168:1157–1162
24. da Silva SA, Lee S-H, Endo T, Bon EPS (2011) Major improvement in the rate and yield of enzymatic saccharification of sugarcane bagasse via pretreatment with the ionic liquid 1-ethyl-3-methylimidazolium acetate ([Emim] [Ac]). Bioresour Technol 102:10505–10509
25. da Silva AS, Inoue H, Endo T, Yano S, Bon EPS (2010) Milling pretreatment of sugarcane bagasse and straw for enzymatic hydrolysis and ethanol fermentation. Bioresour Technol 101:7402–7409

26. Buaban B, Inoue H, Yano S, Tanapongpipat S, Ruanglek V, Champreda V, Pichyangkura R, Rengpipat S, Eurwilaichitr L (2010) Bioethanol production from ball milled bagasse using an on-site produced fungal enzyme cocktail and xylose-fermenting *Pichia* stipitis. J Biosci Bioeng 110:18–25
27. Zheng Y, Pan Z, Zhang R (2009) Overview of biomass pretreatment for cellulosic ethanol production. Int J Agric Biol Eng 2:51–68
28. Binod P, SatyanagalakshmI K, Sindhu R, Janu KU, Sukumaran RK, Pandey A (2012) Short duration microwave assisted pretreatment enhances the enzymatic saccharification and fermentable sugar yield from sugarcane bagasse. Renewable Energy 37:109–116
29. Alvira P, Tomás-Pejó E, Ballestero M, Negro MJ (2010) Pretreatment technologies for an efficient bioethanol production process based on enzymatic hydrolysis: a review. Bioresour Technol 101:4851–4861
30. Laser M, Schulman D, Allen SG, Lichwa J, Antal Jr MJ, Lynd LR (2002) A comparison of liquid hot water and steam pretreatments of sugar cane bagasse for bioconversion to ethanol. Bioresour Technol 81:33–44
31. Lavarack BP, Griffin GJ, Rodman D (2002) The acid hydrolysis of sugarcane bagasse hemicelluloses to produce xylose, arabinose, glucose and other products. Biomass Bioenergy 23:367–380
32. Dekker RFH, Wallis AFA (1983) Enzymatic saccharification of sugarcane bagasse pretreated by autohydrolysis—steam explosion. Biotechnol Bioeng 25:3027–3048
33. Brugnago RJ, Satyanarayana KG, Wypych F, Ramos LP (2011) The effect of steam explosion on the production of sugarcane bagasse/polyester composites. Composites: Part A Appl Sci Manuf 42:364–370
34. Singh RK, Varma AJ, Laxman RS, Rao M (2009) Hydrolysis of cellulose derived from steam exploded bagasse by Penicillium cellulases: comparison with commercial cellulase. Bioresour Technol 100:6679–6681
35. Sendelius J (2005) Steam pretreatment optimisation for sugarcane bagasse in bioethanol production. Master of Science Thesis, Department of Chemical Engineering, Lund University, Sweden
36. Narayanaswamy N, Faik A, Goetz DJ, Gu T (2011) Supercritical carbon dioxide pretreatment of corn stover and switchgrass for lignocellulosic ethanol production. Bioresour Technol 102:6995–7000
37. Sarkar N, Ghosh SK, Bannerjee S, Aikat K (2012) Bioethanol production from agricultural wastes: an overview. Renewable Energy 37:19–27
38. Zheng Y, Lin H-M, Tsao GT (1998) Pretreatment for cellulose hydrolysis by carbon dioxide explosion. Biotechnol Prog 14:890–896
39. Carrasco C, Baudel HM, Sendelius J, Modig T, Roslander C, Galbe M, Hahn-Hägerdal B, Zacchi G, Lidén G (2010) SO2-catalyzed steam pretreatment and fermentation of enzymatically hydrolyzed sugarcane bagasse. Enzyme Microb Technol 46:64–73
40. Prior BA, Day DF (2008) Hydrolysis of ammonia-pretreated sugar cane bagasse with cellulase, β-glucosidase, and hemicellulase preparations. Appl Biochem Biotechnol 146:151–164
41. Chandel AK, Chandrasekhar G, Narasu ML, Rao LV (2010) Simultaneous saccharification and fermentation (SSF) of aqueous ammonia pretreated *Saccharum spontaneum* (wild sugarcane) for second generation ethanol production. Sugar Technol 12:125–132
42. Carvalheiro F, Duarte LC, Gírio FM (2008) Hemicellulose biorefineries: a review on biomass pretreatments. J Sci Ind Res 67:849–864
43. Aita GA, Salvi DA, Walker MS (2011) Enzyme hydrolysis and ethanol fermentation of dilute ammonia pretreated energy cane. Bioresour Technol 102:4444–4448
44. Canilha L, Santos VTO, Rocha GJM, Silva JBM, Giulietti M, Silva SS, Felipe MGA, Ferraz A, Milagres AMF, Carvalho W (2011) A study on the pretreatment of a sugarcane bagasse sample with dilute sulfuric acid. J Ind Microbiol Biotechnol 38:1467–1475
45. Sarrouh B, Silva S, Santos D, Converti A (2007) Technical/economical evaluation of sugarcane bagasse hydrolysis for bioethanol production. Chem Eng Technol 30:270–275

46. Moutta RO, Chandel AK, Rodrigues RCLB, Silva MB, Rocha GJM, Silva SS (2011) Statistical optimization of sugarcane leaves straw hydrolysis into simple sugars by dilute sulfuric acid as a catalyst. Sug Tech. doi:10.1007/s12355-011-0116-y
47. Fonseca BG, Moutta RO, Ferraz FO, Vieira ER, Nogueira AS, Baratella BF, Rodrigues LC, Hou-Rui Z, Silva SS (2011) Biological detoxification of different hemicellulosic hydrolysates using Issatchenkia occidentalis CCTCC M 206097 yeast. J Ind Microbiol Biotechnol 38:199–207
48. Chandel AK, Silva SS, Singh OV (2011) Detoxification of lignocellulosic hydrolysates for improved bioconversion of bioethanol. In: Bernardes MAS (ed) Biofuel production—recent developments and prospects. In Tech, Rijeka
49. Gírio FM, Fonseca C, Carvalheiro F, Duarte LC, Marques S, Bogel-Łukasik R (2010) Hemicelluloses for fuel ethanol: a review. Bioresour Technol 101:4775–4800
50. Chandel AK, Chandrasekhar G, Silva MB, Silva SS (2011) The realm of cellulases in biorefinery development. Crit Rev Biotechnol 32:187–202
51. Wu L, Li Y, Arakane M, Ike M, Wada M, Terajima Y, Ishikawa S, Tokuyasu K (2011) Efficient conversion of sugarcane stalks into ethanol employing low temperature alkali pretreatment method. Bioresour Technol 102:11183–11188
52. Rivera EC, Rabelo SC, dos Reis Garcia D, Filho RM, da Costa AC (2010) Enzymatic hydrolysis of sugarcane bagasse for bioethanol production: determining optimal enzyme loading using neural networks. J Chem Technol Biotechnol 85:983–992
53. Karr WE, Holtzapple T (2000) Using lime pretreatment to facilitate the enzymatic hydrolysis of corn stover. Biomass Bioenergy 18:189–199
54. Wu L, Arakane M, Ike M, Wada M, Takai T, Gau M, Tokuyasu K (2011) Low temperature alkali pretreatment for improving enzymatic digestibility of sweet sorghum bagasse for ethanol production. Bioresour Technol 102:4793–4799
55. Rabelo SC, Carrere H, Filho MR, Costa AC (2011) Production of bioethanol, methane and heat from sugarcane bagasse in a biorefinery concept. Bioresour Technol 102:7887–7895
56. Rocha GJM, Goncalves AR, Oliveira BR, Olivares EG, Rossell CEV (2012) Steam explosion pretreatment reproduction and alkaline delignification reactions performed on a pilot scale with sugarcane bagasse for bioethanol production. Ind Crops Prod 35:274–279
57. Xu F, Sun JX, Liu CF, Sun RC (2006) Comparative study of alkali- and acidic organic solvent-soluble hemicellulosic polysaccharides from sugarcane bagasse. Carbohydr Res 341:253–261
58. Zhao X, Peng F, Cheng K, Liu D (2009) Enhancement of the enzymatic digestibility of sugarcane bagasse by alkali–peracetic acid pretreatment. Enzyme Microb Technol 44:17–23
59. Martín C, Thomsen AB (2007) Wet oxidation pretreatment of lignocellulosic residues of sugarcane, rice, cassava and peanuts for ethanol production. J Chem Technol Biotechnol 82:174–181
60. Azzam AM (1989) Pretreatment of cane bagasse with alkaline hydrogen peroxide for enzymatic hydrolysis of cellulose and ethanol fermentation. J Environ Sci Health 24:421–433
61. Zhao XB, Wang L, Liu DH (2008) Technical note peracetic acid pretreatment of sugarcane bagasse for enzymatic hydrolysis: a continued work. J Chem Technol Biotechnol 83:950–956
62. Neely WC (1984) Factors affecting the pretreatment of biomass with gaseous ozone. Biotechnol Bioeng 26:59–65
63. Diedericks D, Rensburg EV, del Prado García-Aparicio M, Görgens JF (2011) Enhancing the enzymatic digestibility of sugarcane bagasse through the application of an ionic liquid in combination with an acid catalyst. Biotechnol Prog. doi:10.1002/btpr.711
64. Olivier-Bourbigou H, Magna L, Morvan D (2010) Ionic liquids and catalysis: recent progress from knowledge to applications. Appl Catal A: Gen 373:1–56
65. Liu CF, Sun RC, Zhang AP, Ren JL (2007) Preparation of sugarcane bagasse cellulosic phthalate using an ionic liquid as reaction medium. Carbohydr Polym 68:17–25
66. Ma S, Xue XL, Yu SJ, Wang ZH (2012) High-intensity ultrasound irradiated modification of sugarcane bagasse cellulose in an ionic liquid. Ind Crops Prod 35:135–139
67. Zhao X, Cheng K, Liu D (2009) Organosolv pretreatment of lignocellulosic biomass for enzymatic hydrolysis. Appl Microbiol Biotechnol 82:815–827

68. Chen S, Zhang X, Singh D, Yu H, Yang X (2010) Review: biological pretreatment of lignocellulosics—potential, progress and challenges. Biofuels 1:177–199
69. Costa SM, Gonçalves AR, Esposito E (2005) Ceriporiopsis subvermispora used in delignification of sugarcane bagasse prior to soda/anthraquinone pulping. Appl Biochem Biotechnol 121–124:695–706
70. Chandel AK, Calvet G, Giese EC, Reis J, Silva SS (2011) Statistical optimization of dual acid-base pretreatment of sugarcane bagasse improves the enzymatic hydrolysis into fermentable sugars. Workshop on second generation of bioethanol: enzymatic hydrolysis. Campinas Brazil, 24–25 Nov
71. Chandel AK, Antunes FAF, Calvet G, Finotti SB, Marino VF, Silva SS (2011) Hydrolysis of sugarcane bagasse with oxalic acid catalyzed process for ethanol production by *Pichia stipitis* NRRL-Y-7124. 1st Brazilian bioenergy science and technology (BBEST) conference.. Campus da Jordao, Brazil, 14–18 Aug 2011
72. Eggeman T, Elander TR (2005) Process and economic analysis of pretreatment technologies. Bioresour Technol 8:2019–2025
73. Chandel AK, Singh OV, Chandrasekhar G, Rao LV, Narasu ML (2010) Key-drivers influencing the commercialization of ethanol based biorefineries. J Commer Biotechnol 16:239–257
74. Jeoh T, Ishizawa CI, Davis MF, Himmel ME, Adney WS, Johnson DK (2007) Cellulase digestibility of pretreated biomass is limited by cellulose accessibility. Biotechnol Bioeng 98:112–122

Chapter 17
Pre-treatment of Malaysian Agricultural Wastes Toward Biofuel Production

Suzana Yusup, Murni Melati Ahmad, Yoshimitsu Uemura, Razol Mahari Ali, Azlin Suhaida Azmi, Mas Fatiha Mohamad and Sean Lim Lay

Abstract Various renewable energy technologies are under considerable interest due to the projected depletion of our primary sources of energy and global warming associated with their utilizations. One of the alternatives under focus is renewable fuels produced from agricultural wastes. Malaysia, being one of the largest producers of palm oil, generates abundant agricultural wastes such as fibers, shells, fronds, and trunks with the potential to be converted to biofuels. However, prior to conversion of these materials to useful products, pre-treatment of biomass is essential as it influences the energy utilization in the conversion process and feedstock quality. This chapter focuses on pre-treatment technology of palm-based agriculture waste prior to conversion to solid, liquid, and gas fuel. Pre-treatment methods can be classified into physical, thermal, biological, and chemicals or any combination of these methods. Selecting the most suitable pre-treatment method could be very challenging due to complexities of biomass properties. Physical treatment involves grinding and sieving of biomass into various particle sizes whereas thermal treatment consists of pyrolysis

S. Yusup (✉) · M. M. Ahmad · Y. Uemura
Chemical Engineering Department, Universiti Teknologi PETRONAS,
Bandar Seri Iskandar, 31750 Tronoh, Perak, Malaysia
e-mail: drsuzana_yusuf@petronas.com.my

R. M. Ali
Management and Humanities Department, Universiti Teknologi PETRONAS,
Bandar Seri Iskandar, 31750 Tronoh, Perak, Malaysia

A. S. Azmi
Department of Biotechnology Engineering, Faculty of Engineering,
International Islamic University Malaysia,
Jalan Gombak, 53100 Kuala Lumpur, Malaysia

M. F. Mohamad
Biomass Processing Laboratory, Green Technology MOR,
Universiti Teknologi PETRONAS, Bandar Seri Iskandar,
31750 Tronoh, Perak, Malaysia

S. L. Lay
Kawasan Institusi Bangi, PETRONAS Research Sdn. Bhd.,
Lot 3288 & 3289, Off Jalan Ayer Itam, 43000 Kajang, Selangor, Malaysia

Z. Fang (ed.), *Pretreatment Techniques for Biofuels and Biorefineries*,
Green Energy and Technology,
DOI 10.1007/978-3-642-32735-3_17,

and torrefaction processes. Additionally biological and chemical treatment using enzymes and chemicals to derive lignin from biomass are also discussed.

Keywords Pre-treatment · Palm waste · Physical · Thermal · Biological · Chemical

17.1 Introduction

Energy from biomass sources accounts for 11 % of the total world energy supply [1]. Biomass of forestry, agricultural, and municipal wastes has become an alternative source for renewable energy to fulfill global energy demand which currently is sustained by the depleting fossil-based fuels, due to its abundance amount. As the world's largest producer of palm oil, the availability of biomass from the palm oil industry provides an excellent opportunity for Malaysia to produce biofuels [2]. There are more than 3 million hectares of oil palm plantations in Malaysia and each year, about 90 million metric ton of renewable biomass in the form of trunks, fronds, shells, palm press fiber, and the empty fruit bunch (EFB) are produced [2]. Biomass is characterized as a low energy density material, thus converting biomass into gaseous, liquid or solid-derived fuels and chemicals can be challenging. In general, biomass are characterized using several important properties such as calorific value (CV), bulk density, moisture content, ash content, and volatile and fixed carbon (FC) content. A lot of difficulties need to be overcome in order to utilize the biomass as fuel feedstock; one of them is the limitations associated with the biomass fuel characteristics. Pre-treatment can enhance the properties of biomass prior to its effective conversion into fuels and chemicals. In this chapter, four pre-treatment methods including physical, thermal, biological, and chemical treatment are reviewed.

For example, direct comparison of solid biomass with coal, which is still the leading solid fuel for electricity and heat generation, frequently discloses inferior properties of biomass [3]. Typically, biomass has low-energy densities and high moisture content and is more tenacious (due to its fibrous nature). These properties give drawbacks such as lower combustion efficiencies and gasifier design limitations. Compared to coal, biomass varies in many properties; such as heating values, ultimate analysis (amounts of carbon, hydrogen, nitrogen, sulphur, and other impurities), and proximate analysis (FC, volatile material, ash content, and moisture content) [4]. Properties of biomass such as moisture and ash contents, volatile compounds, and particle size impose significant effect on the performance of a gasifier [5, 6]. Thus pre-treatment of biomass is crucial to increase its potentials for subsequent utilization as a solid fuel.

Biomass can be converted to liquid fuel through thermochemical or pyrolysis process. The liquid product is known as bio-oil. Bio-oil is a complex mixture of highly oxygenated compounds [7] and can be burnt efficiently in standard or slightly modified boilers [8, 9] and internal combustion engines [10, 11] at rates similar to those of commercial fuels. However, the combustion occurs at compromised heating values, that is, 40–50 % of that for hydrocarbon fuels. This is due to high water content (15–25 wt%) and high oxygen content (35–40 wt% on dry basis) that is detrimental for ignition. In addition, organic acids, mostly acetic and formic acid, in bio-oil are corrosive to common construction materials [12, 13]. The corrosiveness becomes severe at elevated temperature and with the increase of water content [14]. Moreover, solids (char) in bio-oil can cause clogging in injectors or corrode turbine blades. Over time, physicochemical changes will occur that further degrade the quality of the bio-oil. Presence of moistures will also decrease the biomass heating value. The density and viscosity of the liquid oil increase as the high molecular mass lignin fraction content increases. The increase in viscosity increases the pour point. The decrease in volatile aldehydes and ketones increases the flash point of the liquids.

Fermentation of biomass that is composed of cellulose, hemicellulose, lignin, and proteins, into bioalcohols quite often requires pre-treatment of biomass to break away the sugars contained within the matrix of cellulose fibers in plant cell walls. Pre-treatment also removes lignin from biomass, making it more digestable for enzymatic and microbial hydrolysis of lignocollulosic biomass [15]. Degradation of the lignin and hemicelluloses by the action of white-rot fungi is an aerobic process but there are some bacteria like *Entrobacter lignoluticus* SCF1 and rumen microorganisms with the lignin degradating capability under anaerobic condition [16, 17].

17.2 Pre-Treatment Processes

Pre-treatment methods for biomass prior to conversion into fuels and chemicals can be classified into physical, thermal, biological, and chemical, or any combination of these methods. The selection of appropriate pre-treatment should be based on the aspects to enhance the biomass properties that increase the efficiency of the conversion process.

Biomass contains varying amounts of cellulose (40–60 %), hemicellulose (20–40 %), lignin (10–25 %), and small amount of extractives. These wide range of these fractions lead to different thermal behavior and digestability. Hemicellulose, which is the most reactive compound in biomass, decomposes at relatively low temperature, that is, within the range of 225–325 °C. Meanwhile, cellulose degrades between 305 and 375 °C and lignin decomposes gradually over the temperature range of 250–500 °C [5].

The four methods reviewed in this chapter are physical, thermal, biological, and chemical pre-treatments. Physical and thermal treatments are mainly to remove moisture; hence increase the energy density of that particular biomass. The chemical and biological treatments are performed to de-polymerize lignin.

17.2.1 Physical Pre-treatment

Various physical treatment of biomass prior to its conversion into liquid, solid or gaseous products are available. The main purpose of the selected treatment depends

Table 17.1 Physical pre-treatment for biomass hydrolysis [31]

Technique	Particle size (mm)	Purpose
Harvesting and preconditioning	10–50	–
Chipping	10–30	To reduce heat and mass transfer limitations
Grinding and milling (either before or after chemical pre-treatment)	0.2–2	To reduce size and crystallinity (more effective compared to chipping due to shear forces generated during milling). Milling increases yields of biogas, bioethanol, and biohydrogen [32]
Gamma rays	–	To cleave β-1,4 glycosidic bonds [33], however very expensive, with environmental and safety concerns
	<0.4	Has little effect on rates and yields [34]

on the requirement of the conversion process, that is, to have the biomass at appropriate size for easiness or strength and durability in handling and processing. Specifically biomass conversion can be enhanced via increased digestibility [18] from increased specific surface area and reduced degree of polymerization and cellulose crystallinity [19].

Generally, different biomass may have different density and resistance that may hinder in handling, transportation, and storage. These limitations, however, can be overcome via densification of the biomass to improve its properties such as bulk density, that is, from 40–200 to 600–800 kg/m^3 [20], abrasion resistance, impact resistance, compressive strength, water resistance, and long-term performance [21, 22].

Process for biomass densification includes agglomeration that increases the particle size. This can be performed using pressure [23] in such techniques like extrusion, pelletizing, roll briquetting [24–26], and compaction [20, 22]. Another technique is tumble agglomeration that uses binding agents [23] that chemically or physically adhere to solid surfaces and form a bridge between the biomass particles.

On the other hand, appropriate sizing of biomass can also improve its physical properties for enhanced combustion efficiency via higher burning rates [27]. Full automatic operation and complete combustion in furnaces can also be achieved when the biomass is homogenously densified [28].

Grinder and sieve shaker are used to provide homogeneous particle size for the biomass feedstock. This eases the feeding process of biomass into the gasification system. For more efficient feed preparation, biomass is dried and subsequently ground and sieved to a range of 250–500 μm to enable smooth feeding and minimum fluidization velocity [2, 29].

Moreover, biomass ash may have high inorganic trace elements that can cause slagging, fouling, and corrosion in combustion equipment. However, the quantity of these trace elements such as potassium, sodium, and chlorine content can be reduced via leaching, that is, washing [30].

As for hydrolysis of biomass into fuels, different mechanical size reduction techniques can be applied mainly to reduce cellulose crystallinity. The techniques and its uses are listed in Table 17.1.

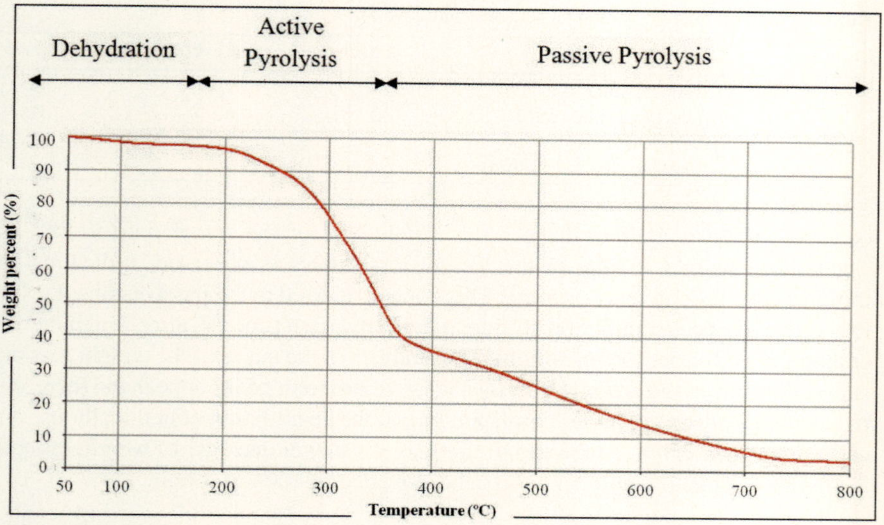

Fig. 17.1 Thermogram of dried OPF

17.2.2 Thermal Pre-treatment

Biomass can be thermally pretreated through either pyrolysis or torrefaction process. Pyrolysis involved breakdown of large complex hydrocarbon molecules of biomass into relatively smaller and simpler molecules of gas, liquid, and char [35]. Torrefaction involves heating of biomass in inert condition at temperature between 200 and 300 °C. Both pyrolysis and torrefaction removes moisture from the biomass followed by decomposition of biomass. Thermally pretreated biomass has better handling and fuel properties. One of the most common disadvantages of biomass compared to coal is the bulk amount of moisture that lowered its heating value and combustion performance.

17.2.2.1 Pyrolysis

Most researchers conducted pyrolysis in the absence of air or in the presence of medium such as water or hydrogen. Pyrolysis process in vacuum at medium heating rate, temperature of 400 °C, and residence time of 2–30 s will result in bio-oil as the product [4, 35]. Figure 17.1 shows the weight loss (%) curve obtained during the pyrolysis of dried oil palm fronds (OPF) under inert atmosphere at a heating rate of 40 °C/min. OPF has been selected as an example among generated waste from oil palm plantation in Malaysia, since it is easily available due to pruning of palm tree. According to this figure, pyrolysis curves of dried OPF follow the usual shape for most of the lignocellulosic materials [36]. During thermal degradation of dried OPF, two distinct pyrolysis zones are observed. Once the loss of water and volatile compounds

Table 17.2 Experimental parameters (factors and levels)

Column	Factors	Level 1	Level 2	Level 3
1	Temperature (°C)	220	250	280
2	Residence time (min)	30	60	90
3	Particle size (mm)	<0.25	0.25–0.50	>0.50
4	Nitrogen flowrate (l/min)	0.1	0.2	0.3

occurred, there is a sharp drop in the weight loss of the samples up to 385 °C. Thereafter, a slight change in the weight loss curves is observed, indicating the initiation of a second reaction zone. This zone is referred to the passive zone. Further loss of weight occurs until 800 °C due to devolatization process, after which there is essentially no further loss of weight. The analysis of the curve of the weight loss rate shows that during the active pyrolysis zone, two different peaks appear and therefore, two decomposition processes corresponding to the degradation of hemicellulose and cellulose are observed. At the end of this stage, a slower decrease of weight loss rate is observed. This loss corresponds to the slow degradation of lignin.

17.2.2.2 Torrefaction Process

Torrefaction is the heating of biomass in the absence of oxygen at temperature lower than 300 °C, or in other words, partial pyrolysis of biomass, to produce uniform product with lower moisture contents and higher energy contents. Prins et al. in their study on torrefaction of dry willow at two different temperatures of 250 °C and 300 °C, observed that more volatiles are obtained at 350 °C with mass yield of 67 % [4].

The mass and energy from the biomass is predominantly conserved in the solid product (torrefied biomass) [1]. Besides, this process is able to depolymerize the long polysaccharide chains, producing a hydrophobic solid product with an increased energy density and consequently increase grindability [3]. Much better fuel quality with higher CV for combustion and gasification applications is obtained through torrefaction [4, 37]. Torrefied materials are commonly used as solid fuel in industries and residences, similar to that of charcoal. However, when performed at temperature higher than 300 °C, the process is no longer referred to as torrefaction, but as flash or rapid pyrolysis that occurs at faster rates.

In this study, the torrefaction was performed in a horizontal tubular furnace flowed with nitrogen; N_2 being inert was used as torrefaction furnace. Four factors that include temperature, residence time, particle size, and inert gas flow rate were studied as listed in Table 17.2. Each was varied at three levels to perform nine sets of experiments following L9 orthogonal array following Taguchi Method. This method was used to analyze the energy density of torrefied products in order to get the optimum results. From the study, temperature contributes the most. Since the process is endothermic and temperature increase will also increase the minimum amount of energy needed to induce the reaction. Residence time and particle size are moderate contributors to the high heating values resulting in energy density and mass loss,

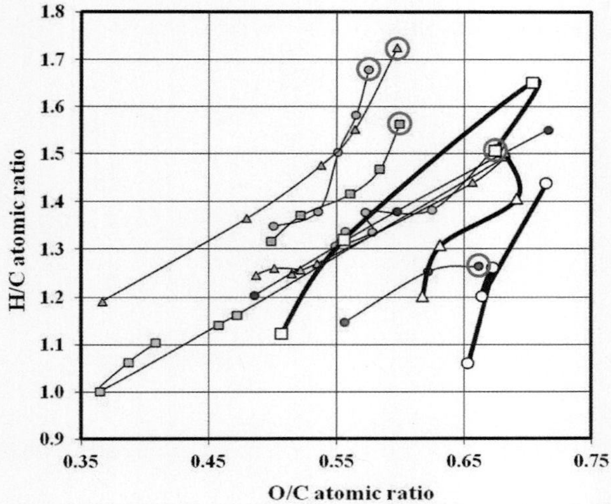

Fig. 17.2 Van Krevelen plot for untorrefied and torrefied biomass. (Reprinted from Ref. [40], Copyright 2011, with permission from Elsevier)

while contribution of nitrogen flow rate is insignificant. For oil palm kernel shell, the optimum condition was found to be at 280 °C and 60 min residence time, while particle sizes have insignificant effect in the process. Torrefaction process is found to be able to overcome some of the above limitations [3, 37, 38]. Torrefaction is able to remove moisture and low weight organic volatile component (carbon monoxide (CO) and carbon dioxide (CO_2)) and increase grindability as well as energy density. O/C ratio of torrefied biomass is lowered through the release of moisture, CO, and CO_2, which remove oxygen from the biomass, resulting in a more efficient gasification [4].

In general, the decomposition of lignocellulosic biomass under inert atmosphere proceeds in three steps; decomposition of hemicellulose (180–300 °C), cellulose (240–400 °C), and lignin (280–550 °C). Torrefaction is the first step, at which hemicellulose decomposes [38]. On the contrary, only a small portion of cellulose decomposes due to its crystallinity compared to hemicellulose. The third component which is lignin decomposes only a little due to its cross-linked structure. During torrefaction, major low molecular weight products include acetic acid, water, formic acid, methanol, lactic acid, furfural, hydroxyacetone, phenol, carbon dioxide and carbon monoxide [39]. Traces of hydrogen and methane are also detected [39]. Water may come from mainly moisture of biomass, and partly from dehydration of holocellulose. Acetic acid and carbon dioxide may come from acetyl group on hemicellulose. Change in the elementary composition during torrefaction may be exhibited using the Van Krevelen plot. In Fig. 17.2, the results extracted from some torrefaction studies are plotted; the H/C atomic ratio against the O/C ratio [40]. The raw biomass was shown by a key with a circle. As torrefaction proceeds, both the ratios, H/C and O/C, decrease in parallel regardless of the type of raw material biomass including beechwood, eucalyptus, canary grass, wheat straw, willow, oil palm empty fruit bunches (OPEFB), oil palm mesocarp fiber, and oil palm kernel

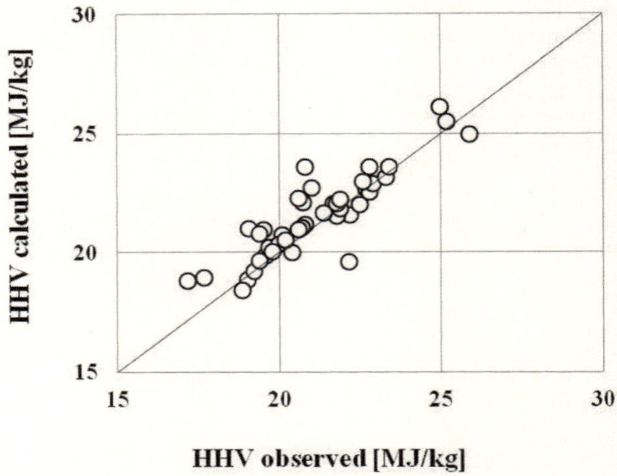

Fig. 17.3 HHV calculated by carbon content vs. HHV observed

shell. This change is mainly due to removal of water and carbon dioxide [39] as discussed previously.

Relationship Between Carbon Content and Calorific Value

The CV of lignocellulosic biomass and lignocellulosic biomass-derived fuels is one of the important fuel properties, which defines the energy density of those fuels. Since the estimation of this value from the elementary composition has been recognized as an important step for performance modeling of biomass and biomass-derived fuels, many correlations have been reported [31–48]. Nonetheless, only one report has been found that correlates the CV of torrefied biomass with its elementary composition [49]. This paper proposed the following equation, by which the CV *high heating value* (*HHV*) (MJ/kg) is well correlated with the elementary composition. HHV also known as gross calorific value (GCV) is defined as the amount of heat released by the unit mass or volume of fuel (initially at 25 °C) once it is combusted [35], as shown in Eq. 17.1 [38]:

$$HHV = 0.4373 \times C - 1.6701 \tag{17.1}$$

where C denotes the weight percent of carbon in the torrefied biomass. The correlation plot is shown in Fig. 17.3.

Effects of Parameters

Temperature

Torrefaction consists of a variety of chemical reactions that lignocellulose undergoes, as described above. The fact, mass yield decreases with torrefaction temperature clearly shows most of the activation energies are positive. Although the activation energy for each reaction has not been reported, the activation energy of the overall reaction has been reported in a few cases: willow biomass to be 76.0 kJ/mol [50] and oil palm empty fruit bunch (OPEFB) to be 37.3 kJ/mol [51]. The difference between these two may attribute to different type of biomass in this case due to different properties of wood and oil palm waste. The second factor is the experimental technique.

Residence Time

In previous papers for torrefaction, the residence time was selected to be between 0.5 and 5 h. The mass yield decreases fast in the first 30 min, and becomes almost constant, specifically below torrefaction temperature of 360 °C. At 280 °C, after 30 min a considerable decrease in mass yield was observed by 3 h [52] and 5 h [53].

Particle Size

Torrefaction is an endothermic reaction. The effect of heat transfer was discussed by Bergman et al. [54]. They proposed an index number, called the Pyrolysis number (py) and Biot number (Bi), as defined in Eqs. 17.2 [39] and 17.3 [40]:

$$py = \frac{\alpha}{k\rho c_p r_p} \tag{17.2}$$

Biot number:

$$Bi = \frac{\alpha r_p}{\lambda} \tag{17.3}$$

where α is the external heat transfer coefficient in W/(m^2 K), k is the reaction rate constant in s^{-1}, ρ is the density in kg/m^3, C_p is the heat capacity in J/(kg K), r_p is the radius in m, and λ is the thermal conductivity in W/(m K) of the biomass particle.

Using the criterion, Jalan and Srivastava estimated that biomass pyrolysis up to 873 K is controlled by the intrinsic pyrolysis reaction when particle sizes are less than 1 mm [55]. Biomass torrefaction is controlled by the intrinsic torrefaction reaction when biomass particle sizes are less than 2 mm, based on the criterion [4, 52]. Larger particle sizes induced secondary cracking reaction due to increase in the resistance towards primary pyrolysis product [35].

Sweep Gas Composition

In academic studies, nitrogen is commonly used as the sweep gas. Industrially, nitrogen is not a practical option due to its cost. In an industrial concept, re-circulation use of torrefaction gas is proposed [54]. From an economic point of view, use of boiler flue gas for torrefaction of oil palm residue is proposed [51]. They investigated the effects of oxygen concentration in torrefaction gas on the mass yield of torrefied biomass.

Advantages of Torrefaction for Solid Fuel Use and Gasification

The advantages of torrefaction are:

1. Higher CV due to partial decomposition and moisture removal [40, 53].
2. Longer shelf life due to moisture removal.
3. Moisture resistant pellets and briquettes due to higher hydrophobicity [56].
4. Enhanced grindability due to less toughness of torrefied biomass [52, 57].

All these advantages are related to solid use of torrefied biomass. At 1,400 °C, the two types of torrefied wood produced more hydrogen (7 %) and carbon monoxide (20 %) than the untorrefied wood [58]. This higher gas yields were attributed to a higher carbon and hydrogen content of the torrefied biomass.

17.2.3 Biological Pre-treatment

Biological pre-treatment is a slow process which can last for few weeks. However, the process has a mild reaction, less energy demand, low chemical usage, low capital cost, less side reaction, and is environment friendly [59, 60]. The process if compared to other pre-treatment processes can be an attractive alternative process due to its green technology. This pre-treatment employs microorganism and their enzymatic mechanism to break down the lignin and liberate cellulose and hemicellulose from the complex lignin. These selective microorganisms which produce oxidative enzyme to break the lignin are from type of fungi and bacteria.

Fungi, the wood rotten microorganism can be divided into three groups according to the morphology of wood decay which are soft-rot ascomycetes or detromycetes, white-and brown-rot basidomycetous. Degradation of the lignin and hemicelluloses by the action of white-rot fungi is an aerobic process but there are some bacteria like *E. lignoluticus* SCF1 and rumen microorganisms with lignin degrading capability under anaerobic condition [16, 17].

These fungi have shown positive effect on delignification process [61, 62]. White- and soft-rot fungi attack both cellulose and lignin while brown-rot fungi mainly attack cellulose and hemicellulose components in wood [62–64]. Some white-rot fungi can selectively delignify (lignin and hemicellulose) and leave enriched cellulose. Depending on types of fungi and wood, the lignin lost, observed, can reach up to 44 % [64]. These microorganisms, associated with lignin-degrading enzyme, consist of mainly two major families of enzymes which are laccase and peroxidase. Bacteria such asactinomycetes, which is a filamentous bacteria belonging to the genus *Streptomyces* are well known degrader that can mineralize up to 15 % of lignin. Others nonfilamentous bacteria mineralize lignin less than 10 % and can degrade the low molecular weight part of lignin [64].

Laccase (benzenediol:oxygenoxidoreductase, EC 1.10.3.2) belongs to the small group of enzymes called the blue copper proteins or the blue copper oxidases that catalyze one-electron oxidations of aromatic amines and phenolic compounds such as phenolic structures of lignin [60, 65]. Laccase, widely distributed in higher plants and fungi, is especially found abundant from white-rot fungi and is also found in insects and bacteria [65]. Fungal laccases have higher redox potential than bacterial or plant laccases (up to +800 mV), and their action seems to be relevant in nature, also finding important applications in biotechonology [65]. Thus, fungal laccases are involved in the degradation of lignin and detoxification of phenols arising during lignin degradation which inhibit fermentation process. Vikineswary et al. studied the production of laccase from sago hampas and OPF parenchyma tissue, which gave higher laccase productivity compared to rubberwood sawdust, using *Pycnoporus sanguineus* [66].

Peroxidase family consists of ligninolytic enzymes such as lignin peroxidase (LiP), manganese peroxidase (MnP), and versatile peroxidases (VP) [60, 62]. Lignin peroxidase (LiP, diarylpropane peroxidase, EC 1.11.1.14), an extracellular lignin olyticperoxidases which is commonly associated with *Phanerochaete chrysosporium* [67]. It has capability of catalyzing the depolymerization of the aromatic polymer lignin and a variety of non-phenolic lignin model compounds in the presence of H_2O_2 [60]. Others such as *Phanerochaete sordida* [68], *Aspergilllusstrains*, and bacteria such as *Acinetobactercalcoaceticus*, *Streptomyces viridosporus* and *Streptomyces lividans* [59] are also producing extracellular LiP.

MnP, hydrogen peroxidase oxidoreductase (EC 1.11.1.13), is able to oxidize Mn(II) to Mn(III) [60, 67]. It has also been reported isolated from *Cunninghamella elegans* [69], *Schizophyllum sp.*, *Ceriporiopsis subvermispora*, *Panus tigrinus*, *Lentinula edodes*, *Nematolomafrowardii*, *Bjerkandera adusta*, *Tinea versicolor*, and *Dichomitus squalens*. VP (EC 1.11.1.16) oxidize Mn(II) and a high redoc-potential aromatic compound as MnP and LiP, respectively [68]. VP is also reported to be produced by fungi from the genera *Pleurotus*, *Bjerkandera*, and *Lepista* and maybe also by *Panus* and *Trametes* species [70].

The enzyme activity and lignin degradation are influenced by several factors which include the type of strain and nutrient composition (i.e., in the case of delignification by using microorganism), enzyme dosage (i.e., in the case of delignification by isolated enzyme), moisture content, pH, aeration, and temperature. These factors can be manipulated to obtain the optimum pre-treatment process. Ahmad Khushairi and Zainol screened factors affecting biological delignification process of oil palm trunk using local oyster mushroom (*Pleurotus ostreatus*) [61]. In their study, temperature contributed the most in delignification process followed by pH. Other studied factors were fungi-to-medium ratio, moisture content, contact time, lighting, and humidity. Interesting to note that, in the study, even though biological delignification is known to be time consuming, the contact time between 2 and 10 days are among the least important compared to others.

In another study, direct delignification with a commercial biocatalyst called laccase was performed. Taguchi method was applied to determine the optimum lignin degradation of OPF from laccase treatment. The effects of laccase dose, pH, pretreatment temperature, and treatment time were investigated. The experiment results of the nine trial conditions with two runs per trial condition are shown in Table 17.3 where the percentage of lignin degradation is exhibited.

Table 17.3 Lignin degradation (%) of laccase treated OPF

Trial	Lignin degradation (%)	
	R_1	R_2
1	10.95	10.82
2	4.63	4.42
3	4.55	4.27
4	9.71	9.40
5	10.05	9.82
6	5.57	5.08
7	5.08	4.80
8	6.58	6.72
9	3.86	3.96

Results in Table 17.4 indicated that factor that influenced lignin degradation is in the following order: pH value > laccase dose > treatment time > treatment temperature. The optimum combination of factors and levels in lignin degradation of OPF may be predicted in general as follow: pH value of 4, laccase dose of 20 mg, treatment time for 2 h, and treatment temperature at 30 °C. However, the experiment results revealed very low percentage of lignin degradation (10–11 %) as shown in Table 17.3.

Syafwina et al. studied three white-rot fungi and found that *Dichomitus squalens* degraded lignin most rapidly compared to *Ceriporiopsis subvermispora* and Pleurotus ostreatuson in OPEFB [71]. After 8 weeks, weight loss of the lignin and holocellulose in beech wood reached 75.9, and 49.9 %, respectively. The fungus also delignified EFB. After 8 weeks, weight loss of lignin and holocellulose in EFB was 25.7 % and 22.8 %, respectively. Hamisan et al. performed chemical pre-treatment and compared to microbial on OPEFB [72]. Microbial pre-treatment using *Phanerochate chrysosporium* was shown to significantly removed the lignin, but it is timely (7 days) compared to chemical pre-treatment (3 h) which is fast reaction but quite harsh to the substrate. Namoolnoy et al. isolated 63 fungal of white-rot fungi; 27 of them showed high activity of laccase, MnP or LiP was selected to culture on OPFs [73]. Only seven isolates could degrade more than 50 % lignin content in OPFs within 30 days of cultivation. The efficiency in lignin degradation of selected fungal isolates seemed to be related to the ability of the fungi to produce more than one ligninolytic enzyme.

Table 17.4 Response table for signal to noise ratios for "Larger is better" option

Level	Laccase dose (mg)	pH	Treatment temperature (°C)	Treatment time (h)
1	15.57	18	17.2	17.51
2	18.01	16	14.8	13.82
3	14.05	13	15.5	16.31
Delta	3.96	4	2.3	3.69
Rank	2.00	1	4.0	3.00

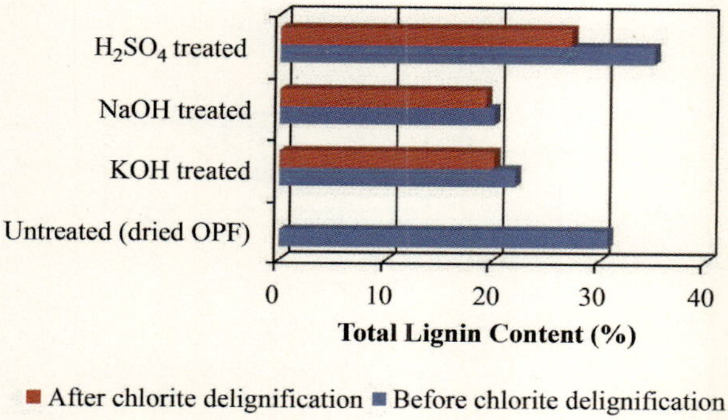

Fig. 17.4 Total lignin content in untreated dried OPF

17.2.4 Chemical Pre-treatment

Chemical pre-treatment such as alkaline (KOH and NaOH) and acid (H_2SO_4) pre-treatments are carried out in this study. Further study is done where three differently pretreated OPF samples were subjected to sodium chlorite delignification. Lignin content in these three pretreated OPF samples are compared before and after sodium chloride delignification. Alkali pre-treatment assists in the hydrolysis of lignin. The best performance is indicated by the lowest percentage of total lignin content remained in the sample after treatment. Study indicated that NaOH pre-treatment coupled with chlorite delignification gave the best performance on lignin removal among the different pre-treatment methods applied, with the lowest total lignin content remained in the sample which accounted for 19.32 % whereby total lignin content in untreated sample was found to have 30.78 % as shown in Fig. 17.4. Chlorite delignification can partially remove lignin from biomass [74]. The chlorite treatment was more effective than the organosolv method, as 2.5 % of lignin was left in chlorite treated barley husks as compared to 3.9 % in organosolv treated barley husks [74].

The treated OPF and untreated OPF is subjected to steam and gasified at temperature higher than 500 °C. For all experiments, biomass sample weight was approximately 5 mg. N_2 was used as inert carrier gas with a constant flow rate of 100 ml/min. The micro vacuum pump (650 mmHg was applied throughout the experiments) was attached to the gas chromatography (GC) to facilitate gaseous product from thermogravimetric analysis (TGA). Steam was generated by super heater up to 400 °C prior to injection in TGA. The system was purged with N_2 gas (100 ml/min) for about 20 min to remove entrapped gases at temperature of 50 °C. All samples were heated at a constant heating rate of 20 °C/min from 50 to 900 °C where it was kept constant for 10 min. To avoid condensation, steam was introduced when temperature inside the TGA reached to 110 °C. The amount of catalyst used was based on biomass-to-catalyst ratio of 3 (mass basis), while steam-to-biomass was kept constant at 1:1 ratio (mass basis). The apparatus used is shown in Fig. 17.5. The biomass steam gasification experiments were performed in a standard TGA (EXSTAR TG/DTA 6300, from SII) and GC (Agilent 7890A, Agilent Technologies) under non-isothermal conditions.

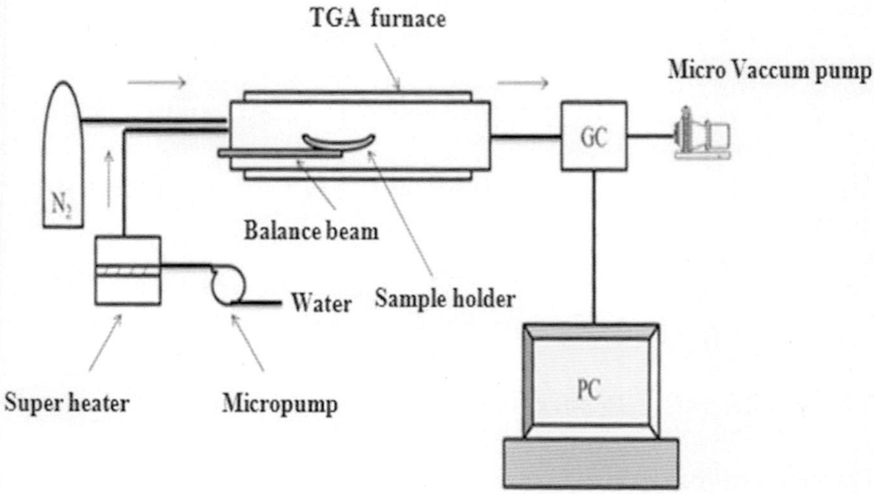

Fig. 17.5 Experimental Apparatus

Results from TGA-GC analysis reveal that NaOH coupled with chlorite treated OPF was able to produce higher yield of H_2 (56.56 vol%) compared to untreated dried OPF (39.48 vol%) as shown in Fig. 17.6. On the other hand, H_2 can be produced at a much lower temperature of 550 °C instead of 850 °C if the OPF is treated with NaOH coupled with chlorite; hence, the chemical pre-treatment has improved the biomass quality.

17.3 Biomass as Fuel

17.3.1 Solid Fuel

The CV of a biomass is a crucial property needed to be considered for its conversion into fuels. The importance of oxygen:carbon (O:C) and hydrogen:carbon (H:C) ratios on the CV of solid fuels can be illustrated using a Van Krevelen diagram. High content of oxygen and H_2 reduces the energy value of the fuel, since energy contained in carbon–oxygen (C–O) and carbon–hydrogen (C–H) bonds are lower than energy in carbon–carbon (C–C) bonds. Materials with low O:C and H:C ratios are usually preferred as fuels for gasification and combustion since they contain more energy. Composition of lignocellulose component of wood is seen to be closely related to coal since wood is the natural precursor for the formation of coal through diverse coal liquefaction processes [4]. The content of alkali metal such as Na, K, Mg, P, and Ca, in biomass is crucial for its thermo-chemical conversion into fuels. Reaction between alkali metals and silica present in ash produces sticky, mobile liquid phase

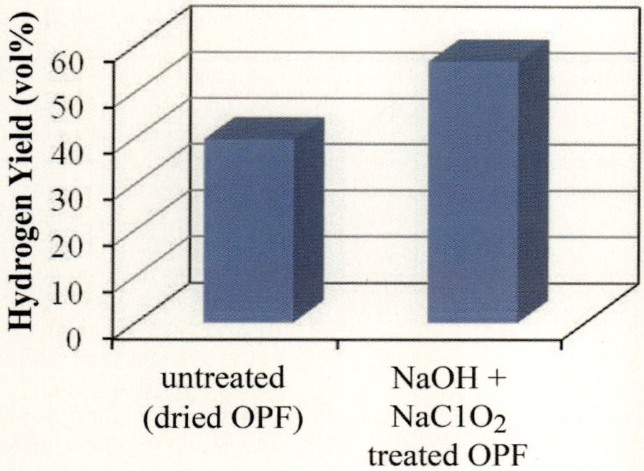

Fig. 17.6 Yield of hydrogen (vol%) produced from untreated and treated OPF

Table 17.5 Malaysian biomass samples

Types of biomass	Biomass sources
Oil palm (OP)	Kernel shell (*PKS*), frond (*OPF*), trunk (*OPT*), mesocarp fiber (*MCF*), empty fruit bunch (*EFB*)
Paddy (P)	Rice husk (*RH*), paddy straw (*PS*)
Sugar cane (SC)	Bagasse (*SCR*)
Coconut (C)	Fiber (*CF*)
Wood industries (W)	Sawdust (*SD*), woodchips (*WC*)
Rubber (R)	Kernel (*RSK*)

which can lead to blockages of airways in furnace. Total silica content may increase significantly due to contamination with soil during harvesting.

In a work performed by Abdullah and Yusup [29], the characterization of eleven samples of Malaysian biomass, listed in Table 17.5, has been performed and evaluated to determine their potential utilization as feedstock for biomass gasification into hydrogen.

In this work, the biomass is physically pre-treated via grinding and sieving. Analyses are carried out on sieved biomass to ensure homogeneous samples. Details on the types and working principle of the analysers used in characterizing the biomass are given in Table 17.6. Table 17.7 shows lists of characteristics of eleven types of biomass properties. The elemental analysis results of this biomass are given in Table 17.8. The composition of the hydrocarbon fuel is expressed in terms of its basic elements except for its moisture (M) and inorganic constituent. This is represented by the ultimate analysis. In comparison, the composition of biomass in terms of its volatile matter, FC, moisture, and ash content is represented by proximate analysis. A typical ultimate and proximate analysis is presented by Eqs. 17.4 and 17.5 [35];

$$\text{Ultimate analysis: } C + H + O + N + S + ASH + M = 100\,\% \tag{17.4}$$

$$\text{Proximate analysis: } VM + FC + M + ASH = 100\,\% \tag{17.5}$$

Table 17.6 Biomass characterization

Analysis	Working principle
Moisture content	A radiator dries the sample while its weight is continuously measured using a precise balance. The moisture content is indicated by the total weight loss
Combustion	It measures the high heating value (HHV) of the sample
Ultimate analysis	CHNS analyzer involves qualitative conversion of the four elements (C, H, N, S) into CO_2, $H_2O_{(g)}$, N_2, and SO_2. Sample is burnt at 1,000 °C in continuous oxygen flow. The components in the product gas are measured using infrared detectors, except for N_2 which is measured by a thermal conductivity detector. In the end, quantitative measurement of C, H, N, and S are recorded, while O content is calculated based on the difference
Proximate analysis	Thermogravimetric analysis (TGA) measures the weight loss of sample as a function of temperature. It provides quantitative measurement of intrinsic moisture, fixed carbon (FC), volatile matter, and ash content of the sample
Alkaline metal content	It measures individual component wavelengths of the fluorescent emission produced when sample is irradiated with x-rays
Bulk density ASTM D1895B	Sample was flowed freely through a funnel into a measuring cylinder. Volume and weight of the filled cylinder were determined and bulk density was measured in kg/m^3

Table 17.7 Characteristics of fuel-proximate analysis

Biomass	Moisture (M) (wt%)	Calorific value, CV (MJ/kg)	Proximate analysis (db, wt%)		
			Ash (db, wt%)	Volatiles, (VM)(db, wt%)	Fixed carbon, (FC) (by diff)
EFB	66.26	18.60	5.50	84.61	9.89
MCF	45.23	19.66	7.10	81.52	11.38
PKS	17.50	20.40	4.10	81.03	14.87
OPF	71.43	15.59	3.90	83.19	12.91
RSK	18.37	15.94	6.10	81.82	12.08
SD	16.30	18.19	5.00	78.80	16.20
SCR	52.20	15.25	3.90	80.19	15.91
RH	13.08	14.79	22.00	59.97	18.03
PS	8.47	13.74	18.30	72.48	9.22
CF	24.51	21.17	6.70	80.24	13.06
CS	18.50	18.18	5.52	75.00	19.48

The C, H, O, N, and S are expressed as the weight percentages of carbon, hydrogen, oxygen, nitrogen, and sulphur, respectively in the fuel.

Biomass moisture content can either be expressed in dry or wet basis as presented in Eqs. 17.6 and 17.7 [35];

$$\text{The dry-basis moisture is: } M_{\text{dry}} = (W_{\text{wet}} - W_{\text{dry}})/W_{\text{dry}} \tag{17.6}$$

$$\text{The wet-basis moisture is: } M_{\text{wet}} = (W_{\text{wet}} - W_{\text{dry}})/W_{\text{wet}} \tag{17.7}$$

Table 17.8 Characteristics of fuels-ultimate analysis

Biomass	Ultimate analysis (daf, wt%)				
	C	H	N	S	O (by diff)
EFB	40.73	5.75	1.40	0.22	92.63
MCF	40.97	5.96	0.77	0.51	92.76
PKS	49.65	6.13	0.41	0.48	92.98
OPF	42.10	5.46	0.70	0.13	93.71
RSK	44.01	6.11	0.58	.03	93.28
SD	43.68	6.65	0.23	0.04	93.08
SCR	42.93	5.82	0.68	0.06	93.44
RH	38.74	5.83	0.55	0.06	93.44
PS	33.48	6.01	1.46	0.15	92.38
CF	45.51	6.02	0.78	0.09	93.11
CS	43.00	6.30	0.75	0.05	92.90

Fig. 17.7 Biomass preference ranking [29]

The wet basis (M_{wet}) and dry basis (M_{dry}) are related as in Eq. 17.8 [35];

$$M_{dry} = M_{wet}/1 - M_{wet} \tag{17.8}$$

The FC content is calculated based on Eq. 17.9 [35];

$$\mathrm{FC} = 1 - \mathrm{M} - \mathrm{VM} - \mathrm{ASH} \tag{17.9}$$

The suitability of biomass as a gasification feedstock is screened using an aggregate matrix as shown in Table 17.9 based on its CV, O:C and H:C ratios and moisture, ash, volatiles, and FC contents [29]. A weighting factor is assigned to each characteristic according its significance for gasification process. The biomass is ranked, based on the scoring performed using the weighting factor. It was found that palm kernel shell (PKS), after being subjected to physical pre-treatment (grinding and sieving) and thermal pre-treatment (oven-drying), is the most preferred biomass among the eleven biomass as the gasification feedstock as shown in Fig. 17.7.

17.3.2 *Liquid Fuel*

On the other hand, as a good liquid fuel, bio-oil from pyrolysis of biomass, has to be homogenous and its properties should not change significantly during the storage [75]. As pyrolysis liquids contain a high number of compounds with various chemical functionalities, the homogenity of the liquid highly depends on the

Table 17.9 Aggregated matrix of Malaysian agricultural waste

Characteristics						
Biomass	Calorific value (CV)	O:C and H:C ratio	Moisture content (M)	Ash	Fixed carbon (FC), and volatiles, (VM)	Total score
	A	B	C	D	E	
EFB	2	2	0	1	3	8
MCF	2	2	0	0	3	9
PKS	3	2	2	2	3	12
OPF	2	2	0	2	3	9
RSK	2	2	2	1	3	10
SD	3	2	2	2	3	12
SCR	2	2	0	2	3	9
RH	2	2	2	0	3	9
PS	2	1	3	0	3	9
CF	2	2	2	1	3	10
CS	3	2	2	2	3	12

A: 0 for less than 10, 1 for less than 10–14, 2 for between 14 and 20, 3 for higher than 20 MJ/kg. (Note that typically biomass has 12–14 MJ/kg), B: 1 if furthest from coal, 2 if near toward coal, 3 if very near towards coal, C: 0 for more than 50, 1 for between 35and 50, 2 for between 10 and 35, 3 for less than 10 wt% moisture, D: 0 for more than 15, 1 for between 6 and 14, 2 for between 3 and 6, 3 for less than 3 wt% ash, E: 0 for less than 10, 1 for between 10 and 25, 2 for between 25 and 50, 3 for more than 50 wt%. (Note that typically almost more than 60 wt% of biomass is volatiles, VM, and fixed carbon, FC)

Table 17.10 Characterization results of pyrolysis liquid from Malaysian empty palm fruit bunch [7]

Properties	Value
Moisture content (%)	50–60
Density ($g{\cdot}cm^{-3}$)	1.2
pH	3
Solid content (%)	0.02–2
Elemental composition (%)	
C	50.36
H	7.83
N	4.45
S	0.16
O	37.21
Calorific value (CV) (MJ/kg)	21.62

complex solubility and reactivity of these chemical compounds. Typically, the pyrolysis liquids are single-phase liquids contaning varying amounts of solids (char). This char sediment gradually settles at the bottom of the barrel forming a thick sludge over time depending on the density difference between the liquid and particles [76]. Phase separation can occur if the total water content exceeds a certain threshold limit, making the liquid usage as fuel questionable unless it can be emulsified before use [75]. The properties of bio-oil such as moisture content, density, pH, and solid content from Malaysian EFB from palm industry was reported in reference [7] as 50–60 %, 1.2 g/cm^3, 3, and 0.02–2 %, respectively. The elemental compositions of the bio-oil are given in Table 17.10 [7].

Table 17.11 Characterization results of liquid fuel from Malaysian empty palm fruit bunch [77]

Properties	Commercial gasoline (RON95)	Bio-gasoline
Boiling point (°C)	60–120	80–120
Density (kg/m^3) (at 15 °C)	719.7–779.68	815.66–851.65
pH	3	
Solid content (%)	0.02–2	
Elemental composition (%)		
C	80–88	60–75
H	12–15	18–23
O	0	0.5–1.2

Table 17.12 Yield of liquid fuel from Malaysian empty palm fruit bunch [77]

Results	Organic liquid product	Bio-gasoline
Yield	91.2	90.84
Standard deviation	0.35–2.57	0–13.55
Optimum operating condition		
Temperature (°C)	350	400
Reaction time (s)	30	900
Catalyst weight (g)	10	30

In this study, the stable single phase mixture of bio-oil contained water ranging from 40 to 60 %. The density of the liquid was 1.2 g/cm^3, which was higher than that of fuel oil, that is, at around 0.85 g/cm^3, and significantly higher than that of the biomass which was at 1.1452 g/cm^3 [8]. This indicated that the liquid had approximately 42 % of the energy content of fuel oil on a weight basis, but 61 % on a volumetric basis. This may impose implications on the design and specification of equipment to process and handle the bio-oil such as pumps and atomizers in boilers and engines. More importantly, the CV of the pyrolysis liquid as determined mathematically was 21.62 MJ/kg, compared to 42–44 MJ/kg for conventional fuel oils. Thus, the liquid produced needs to be upgraded to become the alternative substitute for the existing fuel oils.

For the upgrading/purification methods, the common processes are catalytic cracking, hydrogenation, and steam reforming [77–83]. There are two approaches for catalytic cracking of bio-oil: offline catalytic cracking that utilizes bio-oil as raw material and online catalytic cracking which utilizes pyrolysis vapors as raw material [77, 84–90].

Hew et al. reported an upgrading study on bio-oil from Malaysian EFB using an off line heterogeneous catalytic cracking process [77]. They applied Taguchi L9 method to identify optimum operating condition to upgrade empty fruit oil palm bunch-derived pyrolysis oil to liquid fuel, mainly gasoline or organic liquid vapor. The properties of the bio-gasoline obtained are given in Table 17.11.

The yields and optimal conditions for the upgrading process of the bio-oil into liquid fuels are reported in Table 17.12. The equations used to calculate the yield of organic liquid product (OLP) and the yield of gasoline are as Eqs. 17.10 and 17.11, respectively [77];

$$\text{Yield}_{\text{OLP}} = (\text{Weight}_{\text{OLP}}/\text{Weight}_{\text{bio-oil}}) \times 100\% \quad (17.10)$$

$$\text{Yield}_{\text{gasoline}} = (\text{Weight}_{\text{gasoline}}/12\text{g}) \times 100\% \quad (17.11)$$

17.4 Summary

Pre-treatment process is essential for subsequent biomass conversion into bio-fuel and chemicals. The challenge lies in overcoming the resistivity of plant cells wall to deconstruct due to entanglement of highly crystalline structure of cellulose which is embedded in a matrix of polymer lignin and hemicelluloses.

References

1. Potential Contribution of Bioenergy to the Works Future Energy Demand. IEA Bioenergy, EXCO 2007:02
2. Sumathi S, Chai SP, Mohamed AR (2008) Utilization of oil palm as a source renewable energy in malaysia. Renewable Sustainable Energy Rev 12:2404–2421
3. Bridgeman TG, Jones JM, Shield I, Williams PT (2007) Torrefaction of reed canary grass, wheat straw and willow to enhance solid fuel qualities and combustion properties. Fuel 87:844–856
4. Prins MJ, Ptasinki KJ, Frans JJ, Janssen G (2006) More efficient biomass gasification via torrefaction. Energy 31:3458–3470
5. Rajvanshi AK (1986) Biomass gasification. In: Goswami DY (ed) Alternative energy in agriculture, vol. II. CRC Press, Boca Raton, pp 83–102
6. McKendry P (2002) Energy production from biomass (part 3): gasification technologies. Bioresour Technol 83:55–63
7. Azizan MT, Yusup S, Laziz FD Md, Ahmad MM (2009) Synthesis of bio-oil from oil palm's empty fruit bunch via pyrolysis. In: 3rd WSEAS international conference on renewable energy source (RES '09), University of Cambridge, United Kingdom, 24–26 Feb 2009
8. Shaddix CR, Huey SP (1997) Combustion characteristics of fast pyrolysis oils derived from hybrid poplar. In: Bridgwater AV, Boocock DGB (eds) Developments in thermochemical biomass conversion, Blackie Academic & Professional, London, pp 465–480
9. Gust S (1997) Combustion experiences of flash pyrolysis fuel in intermediate size boilers. In: Bridgwater AV, Boocock DGB (eds) Developments in thermochemical biomass conversion, Blackie Academic & Professional, London, pp 481–488
10. Sipilä K, Oasmaa A, Arpiainen V, Westerholm M, Solantausta Y, Ahnger A, Gros S, Nyrönen T, Gust S (1996) Pyrolysis oils for power plants and boilers. In: Proceedingsof the 9th European bioenergy conference, Copenhagen, 24–27 June
11. Andrews RG, Fuleki D, Zukowski S, Patnaik PC (1997) Results of industrial gas turbine tests using a biomass derived fuel. In: Overend RP, Chornet E (eds) Proceedings of the 3rd biomass conference of the Americas, Elsevier Science Limited, pp 425–436
12. Soltes EJ, Lin J-CK (1984) Hydroprocessing of biomass tars for liquid engine fuels. In: Tillman DA, Jahn EC (eds) Progress in biomass conversion, Academic Press, New York, pp 1–69
13. Jay DC, Sipilä KH, Rantanen OA, Nylund N-O (1995) Wood pyrolysis oil for diesel engines. In: international combustion engine division of the ASME; 1995 fall technical conference, Sept 24–27

14. Aubin H, Roy C (1980) Study on corrosiveness of wood pyrolysis oils. Fuel Sci Technol. Int 8:77–86
15. Chang VS, Holtzapple MT (2000) Fundamental factors affecting enzymatic reactivity. Appl Biochem Biotechnol 84–86:5–37
16. Susmel P, Stefanon B (1993) Aspects of lignin degradation by rumen microorganisms. J Biotechnol 30:141–148
17. DeAngelis KM, D'Haeseleer P, Chivian D, Fortney JL, Khudyakov J, Simmons B, Woo H, Arkin AP, Davenport K, Goodwin L, Chen A, Ivanova N, Kyrpides NC, Mavromatis K, Woyke T, Hazen TC (2011) Complete genome sequence of *Enterobacter lignolyticus* SCF1. StandGenomic Sci 5:69–85
18. Palmowski L, Muller J (1999) Influence of the size reduction of organic waste on their anaerobic digestion. In: IAWQ II. international symposium on anaerobic digestion of solid waste, Barcelona, 15–17 June, pp 137–144
19. Sun Y, Cheng J (2002) Hydrolysis of lignocellulosic materials for ethanol production: a review. Bioresource Technol 83:1–11
20. Mani S, Tabil L, Sokhansanj S (2003) Compaction of biomass grinds—an overview of compaction of biomass grinds. Powder Handl Process 15:160–168
21. Kaliyan N, Vance Morey R (2009) Factors affecting strength and durability of densified biomass products: review. Biomass Bioenergy 33:337–359
22. Yadong Li, Henry Liu (2000) High-pressure densication of wood residues to form an upgraded fuel. Biomass Bioenergy 19:177–186
23. Pietsch W (2002) Agglomeration processes—phenomena, technologies, equipment. Wiley, Weinheim
24. Van Loo S, Koppejan J (2008) The handbook of biomass combustion and co-firing. Earthscan/James & James, Oxford
25. Klass DL (1998) Biomass for renewable energy, fuels, and chemicals. Academic Press, San Diego
26. Bhattacharya SC, Sett S, Shrestha RM (1989) State of the art for biomass densification. Energy Sources 11:161–182
27. Ryu C, Yang YB, Khor A, Yates NE, Sharifi VN, Swithenbank J (2006) Effect of fuel properties on biomass combustion: Part I. Experiments—fuel type, equivalence ratio and particle size. Fuel 85:1039–1046
28. Obernberger I, Thek G (2004) Physical characterisation and chemical composition of densified biomass fuels with regard to their combustion behaviour. Biomass Bioenergy 27:653–669
29. Abdullah SS, Yusup S (2010) Method for screening of Malaysian biomass based on aggregated matrix for hydrogen production through gasification. J Appl Sci 10(24):3301–3306
30. Davidsson K, Korsgren J, Pettersson J, Jaglid U (2002) The effects of fuel washing techniques on alkali release from biomass. Fuel 81:137–142
31. Agbor VB, Cicek N, Sparling R, Berlin A, Levin DB (2011) Biomass pre-treatment: fundamentals toward application. Biotechnol Adv 29:675–685
32. Delgenes JP, Penaud V, Moletta R (2002) Pretreatment for the enhancement of anaerobic digestion of solid water. In: Biomethanization of the organic fraction of municipal solid waste, Chap. 8. IWA Publishing, London, pp 201–228
33. Takacs E, Wojnarovits L, Foldavary C, Hargagittai P, Borsa, J, Sajo I (2000) Effect of combined gamma-irradiation and alkali treatment on cotton-cellulose. Radiat Phys Chem 57:339–403
34. Chang VS, Burr B, Holtzapple MT (1997) Lime pretreatment of switchgrass. Appl Biochem Biotechnol 63–65:3–19
35. Basu P (2010) Biomass gasification and pyrolysis: practical design and theory. Academic Press, New York
36. Meszaros E, Varhegyi G, Jakab E (2004) Thermogravimetric and reaction kinetic analysis of biomass samples from an energy plantation. Energy Fuels 18:497–507
37. Bergman PCA, Kiel JHA (2005) Torrefaction for biomass upgrading No. ECNC-05–103, the Netherlands

38. Demirbas A (2000) Mechanisms of liquefaction and pyrolysis reactions of biomass. Energy Convers Manage 41:633–646
39. Prins MJ, Ptasinski KJ, Janssen FJJG (2006) Torrefaction of wood Part 2. Analysis of products. J Anal Appl Pyrolysis 77:35–40
40. Uemura Y, Omar WN, Tsutsui T, Yusup S (2011) Torrefaction of oil palm wastes. Fuel 90:2585–2591
41. Channiwala SA, Parikh PPA (2002) Unified correlation for estimating HHV of solid, liquid and gaseous fuels. Fuel 81:1051–1063
42. Abe F (1988) The thermochemical study of forest biomass. Bull For For Prod Res Inst 352:1–95
43. RITE-Tokyo Central Laboratory (2005) Development of mix combustion technology of coal and biomass for power plant. Report for global environment protection technology project, FY 2001–2004
44. Demirbas A (2011) Relationships between lignin contents and heating values of biomass. Energy Convers Manage 42:183–188
45. Selvig WA, Gibson IH (1945) Calorific value of coal. In: Lowry HH (ed) Chemistry of coal utilization, vol. 1. Wiley, New York, p 139
46. Seyler CA (1938) Petrology and the classification of coal: Pts I and II. Proc S Wales Inst Engrs 53:254–327
47. Tilman DA (1978) Wood as an energy resources. Academic Press, New York
48. Demirbas A, Gullu D, Caglar A, Akdeniz F (1997) Estimation of calorific values fuels from lignocellulosics. Energy Sources 19:765–770
49. Uemura Y, Wissam NO, Tsutsui T, Subbarao D, Suzana Y (2010) Relationship between calorific value and elementary composition of torrefied lignocellulosic biomass. J Appl Sci 24:3250–3256
50. Prins MJ, Ptasinski KJ, Janssen FJJG (2006) Torrefaction of wood Part 1. Weight loss kinetics. J Anal Appl Pyrolysis 77:28–34
51. Uemura Y, Wissam NO, Othman NA, Suzana Y, Tsutsuit T (2012) Torrefaction of oil palm EFB in the presence of oxygen. Fuel. Available online 19 November 2011
52. Arias B, Pevida C, Fermoso J, Plaza MG, Rubiera F, Pis JJ (2008) Influence of torrefaction on the grindability and reactivity of woody biomass. Fuel Process Technol 89:169–175
53. Almeida G, Brito JO, Perré P (2010) Alterations in energy properties of eucalyptus wood and bark subjected to torrefaction: the potential of mass loss as a synthetic indicator. Bioresource Technol 101:9778–9784
54. Bergman PCA, Boersma AR, Zwart RWR, Kiel JHA (2005) Torrefaction for biomass co-firing in existing coal-fires power stations. ECN-C-05–013
55. Jalan RK, Srivastava VK (1999) Studies on pyrolysis of a single biomass cylindrical pellet—kinetic and heat transfer effects. Energy Convers Manage 40:467–494
56. Felfli FF, Luengo CA, Suárez JA, Beatón PA (2005) Wood briquette torrefaction. Energy Sustainable Dev 9:19–22
57. Phanphanich M, Mani S (2011) Impact of torrefaction on the grindability and fuel characteristics of forest biomass. Bioresource Technol 102:1246–1253
58. Couhert C, Salvador S, Commandré J-M (2009) Impact of torrefaction on syngas production from wood. Fuel 88:2286–2290
59. Sànchez O, Sierra R, Almeciga-Diaz CJ (2011) Delignification process of Agro-industrial wastes an alternative to obtain fermentable carbohydrates for producing fuel. In: Manzanera M (ed) Alternative fuel. InTech, Croatia, pp 111–121
60. Isroi RM, Siti S, Niklasson C, Cahyanto MN, Lundquist K, Taherzadeh MJ (2011) Biological pretreatment of lignocelluloses with white-rot fungi and its applications: a review. Bioresources 6(4):5224–5259
61. Ahmad Khushairi Z, Zainol N (2011) Screening factors affecting biological delignification process of oil palm trunk using local oyster mushroom (*Pleurotus ostreatus*). Int J Chem Environ Eng 2(4). http://www.warponline.org/research_papers/pdf/pdf259.pdf. Accessed 12 May 2012
62. Mohan PR, Ramesh B, Reddy OVS (2012) Biological pretreatment of rice straw by Phanarocheate chrysosporium for the production of celluloses and xylanases using Asperigillus niger Isolate. Res J Microbiol 7(1):1–12

63. Ejechi BO, Obuekwe CO, Ogbimi AO (1996) Microchemical studies of wood degradation by Brown Rot and White Rot Fungi in two tropical timbers. Int Biodeterior Biodegrad 119–122
64. Hatakka A (2001) Biodegradation of lignin. In: Hofricther A, Steinbüchel A (eds) Biopolymer, biology, chemistry, biotechnology, applications, vol. 1, Wiley, New York, pp 129–180
65. Kunamneni A, Ballesteros A, Plou FJ, Alcade M (2007) Fungal laccase—a versatile enzyme for biotechnological applications. In: Méndez-Vilas A (ed) Communicating current research and educational topics and trends in applied microbiology. FORMATEX, Badajoz
66. Vikineswary S, Abdullah N, Renuvathani M, Sekaran M, Pandey A, Jones EBG (2006) Productivity of laccase in solid substrate fermentation of selectedagro-residues by Pycnoporus sanguineus. Bioresource Technol 97:171–177
67. Sugiura M, Hirai H, Nishida T (2003) Puriçation and characterization of a novel lignin peroxidase fromwhite-rot fungus Phanerochaete sordida YK-624. FEMS Microbiol Lett 224:285–290
68. Ruiz-Dueas FJ, Morales M, Garcia E, Miki Y, Jesus M, Martinez MJ, Martinez AT (2009) Substrate oxidation sites in versatile peroxidase and other basidiomycete peroxidases. J Exp Bot 60(2):441–452
69. Roushdy MM, Abdel-Shakour EH, El-Agamy EI (2011) Biotechnological approach for lignin peroxidase (LiP) production from agricultural wastes (Rice Husk) by Cunninghamella elegans. J Am Sci 7(5). http://www.jofamericanscience.org/journals/am-sci/am0705/02_4896am0705_6_13.pdf. Accessed 14 May 2012
70. Pérez-Boada M, Ruiz-Duenas FJ, Pogni R, Basosi R, Choinowski T, Martinez MJ, Piontek K, Martinez AT (2005) Versatile peroxidase oxidation of high redox potential aromatic compounds: site directed mutagenesis, spectroscopic and crystallographic investigation of three long-range electron transfer pathways. J Mol Biol 354:385–402
71. Syafwina HY, Watanabe T, Kuwahara M (2002) Pretreatment of oil empty fruit bunch by white-rot fungi for enzymatic saccharification. Wood Res 89:19–20
72. Hamisan AF, Abd-Aziz S, KamaruddinK (2009) Delignification of oil palm empty fruit bunch using chemical and microbial pretreatment methods. Int J Agric Res 4(8):250–259
73. Namoolnoy P, Phoolphundh S, Wongwincharn A (2011) Biodegradation of lignin in oil palm fronds by white rot fungi. Kasertsart J (Nat Sci) 45:254–259
74. Höije A, Gröndahl M, Tømmeraas K, Gatenholm P (2005) Isolation and characterization of physicochemical and material properties of arabinoxylans from barley husks. Carbohydr Polym 61:266–275
75. Oasmaa A, Leppä mäki E, Koponen P, Levander J, Tapola E (1997) Physical characterisation of biomass-based pyrolysis liquids. Application of standard fuel oil analyses. Technical Research Centre of Finland, Espoo
76. Sarkilahti H (1996) Pyrolyysioljyn suodatus (Filtration of pyrolysis oil). Master thesis, Helsinki University of Technology, Espoo
77. Hew KL, Tamidi AM, Yusup S, Lee KT, Ahmad MM (2010) Catalytic cracking of bio-oil to organic liquid product (OLP). Bioresource Technol 101(22):8855–8858
78. Judit A, Eleni A, Angelos L, Michael S, Meret HN, Aud B, Johan EH, Gisle O (2006) In situ catalytic upgrading of biomass derived fast pyrolysis vapours in fixed bed reactors using mesoporous materials. Microporous Mesoporous Mat 96:93–101
79. Zhu XF, Zheng JL, Guo QX, Zhu QS (2005) Upgrading and utilization of bio-oil from biomass. Eng Sci 7(9):83–88
80. Zhang Q, Chang J, Wang TJ, Xu Y (2006) Progress on research of properties and upgrading of bio-oil. Petrochem Technol 35(5):493–498
81. Guo XY, Yan YJ, Li TC (2004) Influence of catalyst type and regeneration on upgrading of crude bio-oil through catalytical thermal cracking. Chinese J Process Eng 4(1):53–58
82. Babu BV, Chaurasia AS (2002) Modeling for pyrolysis: kinetic and heat transfer effects. Energy Convers Manage 44:2251–2275
83. Twaiq FA, Zabidi NAM, Mohamed AR, Bathia S (2003) Catalytic conversion of palm oil over mesoporous aluminosilicate MCM-41 for the production of liquid hydrocarbon fuels. Fuel Process Technol 1660:1–16

84. Bao WR, Xue XL, Cao Q, Lu JJ, Lu YK (2006) Study on biomass pyrolytic liquid products with MCM-41/SBA-15 as catalyst. J Fuel Chem Technol 34(6):675–679
85. Guo XY, Yan YJ (2006) Analysis of the coke precursor on bio-oil refining catalyst and the regeneration of the deactivated catalyst. J Chem Eng Chin Univ 20(2):222–226
86. Sharma RK, Bakhshi NN (1993a) Catalytic conversion of fast pyrolysis oil to hydrocarbon fuels over HZSM-5 in a dual reactor system. Biomass Bioenergy 5(6):445–455
87. Sharma RK, Bakhshi NN (1993b) Conversion of non-phenolic fraction of biomass derived pyrolysis oil to hydrocarbon fuels over HZSM-5 using a dual reactor system. Bioresour Technol 45(3):195–203
88. Hyun JP, Young-Kwon P, Joo-Sik K, Jong-ki J, Kyung-Seun Y, Jin-Heong Y, Jinh J, Jung MS (2006) Bio-oil upgrading over Ga modified zeolites in a bubbling fluidized bed reactor. Stud Surf Sci Catal 159:553–556
89. Adam J, Antonakou E, Lappas A, Stocker M, Nilsen MH, Bouzga A, Hustad JE, Gisle O (2006) In situ catalytic upgrading of biomass derived fast pyrolysis vapours in a fixed bed reactor using mesoporous materials. Microporous Mesoporous Mater 96(1–3):93–101
90. Nokkosmaki MI, Kuoppala ET, Leppamaki EA, Krause AOI (2000) Catalytic conversion of biomass pyrolysis vapours with zinc oxide. J Anal Appl Pyrolysis 55(1):119–131

Chapter 18
Pretreatment Methods in Biodiesel Production Processes

Ahmed Tafesh and Sobhi Basheer

Abstract Biodiesel has emerged as one of the most growing biofuels to replace diesel fuel. Its preference as one of the most popular alternative fuels was based on its characteristics as it is environment friendly, sustainable, biodegradable, and non-toxic. Biodiesel is mandated by many governments worldwide for incorporation into their diesel supply base. Biodiesel is easily produced through transesterification reactions of vegetable oils (triglycerides). However, current commercial usage of refined vegetable oils for biodiesel production is impractical and uneconomical due to high feedstock cost and priority as food resources.

Low-grade oils, typically waste cooking oils, brown greases, crude corn oils, etc., can be better alternatives; however, the high free fatty acids (FFAs) content in such oils has become the main constrain for those potential feedstocks, and therefore pretreatment methods become necessary to prepare such feedstocks to make biodiesel. The chapter highlights the pretreatment methods to utilize and convert the FFAs from various feedstocks to biodiesel and presents the advantages and limitations of using enzymes and conventional catalysts, distillation, blending, and glycerolysis methods to lower FFAs in the feedstocks. An overview on the current status of biodiesel production, the feedstocks and the FFAs factors are also discussed. With the proper pretreatment methods, the high-FFAs feedstocks can indeed become the next ideal feedstocks for the production of biodiesel.

Keywords Enzymes · Vegetable oil · Biodiesel · Pretreatment · FFA · Triglyceride

18.1 Introduction

Biodiesel production is a globally advancing field, with biodiesel fuel increasingly being used in compression diesel engines to replace diesel fuel which stands at a market value of $200 billion US dollars per year worldwide. The biodiesel, which

A. Tafesh (✉) · S. Basheer
TransBiodiesel Ltd, Nazareth Street 79, P. O. Box 437, Shefar-Am 20200, Israel
e-mail: atafesh@transbiodiesel.com

Z. Fang (ed.), *Pretreatment Techniques for Biofuels and Biorefineries*,
Green Energy and Technology,
DOI 10.1007/978-3-642-32735-3_18,

is becoming one of the most popular alternative and environment-friendly fuels, is mandated by many governments for incorporation into their diesel-supply base. Society's concerns with the environment have made governments, industries, and businesses start to assess how their activities affect the environment. Such biodiesel is a more natural, more sustainable biofuel known to reduce carbon dioxide emission by 78 % when compared to regular diesel and its energy content is 88–95 % that of diesel [1].

Production scale of biodiesel in the European Union in 2006 was estimated to be approximately 4.8 million tons, which constitutes about 77 % of the produced biodiesel in the world followed by the US with 13 %, and the rest of the world to be 10 %. Market growth rate data in Europe was over 50 % in 2006 after growth of 37 % in 2005. It is estimated that biodiesel will reach 3.3 billion gallons by 2022, an increase of 2.5 billion gallons from 2011, where nearly 80 % of the increase comes from three feedstocks: soybean oil (31 %), corn oil (22 %), and palm fatty acid distillate (26 %) [2]. The volume of all biodiesel is growing at a 10 % rate per year and stood at $22 billion market value in 2007, and is estimated to reach $88 billion by 2020 [3, 4].

18.2 Biodiesel Acceptable Standards

The biodiesel acceptable standards around the world normally follow either the American Society for Testing and Materials (ASTM) specification D6751 [5] or the European Specification EN14214 [6]. The variable level of substituted or bound glycerol to fatty acid as tri-, di- and mono-glycerides is summed up and calculated as the total bound glycerin value with an allowable value lower than 0.24 % in biodiesel. The biodiesel specification in Europe EN14214 includes a minimum requirement for the ester content (96.5 %) and individual maximum levels for the mono-, di-, and tri-glycerides. Both specifications limit the level of total glycerin remaining in biodiesel fuel to approximately the same value (0.24 % in ASTM D 6751 and 0.25 % in EN 14214). When measuring the level of free fatty acids (FFAs), the total FFAs should be below 0.25 mg/ml.

18.3 Feedstock's Decomposition

Virgin oil when exposed to moisture, microbial contamination, heating, and light in the presence of air such as in the case of long-term storage or cooking, undergoes a decomposition process such as oxidation and hydrolysis leading to the formation of FFAs and other low- and high-molecular weight hydrocarbons, alcohols, oxidized monomers, dimmers, and trimers. Moisture and lipases excreted by microorganisms promote the hydrolysis of vegetable oil triglycerides to form FFAs, mono-, and di-acylglycerols, which result in increasing of the refining losses directly related to the free fatty acid content of oils and fats (Fig. 18.1). Oxygen and heat cause oxidation

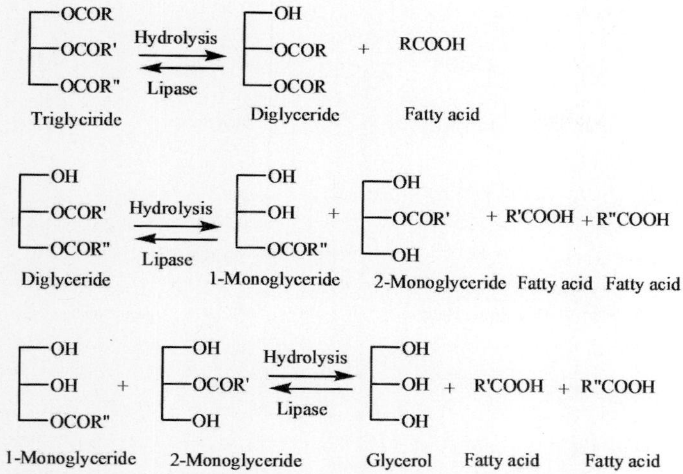

Fig. 18.1 Moisture and lipase-aided decomposition of oil triglycerides

of oil triglycerides which results in the formation of hydroperoxides, leading to volatiles responsible later for the unpleasant odor of spent frying oils. The heat can also initiate cyclization reactions; both intra- and inter-molecular, occurring via electrocyclic Diels–Alder and Ene reactions. Furthermore, deterioration of heated oil used for multiple frying operations, is characterized by the formation of a polymeric dark mass and other low-molecular weight compounds. During frying operations, different compounds which include fatty acids, mono-, di-, and tri-glycerides, low- and high-molecular weight polymeric dark brown materials accumulate in oil. It has been reported that oxidation of oil is mostly responsible for much more of the deterioration of fats and oils than hydrolysis [7].

The decomposed oil loses fatty acids which become detached FFAs, or "used" oil and therefore less expensive feedstock to purchase. In order to increase the overall yield of biodiesel, the high-FFAs oil requires pretreatment processing that means subjecting the oil to acidic conditions, where the FFAs are converted to fatty acid esters and thereby lowered pH to acceptable values in the oil, so that the oil will later be converted to biodiesel by conventional alkaline catalysts (reactions 1 and 3, Fig. 18.2) [8]. The FFAs can also be removed under alkaline conditions (reaction 2, Fig. 18.2) when treated with alkaline reagent such as KOH to yield fatty acid potassium salts or soap which can be removed by water-wash process [9, 10].

The conventional chemical synthesis of biodiesel is typically carried out under alkaline conditions and to a lesser extent under acidic conditions [11, 12]. The alkaline reaction involves transesterification conditions using vegetable oil (triglyceride and to a lesser extent mono- and di-glycerides), methanol, and a chemical base such as sodium hydroxide (NaOH) (reaction 3, Fig. 18.2). The reaction can also be carried out with sodium methylate ($NaOCH_3$) as a catalyst which produces biodiesel and glycerol with improved reaction yields [13–15]. The transesterification of glycerides

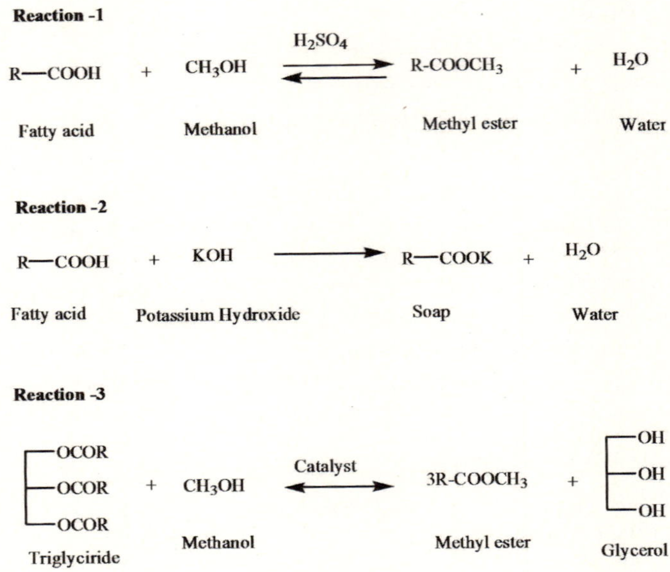

Fig. 18.2 Formation of fatty acid methyl esters (biodiesel) and fatty acid salts from vegetable oil and fatty acids in the presence of methanol and a catalyst

and methanol can also be carried out by acid catalysis to produce fatty acid methyl esters (FAMEs) and glycerol; however, this type of reaction is unfavored due to its low rates and its strict reaction conditions with regards to high temperatures and high excess of methanol.

18.4 Feedstock Factor

The feedstock constitutes 70–95 % of the biodiesel's production costs [16]. The price of biodiesel in the open market stands approximately at $1,300/ton. This means the lower the cost of feedstock which contains higher %FFAs, the bigger the margins and the more profit is made on biodiesel from lower value feedstock. When choosing feedstock, there are several important factors to consider such as quality, price, availability, and ability to meet product specifications "ASTM and EN specs" which can be achieved by choosing proper pretreatment and to some extent post-treatment methods [6]. Understandably, higher quality feedstocks which don't need pretreatment typically are much more costly. Refined, bleached, and deodorized (RBD) vegetable oils are more expensive than crude vegetable oils, both vegetable oils are more expensive than tallow and waste-cooking oil (yellow grease), yellow grease is more expensive than fat trap (brown grease), and so on. However, soybean oil which is very abundant is also more expensive than palm oil simply because of its process ability and inherent characteristics that impacts cloud point (CP) and cold filter plugging point (CFPP) of the biodiesel product [17, 18].

Crude corn oil which is available as byproduct of the ethanol industry typically contains 10–15 % FFAs and is reddish in color, requires pretreatment process. The purification needed is to minimize the sterol, glycosides, waxes, and FFAs to ASTM acceptable levels in the oil in order to make an acceptable transesterification feedstock [19]. The ability to convert FFAs into methyl esters instead of soap increases product yield and reduces feedstock cost per gallon of biodiesel. The high-FFAs crude corn oil is 20 cents cheaper than the soybean oil, and it has a −3 °C CP, nearly 5 °C better than soybean [20]. This is great to the biodiesel consumers, especially since there will be at least twice as much crude corn oil in the market in 2012 compared to 2011 which will also yield biodiesel suitable for cold winters [20].

18.5 FFAs Factor

Crude plant oils typically contain FFAs < 3.0 % and gums in the range of 0.05–0.5 % [21]. The only pretreatment needed for such oil feedstocks is to remove the gums using either conventional degumming methods or using enzymes [22]. Higher FFAs oil feedstocks, such as recovered yellow grease, are no longer viable for the production of biodiesel by the conventional alkaline process. Such feedstocks containing up to 9 % FFAs can be pretreated by different methods such as the SRS (see Sect. 18.6.1) continuous flow acid by SRS Engineering, removal of FFAs by distillation or adsorption, or treated with enzymes (Table 18.1) [27]. Crude corn oil or yellow grease with FFAs >9 % requires acidic pretreatment, enzymes or $Ca(OH)_2$ (Table 18.1). High FFAs oil which is disposed in the drains "brown grease" is a potentially problematic waste stream and it clogs installations in waste-water treatment plants and thus, it adds to the cost of treating effluent [23, 24]. Such potential feedstock contains typically 30 % and up to 100 % FFAs, requires an enzymatic or acid-based esterification methods (Table 18.1).

Biodiesel producers always have to decide on what is the cutoff of the FFAs level in the feedstock and decide how to move forward with biodiesel production. Some producers work with feedstock streams containing FFAs lower than 1 % (virgin oil) where any known technology for making biodiesel can be applied (Table 18.1).

When the feedstock contains up to 3.5 % FFAs, the soap formation during the alkaline reaction will be challenging to work with. One way to deal with it, is blending such oil (FFAs <3.5 %) with a lower FFAs feedstocks (<1 %) to obtain oil feedstocks of FFAs less than 2 %. Also, feedstocks containing FFAs levels below 3 % can be pretreated by adsorbents which extract FFAs from the oil out into the matrix. Spent adsorbents can normally be disposed off in a landfill. Another approach would be to convert the FFAs to their potassium salts and be removed by either water-wash process or centrifugation. When oil feedstocks contain higher than 3.5 % FFAs, pretreatment methods should be applied for the removal of FFAs, simply because the alkaline biodiesel production process converts the FFAs to soaps which eventually result in complicating of the downstream processing of the final products [25].

Table 18.1 Pre- and post-treatment methods of different variable levels of FFAs

FFAs %/ preteatment	<1 %	4–9 %	10–15 %	>15 %	Post-treatment of biodiesel
Feedstock	Crude and refined Soybean palm Rapeseed corn Canola sunflower	Recovered—UCO* Jatropha Algae	Yellow grease Crude corn oil in ethanol plants	Crude oil PFAD** Brown grease Recovered soap	Ion exchange and solid adsorbents for removal of soap, glycerol from BD*** [33] (US Patent 7,635,398)
Method	Adsorbents [31] Magnesium Silicate [36, 37]	Enzyme [40, 41] Steam Stripping continuous flow acid- [27]	Steam stripping [20] Acid-Esteri- fication $Ca(OH)_2$ extraction [24]	Enzymatic Acid-Esteri fication Amberlyst -BD20 (32, 33)	INDION-BF-100 (good for 500 ppm FFAs) [18]
Phosphorus	Purifine PLC [22, 51] Lecitase (Novo) (42) Adsorbent Magneso1600R	Bentonite clay [35]	–	–	–
Sulfur, soaps, sterol glucosides	Adsorbent Magnesol 600R (36)	–	–		Post-treatment distillation-BD Amberlite BD 10 DRY-remove soap [14]
Ca (calcium) Mg (magnesium)	Bentonite clay [35]	–	–	–	–

**UCO*—used cooking oil, ***PFAD*—palm fatty acid distillate, ****BD*—biodiesel

Not all pretreatment methods are suitable for every biodiesel production plant. Several factors will have to be considered such as the process type (alkaline, acidic, heterogeneous, enzymatic, etc.), the type of existing equipment in the plant, the long-term availability of feedstocks used with variable contents of FFAs, and how the pretreatment technology ties into those systems.

18.6 Pretreatment Methods

Pretreatment means the applied stages required in the plant in order to process feedstocks prior to their conversion to biodiesel. Such stages typically involve reducing factors of negative impacts on the biodiesel production process such as water, gums, suspended particles, polymers, and mostly FFAs. Water normally leads to formation of increased concentration of soaps during alkaline transesterification, reacts with the alkaline catalyst sodium methylate to form methanol and sodium hydroxide, and also shifts the equilibrium reaction toward hydrolysis under acid-catalysis conditions. The soaps can solidify or freeze up and clog lines which results in equipment lost time (downtime). One pretreatment method involves reacting caustic soda with FFAs; however, there will be significant yield losses in such pretreatment method. On the other hand, the acid pretreatment doesn't lead to the formation of soaps and therefore will be minimal yield losses. Polymers, gums, and particulates in oil feedstocks impose also negative impacts on the biodiesel production process as they lead to destroying of the catalyst and also implications on phase separation of oil/glycerol phases. The biodiesel industry worldwide has adopted several pretreatment methods that are described below:

- Liquid acid treatment—Pretreatment by esterification of FFAs with a liquid acid catalyst (Sect. 18.6.1).
- Distillation—Removal of FFAs by distillation (Sect. 18.6.2).
- Blending—Blending low FFAs feedstock with higher FFAs feedstock (Sect. 18.6.3).
- Glycerolysis—Glycerol reaction with FFAs (Sect. 18.6.4).
- Acid esterification with solid catalysts—Lower FFA with ion exchange (Sect. 18.6.5).
- Removal of FFA with solid adsorbents (Sect. 18.6.6).
- Pretreatment with enzymes (Sect. 18.6.7).
- Degumming—Removal of gums from crude oils (Sect. 18.6.8).

18.6.1 Liquid Acid-Catalyzed Pretreatment Process of High-FFAs Feedstocks

Feedstocks containing FFAs $>10\,\%$ require pretreatment with a strong acid, such as sulfuric acid, to remove the FFAs by converting the FFAs to their methyl esters, thereby reducing the FFAs level in the feedstock. The acidic process requires high temperature and high excess of methanol in order to achieve high conversions during a reasonable reaction time. Water is produced during the esterification, and as it accumulates in the reaction medium leads to slowing down of the reaction rate. The disadvantage of the acidic process is the need to neutralize the acid at the end of the reaction producing a large amount of salts which moves into the water/methanol phase, generating large amounts of low-grade methanol/water/salts solution. The neutralized oil feedstock is then treated by alkaline catalyst and methanol to produce biodiesel [26].

The biodiesel industry has adopted a liquid acid technology for the conversion of FFAs to FAMEs. SRS Engineering, USA has developed a continuous flow acid esterification system that utilizes methanol and acid to convert FFAs into usable oil feedstocks as opposed to washing with caustic soda which contribute to high yield losses [27]. Such process is energy efficient, prevents soap formation, and converts up to 15 % FFAs feedstock to biodiesel (Table 18.1). SRS further adds to this process a proprietary process of utilizing a modified waste additive to remove excess water from the oil prior to a transesterification process to further prevent the development of soap production [28].

US-based Crown Iron Works (CIW) has established experience in removing FFAs from culinary oils a process known as oil deodorization. The company has adapted this technology for FFAs treatment in biodiesel feedstocks. CIW's biodiesel pretreatment plants are designed for continuous operation to achieve maximum efficiency and safety. The process is similar to a typical oil refining plant used for refining edible oils. Oils or fats containing high amounts of phosphorus and high amounts of fatty acids are degummed (degumming will be further explained later in the chapter) and de-acidified in two steps. The degumming process, which removes the phosphorus, is similar to the standard oil and fat pretreatment process but does not neutralize the fatty acids. After degumming, the fatty acids are removed by distillation under vacuum at a high temperature (>100°C), or by alkaline neutralization followed by centrifugation and then water wash.

This de-acidification process is also known as stripping because a small amount of low-pressure steam is used to strip the fatty acids from oils or fats. Some oils and fats containing low amounts of phosphorus and high amounts of fatty acids are degummed by adding acid along with diatomaceous earth which enhances the adsorption process [29]. The acid is added and reacted in a separate vessel prior to the addition of diatomaceous earth addition step. Phosphorus and fatty acid-discharged oil is equivalent to refined and bleached oil (Fig. 18.3).

18.6.2 Fatty Acid Distillation

When the FFAs value is below 10 %, some companies remove such range of FFAs% by distillation. The distillation of crude fatty acids removes both the low and high boiling impurities typically present in such feedstocks. Fatty acids depending on their

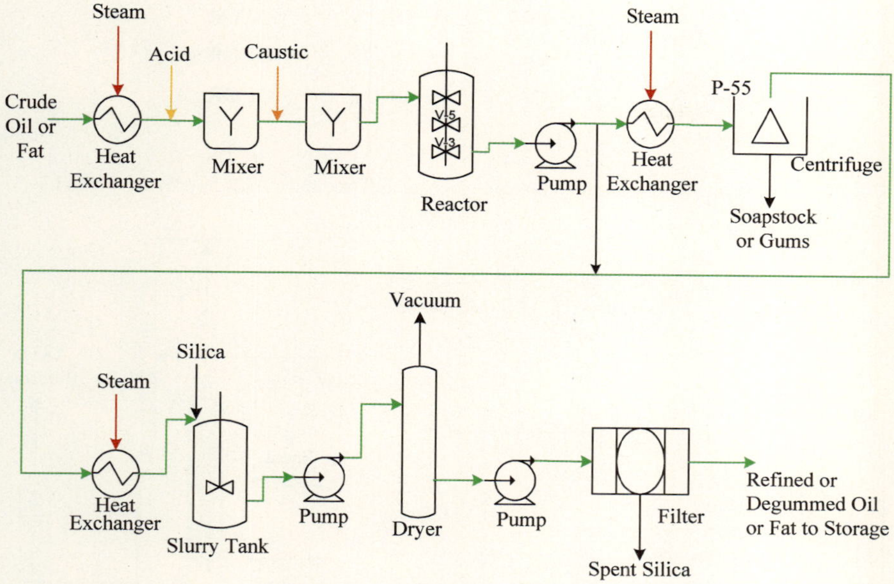

Fig. 18.3 Crown Iron Works standard oil and fat pretreatment flow diagram

degree of saturation are sensitive to heat, oxidation, and corrosion effects. Normally, the higher degree of saturation of FFAs is the higher resistance of FFAs toward heat and oxidation conditions. Distillation of FFAs from oil feedstocks is in general carried out under high vacuum and lower temperatures and with the shortest residence time allowable. Normally, feedstocks are pre-dried and degassed under vacuum and then fed to the distillation unit. Either tripping steam or high vacuum systems are provided to improve circulation and reduce partial pressure, thus lowering the distillation temperature and reducing degradation losses of FFAs [30].

18.6.3 Blending of Feedstocks

It is favored that the feedstock is low FFAs virgin oils which can be blended with lower quality oils without seriously compromising production of high-quality biodiesel. Although this technique is widely practiced by biodiesel producers such pretreatment method doesn'tsolve the long-term problem of low-quality oil and also would not offer any potential solution for the conversion of FFAs to biodiesel.

18.6.4 Glycerolysis of FFAs

Glycerolysis is a process in which the content of FFAs is reduced in feedstocks by combining FFAs with glycerin to create mono-, di-, and tri-glycerides as shown in the process flow diagram of Albemarl Corp (Figs. 18.4 and 18.5) [20]. The glycerin

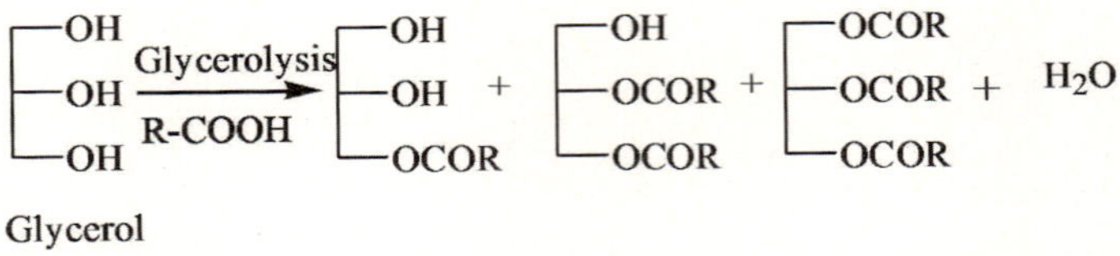

Fig. 18.4 The glycerolysis reaction between glycerol and free fatty acids to produce glycerides

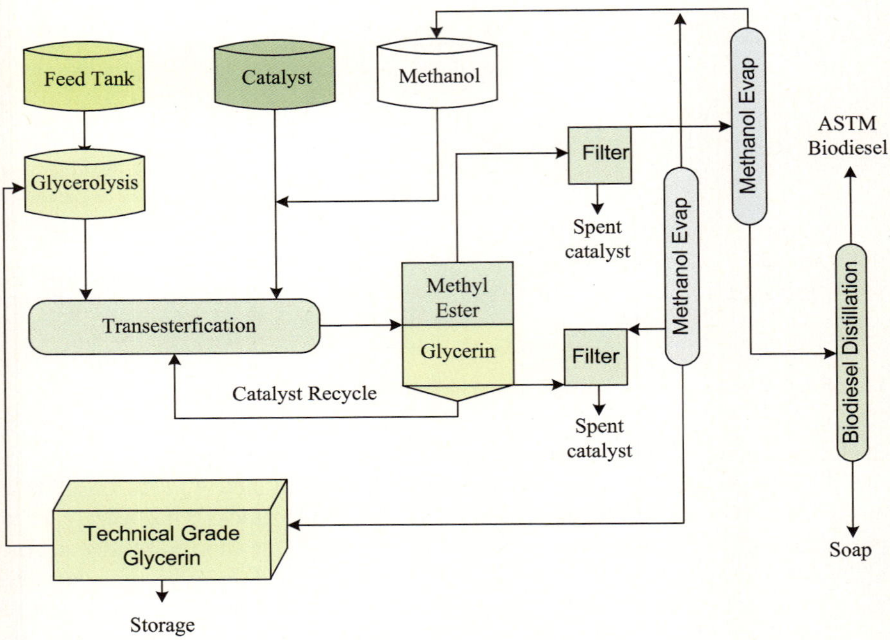

Fig. 18.5 Albemarle conversion of FFAs to mono-, di-, and tri-glyceride

is routed back to a reactor where it is combined with the high FFAs feedstock. The effluent of the glycerolysis process is a low FFAs feedstock, typically less than 1 %, suitable for alkaline transesterification process under normal operating conditions. JatroDiesel, USA, offers a unit for the glycerolysis of acid feedstocks with different levels of FFAs. Glycerolysis with up to 100 % FFAs has been demonstrated at industrial scales while acid esterification process has been used for feedstocks with FFAs below 25 %, and stripping of FFAs has been used with 15 % FFAs or lower, prior to alkaline transesterification process for the production of biodiesel [20].

18.6.5 *Heterogeneous Ion Exchange Resin Catalyzed Reduction of FFAs*

Resin beads containing exotic catalytic sites can also be used as heterogeneous catalysts for lowering the FFAs content in a pretreatment process for feedstocks with high content of FFAs. Resin beads do not dissolve with the reagents as homogeneous catalysts do, making their separation a great deal easier [31]. Furthermore, the use of such resins does not require neutralizing acid, so no water or salts are generated at the end of the biodiesel production process, leading to cleaner biodiesel and glycerol byproduct. Resin beads can be used in both stirred tank (batch and continuous) reactors and in continuous packed column reactors. Currently, one of the most promising resin catalysts for esterification of FFAs is AMBERLYST BD20 from Rohm and Haas (part of the Dow group). Emulsions form upon transesterification of high-FFAs feedstocks whereas a clean separation occurs when the FFAs content in feedstocks is first esterified using AMBERLYST™ BD20 technology [32, 33].

Bayer Technology Services has designed a complete biodiesel production system to get the best performance from this catalyst. The technology converts FFAs into their FAME's. It is a versatile system that can be adopted to deal with any amount of FFAs up to 100 % by adding more reactors in series. Depending on the initial FFAs content allowable in the oil going downstream, a single reactor will deal with oils containing around 5 % FFAs, two reactors push this to around 70 % and three to 100 % conversion. Several resins are available commercially. Rohm and Haas has Amberlite 1R 120 or IRA 900 or Duolite C20 as strong-cation exchange resins; IRA 93SP or Duolite A 378 as weak-anion exchange resins; and the mixed bed polishing resins C20MB and A 101D as strong-cation and anion exchange resins [34].

18.6.6 *Heterogeneous Adsorbent Additives for Removing Free Fatty Acids*

Pretreatment of feedstocks with adsorbents such as magnesium silicates (such as Magnesol 600R, The Dallas Group, USA) were found to be very effective in removing FFAs. D-SOL has been introduced successfully at commercial scales for food frying operations to remove FFAs from frying oil. At 2 % additive concentration, the Magnesol 600R reduced FFAs from 3.8 % to around 1.24 %. When a blend of chicken fat and vegetable oil with FFAs concentration of 1.45 % was tested with Magnesol, all concentrations of the 600R product reduced the FFAs to below 1 %. This means, the 600R is a low-cost solution for reducing FFAs levels of <4.0 % which works out to be about 5 % per gallon per 1% FFAs reduction [35].

Bentonite clay on the other hand, reduced calcium, magnesium, and phosphorus in feedstocks better than the Magnesol 600R. If a plant relies on a proprietary catalyst, the producer is tied-in to one manufacturer. It is possible that as the market matures different catalyst manufactures will offer drop-in alternatives [36].

18.6.7 *Immobilized Enzyme-Catalyzed Reduction of FFAs*

Enzyme is a new biocatalyst to the biodiesel industry. Lipases belonging to the enzyme group of hydrolases are capable of converting FFAs in an esterification reaction with methanol to biodiesel and water byproduct. If used properly, the use

Table 18.2 Advantages of the enzymatic process over the chemical process

Chemical	Enzymatic
Conditions and parameters	
Pre cleaning required (acid)	Pre cleaning required (filtration to 1 μm)
$T = 80–150\,°C$	$T = 10–25\,°C$
FFAs (free fatty acids)	
FFAs reduction required	FFAs reduction not required
FFAs level—0–3 %	FFAs level—0–100 %
Acid use	
Acid esterification process	Acid esterification not required
Sulfuric acid required	Sulfuric acid not required
Soap and caustic	
Caustic required	Caustic not required
Soap formation	No soap formation
Methanol	
Molar ratio—methanol to oil 6–1	Molar ratio—methanol to oil 4–1
Uses large amounts of dry methanol	Uses minimal amounts of (hydrous) methanol
Glycerin	
Low quality of glycerin	High quality of glycerin
Dry wash required	Dry wash not required
Cost	
Production cost per gallon—$ 1.50 USD	Production cost per gallon –$ 1.00 USD

of lipases is cost effective and environment friendly. Lipases can be easily used for lowering the FFAs in different feedstock through esterification with methanol to form FFAs and water [37]. The use of such type of biocatalyst would provide an elegant solution for reducing the environmental impact of yellow grease collected from restaurants, brown grease (>90 % FFAs) and fat collected in municipal and industrial waste-water treatment plants [38, 39].

Most recently TransBiodiesel, Israel has developed and commercialized unique immobilized biocatalysts for the conversion of crude and low-grade feedstocks to biodiesel. The developed biocatalysts are capable of converting any grade of vegetable oil and animal fat to biodiesel with minimal waste products [40, 41]. The biocatalysts would act on any oil feedstock with any level of FFA-containing oil including crude vegetable oils, vegetable oil distillates, yellow and brown greases, and virgin oil, and to reduce their FFAs content to lower than 1 %. These feedstocks with high FFAs levels are much cheaper feedstocks than virgin plant oils (40–60 % cheaper). It is estimated that 20–40 % of the operational costs alone can be saved when dealing with the enzymes developed by TransBiodiesel (www.transbiodiesel.com).

The proposed enzyme technology offers biodiesel manufacturers flexibility in their choice for feedstocks which might contain FFAs in the range of 0–100 %. It allows biodiesel manufacturers to expand their feedstocks selection from expensive virgin oil (approx $1,100/t) to yellow grease ($700/ton) to inexpensive brown-grease feedstock obtained from waste-water treatment plants ($300/t). The major advantages of the enzymatic process over the chemical processes are summarized in Table 18.2.

It has been demonstrated that feedstocks need not be FFAs free in the enzymatic process, and de-hydrated feedstock is not a requirement as in the case of the chemical process. Operating at a relatively low temperature and with no need to neutralize acid, TransBiodiesel's enzymatic process produces remarkably clear biodiesel and high-quality glycerol that needs little refining because enzymes are used at room temperatures (20–30°C) without any other acids or bases.

TransBiodiesel has two main enzymes TransZyme and EsterZyme. TransZyme is an immobilized lipase of high transesterification as well as esterification activity. TransZyme is capable of converting any type of feedstock, including virgin oils, crude plant oils, animal fats, waste-cooking oils, acid oils, and brown grease, regardless of the FFAs content (0–100 %), to form biodiesel through transesterification and esterification processes simultaneously [40, 41]. TransZyme favors more transesterification and esterification than hydrolysis even in the presence of 1–10 % water. TransZyme is also capable of transesterifying phospholipids and wax esters to form biodiesel and free alcohols allowing the use of crude unrefined vegetable oils. Due to the capability of the developed biocatalyst to transesterify phospholipids the overall yield of biodiesel production from crude plant oils would be increased by 1–3 %.

EsterZyme is an immobilized lipase of high esterification activity. It transforms free fatty acid in the presence of methanol (or other alcohols) and under reduced amount of water (preferably below 0.5 %) and glycerol into biodiesel and water byproduct [40, 41]. Furthermore, EsterZyme exhibits relatively high transesterification activity toward partial glycerides and wax esters and lower activity toward triglycerides. The biocatalyst can also be used for lowering the FFAs% in any type of feedstock down to 0–2 % starting from any type of feedstock containing FFAs from 3 % and up to 100 %.

Both enzymes developed by TransBiodiesel are suitable for use in batch and continuous reactors using stirred tank or packed column reactors (Fig. 18.6). While many plants using acid esterification and de-gum their feedstock, TransBiodiesel's technology uses crude feedstock without resorting to de-gumming since gums don't interfere with the enzymatic step.

18.6.8 Degumming

The presence of phospholipids in feedstocks is a serious concern to the biodiesel industry. This is basically due to their emulsifying properties and their negative impact on cold soak filtration times and consequently cold weather fuel performance. When phospholipids are present in the biodiesel alkaline transesterification production process they complicate phase separation of products as they lead to the formation of emulsions which are hard to break. This situation can be detrimental to downstream processing by ion-exchange resin, processing yield, and final product quality. Also, phospholipids pose many problems for the storage and processing of the crude oil, therefore must be removed from oil during refining by a process known as degumming [42, 43].

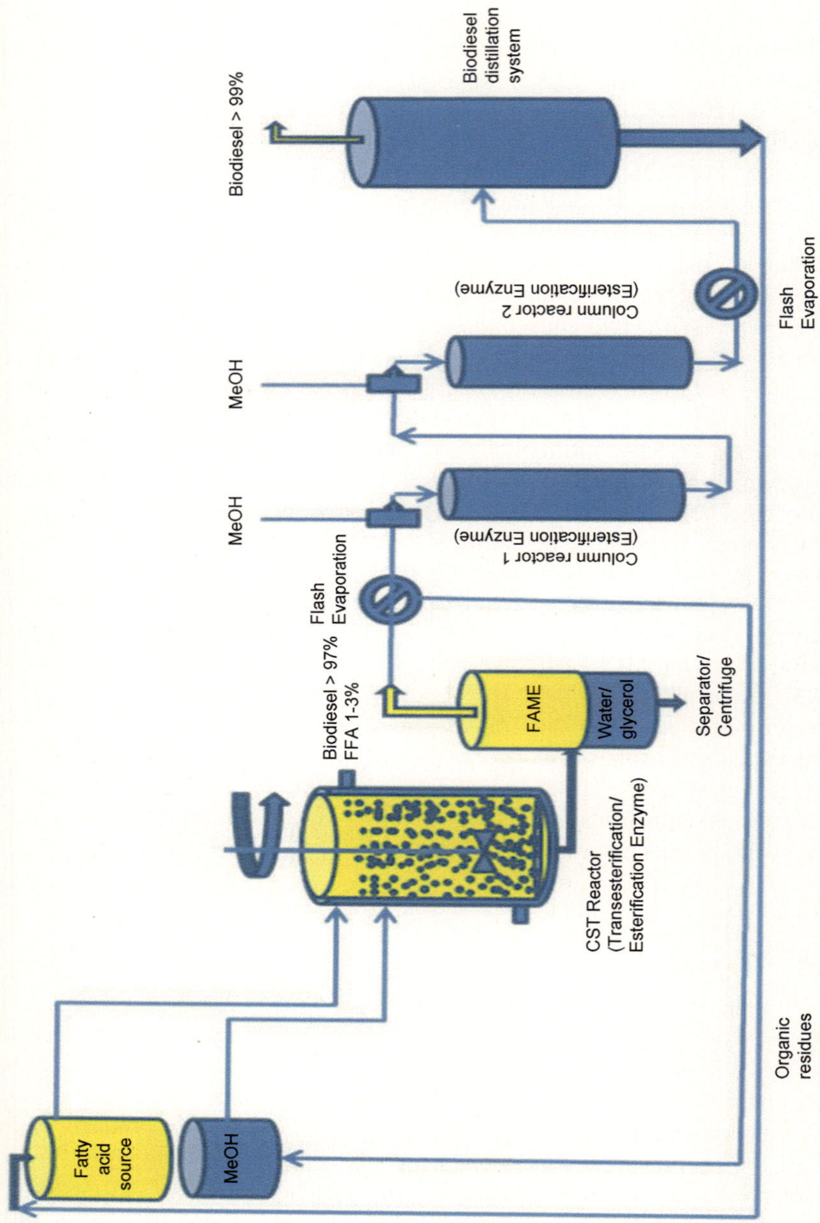

Fig. 18.6 TransBiodiesel's commercial enzymatic process for the production of biodiesel

Lipids obtained by screw pressing (mechanical extruding) and solvent extraction are termed "crude oils" which form deposits or gums upon storage. The chemical nature of these gums consists mainly of phosphatides which entrain oil and meal particles which are formed when the oil absorbs water. Under such conditions gums become oil-insoluble (hydrated phosphatides) which can be readily removed by filtration.

Accordingly, hydrating the gums and removing the hydrated gums before storing the oil can prevent the formation of gum deposits; such treatment is called *water degumming*. Other ingredients, including FFAs, hydrocarbons, ketones, tocopherols, glycolipids, phytosterols, phospholipids, proteins, pigments, and resins, which are oil-soluble or form stable colloidal suspensions in the oil are normally removed from vegetable oils by chemical or physical refining processes which involve the use of phosphoric acid, citric acid, or other degumming substances [44, 47, 48].

Degumming process plays a critical role in the physical refining process of edible oils. Traditional degumming processes, including use of membranes [49], chelating agents [50], enzymes [42, 51, 52], water degumming, acid treatment, and TOP degumming (water degummed oil is heated up to 90–105° C, thoroughly mixed with degumming acid, mixed with dilute caustic and then separated, and washed with water) or "total degumming process," [45, 46] cannot all guarantee the achievement of low phosphorus contents required for physical refining. Such methods are not always optimally suited for all oil qualities because of the high content of non-hydratable phospholipids [53].

The vegetable oil refining industry has also recently experienced the use of free microbial enzymes for degumming of plant oils. Phospholipase A1 (Lecitase Ultra, Novozymes) and Phospholipase C (Purifine, Verenium) are the most prominent enzymes which already have found real industrial applications in oil degumming [42, 51]. Phospholipase A1 catalyzes the hydrolysis of fatty acyl groups at the sn-1 position of phospholipids to form 1-Lyso-phospholipids and a free fatty acid in the first stage, and glycerophospholipids and a free fatty acid in the second hydrolysis stage. The formed 1-Lyso-phopholipids and glycerophospholipids have both increased hydrophilic characteristics which can be easily washed out of oil with mild-acidic water solution.

Phospholipase C cleaves the phosphoric acid ester bondage in phospholipids molecules to form di-glycerides and phosphoryl alcohols. The formed phosphoryl alcohols are hydrophilic molecules which can also be washed easily with water to obtain degummed oils [22].

Both enzymes, Phospholipase A1 and Phospholipase C have been applied at industrial scales in the oil refining industry; however, because of their costs are still not widely used in the oil industry. Similarly, the costs related to the use of both enzymes in degumming of oils feedstocks are economically unaffordable in the biodiesel industry.

It has been demonstrated that the immobilized enzymes, TransZyme and EsterZyme, both developed by TransBiodiesel, are capable of transesterifying phospholipids and methanol to form biodiesel and glycerophospholipids, thereby allowing the use of crude plant oils in the biodiesel production process. While the formed glycerophospholipids are of hydrophilic characteristics, they accumulate in the glycerol/water phase and thereby facilitating the biodiesel downstream processing as well as increasing the biodiesel production yield by 1–2 %.

References

1. Burrell A (ed) (2010) JRC scientific and technical reports impacts of the EU biofuel target on agricultural markets and land use: a comparative modeling assessment. European commission. http://ec.europa.eu/energy/renewables/studies/doc/land_use_change/study_jrc_biofuel_target_iluc.pdf. Accessed 25 June 2012
2. Sims B (2012) Is there sufficient supply of feedstock to support the build-out of a 6 billion gallon biodiesel industry over the next decade. Biodiesel Mag. http://www.biodieselmagazine.com/articles/8296/keeping-pace. Accessed 25 June 2012
3. Bart JCJ, Palmeri N, Cavallaro S (2010) Biodiesel science and technology: from soil to oil. Woodhead, Cambridge
4. Eisentraut A (2010) Sustainable production of second generation biofuels. International Energy Agency. http://www.iea.org/papers/2010/second_generation_biofuels.pdf. Accessed 25 June 2012
5. ASTM (2008) D6751-08 Standard specification for biodiesel fuel blend stock (B100) for middle distillate fuels in annual book of ASTM standards. ASTM International, West Conshohocken
6. European Committee for Standardization (2008) Biodiesel Standard EN14214. Brussels
7. List GR, Wang T, Shukla VKS (2005) Storage, handling, and transport of oils and fats. In: Shahidi F (ed) Bailey's industrial oil and fat products, 6th edn. Wiley, New York
8. Keim GI (1945) Treating fats and fatty oils. US Patent 2,383,601
9. Kawahara Y, Ono T (1979) Process for producing lower alcohol esters of fatty acids. US Patent 4,164,506
10. Jeromin L, Peukert L, Wollman G (1987) Process for the pre-esterification of free fatty acids in fats and oils. US Patent 4,698,186
11. Haas MJ, Michalski PJ, Runyon S, Nunez A, Scott KM (2003) Production of FAME from acid oil, a by-product of vegetable oil refining. J Am Oil Chem Soc 80:97–102
12. Lele S (2006) Biodiesel and jatropha plantation. Agrobios India, Jodhpur
13. Van Gerpen J (2005) Biodiesel processing and production. Fuel Process Technol 86:1097–1107
14. Knothe G, Van Gerpen J, Krahl J (eds) (2005) Biodiesel handbook. AOAC Press, Champaign
15. Canakci M, Van Gerpen J (2001) Biodiesel production from oils and fats with high free fatty acids. Trans ASAE 44:1429–1436
16. Elsolh N (2011) The manufacture of biodiesel from the used vegetable oil. Dissertation, Kassel University
17. Moss P (2012) Feedstock flexibility: key to future profitability. Biodiesel Mag. http://www.biodieselmagazine.com/articles/8281/feedstock-flexibility-key-to-future-profitability. Accessed 25 June 2012
18. Ion Exchange (India) Ltd. Product data sheet of INDION® BF 100. http://www.ionresins.com/pds/BF%20100 %20PDS.pdf. Accessed 25 June 2012
19. Fjerbaek L, Christensen KV, Norddahl B (2009) A review of the current state of biodiesel production using enzymatic transesterification. Biotechnol Bioeng 102:1298–1315
20. Voegele E (2012) A critical component. Biodiesel Mag. http://biodieselmagazine.com/articles/8295/a-critical-component. Accessed 25 June 2012
21. Leung DYC, Wu X, Leung MKH (2010) A review on biodiesel production using catalyzed transesterification. Appl Energy 87:1083–1095
22. Hitchman T (2009) Purifine® PLC industrial application in oil degumming and refining soybean phospholipid. Oil Mill Gazetter 115:2–4

23. Gold M (2001) Restaurants stuck with a problem. Los Angeles Times. http://articles.latimes.com/2001/jan/10/local/me-10620. Accessed 25 June 2012
24. Tunc MF, Gurbuz H, Türkay SZ (2010) Non-evaporative solvent recovery step in deacidification of used frying oil as biodiesel feedstock by methanol extraction. J Am Oil Chem Soc 87:195–203
25. Li Y, Zhang X, Sun L (2010) Fatty acid methyl esters from soapstocks with potential use as biodiesel. Energy Convers Manage 51:2307–2311
26. Ataya F, Dubé MA, Ternan M (2007) Acid-catalyzed transesterification of canola oil to biodiesel under single- and two-phase reaction conditions. Energy Fuel 21:2450–2459
27. SRS Engineering Corp (2012) Feedstock pretreatment—high FFA pretreatment. http://www.srsbiodiesel.com/HighFFAPretreatment.aspx. Accessed 25 June 2012
28. Thane P (2012) Pretreatment biofuels. Biofuels Int 6:59–63
29. Silverson Machines Inc. Refining of edible oils. http://edge.silverson.com/assets/PDFs/AppReports/Food/FEdibleOils.pdf. Accessed 25 June 2012
30. Lurgi AG (1991) Fatty acid technology, technical brochure no. 197e/3.91/30, Lurgi AG, Frankfurt
31. Lam MK, Lee KT, Mohamed AR (2010) Homogeneous, heterogeneous and enzymatic catalysis for transesterification of high free fatty acid oil (waste cooking oil) to biodiesel: a review. Biotechnol Adv 28:500–518
32. Rohm and Haas Company (2009) AMBERLYST^{TM} BD20—solid catalyst FFA esterification technology. https://dowanswer.custhelp.com/ci/fattach/get/23481/. Accessed 25 June 2012
33. Berrios M, Skelton RL (2008) Comparison of purification methods for biodiesel. Chem Eng J 144:459–465
34. De Dardel F (1989) Glycerin purification. Rohm and Haas, Paris
35. Department of biological and agricultural engineering, University of Idaho. using absorbents to lower FFA level in Biodiesel feedstocks, Biodiesel tech notes. http://www.uiweb.uidaho.edu/bioenergy/Adsorbents.pdf. Accessed 25 June 2012
36. Kotbara R (2011) 4 feedstock pretreatment adsorbents tested, 1 works. Biodiesel Mag. http://www.biodieselmagazine.com/blog/article/2011/11/4-feedstock-pretreatment-adsorbents-tested-1-works. Accessed 25 June 2012
37. Bertram B, Abrams C, Kauffman J (2002) Adsorbent filtration system for treating used cooking oil or fat in frying operations. US Patent 6,368,648
38. Ranganathan SV, Narasimhan SL, Muthukumar K (2007) An overview of enzymatic production of biodiesel. Bioresour Technol 99:3975–3981
39. Dorado MP, Ballesteros E, López FJ, Mittelbach M (2004) Optimization of alkali-catalyzed transesterification of *Brassica carinata* oil for biodiesel production. Energy Fuel 18;77–83
40. Basheer S, Haj M, Kaiyal M (2010) A robust multi-enzyme preparation for the synthesis of fatty acid alkyl esters. US Patent 7,790,429
41. Basheer S et al (2010) Modified immobilized enzymes of high tolerance to hydrophilic substrates in organic media. US Patent Application 20,100,209,982
42. Dixit L, Kanakraj S (2010) Enzymatic degumming of feedstock's (vegetable oil) for biodiesel—a review. J Eng Sci Manage Edu 3:57–59
43. Erickson DR, Pryde EH, Brekke OL, Mounts TL, Falb RA (eds) (1980) Handbook of soy oil processing and utilization. American Soybean Association, St. Louis, and American Oil Chemists' Society, Champaign
44. Verleyen T, Sosinska U, Ioannidou S, Verhe R, Dewettinck K, Huyghebaert A, De Greyt W (2002) Influence of the vegetable oil refining process on free and esterified sterols. J Am Oil Chem Soc 79:947–953
45. Indira TN, Hemavathy J, Khatoon S, Gopala Krisna AG, Bhattacharya S (2000) Water degumming of rice bran oil: a response surface approach. J Food Eng 43:83–90
46. Van Nieuwenhuyzen W, Tomás MC (2008) Update on vegetable lecithin and phospholipid technologies. Eur J Lipid Sci Tech 110:472–486
47. Careli AA, Brevedan MIV, Crapiste GH (1997) Quantitative determination of phospholipids in sunflower oil. J Am Oil Chem Soc 74:511–514

48. Szydlowska-Czerniak A (2007) MIR spectroscopy and partial least-squares regression for determination of phospholipids in rapeseed oils at various stages of technological process. Food Chem 105:1179–1187
49. Ochoa N, Pagliero C, Marchese J, Mattea M (2001) Ultrafiltration of vegetable oils: degumming by polymeric membranes. Sep Purif Technol 22:417–422
50. Choukri A, Kinany MA, Gibon V, Tirtiaux A, Jamil S (2001) Improved oil treatment conditions for soft degumming. J Am Oil Chem Soc 78:1157–1160
51. Voegele E (2011) Verenium markets biodiesel pretreatment. Biodiesel Mag. http://biodieselmagazine.com/articles/8224/verenium-markets-biodiesel-pretreatment-enzyme. Accessed 25 June 2012
52. Yang B, Wang Y, Yang J (2006) Optimization of enzymatic degumming process for rapeseed oil. J Am Oil Chem Soc 83:653–658
53. Copeland D, Belcher MW (2005) Vegetable oil refining. W. Int. Cl. C11B3/00. US Patent 6,844,458

Chapter 19
Organosolv Pretreatment of Pine Sawdust for Bio-ethanol Production

Chunbao (Charles) Xu, Liao Baoqiang and Wei Shi

Abstract This chapter presents some recent research results on pretreatment of pine sawdust for bio-ethanol production, using organosolv extraction-based methods, combined with other methods including ultrasonic treatment, and sodium hydroxide treatment, and enzymatic hydrolysis of the pretreated pine sawdust samples. The pretreatment efficiency (PE) and delignification efficiency (DE) of various pretreatment methods were studied. All the pretreatment methods, in particular the organosolv extraction, resulted in significant removal of lignin and hemicellulose. The results indicated that the combination of three pretreatment methods (organosolv extraction + ultrasound + NaOH) achieved the best PE (61.6 % $\pm$ 1 %) and DE (86.4 % $\pm$ 3 %). Enzymatic hydrolysis of pine samples treated with different pretreatment methods was comparatively studied. Glucose yields, total sugar yields, and total weight loss were obtained under various enzyme loadings (0~15.6 FPU cellulase) and reaction times (up to 48 h). The maximum glucose yield and the maximum total sugar yield were 5.8 % and 7.1 %, respectively, for un-pretreated raw pine samples, compared with 19.3 % and 22.40 % for the (organosolv extracted + ultrasound + NaOH) treated samples.

Keywords Softwood · Jack pine · Pretreatment · Organosolv extraction · Ultrasound Sodium hydroxide · Pretreatment efficiency · Delignification efficiency · Enzymatic hydrolysis · Glucose

19.1 Introduction

Fossil fuels, mainly coal, petroleum and natural gas, account to more than 80 % of the primary energy consumption in the world. The burning of fossil fuels emits around 21.3 billion tonnes of greenhouse gases (GHGs) annually. As such, it is

C. Xu (✉)
Department of Chemical and Biochemical Engineering, Western University, London, ON, N6A 5B9, Canada
e-mail: cxu6@uwo.ca

B. Liao · W. Shi
Department of Chemical Engineering, Lakehead University, Thunder Bay, Ontario P7B 5E1, Canada

Z. Fang (ed.), *Pretreatment Techniques for Biofuels and Biorefineries*, Green Energy and Technology,
DOI 10.1007/978-3-642-32735-3_19,

strategically pivotal to pursue alternative and renewable energy sources due to the rapidly increasing demand for energy, the depleted fossil resources and the growing concerns over climate changes and energy security. The pursuit of bio-ethanol as an alternative energy source has attracted increasing interest recently.

As a typical bio-fuel, bio-ethanol is considered carbon neutral since the carbon dioxide released from combustion of ethanol produced from renewable lignocellulosic materials is the CO_2 sequestered by the plants during their growth. Bio-ethanol can be used in various ways for energy and chemicals, while more commonly as a blended fuel in gasoline. Nowadays, all gasoline engines can use up to 10 wt% ethanol blended fuel without any need of engine modification [1, 2]. However, the main challenges for commercialization of the bio-ethanol technologies (particularly for the cellulosic ethanol) may be the combination of the low cost of conventional energy resources and the high biomass processing cost [1]. Compared with the conventional starch-based bio-ethanol manufacture, production of cellulosic ethanol using non-food lignocellulosic feedstock is advantageous as it does not compete with the food industry for feedstocks. Typical feedstock for cellulosic bio-ethanol production includes crop residues, grasses, forest biomass and waste, such as sawdust and wood chips [1, 3]. Softwoods are the dominant wood species in North America. Softwoods, for example, pine and spruce, contain about 40–45 wt% cellulose, 20–25 wt% hemicelluloses, and 25–30 wt% lignin. Because of the structural characteristics of woody biomass (large polymeric molecules and high crystallinity) of the cell wall and the presence of hemicellulose and lignin, enzymatic hydrolysis of cellulose into glucose for bio-ethanol production is challenging due to its low accessibility to enzymes. As a result, pretreatment of lignocellulosic biomass to loosen the cellulose crystalline structures and increase its accessibility is necessary, important, and the key in the bioconversion process. For instance, Zhu and his team in US Forest Service of Forest Products Laboratory reported a very high enzymatic hydrolysis efficiency for softwood species (e.g., spruce, red pine) with combined dilute acid and sulfite pretreatment, or hot water, dilute acid or sulfite pretreatment, combined with disk-milling pretreatment [4–6]. With sulfite pretreatment using 8–10 % bisulfite and 1.8–3.7 % sulfuric acid on oven dry wood at 180 °C for 30 min, enzymatic hydrolysis of spruce chips led to more than 90 % cellulose conversion at an enzyme loading of about 14.6 FPU cellulase plus 22.5 CBU β-glucosidase per gram of od substrate after 48-h hydrolysis [4]. Enzymatic hydrolysis of lignocellulosic materials into fermentable sugars can achieve a glucose yield from 10 % to 60 %, depending on the type of lignocellulosic materials [7]. There are a number of different pretreatment methods that can be generally classified into the following types: physical pretreatment, physical-chemical pretreatment, chemical pretreatment, and biological pretreatment, whose details will be overviewed as follows.

19.1.1 Physical Pretreatment

Physical pretreatment such as comminution of lignocellulosic materials through a combination of chipping, grinding, and milling can be applied to reduce cellulose crytallinity [3]. Zhu et al. [5, 8] demonstrated the effectiveness of different millings methods as pretreatment approaches in enzymatic hydrolysis. These studies demonstrated the effects of chemical pretreatment (hot water, dilute acid, and sulfite) and disk-milling conditions on the efficiency of enzymatic cellulose saccharification of softwood as well as on energy consumption for size-reduction. Yeh et al. [9] applied a media mill (0.94 kW) for milling cotton particles to 0.3 mm, and the yield of glucose was significantly increased by five-folds or more. Ball milling was found to be effective in reducing cellulose crystallinity of lignocellulosic materials such as woody biomass [10]. For the physical pretreatment method, energy consumption is always an undesirable issue. Obviously, the power requirement for mechanical comminution of lignocellulosic materials is determined by the final particle size and the biomass characteristics. It was estimated that the energy input for comminution can be lower than 30 kW h/t of biomass if the final particle size is in the range of 3–6 mm, while in most cases the energy consumption could be even higher than the theoretical energy content of the biomass treated, limiting the economic viability of the physical pretreatment method for cellulosic ethanol production [11].

19.1.2 Physical–Chemical Pretreatment

Physical–chemical pretreatment methods include steam explosion, ammonia fiber explosion (AFEX), microwave pretreatment, and ultrasonic pretreatment, etc. Steam explosion is the most commonly used physical–chemical method for the pretreatment of lignocellulosic materials in particular for hardwood and agricultural biomass [12]. In this method, biomass is treated with high-pressure saturated steam and then the pressure is suddenly reduced such that the biomass undergoes an explosive decompression. Steam explosion is typically carried out at 160–260 °C and 0.69–4.83 MPa for several seconds to a few minutes before the materials are exposed to atmospheric pressure [13]. Like a hydrothermal pretreatment, steam explosion removes hemicellulose mainly by autohydrolysis and partially by chemical effects and mechanical forces. The high temperature and pressure promote the acetyl groups present in hemicellulose to be automatically hydrolyzed to acetic acid; on the other hand, the water may act as an acid under such high-temperature condition. All of these acids formed in the steam explosion process could thus hydrolyze hemicellulose. Removal of hemicellulose exposes the cellulose surface and increases enzyme accessibility to the cellulose microfibrils [14]. In the stream explosion process, lignin can also be removed to a certain extent, but is redistributed on the fiber surfaces as a result of melting and depolymeriation/repolymerization reactions [15]. The removal and redistribution of hemicellulose and lignin could swell the pretreated sample and increase its accessible surface area [7]. The main drawback of this method is that many enzyme-inhibitors are produced in the pretreatment. For example, the pentoses and hexoses formed from the hydrolyzed hemicellulose and cellulose can be further degraded to furfural, 5-hydroxymethylfurfural (HMF), levullinic acid, and formic acid, which would deactivate the enzymes used in the consecutive enzymatic hydrolysis process. In the AFEX pretreatment process, lignocellulosic materials are treated with liquid ammonia at the temperature between 60 °C and 100 °C under high pressure for a certain period of time, normally 10 min, before the pressure is released swiftly, which would cause mechanical explosion of the materials from the inside. Recycling of ammonia in the system after the pretreatment is economically feasible due to the high volatility of ammonia at atmospheric pressure [16]. The AFEX process was employed for the pretreatment of a variety of lignocellulosic materials such as alfalfa, wheat straw, wheat chaff, barley straw, corn stover, rice straw, municipal solid waste, softwood newspaper, kenaf newspaper, coastal Bermuda grass, switchgrass, aspen chips, and bagasse [17, 18]. AFEX pretreatment does not remove much hemicellulose and lignin, but it can decrease the cellulose crystallinity, disrupt the lignin–carbohydrate linkages and remove the acetyl groups from hemicellulose [19]. After pretreatment, the enzymatic digestibility of lignocellulosic materials can be increased. Moniruzzaman et al. [20] achieved more than 80 % of the theoretical sugar yield from corn fiber pretreated using AFEX at 90 °C, an ammonia-to-corn fiber mass ratio of 1:1 and 200 psi for 30 min. Superior to the steam-explosion pretreatment where many inhibitors are formed from the degradation of hemicellulose and cellulose, the AFEX pretreatment is likely more advantageous as no toxic byproducts are formed except for some phenolic fragments from lignin [3]. However, the AFEX process was not very effective for biomass with higher lignin content such as newspaper and woody biomass [12, 13].

Microwave pretreatment may be considered a physico-chemical process since it involves both non-thermal and thermal effects. The microwave method has proved to be effective for improving enzymatic hydrolysis of many agricultural residues/biomass such as rice straw and wheat straw [21, 22]. Ma et al. [22] systematically optimized the pretreatment conditions for rice straw hydrolysis, and studied the effects of microwave intensity, irradiation time, and substrate concentration on the hydrolytic conversion of cellulose and hemicellulose. It was believed that microwave and microwave-based pretreatment could hydrolyze hemicellulose and solubilize lignin. Moreover, the thermal and non-thermal effects arising from heating could enlarge the pore size of the lignocellulosic materials, and enhance the accessibility of cellulose to enzymes [23]. Ultrasonic pretreatment could be another promising method for the removal of hemicellulose and lignin, and it has been widely used in the extraction of hemicellulose and enzyme proteins from biomass and organic waste materials such as biosludge [24, 25], although there is very limited published literature with respect to the effects of ultrasonic pretreatment on the glucose yield and carbohydrate conversion during enzymatic hydrolysis of lignocellulosic materials. Yachmenev et al. [26] reported that saccharification of cellulose was enhanced considerably by ultrasonic pretreatment. The increased enzymatic hydrolysis yields after ultrasonic pretreatment could be explained by the effects of ultrasonic pretreatment: cracking of the cell wall, dislocation of the secondary wall of the middle layer of the cell wall, and exposure of the middle layer to enzymes [27]. Yu et al. [28] investigated the effects of ultrasonic pretreatment on enzymatic hydrolysis of rice hull using 250 W, 40 kHz at 25 °C for different period of time ranging from 10 min to 60 min. Their results showed that the yields of total sugar and glucose increased from 11.7 % to 10.9–16.3 % and 15.8 %, respectively, via ultrasonic pretreatment for 30 min. In summary, ultrasonic treatment can cause cavitation, crash the cell wall structure, and provide more accessible surfaces in the substrate. Ultrasonic pretreatment can thus be a potential method for pretreatment of lignocellulosic materials due to its lower energy consumption. Due to the limited research in this respect, it is of interest to examine the effects of ultrasonic pretreatment on enzymatic hydrolysis of lignocellulosic materials, in particular, when combining it with other chemical pretreatment approaches such as organosolv and alkaline methods.

19.1.3 Chemical Pretreatment

Chemical pretreatment includes acid pretreatment, alkaline pretreatment, and organosolv pretreatment as detailed below. Acid pretreatment can be performed with a concentrated acid such as hydrochloride acid, sulfuric acid, phosphoric acid, and nitric acid, or a dilute acid (such as dilute sulfuric acid). Although concentrated acid proved to be very effective for the hydrolysis of hemicellulose, it is less attractive for practical application to ethanol production, due to the significant formation of high concentration of inhibitory compounds (such as furfural, HMF, etc.). In addition, pretreatment using concentrated acid under a high temperature (>180 °C) may cause the decomposition of cellulose. Moreover, a strong acid at a high concentration would corrode the reactor and cause acid recovery issues, which would lead to high operational and maintenance costs [29]. Unlike the concentrated acid pretreatment, dilute acid pretreatment has been successfully developed for the pretreatment of lignocellulosic materials under relatively moderate experimental conditions. It can be used either for pretreatment of lignocellulosic materials before enzymatic hydrolysis or for direct hydrolysis of the biomass into fermentable sugars. Typically, dilute acid pretreatment can achieve a high reaction rate and obtain nearly 100 % hemicellulose removal under the optimal conditions of 180 °C for a short reaction time (e.g., 5 min) or at 120 °C for a comparatively long reaction time (e.g., 30–90 min). Due to the possible decomposition of cellulose and formation of inhibitors at severe conditions, many studies have focused on conditions at lower temperatures. For example, Guo et al. [30] attained a high yield of xylose (70 % and 75 %) from silvergrass when treated with 1 %, 2 %, and 3 % dilute sulfuric acid at 121 °C for 30 min. The dilute acid pretreatment process also produced glucose at a yield of 10 %, and converted more than 90 % of the acetyl groups into acetic acid and 4–5 % of furfural, which are the inhibitors for the fermentation process. Dilute sulfuric acid hydrolysis was performed on an energy crop giant reed (*Arundo donax* L.), and the monomeric xylose recovery 94 % was achieved under optimized hydrolysis conditions (1.27 % acid concentration, 141.6 °C and 36.4 min) [31]. However, the highest hemicellulosic sugars recovery during pretreatment does not necessarily lead to the highest enzymatic hydrolysis efficiency [32, 33].

Alkalis or alkaline solutions such as NaOH and $Ca(OH)_2$ have been applied for effectively removing lignin and hemicellulose from lignocellulosic materials in the pulping processes. They also proved to be effective reagents in pretreating lignocellulosic materials to increase the accessibility of enzymes to cellulose for bio-ethanol production. Compared with the steam-explosion approach and acid pretreatment as introduced previously, alkaline pretreatment can be carried out under ambient or mild conditions. In addition, unlike the acid pretreatment with the issues of corrosion and inhibitory product formation, alkaline pretreatment causes less sugar degradation to form toxic compounds, and many caustic salts may be recovered or regenerated to achieve better process economics. However, a drawback of alkaline pretreatment may be that it normally needs a longer pretreatment time ranging from hours to days. Alkaline pretreatments can increase cellulose digestibility, and they are more effective for the solubilization and removal of lignin, while having minimum effects on cellulose and hemicellulose [34]. The mechanism of alkaline hydrolysis is believed to involve saponification of intermolecular ester linkages in the xylan of hemicellulose and lignin. Alkaline pretreatment using sodium hydroxide has been widely reported in a number of publications. NaOH treatment of lignocellulosic materials was found to cause swelling, an increase in internal surface area, decrease in crystallinity and degree of polymerization of cellulose, separation of the structural linkages between lignin and carbohydrates, and disruption of the lignin structure [32]. Zhao et al. [35] compared the effects of pretreatments using mild sulfuric acid and NaOH on enzymatic hydrolysis of crofton weed stem, and they concluded that the optimal pretreatment conditions using NaOH was at 110 °C for 120 min with NaOH loading of 10 wt% and a liquid-to-solid ratio of 6:1 (w/w). Under these optimal conditions, 30 % of the raw material and 25 % of the lignin were removed in the pretreatment. Compared to mild sulfuric acid pretreatment, the NaOH pretreatment obtained a higher enzymatic conversion ratio of cellulose. Xu et al. [36] employed NaOH to pretreat switchgrass and achieved 45 g of total reducing sugars per 100 g of raw biomass with the glucan and xylan conversion ratios as high as 74.4 % and 62.8 %, respectively, under optimal conditions (e.g., 50 °C, 12 h, and 1.0 % NaOH), which were nearly four times that of the untreated biomass.

The addition of an oxidative agent, such as H_2O_2, to alkaline pretreatment can improve the performance by favoring lignin removal. Alkaline peroxide treatment is an effective method. The typical procedure of this method is as follows: The lignocellulosic materials are soaked in an alkaline solution at optimal pH 11–12 (pretreatment efficiency may change at different pH values), then low concentration H_2O_2 is added to the solution at a low temperature normally 30–70 °C for typically several hours. The ability of H_2O_2 to delignify biomass is attributed to its ability to react with several carbonyl-containing structures in lignin, which can be explained through the reactions of the hydro-peroxide anion (HOO^-), formed in an alkaline medium according to the following equilibrium [37]:

$$H_2O_2 + HO^- \leftrightarrow HOO^- + H_2O, \quad pKa = 11.6 \text{ at } 25\,^\circ C$$

This anion is a strong nucleophile that preferentially attacks ethylenic and carbonyl groups present in lignin. As a consequence, such chromophores as quinines, cinnamaldehyde and ring-conjugated ketones could be converted to nonchromophoric species in the alkaline solution [38]. On the other hand, the H_2O_2 itself is unstable under alkaline conditions and can readily decompose as the temperature is increased. At the same time, catalyzed by some metal ions such as manganese, iron, and copper presence in the biomass, the H_2O_2 would release more active radicals, for example, $HO^{\bullet}$ and O^{2-}, to facilitate the delignification of biomass. Sun et al. [39] reported the delignification efficiency of rye straw using H_2O_2 at different temperatures from 20 °C to 70 °C with 2 % H_2O_2 at pH 11.5 for 12 h. The maximum delignification efficiency was calculated to be 87.8 % at 70 °C and correspondingly 71.9 % of hemicellulose was removed as well. Saha and Cotta [40] pretreated rice hull using 7.5 % (v/v) H_2O_2 at pH 11.5, 35 °C for 24 h and reached a theoretical sugar yield of 90 %, where no measurable furfural and HMF were detected in the process.

19.1.4 Organosolv Pretreatment

Organosolv pretreatment is another promising pretreatment method suitable for enzymatic hydrolysis of lignocellulosic materials. In general, it uses an organic or aqueous organic solvent to remove or cleave the linkages between lignin, hemicellulose, and cellulose. In this process, lignocellulosic materials are mixed with an organic solvent with or without water and heated up to a high temperature, typically 150–250 °C, and high pressure, although some literature studies revealed that it can also be operated under a mild condition. An acid-free organosolv process can overcome the problems caused by acid catalyzed pretreatment. Ethanol–water solvent was found to be an effective pretreatment method for bio-ethanol production from *Liriodendron tulipifera*. Although relative lignin contents were above 20 %, enzymatic conversion increased significantly to 65 % by organosolv pretreatment with sodium hydroxide [41]. A mechanism has been proposed that stipulates that under high temperature and pressure, organic reagents can cleave the bonding of different functional groups (such as the α-O-4 and β-O-4 ethers) in lignin and fragment larger lignin molecules into smaller ones. It also accompanies the formation of acetic acid or other acids or byproducts from the acetyl groups in hemicellulose under severe conditions. These acids can act as catalysts accelerating the rapture of lignin–carbohydrate complex [7]. At the same time, water acts like a medium to dissolve these products. At the end of the reaction, it produces three phases (oily phase, water soluble phase, and solid residue). Cellulose can be recovered in the solid phase and hemicellulose and lignin fractions can be obtained after extraction or distillation of the liquid phase [42]. Many organic solvents such as alcohols, esters, ketones, glycols, organic acids, phenols, and ethers have been tested as the organosolv pretreatment solvents for the pulping processes and enzymatic bio-ethanol production [43]. From an economic perspective, the use of low-molecular weight alcohols such as ethanol and methanol is more favorable as these solvents are easy for recovery due to their lower boiling points. Pasquini et al. [44] extracted lignin using ethanol–water mixtures and CO_2 from sugar cane bagasse and *Pinus* tarda wood chips, and achieved a high delignification efficiency of 93.1 % for *P. tarda* wood chips and 88.4 % for sugar cane bagasse. Using hot aqueous 60 % ethanol and 1.25 % H_2SO_4 to pretreat poplar chips at 180 °C for 60 min before enzymatic hydrolysis, Pan et al. [42] obtained 74 % of lignin removal, and 82 % of total cellulose converted to glucose. In a study by Araque et al. [45], organosolv pretreatment of woody biomass using acetone-water 1:1 (w/w) at pH 2.0 and 195 °C for 5 min also resulted in 71.8 % total sugar yields in the enzymatic hydrolysis.

Softwood is among the most difficult biomass feedstocks for enzymatic hydrolysis due to its high lignin content and high cellulose crystallinity. For enzymatic hydrolysis of softwood, the pretreatment methods discussed above have demonstrated varying degrees of success, so continued research efforts in the area of pretreatment are needed. The primary goal of this study was thus to investigate various pretreatment methods for softwood biomass (pine sawdust) to fermentable sugars (glucose, xylose, etc.). Specific objectives of this study are to (1) pretreat pine sawdust using various pretreatment methods including organosolv extraction, ultrasound, and NaOH pretreatment, and (2) evaluate enzymatic hydrolysis of the pretreated pine sawdust.

19.2 Experimental

19.2.1 Materials

The *Pinus bank siana* (Jack pine) sawdust was supplied by a local lumber mill (Northern Wood Ltd). The sawdust was grounded with a Wiley mill and screened using a 20-mesh to a particle size of 0.75 mm for the experiments. The particles were dried in an oven at 105 °C for 12 h before use. The results of proximate and ultimate analysis of the pine sample and the chemical compositions of the ash from pine wood samples are summarized in Table 19.1. Wherein, the proximate analysis (ash, volatile matters, and fixed carbon contents) was performed in accordance with the standard of ASTM D 5142 by thermogravimetric analysis (TGA), and the results are reported in Table 19.1 on a dry basis. The ultimate analyses or elemental compositions were conducted according to ASTM D 5373 on a CHNS elemental analyzer, while the oxygen content was calculated by difference on a dry basis (including the ash content). The cellulose, hemicelluloses, and lignin content were determined by the Analytical Laboratory of FPInnovations, Montreal, Canada. The samples were extracted with acetone to obtain extractive-free test specimens. Carbohydrates were determined according to the Technical Association of the Pulp and Paper Industry (TAPPI) test method T249 cm-85 and the acid-soluble and acid-insoluble lignin was determined according to the TAPPI test method T222 cm-88. The results are summarized in Table 19.2.

Table 19.1 Proximate and ultimate analysis results of the pine wood sample and chemical compositions of the ash from the wood sample

	Proximate analysis (wt%) (d.b.[a])			Ultimate analysis (wt%) (d.b.[a])				
	VM	FC	Ash	C	H	N	S	O[b]
Pine wood sample	81.52	18.31	0.17	53.3	6.2	0.1	0.1	40.3
Major elements in the ash, ppmw (d.b.)[c]								
	Na	K	Mg	Ca	Mn	Fe	Zn	Al	Si
Ash from the sample	7	114	100	440	20	9	10	16	3

[a]On a dry basis
[b]By difference
[c]Determined by ICP-AES

Table 19.2 Carbohydrates and lignin contents (wt%) in the pine sawdust sample

Acid-insoluble lignin	28.2
Acid-soluble lignin	0.22
Carbohydrates	
Arabinan	1
Xylan	3.5
Mannan	11.4
Galactan	1.5
Glucan	44.7
Acetone extractives	5.95
Total lignin	28.4
Cellulose	40.2
Hemicelluloses	21.9

19.2.2 Pretreatment of Pine Sawdust

Pretreatment of pine sawdust was conducted using various methods, including organosolv extraction, organosolv extraction followed by ultrasonic treatment, organosolv extraction followed by NaOH treatment, and combined organosolv extraction, ultrasonic treatment and NaOH treatment in sequence, to remove lignin and/or hemicellulose and disrupt the textual structure of the pine sample to increase the accessibility of cellulose to enzymes.

19.2.2.1 Organosolv Extraction

Pine sawdust (20 mesh) was dried in an oven at 105 °C for 12 h before use. The experiment was carried out in a 1 L stirred autoclave reactor under nitrogen atmosphere. A sample of 50 g was weighed and mixed with an ethanol–water solution (1:1 v/v) at a biomass-to-solvent ratio of 1:10 (w/v). The reaction temperature was set at 190 °C and the stirring speed was controlled at 400 ± 20 rpm. The initial nitrogen pressure was 300 psi in the reactor and the maximum pressure during the reaction was 700 psi. After a reaction time of 4 h, the reactor was cooled down rapidly using cooling water. The liquid and solid phases in the reactor were then separated via filtration. The liquid phase was used to recover hemicellulose and lignin. The dried solid phase was considered to be cellulose-rich residue, which was divided into four parts and subjected to further treatments as described below. One part of the residue was further treated with ultrasound and NaOH in sequence and was designated as pine sample with (organosolv + ultrasound + NaOH) pretreatment. Another part was treated with ultrasound only and was denoted as pine sample with (organosolv + ultrasound) pretreatment. The third part was treated with NaOH only denoted as pine sample with (organosolv + NaOH) pretreatment. The fourth part without any further treatment and was designated as pine samples with (organosolv) pretreatment. For comparison, the raw (untreated) pine sawdust sample was used as a reference for enzymatic hydrolysis. Detailed experimental conditions and methods for the ultrasonic and NaOH treatment are described as follows.

19.2.2.2 Organosolv Extraction Followed by Ultrasonic Treatment

The solid residue from the organosolv treatment was weighed and placed in a 500 mL beaker with distilled water (5 g, 200 mL water). The mixture was treated with ultrasound (Fisher Scientific Ultrasonic Cleaner, 100 W, 42 kHz output) for 3 h at room temperature. After that, the solid phase was separated from the liquid phase by filtration. The collected solid sample with (organosolv + ultrasonic) pretreatment was then used for enzymatic hydrolysis.

19.2.2.3 Organosolv Extraction Followed by NaOH Treatment

Five grams of solid residue from the organosolv treatment was placed in a 500 mL beaker with the addition of 250 mL 1 N NaOH. The NaOH treatment was carried out at 70 °C in a water bath shaker at 100 rpm for 3 h. After filtration, the solid residue was collected for enzymatic hydrolysis.

19.2.2.4 Organsolv Extraction Followed by Ultrasonic and NaOH Treatment

The solid residue from the organosolv treatment was treated with ultrasound and NaOH in sequence at the same reaction conditions as described above. The solid was recovered after filtration and used in the enzymatic hydrolysis process.

19.2.3 Enzymatic Hydrolysis

Pretreated pine samples and raw pine sawdust were hydrolyzed using cellulase (*Trichoderma reesei,* from Sigma Alderich) supplemented with β-*glucosidase* (Novozym 188) at a ratio of 1 FPU:2 IU, where the activities of cellulase and β-*glucosidase* are expressed in Filter Paper Unit (FPU) and International Units (IU), respectively. The hydrolysis was performed in a water bath shaker at 100 rpm and a 50 mL flask using 0.10 g cellulosic substrate and in 25 mL 0.1 M sodium citrate buffer solution (pH of 4.8). The temperature was set at 50 °C. In this study, the effects of different pretreatment methods (organosolv, organosolv + ultrasound, organosolv + NaOH, and organosolv + ultrasound + NaOH), hydrolysis time (0–72 h), and enzyme doses (0–16 FPU) on glucose yield, total sugar yield, and total weight loss were examined. After the reaction was completed, the liquid solution was separated from residual solids by filtration. The filtrate was stored at −20 °C in a freezer for further analysis for sugar contents. The weight of the residual solids after the enzymatic hydrolysis was determined. At last duplicate tests were performed for each condition of the enzymatic hydrolysis experiments.

The pretreatment efficiency, delignification efficiency, and reducing sugar (glucose) yield and total sugar yield are calculated as Eqs. 19.1–19.4:

$$\text{Pretreatment efficiency (PE)} = \left(1 - \frac{\text{Mass of residual solids}}{\text{Mass of raw materials}}\right) \times 100\,\% \quad (19.1)$$

$$\text{Delignification efficiency (DE)} = \left(1 - \frac{\text{Mass of residual lignin}}{\text{Mass of original lignin}}\right) \times 100\,\% \quad (19.2)$$

$$\text{Glucose yield (wt\%)} = \frac{\begin{array}{c}\text{Mass of reducing glucose}\\ \text{from hydrolysis}\end{array}}{\begin{array}{c}\text{Mass of cellulsoe and hemicellulose}\\ \text{in the substrate}\end{array}} \times 100\,\% \quad (19.3)$$

$$\text{Total sugar yield (wt\%)} = \frac{\begin{array}{c}\text{Mass of total sugar from the}\\ \text{hydrolysis}\end{array}}{\begin{array}{c}\text{Mass of cellulsoe and}\\ \text{hemicellulose in the substrate}\end{array}} \times 100\,\% \quad (19.4)$$

The glucose concentration after enzymatic hydrolysis was determined using the dinitrosalicylic acid (DNS) assay method [46]. Using glucose solution as the standard, a standard calibration curve was prepared for each batch of measurements. The total sugar concentration after the enzymatic hydrolysis was determined using the anthrone reagent method, a colorimetric method [47], using glucose as the standard. The Klason lignin contents of the raw and pretreated pine sawdust were determined by complete hydrolysis performed in accordance with a modified TAPPI standard method, where 0.3 g of the lignocellulosic material was treated with 3 mL of 72 wt% sulfuric acid at 30 °C for 60 min, followed by a dilution with 57 mL water and pressure cooking at 120 °C for 40 min.

19.2.4 Feedstock and Treatment Residue Characterization

Ultraviolet–visible spectroscopy or ultraviolet-visible spectrophotometer (UV-Vis or UV/Vis) was employed to determine the characteristic absorption of the sawdust samples prior to and after the treatment. The analysis was performed on a CARY-5E-UV-VIS-NIR spectrophotometer. The UV wavelength range was setting from 240 to 380 nm. X-ray diffraction (XRD) is a method of determining the arrangement of component structure within a crystal. The carbon crystalline structure of the pine sawdust before and after pretreatment was determined by XRD using Rigooku DMAX-RB diffractometer operated at 45 kV and 50 mA. Following the method of Kim and Holtzapple [48] as well as Liu et al. [49], the carbon crystallinity index (CrI) was calculated with the Eq. 19.5:

$$CrI(\%) = \frac{I_{002} - I_{am}}{I_{002}} \times 100\ \% \tag{19.5}$$

where I_{002} is the intensity of the crystalline area of (0 0 2) plane at $2\theta = 22.4\,°$; I_{am} is the intensity of the amorphous region at $2\theta = 18.7\,°$. It shall be noted that I_{002} includes both crystalline and amorphous intensity (background). The scanning electron microscope (SEM) was used to characterize the morphological structure change in pine sawdust before and after pretreatment. All solid samples were milled to 20 mesh in particle size, mounted with gold coating before the SEM scanning. The images were taken by a JEOL5900LV SEM, Oxford INCA microanalysis full system with 4 "×5" analytical stage.

19.3 Results and Discussion

The results of the pine sawdust pretreatment using different methods including organosolv extraction, followed by ultrasonic and/or NaOH treatment, are discussed below. The organosolv extraction pretreatment aimed to remove or cleave the linkages between lignin, hemicellulose, and cellulose using organic solvent, loosen the cellulose fibrils and decrease the degree of polymerization. Ultrasonic treatment was employed for the rupture of the cell wall, dislocation of the secondary wall of the middle layer of the cell wall to expose and fibrillate the middle layer. NaOH treatment was used to remove hemicellulose and residual lignin. It is believed that NaOH treatment could lead to saponification of intermolecular ester bonds of hemicelluloses and lignin. The pretreated pine sawdust samples are classified into four groups: Group 1—organosolv extraction; Group 2—(organosolv extraction + ultrasonic) treatment; Group 3—(organosolv extraction + NaOH) treatment; and Group 4—(organosolv extraction + ultrasonic + NaOH) treatment. Pretreatment efficiency (PE) and delignification efficiency (DE) were determined after the pretreatment in accordance with Eqs (19.2) and (19.3). The raw and pretreated pine sawdust samples were further characterized using Fourier transform infrared spectroscopy (FTIR), XRD, and SEM to examine changes in the characteristic functional groups, crystalline structure, and cell wall structure of the pine sawdust before and after the treatments.

19.3.1 Pretreatment Efficiency and Delignification Efficiency

The results of PE and DE after each pretreatment are summarized in Table 19.3.

For samples after organosolv extraction (Group 1), the PE and DE were 51.4 % and 76.5 %, respectively. The results indicated that organosolv extraction had a significant effect on delignification. The delignification efficiencies are comparable or slightly higher than those reported in previous studies [42, 50]. The higher PE and higher DE might be accounted for by the slightly higher temperature (by 20 °C) employed in this work. Pasquini et al. [51] found that both temperature and pressure can affect PE and DE: A higher temperature and pressure could lead to an increase in the PE and DE. In a study by Pasquini et al., the DE was on the order of 93.1 % for *P. tarda* wood chips when 16.0 MPa and 190 °C were employed. This process also leads to the formation of acetic acid or other carboxylic acids that can act as catalysts accelerating the rupture of lignin–carbohydrate polymers in the presence of water.

For Group 2 samples (with organosolv extraction followed by ultrasonic treatment), slightly improved PE (53.3 %) and PE (77.2 %) were obtained. The results suggest that the ultrasonic treatment did affect PE and DE, although the effect was not significant. This could be caused by the decreased particle size of pine sawdust after ultrasonic treatment. A decrease in particle size would lead to an increase in specific surface area and thus release of hemicellulose and lignin. Ultrasonic treatment has been widely used to enhance the extraction of hemicellulose in alkaline solutions by introducing violent cavitation. In this study, the ultrasonic treatment of pine sawdust accounted for only 3–5 wt% weight loss, due to the fact that distilled water rather than alkaline solution was used as liquid phase in the ultrasonic treatment. Extractives from hemicellulose and cellulose could not be dissolved easily in water. The comparable DE values with and without ultrasonic treatment (76.5 % vs. 77.2 %) suggest that ultrasonic treatment was not very effective for removing lignin.

For Group 3 samples (after organosolv extraction followed by NaOH treatment), the PE (57.7 %) and DE (81.5 %) were significantly higher, implying the high efficiency of NaOH treatment in the extraction of hemicellulose and lignin. When the organosolv extracted solid was added into NaOH solution, its color turned to dark brown, and the mixture was much finer and more viscous than it was before the treatment. This could be explained by the fact that the NaOH treatment of lignocellulosic materials can cause swelling which leads to an increase in internal surface area, a decrease in the degree of polymerization, a decrease in crystallinity, and disruption of the lignin structure [13, 32].

It is not surprising to observe that organosolv extraction followed by ultrasonic and NaOH treatment in sequence (Group 4) led to the highest PE (61.1 %) and DE (86.4 %), approximately 10 %, respectively higher than that for the Group 1 pretreatment, as shown in Table 19.3.

Table 19.3 Pretreatment efficiency and delignification efficiency of pine sawdust

	Group 1	Group 2	Group 3	Group 4
Step 1[a]	Ethanol:water (1:1), 190 °C, 700 psi, 4 h	Ethanol:water (1:1), 190 °C, 700 psi, 4 h	Ethanol:water (1:1), 190 °C, 700 psi, 4 h	Ethanol:water (1:1), 190 °C, 700 psi, 4 h
Step 2	N/A[b]	Ultrasound (100 W), 40 kHz, 25 ° C, 3 h	NaOH (1 mol/L), 70 °C,100 rpm, 3 h	Ultrasound (100 W), 40 kHz, 25 °C, 3 h
Step 3	N/A	N/A	N/A	NaOH (1 mol/L), 70 °C, 100 rpm, 3 h
PE	51.4 ± 2 %	53.3 ± 1 %	57.7 ± 1.1 %	61.1 ± 1 %
DE	76.5 ± 3 %	77.2 ± 2.6 %	81.5 ± 3 %	86.4 ± 3 %

[a] Steps indicated different pretreatment methods taken
[b] *N/A* not applied

19.3.2 Characterization of Pretreated Pine Sawdust Samples

Effect of pretreatment (organosolv extraction, ultrasonic treatment, and NaOH treatment) on the microstructure of pine sawdust was analyzed using SEM. Comparison of the changes in microstructure of pine sawdust before and after the organosolv extraction pretreatment is shown in Fig. 19.1a–d. The SEM pictures clearly show that the microstructure of pine sawdust changed after pretreatments. Compared with the nicely presented cell wall structure in the raw pine sawdust, the pretreatments led to a significant swelling of the raw pine sawdust and caused disorder in the cell structure. After pretreatment, the cell wall seems to be destroyed. The original cell wall was twisted and disordered, suggesting that significant morphological changes occurred in the process. Similarly, the raw pine sawdust sample exhibited rigid and highly ordered fibrils. The fibers of pine sawdust samples after pretreatment were distorted and twisted. The microfibrils were also isolated from the initial connected structure and made fully exposable, thus increasing the external surface area and the porosity of the biomass structure.

From the XRD measurement of pine sawdust before and after the various types of pretreatments, there are three peaks occurring at 2θ of 14.6 °, 16.5 °, and 22.4 ° with varying intensities. These peaks correspond to cellulose XRD, corresponding to cellulose's crystalline planes of ($1\bar{1}0$), (110) and (220), respectively [52]. Compared with the un-treated raw material, the cellulose peaks all increase considerably in the pretreated pine sawdust samples. This demonstrates that the concentration of cellulose in the residues after pretreatment increases owing to the removal of hemicellulose and lignin in the pretreatment processes. The intensity of the cellulose peaks in sawdust samples increases with an increase in the PE and DE, which is expected as the cellulose content in the pretreated sawdust samples increased with an increase in the PE and DE. The *CrI* calculated from the Eq. 19.5 is 43 %, 63 %, 55 %, 64 %, and 65 %, respectively, for the untreated sawdust sample and the pretreated samples with the organosolv extraction, organosolv + NaOH, organosolv + ultrasound and organosolv + NaOH + ultrasound. As a common observation from chemical or

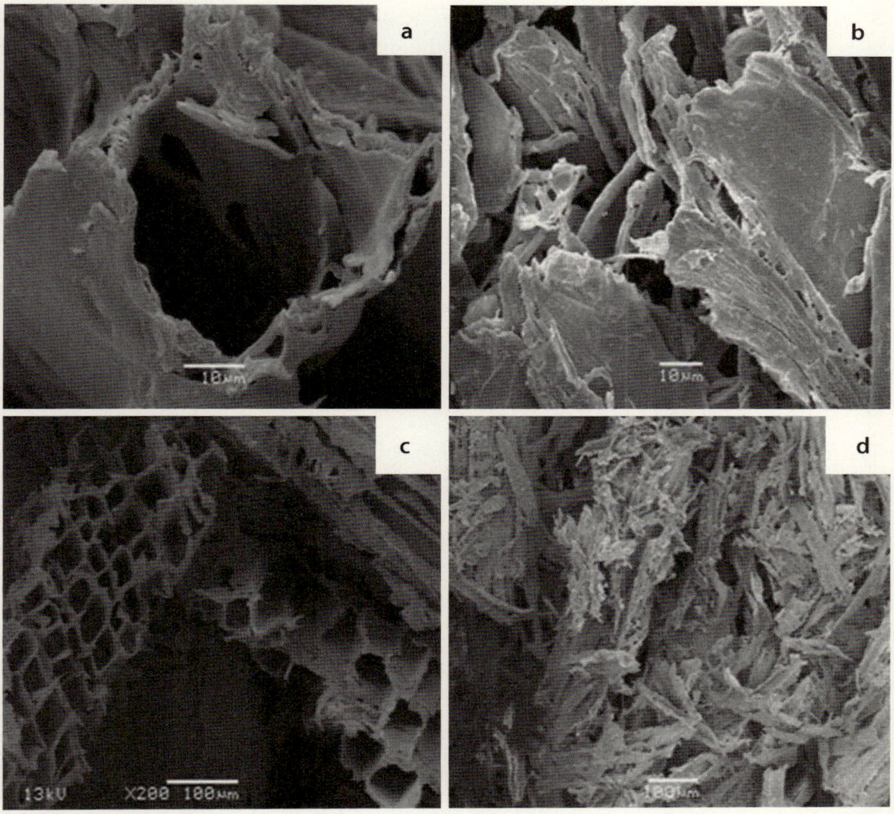

Fig. 19.1 Comparison of the changes in microstructure of pine sawdust before and after the organosolv extraction: **a** a single cell before treatment; **b** a single cell after the organosolv extraction treatment; **c** pine wood structure before treatment; **d** pine wood structure after the organosolv extraction treatment

organosolv pretreatment of biomass [32, 49], crystallinity of the pretreated samples increases due to the removal of hemicelluloses and lignin. However, it should be noted that an increase in crystallinity of pretreated material was not expected to negatively affect the product yield during enzymatic hydrolysis of biomass [32, 49].

The functional groups of the raw and pretreated sawdust samples, as well as pure cellulose and pure lignin were characterized using the FTIR technique (while the IR spectra are not included in this chapter). The FTIR results show that the pretreated sawdust samples are very similar to the peak patterns of the pure cellulose in the whole wavenumber range, simply because the cellulose content is high in pretreated sawdust samples when compared to the raw pine sawdust. The absorption bands at 1,275 cm^{-1} and 1,516–1,700 cm^{-1} correspond to the functional groups of lignin. The absorption band at 1,275 cm^{-1} is related to vibrations of guaiacyl rings and the absorption bands at 1,516–1,700 cm^{-1} ascribe to aromatic ring vibrations. The pure lignin and raw pine sawdust materials possess stronger signal at these absorption band, due to the higher content of lignin. The pretreated pine sawdust had a weaker signal at these absorption bands, because the majority of lignin was removed by the pretreatment. The absorption band at 1,711 cm^{-1} is related to carbonyl absorption in hemicelluloses. The raw and pretreated pine sawdust samples all showed absorption at 1,711 cm^{-1}, implying that the pretreatment were unable to remove all hemicellulose in the raw materials, as also reported by Sreenivasan et al. [53]. In summary, the FTIR measurement indicated that the pretreatment was more efficient for removing of lignin than hemicellulose and cellulose from the pine sawdust.

19.3.3 Enzymatic Hydrolysis of Pretreated and Untreated Pine Sawdust

The raw and pretreated pine sawdust samples were hydrolyzed using enzymes to produce sugar for biofermentation. Effects of reaction time and enzyme dose on the glucose yield, total sugar yield, and weight loss of the raw and pretreated pine sawdust were systematically studied. The results of enzymatic hydrolysis are presented as follows.

Figure 19.2 shows the effect of reaction time on the glucose yield under two different enzyme doses. The general trends observed in Fig. 19.2a (7.67 FPU) are similar to those in Fig. 19.2b (11.76 FPU). In the first 12 h of operation, the glucose yield increased significantly with time for all samples. After that, the glucose yield leveled off for 12–48 h of reaction. This could be explained by the fact that the enzymatic hydrolysis rate, especially the initial hydrolysis rate, strongly depends on the initial extent of enzyme adsorption and the effectiveness of the adsorbed enzymes [54]. At the beginning of the hydrolysis reaction, there were ideally a maximum number of active binding sites on the surface of the substrate, and enzymes could be fully absorbed onto the area. At this time, the hydrolysis rate could be considered the fastest. After a certain period of time, the process of enzymes adsorption and desorption reached a saturation point. Additionally, the production of cellobiose and glucose, which have been considered inhibitory to enzymatic hydrolysis, might be accumulated and inhibited enzymatic hydrolysis. As shown in Fig. 19.2, there are obvious differences in the glucose yield among pine sawdust samples pretreated using different methods. From Fig. 19.2, it is clear that the glucose yield increased with an increase in the enzyme dose (Fig. 19.2a vs. b). The raw pine sawdust (untreated) only had 5–6 % glucose yield while pine sawdust treated with (organosolv + ultrasound + NaOH) contributed to nearly 16–18 % glucose yield (Fig. 19.2b), which is more than three times the yield obtained with the raw pine sawdust sample. This can be explained by that the pretreatment removed lignin, loosened the cellulose crystalline structures and increased its accessibility to enzyme. The order of glucose yield positively correlated to the order of PE and DE, as shown in Fig. 19.4. In other words, a higher PE and DE led to a higher glucose yield. The results suggest that pretreatment did play an important role in enzymatic hydrolysis [50, 55].

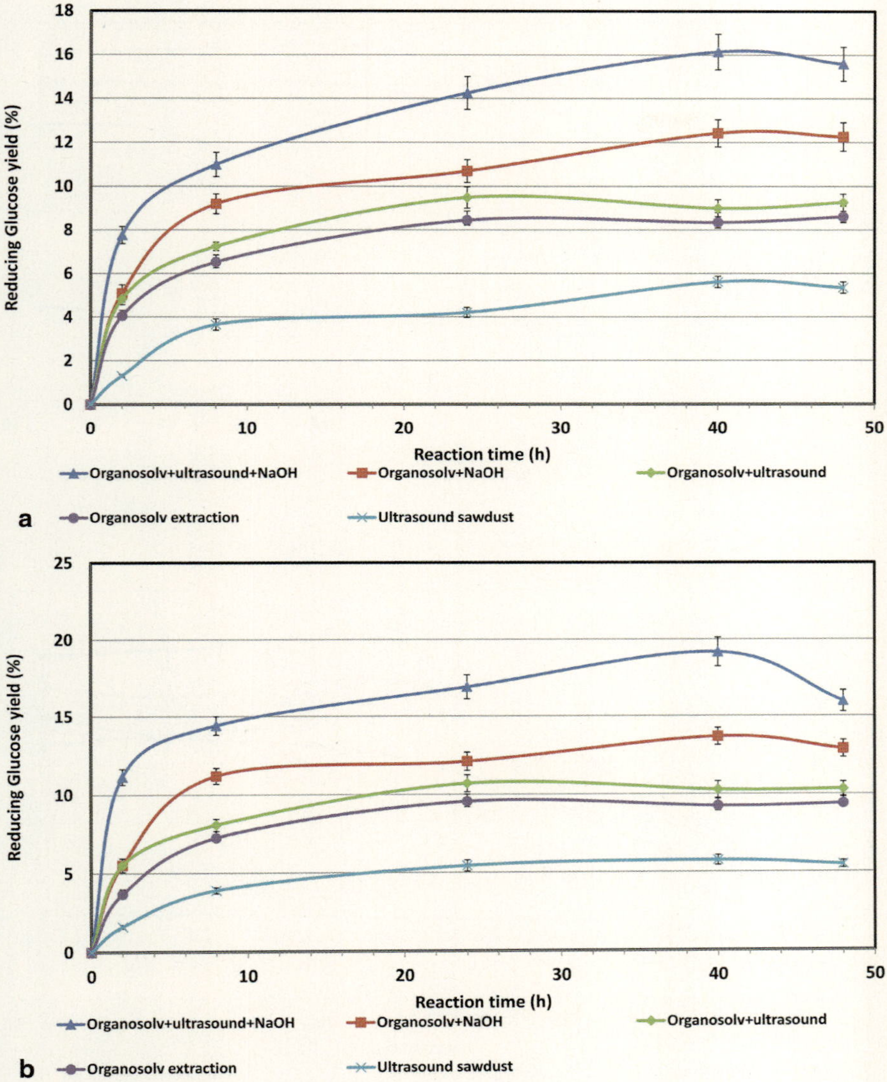

Fig. 19.2 Glucose yield from various pretreated pine sawdust samples at 50 °C in sodium citrate solution, pH 4.8 and 100 rpm shaking speed: **a** 7.67 FPU; **b** 11.76 FPU

Admittedly, the maximum glucose yield from this work was lower than those reported in some other studies, for example by Sannigrahi et al. [50] and Palonen et al. [55] using different pretreatment methods. Sannigrahi et al. [50] obtained a nearly 70 % of sugar yield when 65 % ethanol/water solution containing 1.1 % sulfuric acid was used as the pretreatment reagent. The addition of acid could lead to

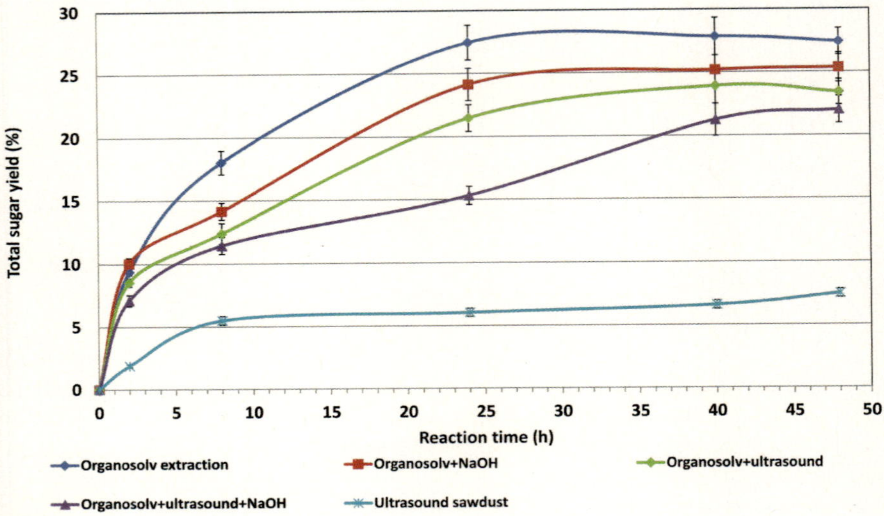

a

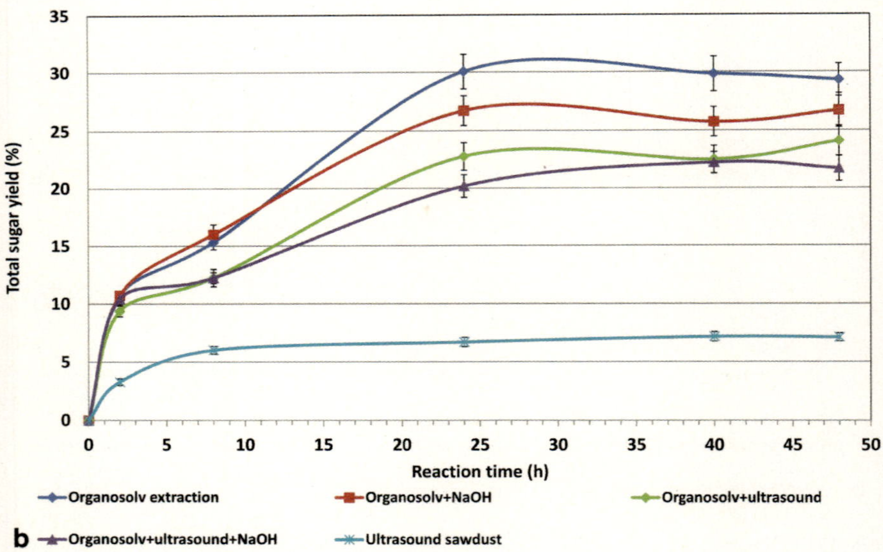

b

Fig. 19.3 Total sugar yield from various pretreated pine sawdust samples at 50 °C in sodium citrate buffer solution, pH 4.8, and 100 rpm shaking speed: **a** 7.67 FPU; **b** 11.76 FPU

a higher pretreatment efficiency not only for lignin but also for hemicellulose. The higher lignin and hemicellulose removal efficiency led to the higher glucose yield.

The glucose yields achieved in this study for Jack pine sawdust pretreated with the combination of (organosolv + ultrasound + NaOH) methods are lower than those reported in the work of Zhu et al. [4–6] where very high percentage of cellulose conversion (up to 90 %) was reported for hardwood and softwood samples with combined pretreatment methods of dilute acid and sulfite and disk milling [4–6]. However, compared with the disk milling pretreatment, a mechanical pretreatment approach, the organosolv-based pretreatment approaches may be advantageous in terms of pretreatment energy costs and the overall economics of the process. An acid-free organosolv process can overcome the problems caused by acid-catalyzed pretreatment. It shall be noted that the effect of the organosolv-based pretreatment on cellulose conversion can be greatly improved by combining it with dilute acid pretreatment, as demonstrated by Sannigrahi et al. [50]. Moreover, one of the striking advantages of the organosolv-based pretreatment approaches is that lignin of high purity is generated as a byproduct from the organosolv-based pretreatment processes, which can be utilized as a highly valuable feedstock for the production of phenolic resins and adhesives and biophenols [56].

Figure 19.3a and 19.3b shows the total sugar yield as the function of hydrolysis reaction time. Similar to the results obtained for glucose yield as dicussed above, for the sample pretreated with organo + ultrasound + NaOH, the total sugar yield increased to approximately 27 % in the initial 24 h and then reached the maximum yield of about 30 %. In the next 24 h, there was no significant increase in total sugar yield. The total sugar yields of other pretreated pine sawdust samples were lower but followed the same trend. Interestingly, in contrast to the glucose yield, a higher PE and DE led to a lower total sugar yield (Fig. 19.4). This was probably due to the fact that the higher removal efficiency of hemicellulose at a higher PE resulted in a lower content of hemicellulose in the solid residues, which decreased the formation of non-glucose carbohydrates (such as xylose). From the Fig. 19.4, the effects of enzyme dose on the total sugar yield are less significant (with approximately 5 % differecne in the yield).

19.4 Conclusions

1. Various pretreatment methods (organosolv extraction, followed by ultrasonic and/or NaOH treatment) resulted in a significant removal of lignin and hemicellulose.
2. The observations from SEM, FTIR, and XRD clearly demonstrate that the pretreatment led to disordered cell wall, twisted and exposed inner structure, reduced lignin and hemicelluloses contents, accompanied by increased cellulose content in the solid residues after treatment.
3. Pretreatment of pine sawdust samples had a significant impact on the glucose yield and total sugar yield. The treatment with different methods produced two-tothree-fold increase in the glucose and total sugar yields. The maximum glucose and total sugar yields were 5.8 % and 7.1 %, respectively for the raw pine sample, but they were increased to 9.6 % and 30.1 % with organosolv pretreatment, 10.7 % and 24.1 % with organosolv + ultrasound pretreatment, 13.6 %

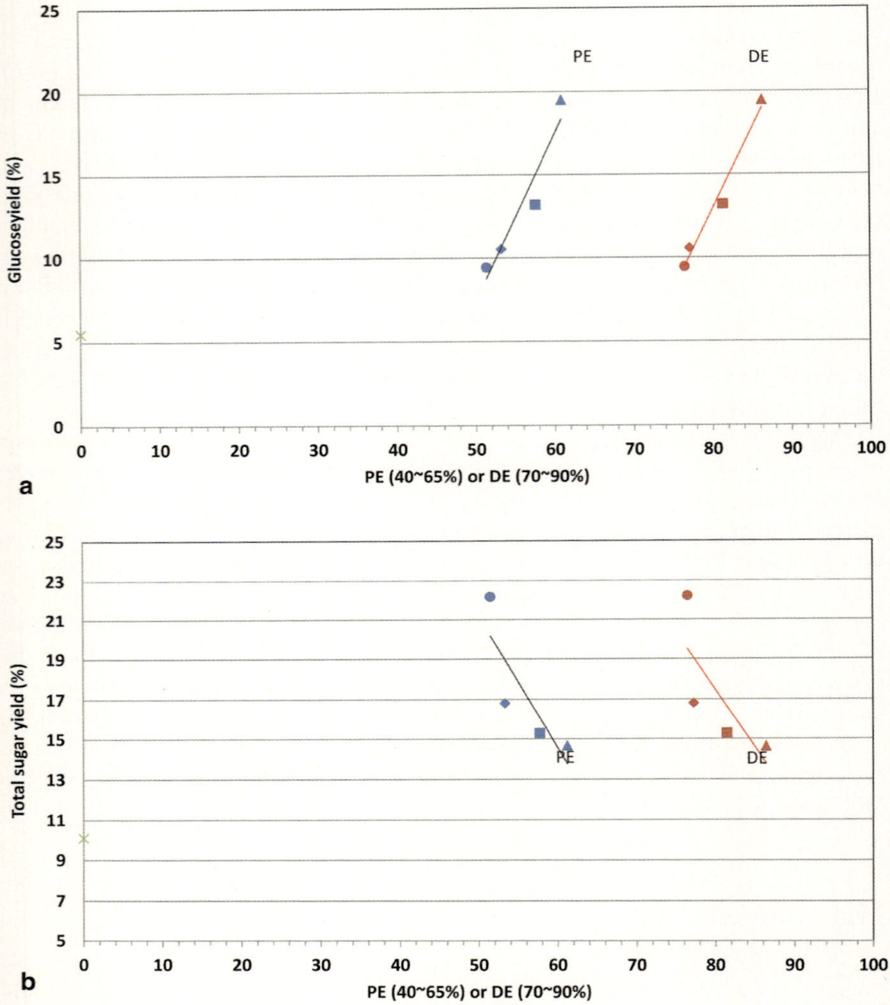

Fig. 19.4 Effects of PE and DE on glucose yield (**a**) and total sugar yield (**b**): • organosolv extraction; ♦ organosolv + ultrasound; ■ organosolv + NaOH; ▲ organosolv + ultrasound + NaOH; * untreated sawdust

and 26.8 % with organosolv + NaOH pretreatment, and 19.3 % and 22.4 % with organosolv + ultrasound + NaOH pretreatment.

4. In enzymatic hydrolysis of the pretreated pine sawdust, increasing PE and DE led to an increase in glucose yield, but a decrease in total sugar yield.

Acknowledgments The authors are grateful for the financial support from the Ontario Ministry of Energy and Ontario Centers of Excellence (OCE) through the Atikokan Bioenergy Research Center (ABRC) program. The authors would also like to acknowledge Natural Sciences and Engineering Research Council of Canada (NSERC) for the Discovery Grants.

References

1. Prasad S; Singh A, Joshi HC (2007) Ethanol as an alternative fuel from agricultural, industrial and urban residues., Resour Conserv Recycl 50:1–39
2. Gonzalez-Garcia S, Gasol CM, Gabarrell X, Rieradevall J, Mcreira MT, Feijoo G (2009) Environmental aspects of ethanol-based fuels from *Brassica carinata*: a case study of second generation ethanol. Renew Sustainable Energy Rev 13(9):2613–2620
3. Sun Y, Cheng J (2002) Hydrolysis of lignocellulosic materials for ethanol production: a review. Bioresour Technol 83(1):1–11
4. Zhu JY, Pan XJ, Wang GS, Gleisner R (2009) Sulfite pretreatment (SPORL) for robust enzymatic saccharification of spruce and red pine. Bioresour Technol 100(8): 2411–2418
5. Zhu W, Zhu JY, Gleisner R, Pan XJ (2010) On energy consumption for size-reduction and yields from subsequent enzymatic saccharification of pretreated lodgepole pine. Bioresour Technol 101(8):2782–2792
6. Zhu JY, Gleisner R, Scott CT, Luo XL, Tian S (2011) High titer ethanol production from simultaneous enzymatic saccharification and fermentation of aspen at high solids: a comparison between SPORL and dilute acid pretreatments. Bioresour Technol 102(19): 8921–8929
7. Duff SJB, Murray WD (1996) Bioconversion of forest products industry waste cellulosics to fuel ethanol: a review. Bioresour Technol 55(1):1–33
8. Zhu JY, Wang GS, Pan XJ, Gleisner R (2009) Specific surface to evaluate the efficiencies of milling and pretreatment of wood for enzymatic saccharification. Chem Eng Sci 64(3):474–485
9. Yeh An-I, Huang Yi-C, Chen SH (2010) Effect of particle size on the rate of enzymatic hydrolysis of cellulose. Carbohydr Polym 79:192–199
10. Hideno A, Inoue H, Tsukahara K, Fujimoto S, Minowa T, Inoue S, Endo T, Sawayama S (2009) Wet disk milling pretreatment without sulfuric acid for enzymatic hydrolysis of rice straw. Bioresour Technol 100: 2706–2711
11. Cadoche L, Lopez GD (1989) Assessment of size reduction as a preliminary step in the production of ethanol from lignocellulosic wastes. Biol Wastes 30:153–157
12. Millan MJD (1994) Pretreatment of lignocellulosic biomass. In: Himmel ME, Baker JO, Overend RP (eds) Enzymatic conversion of biomass for fuels production. American Chemical Society, Washington, DC, pp 292–324
13. Kumar P, Barrett DM, Delwiche MJ, Stroeve P (2009) Methods for pretreatment of lignocellulosic biomass for efficient hydrolysis and biofuel production. Ind Eng Chem Res 48(8): 3713–3729
14. Kabel MA, Bos G, Zeevalking J, Voragen AG, Schols HA (2007) Effect of pretreatment severity on xylan solubility and enzymatic breakdown of the remaining cellulose from wheat straw. Bioresour Technol 98: 2034–2042
15. Li J, Henriksson G, Gellerstedt G (2007) Lignin depolymerization/repolymerization and its critical role for delignification of aspen wood by steam explosion. Bioresour Technol 98:3061–3068
16. Eymouri F, Laureano-Perez L, Alizadeh H, Dale.BE (2005) Optimization of the ammonia fiber explosion (AFEX) treatment parameters for enzymatic hydrolysis of corn stover. Bioresour Technol 96:2014–2018
17. Holtzapple MT, Lundeen JE, Sturgis R (1992) Pretreatment of lignocellulosic municipal solid waste by ammonia fiber explosion (AFEX). Appl Biochem Biotechnol 34/35:5–21
18. Vlasenko EY, Ding H, Labavitch JM, Shoemaker SP (1997) Enzymatic hydrolysis of pretreated rice straw. Bioresour Technol 59:109–119
19. Chandra RP, Bura R, Mabee WE, Berlin A, Pan X, Saddler JN (2007) Substrate pretreatment: the key to effective enzymatic hydrolysis of lignocellulosics. Adv Biochem Eng Biotechnol 108:67–93
20. Moniruzzaman M, Dale BE, Hespell RB, Bothas RJ (1997) Enzymatic hydrolysis of high-moisture corn fiber pretreated by AFEX and recovery and recycling of the enzyme complex. Humana Press, New York, pp 1–2 and 113–127

21. Zhu S, Wu Y, Yu Z, Liao J, Zhang Y (2005) Pretreatment by microwave/alkali of rice straw and its enzymic hydrolysis. Process Biochem 40:3082–3086
22. Ma H, Liu W-W, Chen X, Wu Y, Yu Z (2009) Enhanced enzymatic saccharification of rice straw by microwave pretreatment. Bioresour Technol 100:1279–1284
23. Banik S, Bandyopadhyay S, Ganguly S (2003) Bioeffects of microwave—a brief review. Bioresour Technol 87(22):155–159
24. Sun RC, Tomkinson RC (2002) Characterization of hemicelluloses obtained by classical and ultrasonically assisted extractions from wheat straw. Carbohydr Polym 50(3):263–271
25. Yin X, Han P, Lu X, Wang Y (2004) A review on the dewaterability of bio-sludge and ultrasound pretreatment. Ultrason Sonochem 11(6):337–348
26. Yachmenev V, Condon B, Klasson T, Lambert A (2009) Acceleration of the enzymatic hydrolysis of corn stover and sugar cane bagasse celluloses by low intensity uniform ultrasound. J Biobased Mater Bioenergy 3:25–31
27. Tang AM, Liang WZ (2000) Studies on the structure changes of fast-growing wood fiber treated by ultrasonic wave. Tech Acoust 19:78–85
28. Yu J, Zhang J, He J, Liu Z, Yu Z (2009) Combinations of mild physical or chemical pretreatment with biological pretreatment for enzymatic hydrolysis of rice hull. Bioresour Technol 100:903–908
29. Alvira P, Tomas-Pejo E, Balleserros M, Negro MJ (2010) Pretreatment technologies for an efficient bioethanol production process based on enzymatic hydrolysis: a review. Bioresour Technol 101:4851–4861
30. Guo GL, Chen WH, Chen WH, Men LC, Hwang WS (2008) Characterization of dilute acid pretreatment of silvergrass for ethanol production. Bioresour Technol 99:6046–6053
31. Shatalov AA, Pereira H (2012) Xylose production from giant reed (*Arundo donax* L.): modeling and optimization of diluted acid hydrolysis. Carbohydr Polym 87(1):210–217
32. Taherzadeh MJ, Karimi K (2008) Pretreatment of lignocellulosic wastes to improve ethanol and biogas production: a review. Int J Mol Sci 9(9):1621–1651
33. Cara C, Ruiz E, Oliva JM, Saez F, Castro E (2008) Conversion of olive tree biomass into fermentable sugars by dilute acid pretreatment and enzymatic saccharification. Bioresour Technol 99:1869–1876
34. Carvalheiro F, Duarte LC, Girio FM (2008) Hemicellulose biorefineries: a review on biomass pretreatments. J Sci Ind Res 67:849–864
35. Zhao X, Zhang L, Liu D. (2007) Comparative study on chemical pretreatment methods for improving enzymatic digestibility of crofton weed stem. Bioresour Technol 99:3729–3736
36. Xu J, Cheng JJ, Sharma-Shivappa RR, Burns JC (2010) Lime pretreatment of switchgrass at mild temperatures for ethanol production. Bioresour Technol 101:2900–2903
37. Sun RC, Sun XF (2002) Fractional and structural characterization of hemicelluloses isolated by alkali and alkaline peroxide from barley straw. Carbohydr Polym 49(4):415–423
38. Sun RC, Fang JM, Tomkinson J (2002) Delignification of rye straw using hydrogen peroxide. Ind Crops Prod 12:71–83
39. Sun RC, Tomkinson J, Wang S, Zhu W (2000) Characterization of lignins from wheat straw by alkaline peroxide treatment. Polym Degrad Stab 67:101–109
40. Saha BC, Cotta MA (2007) Enzymatic saccharification and fermentation of alkaline peroxide pretreated rice hulls to ethanol. Enzyme Microb Technol 41:528–532
41. Koo B-W, Kim H-Y, Park N, Lee S-M, Yeo H, Choi I-G (2011) Organosolv pretreatment of *Liriodendron tulipifera* and simultaneous saccharification and fermentation for bioethanol production. Biomass Bioenergy 35(5):1833–1840
42. Pan X, Xie D, Yu RW, Lam D, Saddler J (2007) Pretreatment of lodgepole pine killed by mountain pine beetle using the ethanol organosolv process: fractionation and process optimization. Ind Eng Chem Res 46:2609–2617
43. Zhao X, Cheng K, Liu D (2009) Organosolv pretreatment of lignocellulosic biomass for enzymatic hydrolysis. Appl Microbiol Biotechnol 82:815–827
44. Pasquini D, Pimenta MTB, Ferreira LH (2005) Extraction of lignin from sugar cane bagasse and *Pinus tarda* wood chips using ethanol-water mixtures and carbon dioxide at high pressure. J Supercrit Fluids 36:31–39

45. Araque E, Parra C, Freer J, Contreras D, Rodríguez J, Mendonc R, Beza J (2008) Evaluation of organosolv pretreatment for the conversion of *Pinus radiata* D. Don to ethanol. Enzyme Microb Technol 43:214–219
46. Bailey MJ (1988) A note on the use of dinitrosalicylic acid for determining the products of enzymatic reactions. Appl Microbiol Biotechnol 29:494–496
47. Dubois M, Gilles K, Hamilton JK, Rebers PA, Smith FA (1951) Colorimetric method for the determination of sugars. Nature 168:167. doi:10.1038/168167a0
48. Kim S, Holtzapple MT (2006) Effect of structural features on enzyme digestibility of corn stover. Bioresour Technol 97:583–591
49. Liu K, Lin X, Yue J, Li X, Fang X, Zhu M, Lin J, Qua Y (2010) High concentration ethanol production from corncob residues by fed-batch strategy. Bioresour Technol 101:4952–4958
50. Sannigrahi P, Ragauskas AJ, Miller SJ (2010) Effects of organosolv pretreatment and enzymatic hydrolysis on cellulose structure and crystallinity in Loblolly pine. Carbohydr Res 345:965–970
51. Pasquini D, Pimenta MTB, Ferreira LH (2005) Extraction of lignin from sugar cane bagasse and *Pinus tarda* wood chips using ethanol-water mixtures and carbon dioxide at high pressure. J Supercrit Fluids 36:31–39
52. Xu C, Etcheverry T (2008) Hydro-liquefaction of woody biomass in sub- and super-critical ethanol with iron-based catalysts. Fuel 87:335–345
53. Sreenivasan S, Iyer PB, Iyer KRK (1996) Influence of delignification and alkali treatment on the fine structure of coir fibres (Cocos Nucifera). J Mater Sci 31:721–726
54. Fan LT, Lee YH (1982) Kinetic studies of enzymatic hydrolysis of insoluble cellulose: analysis of the initial rates. Biotechnol Bioeng 24;2383–2406
55. Palonen H, Thomsen AB, Tenkanen M, Schmidt AS, Viikari L (2004) Evaluation of wet oxidation pretreatment for enzymatic hydrolysis of softwood, vol 117. Humana Press, New York, pp. 1–17
56. Wang M, Leitch M, Xu C (2009) Synthesis of phenol-formaldehyde resol resins using organosolv pine lignins. Eur Polym J 45:3380–3388

Printed by Printforce, the Netherlands